杨叔子科技论文选(上)

涂又光 题

杨叔子

图书在版编目(CIP)数据

杨叔子科技论文选(上、下)/杨叔子.—武汉:华中科技大学出版社,2012.9
ISBN 978-7-5609-8363-9

Ⅰ.杨… Ⅱ.杨… Ⅲ.机械学-文集 Ⅳ.TH11-53

中国版本图书馆 CIP 数据核字(2012)第 209443 号

杨叔子科技论文选(上、下) 杨叔子

策划编辑:熊新华　杨　玲
责任编辑:包以健
责任校对:张　琳
责任监印:周治超
出版发行:华中科技大学出版社(中国·武汉)
　　　　　武昌喻家山　邮编:430074　电话:(027)81321915
录　　排:华中科技大学惠友文印中心
印　　刷:湖北新华印务有限公司
开　　本:710mm×1000mm　1/16
印　　张:38　插页:4
字　　数:675 千字
版　　次:2012 年 9 月第 1 版第 1 次印刷
定　　价:100.00 元(含上、下册)

本书若有印装质量问题,请向出版社营销中心调换
全国免费服务热线:400-6679-118　竭诚为您服务
版权所有　侵权必究

序

  这不仅仅是我们的恩师、中国科学院院士杨叔子先生的一本科技论文选,更是我们这些弟子的一个真诚心愿。

  2008年9月,杨叔子先生满75周岁,在汉的弟子按一年前的商定,前来给先生过生日。尽管原打算活动只在小范围内举行,只通知在武汉地区先生的弟子,但生日那天,却有许多外地包括北京的已知晓此事的弟子也纷纷赶来祝贺,知晓而不能来的,例如,在天津、广州、上海、长沙、深圳乃至英国的也打电话来祝贺。祝贺欢庆之余,很多弟子提出这次生日应有一个纪念性的东西,除了必需的集体合影外,最有意义的大概应是先生著作的结集出版。

  先生从事教育工作50多年来,不仅作为学者发表科技学术论文、教育学术论文过千篇,出版各类著作、教材几十部,随笔散文与序数百篇,至于信函就以千计了,而且,作为诗人创作的诗词作品也有近千首,目前能收集到的有700多首。同时,先生作为导师培养各类学生几百名,培养形式多种多样,包括以指导毕业设计、指导攻读硕士或博士学位、指导博士后流动站研究工作或长期一起从事教学改革试验等形式所培养的,其中获得博士学位的学生逾百名,学生的年龄跨度从二十几岁至六十几岁,现在从事的职业几乎遍布各行业。我们这些弟子同先生情感深厚。先生常讲:"我所取得的进步,我所作出的贡献,我所获得的荣誉,无不沉淀着我的学生的辛勤劳动与珍贵心血。"我们这些弟子也常讲:"我们今天能有思路,有奔头,挑大梁,作贡献,无不凝聚着先生道德文章的深刻影响。"是的,学生无时不在先生的心头,先生也无时不在学生的心头,师生一同走过的如歌岁月与风雨征程,永在大家心中。

  这次生日聚会,我们这些弟子纷纷表示一种遗憾,由于多种原因,没有一个弟子完整地读过,甚至完整地见过先生的所有著作、论文和诗作等,因此弟子纷纷提议能否将先生的论文、诗作等结集出版;这不仅可给大家留作永远的纪念,而且也有利于教育与科研工作的开展以及学术交流与思想交流。这一提议立即得到所有与会弟子的积极响应,并委托目前仍在华中科技大学工作的师兄弟从他们中推选若干位负责筹备此事。这一工作开始后,才发现这是一项很大的工程,要将所有的论文、诗作等结集出版,工作量将十分浩大,何况先生原本就不同意这项工作,他一再讲:"已经发表了,不必结集出版了!何况其中不成熟部分乃至谬误之处还不少,这么出版,还可能害人呢!还有,我所写的诗词,对诗词行家里手而言,还是槛外人、门外汉之作呢!"大家做了很多工作,先生才勉强同意,但前提是事情一定要办得简单,出版后对大家、对工作要有好处,只出选集,选几分之一乃至十几分之一出版就足够了!为此,经过反复讨论,决定尊重先生的意愿,尽量将事情简化,并分为四个部分结集出版,分别是科技类文选、教育类文选、诗词及有关论文选,以及散文、序、信函文选。按照我们的计划,准备用四年左右的时间,即先生八十华诞前后出齐。

按此计划,2009 年 9 月出版了《杨叔子槛外诗文选》,选编了改革开放 30 年间(1978 年至 2008 年)诗词 200 首和有关的诗教论文 7 篇。2010 年 11 月出版了《杨叔子教育雏论选》,选编了从 1981 年至 2010 年 30 年间有关教育教学的论文 90 篇。这本今年 9 月即将出版的《杨叔子科技论文选》,选编了从 1979 年至 2012 年间有关科技学术论文 48 篇,其中英文 11 篇。

先生 1956 年毕业于原华中工学院机械制造本科专业后,即留校任教,并从事科研工作。1956 年至 1957 年约一年在哈尔滨工业大学进修;1964 年至 1965 年整一年在上海机床厂结合劳动,对磨床磨削开展试验研究;1981 年至 1982 年整一年在美国 Wisconsin 大学(Madison)师从美籍华人吴贤铭教授开展"时间序列分析及其在工程中应用"的研究。先生所教课程固然以机械制造专业课程为主,由于客观需要也教过高等数学、复变函数、变分法、材料力学、理论力学、机械设计、控制工程、时间序列分析等课程;而研究工作一直结合教学工作并瞄准国家需要与科技前沿来开展;开始主要从事有关机床本身的研究,特别从 20 世纪 80 年代以来,先生以高瞻远瞩的眼光与敏锐的思维,紧紧把握世界科技动态、机械制造发展趋势与我国建设需要，努力引领潮流，立足于机械工程，结合控制论、信息论、系统论，开展机械工程特别是机械制造同新兴科技相结合的交叉领域的研究，诸如时间序列分析及其工程应用、信号处理、无损检测新技术、机电设备诊断、先进制造技术、智能制造与网络制造等，先生同他的团队以及研究生们一起，取得了一系列重要的科研成果，培养了一大批研究生。这些重要科研成果取得时，有的无愧于国际前沿水平，有的开拓了国内新的研究领域，有的解决了生产或工程中的重大难题，有的有力推进了有关方面的基础研究，不少对国内有关研究产生了重大影响，不少对教学内容的改革与提高作出了重要贡献并一直影响到现在。先生的学术造诣很高。先生所培养的并取得学位的博士早已超过百人。当然，也包括我们这些执笔写这个序言的、已经成为教授的弟子们。

这次为了选编《杨叔子科技论文选》，以史铁林同志为主，在研究生协助下，弟子们多方从网上下载了近千篇有先生署名的中外学术期刊上发表的学术论文，再加上以往没有电子版的学术论文，篇数过千。论文怎么选编？我们与先生一起，确定以下三条原则。

第一，"厚今薄古"。科技发展日新月异，以往有的论文或多或少甚至基本上无参考价值，对此类论文基本上不选。

第二，"学术认可"。在下载中，我们尽可能取得对本篇论文的学术评价，主要看其点击次数与引用次数，点击或引用次数不多的不选。

第三，"全面兼顾"。其一，论文领域较广，为了反映先生多方面的工作，论文不宜过分集中在一两个方面，如设备诊断类论文，未超过所选论文总数的1/4。其二，论文也不宜集中在少数年份，而应反映先生学术经历特别是改革开放以来的学术经历。所以自1980年至今，大体上每年少则1篇，多则不超过4篇。其三，先生还强调，论文的作者尽可能全面地反映团队的成员，他们都作出了贡献，不要选编的结果遗漏了他们。当然，也绝不会"削足适履"。

此外，先生认为专业学术论文，专业性很强，不宜多选，有代表性即可。先生还认为，从1993年起，他担任行政职务，主要精力不在专业上，学术论文绝大部分也不是他执笔的,也不宜过多选。从而我们只选取了48篇。

本文集有关论文的下载、打印、图与曲线的重新描绘以及校对等,都得到了研究生的帮助,诸如王珂、张嘉琪、蒋淑兰、饶和昌、张财胜等40多位同学,为本文集的出版作出了贡献,我们对他们的工作表示衷心的感谢!

我们还要感谢华中科技大学出版社真诚、积极、认真和热情的态度,在高质量出版了先生的两本文集后,这次又本着出好出快、让我们满意、让读者满意的态度,高质量地完成了这本科技论文选的编辑工作。我们向华中科技大学出版社以及为本文集出版工作付出辛勤劳动的各位同志致以最真诚的敬意!

经过近一年的工作,这本文集即将付梓了。我们都感到由衷的高兴。希望这本文集能对我们大家真正地有所触动,因为其中有我们所做的工作,有我们岁月的留痕,有我们心血的结晶,同时,也凝有我们同先生深厚的师生情谊;让我们多些思考、少些浮躁,时刻牢记我们都是老师了,牢记教书育人这一首要的职责和使命,爱我们的学生,敬我们的事业,在伟大的改革开放的新征程中,为我国的高等教育事业尽到自己应尽的义务,作出自己的贡献,为我国高等教育更加科学地更加"稳中求进"地发展、我国早日成为高等教育强国添砖加瓦。我们相信:这是先生对我们的最大要求和先生的最大心愿,也是和先生一起献给党的十八大的最好礼物。

华中科技大学教授

史铁林 吴 波 吴昌林 李 斌
赵英俊 易传云 康宜华 何岭松
李锡文 黄其柏 胡友民 管在林
刘世元 武新军 廖广兰 等

2012年9月

(农历壬辰年)

编辑凡例

一、据不完全统计，从 1979 年至 2012 年间，中外学术期刊上公开发表的有杨叔子署名的学术论文多达 1000 余篇。编选者以厚古薄今、全面兼顾为原则选取了引用率较高的有代表性的科技类学术论文共计 48 篇构成《杨叔子科技论文选》。

二、本文集所收录的文章，涵盖了先进制造技术，机电设备诊断，机床刚度、振动、噪声、无损检测、时序分析、信号处理以及有关网络技术等内容。它们共同构成了一个思想整体，集中反映了合乎世界科技发展潮流的先进制造技术的发展与趋势。

三、本文集所收录的文章，各篇独立成文。由于其写作的时间不同，论述的重点有异，因此，不另行分类，以发表的时间顺序编排。

四、本文集所收录的文章大部分保留了历史原貌，即各篇文章内各级标题格式基本遵从公开发表时原学术期刊的体例。对少部分文章中少量文字与个别字句错漏之处作了删改。

五、本文集所收录的文章均经过作者本人审阅。华中科技大学机械学院史铁林教授负责文集的选编、整理工作。华中科技大学机械学院研究生王珂、张嘉琪、蒋淑兰等 40 多位同学做了论文的收集、初录及校稿工作。华中科技大学机械学院 Frontiers of Mechanical Engineering 编辑部陈惜曦同志提供了所有论文的复印件和扫描件。华中科技大学机械学院博士后何锐波同志对论文中的部分图与曲线做了重新描绘工作。

目录

三支承主轴部件静刚度的分析与讨论　/1

δ 函数在机械制造中的应用　/19

A Study of the Static Stiffness of Machine Tool Spindles　/30

平稳时间序列的数学模型及其阶的确定的讨论　/57

时序建模与系统辨识　/64

金属切削过程颤振预兆的特性分析　/74

机械设备诊断学的探讨　/86

灰色预测和时序预测的探讨　/97

Quantitative Wire Rope Inspection　/106

钢丝绳断丝定量检测的原理与实现　/116

复杂系统诊断问题的研究　/126

Plant Condition Recognition—A Time Series Model Approach　/136

Space-domain Feature-based Automated Quantitative Determination of Localized Faults in Wire Ropes　/150

基于深知识的多故障两步诊断推理　/168

机床切削系统的强迫再生颤振与极限环　/176

机械设备诊断学的再探讨　/185

机械设备诊断策略的若干问题探讨　/194

Forced Regenerative Chatter and its Control Strategies in Machine Tools　/201

智能制造技术与智能制造系统的发展与研究　/212

两类小波函数的性质和作用　/220

Intelligent Prediction and Control of a Leadscrew Grinding Process Using Neural Networks　/233

大直径钢丝绳轴向励磁磁路的研究　/244

金属切削机床切削噪声的动力学研究　/250

BP 网络的全局最优学习算法　/263

基于神经网络的结构动力模型修改和破损诊断研究　/268

A CORBA-based Agent-driven Design for Distributed Intelligent Manufacturing Systems /275

内燃机气缸压力的振动信号倒谱识别方法 /295

A Novel Co-based Amorphous Magnetic Field Sensor /302

基于高阶统计量的机械故障特征提取方法研究 /312

基于因特网的设备故障远程协作诊断技术 /318

磁性无损检测技术中磁信号测量技术 /323

分布式网络化制造系统构想 /332

网络化制造与企业集成 /342

磁性无损检测技术中的信号处理技术 /351

Intelligent Machine Tools in a Distributed Network Manufacturing Mode Environment /362

AR 模型参数的 Bootstrap 方差估计 /387

虚拟制造系统分布式应用研究 /392

Wigner-Ville 时频分布研究及其在齿轮故障诊断中的应用 /400

先进制造技术及其发展趋势 /409

Feature Extraction and Classification of Gear Faults Using Principal Component Analysis /420

制造系统分布式柔性可重组状态监测与诊断技术研究 /434

基于 Markov 模型的分布式监测系统可靠性研究 /444

再论先进制造技术及其发展趋势 /456

基于神经网络信息融合的铣刀磨损状态监测 /464

以人为本——树立制造业发展的新观念 /472

走向"制造-服务"一体化的和谐制造 /480

Kinematic-parameter Identification for Serial-robot Calibration Based on POE Formula /491

高端制造装备关键技术的科学问题 /524

后记 /536

论文附录 /541

三支承主轴部件静刚度的分析与讨论[*]

杨叔子

在机床设计中,有时由于结构上的限制的需要,主轴部件采用三支承结构。从力学观点来看,在主轴部件静刚度上,三支承主轴部件有什么特点?有关参数如何选择才能使主轴部件静刚度为最大?这是本文所要研究的主要问题。文中所述的一些看法有待深入研究和值得进一步商榷。

本文是文献[1]的继续,文献[2]后部分的改写与补充,并对其中某些公式由校对上疏忽而造成的错误,作了更正。

一、一般原理与计算公式

本文采用《结构力学》中"影响系数"的概念,来研究主轴部件的静刚度问题。所谓"影响系数"是指在弹性杆系 K 处作用一单位载荷时在 i 处引起的位移值。此值成为 k 处对 i 处的影响系数。当载荷为力,位移分别为线位移与角位移(转角)时,影响系数相应地记为 α_{ik} 与 β_{ik};当载荷为力偶矩,位移分别为线位移与角位移时,则相应地记为 γ_{ik} 与 δ_{ik}。

影响系数的重要特性是 $\alpha_{ik}=\alpha_{ki}$,$\beta_{ik}=\gamma_{ik}$(或 $\beta_{ki}=\gamma_{ik}$),$\delta_{ik}=\delta_{ki}$,这就是位移(影响系数)互等原理。

如同文献[1]、[2]所指出,要全面衡量主轴部件的静刚度,必须综合地考虑切削力与传动力的作用,即必须考虑到引起主轴轴端位移的力,这个力不仅有切削力引起的作用在轴端的力 P,而且还有切削力引起的作用在轴端的力偶矩 M,由切削力引起的作用在主轴某一部位的传动力 Q。在三支承主轴部件中,它们引起

---

[*] 赵星、王治藩、彭茂竞、吴雅、吴凤鸣、黄荔、周泽耀参加了本文举例计算与有关部分的讨论。

的轴端位移分别为

$$y_{s\cdot 3}^{P} = P\alpha_{ss\cdot 3} \tag{1}$$

$$y_{s\cdot 3}^{M} = M\gamma_{ss\cdot 3} \tag{2}$$

$$y_{s\cdot 3}^{Q} = Q\alpha_{sQ\cdot 3} \tag{3}$$

式中,注脚"s"表示主轴轴端,"3"表示三支承主轴部件的有关参数;而注脚后不加"3"的,则表示两支承主轴部件有关参数。正如文献[1]、[2]指出那样,一般不能选择这么一种主轴部件结构与参数,使轴端位移的矢量和 $\mathbf{y}_{s\cdot 3} = \mathbf{y}_{s\cdot 3}^{P} + \mathbf{y}_{s\cdot 3}^{M} + \mathbf{y}_{s\cdot 3}^{Q}$ 为最小,但可力求 $\alpha_{ss\cdot 3}$、$\gamma_{ss\cdot 3}$、$\alpha_{sQ\cdot 3}$ 尽可能小,力求 $y_{s\cdot 3}^{P}$、$y_{s\cdot 3}^{M}$、$y_{s\cdot 3}^{Q}$ 在对加工精度具有决定性影响的敏感方向上相互抵消一部分。

今求 $\alpha_{ss\cdot 3}$、$\gamma_{ss\cdot 3}$、$\alpha_{sQ\cdot 3}$ 的一般算式。

如图 1 所示,设第 i 个支承的刚度为 C_i,在 P 作用下,其弹性变形为 δ_i,其支反力为 R_i。如规定 $P\alpha_{is}$ 朝某一方向为正,$R_i\alpha_{ii}$ 朝同它相反的方向为正,则对主轴而言,在第 i 支承处的位移显然有下列关系式:

$$P\alpha_{is} - R_i\alpha_{ii} = \delta_i \tag{4}$$

$$R_i = C_i\delta_i \tag{5}$$

联解上列两式,得

$$R_i = \frac{\alpha_{is}}{\dfrac{1}{C_i} + \alpha_{ii}} \cdot P \tag{6}$$

显然,α_{ii} 必定是正值。如 P 为正时,若求得的 R_i 为负,则 R_i 与 P 相同;反之亦然。

如图 1 所示,当以 R_i 代替第 i 支承(图中以 R_2 代替中间支承 2),则三支承结构转化为两支承结构,其上作用力 P 与 R_i 正反向规定同上,故

$$y_{s\cdot 3}^{P} = P\alpha_{ss} - R_i\alpha_{si} \tag{7}$$

将式(6)代入上式,得

$$y_{s\cdot 3}^{P} = P\left[\alpha_{ss} - \frac{\alpha_{is}}{\dfrac{1}{C_i} + \alpha_{ii}} \cdot \alpha_{si}\right] \tag{8}$$

因 $\alpha_{is} = \alpha_{si}$,并记 $1/C_i + \alpha_{ii}$ 为 α_{ii}^{d},称当量影响系数,则有

$$y_{s\cdot 3}^{P} = P(\alpha_{ss} - \alpha_{is}^{2}/\alpha_{ii}^{d}) \tag{9}$$

$$\alpha_{ss\cdot 3} = \alpha_{ss} - \alpha_{is}^{2}/\alpha_{ii}^{d} \tag{10}$$

上列公式中,各影响系数的具体算式如下:

$$\alpha_{1s} = \alpha_{s1} = \frac{l(l+a)}{C_2(l-l_1)^2} + \frac{l_1(l_1+a)}{C_3(l-l_1)^2} + \frac{l_1(2ll_1+2la-l_1a)}{6EI} \tag{a}$$

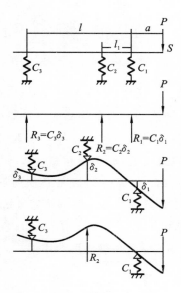

图 1

$$\alpha_{2s} = \alpha_{s2} = \frac{(l+a)(l-l_1)}{C_1 l^2} - \frac{l_1 a}{C_3 l^2} - \frac{l_1 a(l-l_1)(2l-l_1)}{6EIl} \tag{b}$$

$$\alpha_{3s} = \alpha_{s3} = -\frac{(l_1+a)(l-l_1)}{C_1 l_1^2} - \frac{la}{C_2 l_1^2} + \frac{l_1 a(l-l_1)}{6EI} \tag{c}$$

$$\alpha_{11} = \frac{1}{C_1} + \frac{1}{C_2}\left(\frac{l}{l-l_1}\right)^2 + \frac{1}{C_3}\left(\frac{l_1}{l-l_1}\right)^2 + \frac{ll_1^2}{3EI} \tag{d}$$

$$\alpha_{22} = \frac{1}{C_1}\left(\frac{l-l_1}{l}\right)^2 + \frac{1}{C_2} + \frac{1}{C_3}\left(\frac{l_1}{l}\right)^2 + \frac{l_1^2(l-l_1)^2}{3EIl} \tag{e}$$

$$\alpha_{33} = \frac{1}{C_1}\left(\frac{l-l_1}{l}\right)^2 + \frac{1}{C_2}\left(\frac{l}{l_1}\right)^2 + \frac{1}{C_3} + \frac{l(l-l_1)^2}{3EI} \tag{f}$$

上式中，E 为材料的弹性模量，钢的 E 为 2.1×10^5 牛顿/毫米^2；I 为主轴在支承间的平均断面惯性矩，对实心主轴 $I = \frac{\pi}{64} D^4$；D 为支承间主轴平均直径；l、l_1、a 如图 1 所示。

要注意的是 α_{ss}，在应用式(7)～式(10)时，α_{ss} 是指去掉第 i 个支承后的两支承主轴部件的轴端影响系数。

去掉第 1 支承后，有

$$\alpha_{ss} = \frac{(l_1+a)^3}{3EI_{l_1+a}} + \frac{(l-l_1)(l_1+a)^2}{3EI_{l-l_1}} + \frac{1}{C_2}\left(1 + \frac{l_1+a}{l-l_1}\right)^2 + \frac{1}{C_3}\left(\frac{l_1+a}{l-l_1}\right)^2 \tag{g}$$

去掉第 2 支承后，有

$$\alpha_{ss} = \frac{a^3}{3EI_a} + \frac{la^2}{3EI} + \frac{1}{C_1}\left(1+\frac{a}{l}\right)^2 + \frac{1}{C_3}\left(\frac{a}{l}\right)^2 \tag{h}$$

去掉第 3 支承后，有

$$\alpha_{ss} = \frac{a^3}{3EI_a} + \frac{l_1 a^2}{3EI_{l_1}} + \frac{1}{C_1}\left(1+\frac{a}{l_1}\right)^2 + \frac{1}{C_2}\left(\frac{a}{l_1}\right)^2 \tag{i}$$

式中，I 的角标表示 I 是主轴该部分的平均断面惯性矩。由位移互等原理可推论或由式(10)可知，R_i 引起的轴端位移 $R_i\alpha_{si}$ 必与 P 引起的轴端位移 $P\alpha_{ss}$ 反向，即式(7)～(10)中前后两项必均为正值，故两项相减，$\alpha_{ss\cdot 3} < \alpha_{ss}$，主轴部件刚度得到提高。只有在 $\alpha_{is}=0$ 时，$\alpha_{ss\cdot 3}=\alpha_{ss}$。$\alpha_{is}=0$ 是指在 P 作用下(如图 2 所示)，主轴不发生位移处(即所谓"节点"j)的影响系数。根据位移互等原理，在节点 j 处作用外力，则外力引起的轴端位移也为零，$\alpha_{si}=\alpha_{is}=0$。所以，在节点处装上第三支承，必不起加强但也决不会起减弱主轴部件静刚度的作用。

图 2

同理，可得出在 M 作用下的计算公式：

$$R_i = \frac{\gamma_{is}}{\frac{1}{C_i}+\alpha_{ii}} \cdot M \tag{11}$$

$$y_{s\cdot 3}^M = M\left(\gamma_{ss} - \frac{\gamma_{is}}{\alpha_{ii}^d} \cdot \alpha_{si}\right) \tag{12}$$

$$\gamma_{ss\cdot 3} = \gamma_{ss} - \frac{\gamma_{is}}{\alpha_{ii}^d} \cdot \alpha_{si} \tag{13}$$

从位移互等原理可推论，式(12)、(13)中前后两项必均为正值，故两项相减，$\gamma_{ss\cdot 3} < \gamma_{ss}$。只有在 $\gamma_{is}=0$ 或 $\alpha_{si}=0$ 时，$\gamma_{ss\cdot 3}=\gamma_{ss}$。但是，$\gamma_{is}=0$ 时，第三支承处无反力；$\alpha_{si}=0$ 时，第三支承处有反力，不过此反力不引起轴端位移。

同理，可得出在 Q 作用下的计算公式：

$$R_i = \frac{\alpha_{iQ}}{\frac{1}{C_i}+\alpha_{ii}} \cdot Q \tag{14}$$

$$y_{s\cdot 3}^Q = Q\left(\alpha_{sQ} - \frac{\alpha_{iQ}}{\alpha_{ii}^d} \cdot \alpha_{si}\right) \tag{15}$$

$$\alpha_{sQ\cdot 3} = \alpha_{sQ} - \frac{\alpha_{iQ}}{\alpha_{ii}^d} \cdot \alpha_{si} \tag{16}$$

由位移互等原理分析表明,除 Q 作用在悬伸部分外,式(15)、(16)中前后两项的符号不一定相同,故与 α_{sQ} 比较, $\alpha_{sQ\cdot 3}$ 可能增大,可能减小,可能相等,可能反号。显然,在 $\alpha_{iQ}=0$ 或 $\alpha_{si}=0$ 时, $\alpha_{sQ\cdot 3}=\alpha_{sQ}$。但是, $\alpha_{iQ}=0$ 时,第三支承处无反力; $\alpha_{si}=0$ 时,第三支承处有反力,不过此反力不引起轴端位移。

有关影响系数的具体算式可参考《材料力学》教材或文献[3]、[4]。

二、最佳参数的选择

根据式(6)、(9)、(10),并代入影响系数的具体算式(a)～(i),研究在 P 作用最佳参数的选择,可得下列结果。

1. 关于支承反力

α_{1s} 必定为正,且 $\alpha_{1s}>|\alpha_{2s}|$, $\alpha_{1s}>|\alpha_{3s}|$,故 R_1 永远为正,即与 P 反向,并且最大,同时前支承对加强主轴部件静刚度作用最大。R_2、R_3 可为正,可为负,可为零,当然不可同时为正或为零。文献[11]理论计算与试验结果都证明了这点。

2. 关于支承刚度 C_i

C_i 越大, α_{ii}^d 也越小, $\alpha_{ss\cdot 3}$ 也越小,若 C_i 全为 ∞ 时,即得刚性支承梁,则式(10)化为

$$\alpha_{ss\cdot 3} = \frac{a^3}{3EI_a} + \frac{l_1 a^2}{3EI}\left(1 - \frac{l_1}{4l}\right) \tag{17}$$

再当 $l_1 \to 0$ 时,即得具有固定端悬臂梁的轴端影响系数 $a^3/(3EI_a)$,此即 $\alpha_{ss\cdot 3}$ 不可能小过的极限值。理论分析表明, $a^3/(3EI_a)$ 也是 n 支承梁 $\alpha_{ss\cdot n}$ 不可能小过的极限值。

分析式(9)、(b)及(e)可知,当中支承紧靠前支承($l_1 \to 0$)或后支承($l_1 \to l$)时,如果紧靠的两个支承中,有一个刚度很小,有一个刚度很大,同时另一支承的刚度也很大,那么,这个刚度小的支承的反力很小,即其作用很小。推而广之,同一支承中的两个轴承,刚度也不宜相差过大。

3. 关于悬伸长度 a 与主轴直径

同两支支承主轴部件一样,悬伸越短越好,直径越大越好。

4. 关于最佳支承距

从理论上讲,保证 $\alpha_{ss\cdot3}$ 极小值的 l 与 l_1 应从 $\partial \alpha_{ss\cdot3}/\partial l = 0, \partial \alpha_{ss\cdot3}/\partial l_1 = 0$ 这一组方程联立解出,但这太烦琐了。当然可用电子计算机求解。但在一般情况下,取 $l = l_0$ 或 $l_1 = l_0$(l_0 为两支承主轴部件最佳支承距[1,2])就可以了。因为分析式(10)可知,去掉第 2 支承代以 R_2,且取 $l = l_0$,或去掉第 3 支承代以 R_3,且取 $l_1 = l_0$,α_{ss} 均达极小值,但 $\alpha_{ss}^{i2}/\alpha_{ii}^{d}$ 除 $\alpha_{is} = 0$ 外一般不会是极小值,因此,所得到的 α_{ss} 仍是较小的。下面的举例计算也证明了这点,而且还表明了对待三支承结构应从两方面评价:

① 当前中支承距或前后支承距为 l_0 时,后支承或中支承对加强静刚度的作用是小的;

② 当 l 远大于 l_0 时,如果增加中支承,并使 $l_1 = l_0$,那么中支承对加强静刚度的作用十分巨大。

这就是说,从静刚度观点看来,如果结构上没有什么限制,前后支承距可任选,可将 l 取为 l_0,那么,采用三支承结构的好处不大;如果结构上的原因使得 l 远大于 l_0,采用三支承结构并正确选定 l_1,就十分必要。

在选择 $l_1 = l_0$ 后,在 P 的作用下,如无后支承,中支承以后的主轴尾部将与原轴线平行[1,2];如 C_2 也很大时,这就意味着尾部偏离原轴线很小。根据位移互等原理,后支承位置可任意选择,其影响很小。

当选择 $l = l_0$ 后,显然,中支承不宜选在节点上,而宜于偏离节点并向前支承靠近。

通常采用三支承结构是由于 l 较大所致。因此,在给定 l 后,要想精确地确定 l_1 的大小,需在 $l_1 = l_0$ 附近作出 $\alpha_{2s}^2/\alpha_{22}^d$ 的函数图像,求出与其极大值相应的 l_1(详见后面算例及图 3)。

当只有 M 作用时,显然应选 $l_1 = l_0^M$ 或 $l = l_0^M$(l_0^M 是 M 作用时两支承主轴部件的最佳支承距),$l_0^M < l_0$[1][2]。

对作用在悬伸部分的 Q 而言,自然应选取 $l_1 = l_0^Q$ 或 $l = l_0^Q$(l_0^Q 是 Q 作用在悬伸部分时两支承主轴部件的最佳支承距),$l_0^Q > l_0$[1][2]。

如果 P、M 及作用在悬伸部分的 Q 同时作用,则应综合考虑,以选择最佳支承距 l_t;显然,$l_0^M < l_t < l_0^Q$。但一般仍可选 $l_t \approx l_0$。

应指出,文献[5]推荐取 $l - l_0 = 0.58l$,是错误的,但这数据至今还为一些文献所引用。尽管文献[5]讨论的是刚性支承的三支承梁的最佳支承距,但在推导此最佳支承距的过程中,弄错了材料力学公式。

5. 关于传动件的位置

传动件的位置即 Q 的位置。

当 Q 作用在悬伸部分时,由于悬伸部分无节点,故参数选择与 P 作用时的类似,最佳支承距的选择才能如上述。

当 Q 作用在支承间或尾部时,显然应力求 $\alpha_{Q.3}=0$。这就要求解 $\alpha_{Q.3}=0$ 的方程,求出 Q 的位置。但这很麻烦,建议根据式(16)作出函数图像来求:$\alpha_{Q.3}=\alpha_{Q.3}(b)$。$Q$ 作用在支承间时,b 是 Q 距前支承的距离;Q 作用在尾部时,b 是 Q 距后支承的距离。此时,是以 R_2 取代中支承来应用式(16),式(16)中的影响系数可参考文献[1]进行计算。

分析表明,在正确选择支承距的条件下,转动件在尾部时一般应靠近后支承安装;而当转动件在支承间时,如 $l_1=l_0$,则靠近由前中支承确定的节点 j_{12} 安装,如 $l_1=l_0$ 时,则靠近中支承(前后支承所确定的节点 j_{13} 同中支承之间)安装。

三、计算举例

[例一] 已知某一 $\phi400$ 毫米卧式车床,采用三支承主轴部件,它的有关参数是:$D_a=115$ 毫米,$D_l=88$ 毫米,$a=118$ 毫米,$l=640$ 毫米,$l_1=375$ 毫米,$C_1=150$ 千克/微米,$C_2=70$ 千克/微米,$C_3=100$ 千克/微米。分析采用中支承的作用及有关问题。

分析计算(不考虑 M 的作用):

(1) 如果无中间支承,则

$$\alpha_{ss}=\frac{a^3}{3EI_a}+\frac{la^2}{3EI}+\frac{1}{C_1}\left(1+\frac{a}{l}\right)^2+\frac{1}{C_3}\left(\frac{a}{l}\right)^2$$

$=(0.304\times10^{-2}+4.784\times10^{-2}+0.928\times10^{-2}+0.035\times10^{-2})$ 微米/千克

$=6.05\times10^{-2}$ 微米/千克

显然,第二项(简支梁本身变形所引起的轴端位移)占绝大部分,几达 80%,这就是由于 l 太长的缘故。今按给定的 l_1 增加中支承,则有下列结果:

$\alpha_{2s}=-4.91\times10^{-2}$ 微米/千克(负号表示位移向上);

$\alpha_{22}^d=10.92\times10^{-2}$ 微米/千克;

$\alpha_{2s}/\alpha_{22}^d=-0.449$(表示 R_2 向下);

$\alpha_{2s}^2/\alpha_{22}^d = 2.19 \times 10^{-2}$ 微米/千克；

$\alpha_{ss\cdot3} = \alpha_{ss} - \alpha_{2s}^2/\alpha_{22}^d = 3.86$ 微米/千克。

$\alpha_{ss\cdot3}$ 较 α_{ss} 减少 35% 以上。如将 l_1 取为 $l_0 = 233$ 毫米[1]，则可算得 $\alpha_{2s}^2/\alpha_{22}^d = 2.58$ 微米/千克，故

$\alpha_{ss\cdot3} = \alpha_{ss} - \alpha_{2s}^2/\alpha_{22}^d = 3.47$ 微米/千克，此 $\alpha_{ss\cdot3}$ 较上 $\alpha_{ss\cdot3}$ 减少 10%，较 α_{ss} 减少 42% 以上。

(2) 为寻求 l_1 的最佳值与 R_2 (或 $\alpha_{2s}/\alpha_{22}^d$) 的变化情况，按式(b)、(e)计算可得图 3。图 3 表明：

① l_1 的最佳值不明显，如图 3(d)，在 200~275 毫米之间，约在 230 毫米附近，与 l_0 的值极为接近；

② 如图 3(c)所示，R_2 的最小值约在 $l_1 = 370$ 毫米处 (此时 $\alpha_{2s}^2/\alpha_{22}^d = 2.16$ 微米/千克)，R_2 的最大值 l_1 约在 110 毫米、540 毫米附近 (此时 $\alpha_{2s}^2/\alpha_{22}^d$ 值分别为 2.07、1.54)可见，R_2 最大值并非 l_1 最佳值处，它的 $\alpha_{2s}^2/\alpha_{22}^d$ 值甚至比 R_2 最小值处的还小；

③ 如图 3(a)所示，只有前后支承时，存在节点，节点距前支承约 17.5 毫米，这与从文献[1]中查线图的结果相符；

④ 如将 C_2 改为 100 千克/微米，C_3 改为 70 千克/微米，$\alpha_{ss\cdot3}$ 将有所下降，但很不明显，这是因为中、后支承对加强刚度的作用远较前支承为小。

[例二] 已知某一 $\phi 400$ 毫米普通车床，采用两支承主轴部件，它的有关参数如下：$D_a = 115$ 毫米，$D_l = 88$ 毫米，$a = 118$ 毫米，$l = l_0 = 221$ 毫米，$C_1 = 150$ 千克/微米，$C_2 = 100$ 千克/微米。今增加 $C = 70$ 千克/微米的第三支承，分析它对加强刚度的作用。

分析计算(不考虑 M 的作用)：

(1) 如无第三支承时，则

$$\alpha_{ss} = \frac{a^3}{3EI_a} + \frac{al^2}{3EI} + \frac{1}{C_1}\left(1 + \frac{a}{l}\right)^2 + \frac{1}{C_2}\left(\frac{a}{l}\right)^2$$

$= (0.304 \times 10^{-2} + 1.765 \times 10^{-2} + 1.510 \times 10^{-2} + 0.257 \times 10^{-2})$ 微米/千克

$= 3.836 \times 10^{-2}$ 微米/千克

此时的 α_{ss} 比例一中 $l_1 = 375$ 毫米，$l = 640$ 毫米时的 $\alpha_{ss\cdot3}$ 还略小，即只是支承距选择不同，而其他参数相同时，这种两支承主轴部件比上面三支承主轴部件的静刚度还略高，但低于改用 $l_1 = 233$ 毫米时的刚度。

(2) 为分析第三支承的作用，将第三支承 i 从与前支承重合处起变化至距前支承 1000 毫米处止，据式(b)、(c)、(e)、(f)，以 R_i 代替第三支承 i，绘出图 4。图 4

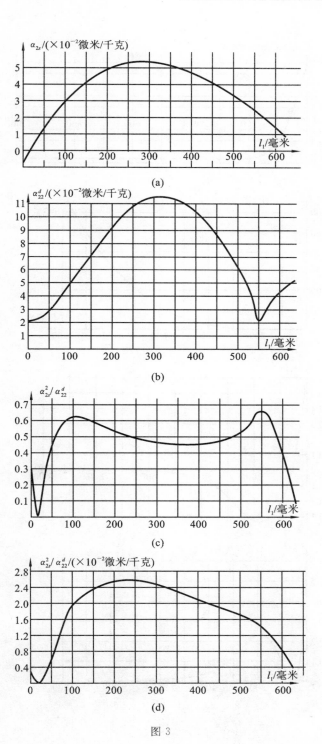

图 3

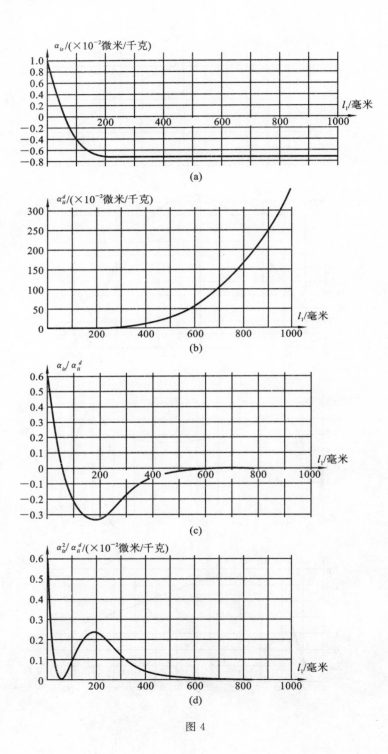

图 4

表明：

① 第三支承对加强刚度的作用并不显著，在 $l_1=62$ 毫米处为节点。

② 第三支承最佳位置在前支承处，此时 $\alpha_{is}^2/\alpha_{ii}^d$ 大于 0.5×10^{-2} 微米/千克，$\alpha_{ss\cdot 3}$ 较 α_{ss} 减少约 13% 以上；其次，是在 $l_1=190$ 毫米附近，$\alpha_{is}^2/\alpha_{ii}^d=0.24\times10^{-2}$ 微米/千克，但 $\alpha_{ss\cdot 3}$ 较 α_{ss} 减少不及 7%。

③ 第三支承移到主轴尾部后，$\alpha_{is}^2/\alpha_{ii}^d$ 一直下降，当距前支承达 500 毫米以上时，$\alpha_{is}^2/\alpha_{ii}^d\to 0$。

④ 第三支承对轴端的影响系数 α_{si} 从 221 毫米处起就是常数 0.715 微米/千克，这是自然的。因为对 $l=l_0$ 的主轴部件，其尾部平行原轴线[1][2]，α_{is} 为常数。

⑤ 第三支承在 $l_1\to 180$ 毫米处，$\alpha_{2s}/\alpha_{22}^d$ 最大，反力最大。

应指出，当前支承刚度为 $C_1+C=250$ 千克/微米时，如取此时的最佳支承距，α_{ss} 将进一步下降。

以上两个例题充分表明，上节关于对三支承结构从两个方面的评价是正确的，关于最佳支承距的选取是能够满足一般要求的。

四、考虑支承反力矩时主轴部件刚度的计算

研究支承反力矩 M_c 对三支承主轴部件静刚度的影响，如同对两支承主轴部件一样，主要是研究推力支承的位置、径向支承的支承距、传动件的位置如何选择，以保证 $\alpha_{ss\cdot 3}^{M_c}$、$\gamma_{ss\cdot 3}^{M_c}$、$\alpha_{Q\cdot 3}^{M_c}$ 为最小。

如图 5 所示，在 P 作用下，弹性支承 W 产生反抗主轴断面转动的反力矩 M_c。支承 W 可以是单独的推力支承，也可以是能产生反力矩的承受径向力的支承。显然，后一种支承可抽象为一个径向支承与一个推力支承 W，这时支承 W 的位置当然与径向支承的位置重合。

如同推导式(6)、(7)、(10)一样，当以 R_i 代替第 i 支承，以 M_c 代替支承 W，则具有反力矩的第三支承结构化为两支承结构，其上作用载荷 P、R_i、M_c，采用类似的正反方向的规定，列出第 i 支承处主轴位移的关系式、支承 W 处主轴转角的关系式、主轴轴端位移的关系式，并设支承 W 的角柔度为 K_w，则可求得计算 M_c、R_i 及 $\alpha_{ss\cdot 3}^{M_c}$ 的矩阵形式公式如下：

$$\begin{bmatrix} M_c \\ R_i \end{bmatrix} = P \begin{bmatrix} K_w+\delta_{ww} & \beta_{wi} \\ \gamma_{iW} & \dfrac{1}{C_i}+\alpha_{ii} \end{bmatrix}^{-1} \begin{bmatrix} \beta_{Ws} \\ \alpha_{is} \end{bmatrix} \qquad(18)$$

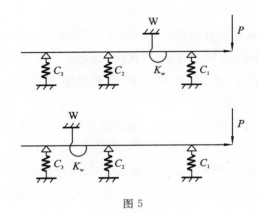

图 5

$$\alpha_{ss\cdot 3}^{M_0} = \alpha_{ss} - \begin{bmatrix} \gamma_{sW} & \alpha_{si} \end{bmatrix} \begin{bmatrix} K_w + \delta_{ww} & \beta_{Wi} \\ \gamma_{iW} & \frac{1}{C_i} + \alpha_{ii} \end{bmatrix}^{-1} \begin{bmatrix} \beta_{Ws} \\ \alpha_{is} \end{bmatrix} \tag{19}$$

式(18)式中 2×2 阶的逆阵如下计算：

$$\begin{bmatrix} K_w + \delta_{ww} & \beta_{Wi} \\ \gamma_{iW} & \frac{1}{C_i} + \alpha_{ii} \end{bmatrix}^{-1} =$$

$$\begin{bmatrix} \frac{1}{C_i} + \alpha_{ii} & -\beta_{Wi} \\ -\gamma_{iW} & K_w + \delta_{ww} \end{bmatrix} \bigg/ \begin{vmatrix} K_w + \delta_{ww} & \beta_{Wi} \\ \gamma_{iW} & \frac{1}{C_i} + \alpha_{ii} \end{vmatrix}$$

同理,可求得在 M、Q 作用下的算式：

$$\begin{bmatrix} M_c \\ R_i \end{bmatrix} = M \begin{bmatrix} K_w + \delta_{ww} & \beta_{Wi} \\ \gamma_{iW} & \frac{1}{C_i} + \alpha_{ii} \end{bmatrix}^{-1} \begin{bmatrix} \delta_{Ws} \\ \gamma_{is} \end{bmatrix} \tag{20}$$

$$\gamma_{ss\cdot 3}^{M_0} = \gamma_{ss} - \begin{bmatrix} \gamma_{sW} & \alpha_{si} \end{bmatrix} \begin{bmatrix} K_w + \delta_{ww} & \beta_{Wi} \\ \gamma_{iW} & \frac{1}{C_i} + \alpha_{ii} \end{bmatrix}^{-1} \begin{bmatrix} \delta_{Ws} \\ \gamma_{is} \end{bmatrix} \tag{21}$$

$$\begin{bmatrix} M_c \\ R_i \end{bmatrix} = Q \begin{bmatrix} K_w + \delta_{ww} & \beta_{Wi} \\ \gamma_{iW} & \frac{1}{C_i} + \alpha_{ii} \end{bmatrix}^{-1} \begin{bmatrix} \beta_{WQ} \\ \alpha_{iQ} \end{bmatrix} \tag{22}$$

$$\alpha_{sQ\cdot 3}^{M} = \alpha_{sQ} - \begin{bmatrix} \gamma_{sW} & \alpha_{si} \end{bmatrix} \begin{bmatrix} K_w + \delta_{ww} & \beta_{Wi} \\ \gamma_{iW} & \frac{1}{C_i} + \alpha_{ii} \end{bmatrix}^{-1} \begin{bmatrix} \beta_{WQ} \\ \alpha_{iQ} \end{bmatrix} \tag{23}$$

当支承 W 在前支承或在中、后支承时,只要将脚标 W 换成 1 或 2、3 即可。当求反力 R_1 或 R_2、R_3 时,只要将脚标 i 换成 1 或 2、3 即可;当然,影响系数 α_{ss}、γ_{ss}、α_{sQ} 的各自的具体算式随 R_i 的脚标不同而不同。

应指出,也可以只用 M_c 代替支承 W,将有 M_c 的三支承结构转化为无 M_c 时的三支承结构,则在 P 作用下,可求得类似式(6)、(10)的算式:

$$M_c = \frac{\beta_{ws\cdot 3}}{K_w + \delta_{ww\cdot 3}} P \tag{j}$$

$$\alpha_{ss\cdot 3}^{M_0} = \alpha_{ss\cdot 3} - \frac{\beta_{ws\cdot 3}}{K_w + \delta_{ww\cdot 3}} \gamma_{sw\cdot 3} \tag{k}$$

也可以只用 R_i 代替第 i 支承,将有 M_c 的三支承结构转化成有 M_c 的两支承结构,也可求得类似式(6)、(10)的算式:

$$R_i = \frac{\alpha_{is}^{M_0}}{\frac{1}{C_i} + \alpha_{ii}^{M_0}} P \tag{l}$$

$$\alpha_{ss\cdot 3}^{M_0} = \alpha_{ss}^{M_0} - \frac{\alpha_{is}^{M_0}}{\frac{1}{C_i} + \alpha_{ii}^{M_0}} \alpha_{si}^{M_0} \tag{m}$$

其实,式(18)正是式(j)、(l)的联合展开式的矩阵形式,式(19)则是式(k)或式(m)的展开式的矩阵形式,式(18)、(19)直接用两支承时的影响系数计算,而式(j)~(m)则用少一个支承时的影响系数来表达多一个支承时的影响系数,物理概念较具体。研究问题较方便,而计算较繁,最后还要化成式(18)、(19)的展开式。

以上讨论了 P 作用时的情况,对于 M、Q 作用时的情况,完全可得类似于式(11)、(13)和式(14)、(16)的形式与结论。

分析以上各式,得到下列结论。

(1) 从式(k)可知,对三支承结构,有 M_c 的轴端影响系数 $\alpha_{ss\cdot 3}^{M_0}$ 必不大于无 M_c 的轴端影响系数 $\alpha_{ss\cdot 3}$,因为式(k)中前后两项必为正(后项最小为零,不可能为负,因为 $\beta_{ws\cdot 3} = \gamma_{sw\cdot 3}$),两项要相减。从式(m)可知,对有 M_c 的两支承结构,增加第三支承,$\alpha_{ss\cdot 3}^{M_0}$ 必不大于 $\alpha_{ss}^{M_0}$;只有当后项为 0 时,两者才相等。广而言之,对 P、M 与作用在悬伸部分的 Q 而言,M_c 的存在一般必能减小轴端影响系数,加强主轴部件静刚度。

(2) 最佳支承距的选择可按无 M_c 时情况处理,这个结论是很自然的。取 $l_1 = l_0$ 或 $l = l_0$,在式(k)中,前项已是十分接近极小值(见上节),后项一般不为零(支承 W 在前支承时就决不为零),前项再减后项,$\alpha_{ss\cdot 3}^{M_0}$ 就会更小。当然,对 M 而言,则取 $l_1 = l_0^M$ 或 $l = l_0^M$,对作用在悬伸部分的 Q 而言,则取 $l_1 = l_0^Q$ 或 $l = l_0^Q$。如 P、M 及作

用在悬伸部分的 Q 同时作用,则应综合考虑,选取 l_1 或 l 为 l_t。一般可取 $l_t \approx l_0$。

(3) 支承 W 的作用在前支承远优于在中、后支承,且当 $K_1 \to 0$,而 $C_1 \to \infty$ 时,轴端影响系数将近于其极限值,即近于具有固定端的悬臂梁的轴端影响系数,而与 l_1、l、C_2、C_3 及支承间主轴直径几乎无关。

(4) 传动力 Q 的位置,当 Q 作用在悬伸部分时已如上述;当 Q 作用在支承间或尾部时,力求作用在节点上。节点的求法可令 $\alpha_{Q}^{M_0}{}_{.3} = 0$,或作出 $\alpha_{Q}^{M_0}{}_{.3} = \alpha_{Q}^{M_0}{}_{.3}(b)$ 的函数图像,求出节点的位置,b 为 Q 距有关支承的距离。但可以指出,当支承 W 在前支承时,与无 M_c 时比较:支承间的节点更靠近中支承[1],传动件也更应靠近中支承;尾部可能没有($l_1 = l_0$ 时)或一般没有($l = l_0$ 时)节点,传动件一般宜靠近后支承。

五、关于支承孔不同轴问题的一些分析

首先讨论无 M_c 时三支承的情况,然后考虑有 M_c 时的情况。要讨论的主要问题是支承孔的不同轴对主轴部件受力与变形的影响及解决办法。

图 6 中,"△"表示支承孔的位置,"▲"表示支承在主轴装配后的实际位置。设以两个支承孔所决定的轴线为基线 L,第 i 个支承孔偏离此基线为 Δ_i(在图 6 (a)、(c)中,Δ_1、Δ_3 应垂直于 L,但因 Δ_i 很小,故可取为垂直于 1、3 孔中心的联线),装配后支承的变形为 δ_i,支承 i 的反力为 R_i^{Δ}。R_i^{Δ} 必与 Δ_i 的偏离方向相反,故令 Δ_i 朝某一方向为正,则 R_i^{Δ} 朝与此相反的方向为正。与前面的分析一样,当以 R_i^{Δ} 代

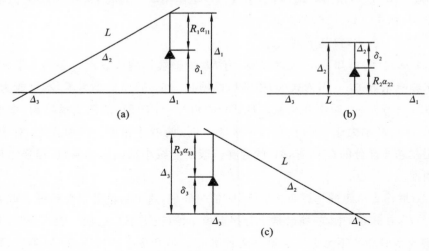

图 6

替支承 i，将三支承结构转化为两支承结构时，则对主轴在第 i 支承处的位移而言，有如下关系式：

$$\Delta_i - R_i^\Delta \alpha_{ii} = \delta_i \tag{24}$$

$$R_i^\Delta = C_i \delta_i \tag{25}$$

联解上列两式，可得

$$R_i^\Delta = \frac{\Delta_i}{\frac{1}{C_i} + \alpha_{ii}} \quad \text{或} \quad R_i^\Delta = \Delta_i / \alpha_{ii}^d \tag{26}$$

显然，对基线而言，Δ_2 与 Δ_1、Δ_3 的方向是相反的，R_2^Δ 与 R_1^Δ、R_3^Δ 的方向也是相反的。应指出，式(26)完全与式(6)、(11)、(14)相呼应，只要考虑到 Δ_i 相当于 $P\alpha_{is}$、$M\gamma_{is}$、$Q\alpha_{iQ}$ 而已。

反力 R_i^Δ 引起轴端相对于基线的位移为

$$y_s^{R_i^\Delta} = R_i^\Delta \alpha_{si} \tag{27}$$

分析式(26)、(27)可知：

(1) $y_s^{R_i^\Delta} \propto R_i^\Delta \propto \Delta_i$，所以，不同轴度要小。

(2) 任意一个支承的刚度 $C_j (j=1,2,3)$ 如增加，则所有支承的反力 $R_i^\Delta (i=1,2,3)$ 都增加，当 C_1、C_2、$C_3 \to \infty$ 时，R_i^Δ 达最大值，此即为刚性支承时的情况，即

$$R_1^\Delta = \frac{3EI}{ll_1^2} \cdot \Delta_1 \quad R_2^\Delta = \frac{3EIl}{l_1^2 (l-l_1)^2} \cdot \Delta_2$$

$$R_3^\Delta = \frac{3EI}{l(l-l_1)^2} \cdot \Delta_3 \tag{28}$$

(3) 上一点的另一面，任意一个支承的刚度如 C_j 下降，则所有支承的反力 R_i^Δ 都下降。显然，当某一 $C_j \to 0$，则 $\frac{1}{C_j} + \alpha_{ii} \to \infty$，所有 $R_i^\Delta \to 0$。这就是说，将某一支承中的轴承采用间隙配合（当然，此间隙应能抵消 Δ_i 的影响，即轴承直径上的间隙应不小于 $2\Delta_i$ 时），则主轴部件在装配后由 Δ_i 引起的支承反力均为零。其实，这是不言而喻的，当三个支承中有一个支承存在足够的间隙时，它就不起支承作用，从而就成为两支承主轴部件，也谈不上有什么反力 R_i^Δ 了。

显然，不能将前支承这个极为关键的支承在装配时给予间隙，C_1 是应大力加强的。因此，当取 $l_1 = l_t$ 时，后支承要采用间隙配合，当取 $l = l_t$ 时，中支承采用间隙配合。这样，不但主轴部件刚度是大的，而且在工作过程中，第三支承起了作用后，主轴部件刚度的变化也是小的。

现分析后支承有初间隙 $2\Delta_3$ 时，外力 P 达到多大，后支承才开始工作。显然，

P 与 Δ_3 空间位置关系不同,情况也就不同,今讨论以下两种极端的情况。

(1) 当 P 与 Δ_3 处于同一平面,且方向相同(见图7(a)),$l_1 = l_0$ 时,则 P 一作用,后支承就开始工作,这是所希望的。

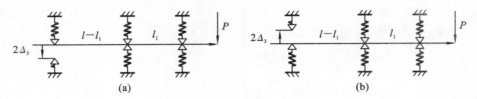

图 7

(2) 当 P 与 Δ_3 处于同一平面,但方向相同反(见图7(b)),当 $l_1 = l_0$,则因尾部平行于原轴线,其间偏离的距离就等于后支承变形的大小 $Pa/(C_2 l_0)$,从而可得:

$$Pa/(C_2 l_0) \geqslant 2\Delta_3 \quad \text{或} \quad P \geqslant 2C_2 l_0 \Delta_3/a \tag{29}$$

时,后支承方开始工作。

当 P 与 Δ_3 不在同一平面,则只要将 P 分解为与 Δ_3 同一平面的和与 Δ_3 相互垂直的两个分量来考虑。如 P 与 Δ_3 成 α 角,则当 $P\cos\alpha$ 与 Δ_3 同向时,P 一作用,后支承也就开始工作。

现考虑有 M_c^Δ 时的情况,如同推导式(18)、(26)一样,当以 R_i^Δ 代替第 i 支承,以 M_c^Δ 代替支承 W,则三支承结构转化为两支承结构,主轴受有载荷 R_i^Δ、M_c^Δ,从而可求得

$$\begin{bmatrix} M_c^\Delta \\ R_i^\Delta \end{bmatrix} = \begin{bmatrix} K_W + \delta_{WW} & \beta_{Wi} \\ \gamma_{iW} & \dfrac{1}{C_i} + \alpha_{ii} \end{bmatrix}^{-1} \begin{bmatrix} 0 \\ \Delta_i \end{bmatrix} \tag{30}$$

从式(30)可明显看到:

(1)支承 W 的存在,不但产生了 M_c^Δ,而且改变了 R_i^Δ 的大小,不过,M_c^Δ、R_i^Δ 仍然正比于 Δ_i;

(2)但是,只要有一个支承刚度为零,即在装配时若有足够的初间隙,则由式(25)可知,$R_i^\Delta = 0$,从而由式(30)可知,Δ_i 不起作用,$M_c^\Delta = 0$。这一点完全与无支承 W 时的相同。

六、小　　结

1) 三支承结构与两支承结构比较。

(1) 在 P、M 与作用在悬伸部分的 Q 的作用下会出现以下情况:

① 如果是在原两支承的结构基础上,增加第三支承,只要第三支承不是增加在节点上,对 M 而言,或者不是增加在 $\gamma_{is}=0$ 处,就一定能减小轴端影响系数,提高主轴部件刚度;

② 特别是当前后支承距远大于两支承结构的最佳支承距 l_i 时,增加中间支承,只要不是增加在节点上,对 M 而言,或者不是增加在 $\gamma_{is}=0$ 处,便能够增加而且能显著增加主轴部件刚度;

③ 但是当前后支承距在 l_i 附近时,增加第三支承,对增加主轴部件刚度的效果是小的;

④ 而当第三支承增加在节点上时,对 M 而言,或者增加在 $\gamma_{is}=0$ 处时,是不会增加,但也决不会减小主轴部件静刚度的;

⑤ 如果三支承结构与两支承结构的支承距都是任选的,而其他参数是相同的,则需根据两者支承距的情况才能断定何者刚度为高。

(2) Q 作用在支承间或尾部时,增加第三支承,可能提高,可能减小,也可能不改变主轴部件刚度。

2) 采用三支承结构时,无论有无支承反力矩 M_c 的存在,可根据结构条件,将前中支承距 l_1 或前后支承距 l 选为 l_t(一般可取($l_t \approx l_0$),而第三支承应偏离节点。

3) 支承反力矩 M_c 的存在(即只要支承 W 不是装在 β_{wS}、δ_{wS}、β_{wQ} 等转角为零处),则对 P、M 与作用在悬伸部分的 Q 而言,一定能减小轴端影响系数,提高主轴部件刚度。

4) 支承 W 在前支承远优于在后支承;且当支承 W 与前支承的刚度极大时,主轴轴端影响系数几乎与支承距 l、l_1,中、后支承刚度 C_2、C_3 以及支承间的主轴直径无关;而且无论 P、M、Q 如何作用,主轴轴端影响系数接近于其极限值,即具有固定端悬臂梁的端部影响系数[1]。

5) 传动件的位置应力求与节点的位置相重合,而节点的位置可按 $\alpha_{sQ.3}=0$ 或 $\alpha_{sQ.3}^{M_0}=0$ 来决定。但可指出:

① 悬伸部分无节点,故传动件宜靠近前支承安装;

② 支承间有节点,当 $l_1=l_0$ 时,节点接近由前中支承决定的节点 j_{12};当 $l=l_0$ 时,节点接近中支承;

③ 尾部往往无节点,故传动件一般宜靠近后支承安装。

6) 支承孔的不同轴,在装配主轴部件时势必引起支承反力 R_i^A 与反力矩 M_c^A,其值与不同轴度成正比,并随支承刚度的增加而增加。消除这一不利影响的措施,是将同前支承距离不为最佳支承距的支承在装配时给予初间隙,即轴承直径

上的间隙不小于 $2\Delta_i$。

7) 如同两支承结构一样,设计主轴部件时,应综合考虑 P、M、Q 的作用,来决定有关参数。

参 考 文 献

[1] 杨叔子. 机床两支承主轴部件静刚度的分析与计算.《机床》,1979 年,第 3 期.

[2] 杨叔子. 机床主轴部件静刚度的分析与计算.《华中工学院学报》,1978 年,第 1 期.

[3] 华中工学院机床教研室.《金属切削机床》,1973 年.

[4] л. Вейц, В. К. Догдоагский, В. И. Черяев.《Вынуждеггые Колощнх Станках》. Мащиностроение,1973.

[5] 后藤 渡边. 旋盘仕上面精度の向上策.《机械技术》,Vol. 12,No2,1964.

[6] J. Koch. Die "Bezongene Starrheit"—Die Bestimmung der Hauptspindel von Werkzeugmaschinen aufgrund einer Neuen Starrheits-betrachtung.《Machinenmarket》,Jg. 75,Nr19,1969.

[7] 汤本诚治,宫川磐. 普通旋盘の仕样汇关すろ调查.《精密机械》,37,卷 7 号,1971.

[8] H. Regab. *Dynamc Behaviour of Machine Tool Spindles Mounted in Three Bearings.* 《Machinery and Production Engeering》,Vol. 124,No3193. 1974.

[9] K. Parsiegla. *Die Verlagerung Hydrostatisch Gelagerter Spindeln bei Verwendung eines Taschen-Axiallagers.* Industrie-Anzeiger,Jg. 98,Nr. 67,1976.

[10] R. Zdenkovic, V. Dukovski. Die Steifigkeit von Werkzeug maschinenspindeln unter Ein-wirkung der Arbeitskraft.《Werkstatt und Betrieb》,Jg. 111,Nr. 2,1978.

[11] 广东师范学院数力系,广州机床研究所静压室. 多支承静压轴承主轴系统刚度的计算与试验.《机床与液压》,1978 年,第 2 期.

(原载《制造技术与机床》1979 年第 9 期)

δ 函数在机械制造中的应用

杨叔子　师汉民

提　要　本文从物理本质出发,归纳了 δ 函数在机械制造中某些较为典型的应用,并给出了若干理论解释。同时,进一步阐述了 δ 函数对于机械系统分析的意义,指出这一概念的引用沟通了周期函数的频谱分析与非周期函数的频谱分析、分布量与集中量、连续量与间断量以及过程激励与初始激励之间的联系。

一

δ 函数近十几年来逐渐进入机械制造领域,以其奇特而有用的性质为进行动态测试与数据分析提供了一些有效的理论与方法。这种函数在机械制造领域中的应用,就其物理本质来讲,可归纳为以下四个主要方面。

(一) 用冲击激振研究机床或其他机械结构的动态性能

对系统采用冲击激振,就是以 δ 函数作为输入,测出系统的脉冲响应,从而掌握机床在外力 $f(t)$ 作用下的响应规律 $f(t)*h(t)$,$h(t)$ 为系统的单位脉冲响应。

从频域方面看,δ 函数的 Fourier 变换为 $F[\delta(t)]=1$,这表明 $\delta(t)$ 中含有从 $-\infty$ 到 ∞ 的各种频率的谐波,各谐波的幅值均为 1,彼此间无相移。这样,δ 函数的"一击"就相当于同时进行了所有频率的谐波激振,足以激发系统的所有模态。并且,$\delta(t)$ 同等地供给各阶模态以能量,这就不仅可测出各个固有频率,而且可辨别各阶模态作用的主次。

从时域方面看,已知一个 n 阶系统的单位脉冲响应为[1]

$$h(t) = \sum_{j=1}^{q} a_j e^{-p_j t} + \sum_{l=1}^{r} e^{-\xi_l w_l t}\left[b_l\cos(\omega_l\sqrt{1-\xi_l^2}\,t) + c_l\sin(\omega_l\sqrt{1-\xi_l^2}\,t)\right]$$

式中,$q+2r=n$;a_j、b_l、c_l 为常系数;ω_l、ξ_l 分别表示系统中第 l 个二阶振荡环节的无阻尼固有频率与阻尼率。显然,上式中除了 $a_j e^{-p_j t}$ 这种指数衰减项外,就是各阶模态的组合,这与频域中的分析是一致的。

现已有较成熟的冲击激振试验技术,其理论根据正在于此。

(二) 用白噪声作为输入研究机床等系统或机械加工等过程的动态性能

采用白噪声这一随机信号作为输入,既便于在线测试,又具有较强的抗干扰能力。只要对输入和输出信号进行相关分析,就可求得系统的单位脉冲响应的最佳估计 $\hat{h}(t)$。

已知 $\hat{h}(t)$ 应满足以下 Wiener-Hopf 方程。

$$R_{xy}(\tau) = \int_0^\tau \hat{h}(\lambda) R_{xx}(\tau-\lambda) d\lambda \tag{1}$$

式中:$R_{xy}(\tau)$ 是输入 $x(t)$ 与输出 $y(t)$ 的互相关函数;$R_{xx}(\tau)$ 是输入 $x(t)$ 的自相关函数。已知 $R_{xy}(\tau)$ 与 $R_{xx}(\tau)$,求解积分方程式(1)即可得到 $\hat{h}(\tau)$。为简化求解,可选取输入 $x(t)$ 为白噪声。因为白噪声的自相关函数为 δ 函数,即

$$R_{xx}(\tau-\lambda) = K\delta(\tau-\lambda) \tag{2}$$

式中,K 是比例系数。

以式(2)代入式(1),得

$$R_{xy}(\tau) = \int_0^\tau K\hat{h}(\lambda)\delta(\tau-\lambda) d\lambda$$

再根据 δ 函数的"筛选性质",立即解出

$$\hat{h}(\tau) = R_{xy}(\tau)/K \tag{3}$$

即在输入为白噪声时,系统的单位脉冲响应与输入输出间的互相关函数成简单的比例关系,基于这个道理研制出的随机激振装置,已经实际用于研究机床的动态性能。如果考虑到采用伪随机序列信号作为输入,仍然利用其相关函数与 δ 函数的关系,则可使测试更为方便。

在离散系统方面,同样可用白噪声作为输入,测出系统的单位脉冲响应的最佳估计 $\hat{h}(n)$。现以一个工艺系统为例[3],在宽砂轮无心磨床上磨削时,毛坯尺寸(输入)与工件尺寸(输出)之间的关系应符合离散的 Wiener-Hopf 方程:

$$R_{xy}(n) = \sum_{m=0}^n \hat{h} R_{xx}(n-m) \tag{4}$$

式中:$R_{xy}(n)$ 是毛坯尺寸与工件尺寸之间的互相关函数;$R_{xx}(n)$ 是毛坯尺寸本身的

自相关函数。如将毛坯尺寸变化规律取为白噪声(将上道工序加工出的零件次序打乱,或采用料斗进行随机上料等),则有

$$R_{xx}(n-m) = \sigma_x^2 \delta(n-m) \tag{5}$$

式中,σ_x^2 为毛坯尺寸的方差;$\delta(n-m)$ 为 δ 函数的离散形式。当 $n-m=0$ 时,$\delta(n-m)=1$,否则为 0。这就是 $\delta(n-m)$ 的筛选性质。

以式(5)代入式(4),并利用 $\delta(n-m)$ 的"筛选性质",可将式(4)解出,得

$$\hat{h}(n) = R_{xy}(n)/\sigma_x^2 \tag{6}$$

$\hat{h}(n)$ 是对机床加工过程进行动态控制或动态补偿的重要技术参数。

(三) 从被噪声所污染的随机信号中检测出周期信号

在这方面要比前两方面以更深刻的方式应用 δ 函数。在(一)中,输入本身是 δ 函数,在(二)中,输入的自相关函数是 δ 函数,而此处输入信号的自相关函数的 Fourier 变换(即功率谱)才是 δ 函数。其分析步骤如下。

第一步,进行相关分析。凡包含有周期成分的任何原始信号的自相关函数仍然含有相应频率的谐波成分。例如,含有正弦波成分 $A\sin(\omega_1 t+\theta)$ 的某信号 $x(t)$ 的自相关函数 $R_{xx}(\tau)$ 中,仍含有形如 $\frac{A^2}{2}\cos(\omega_1\tau)$ 的项。而且,原始信号中各种频率的谐波成分是各自独立地进入自相关函数的,而与其他频率的谐波成分或噪声的存在无关[2]。

第二步,将所得的自相关函数进行 Fourier 变换,化为功率谱。由于

$$F[\cos(\omega_1\tau)] = \pi[\delta(\omega-\omega_1) + \delta(\omega+\omega_1)] \tag{7}$$

如果原始信号中含有频率为 ω_1 的正弦波,则在功率谱图形上 $\pm\omega_1$ 处必有尖峰出现,如图 1 所示。即使原始信号中正弦波很弱,难于发现,但在高分辨率的功率谱中,在相应的频率处尖峰也十分陡峭。这一点,可以用于检测混杂在随机信号中的周期成分,例如,在分析加工表面粗糙度的影响因素、分析齿轮加工机床传动误差的来源、分析工作设备所受干扰的性质中,以及在类似的数据资料的分析处理中都是很有用的。

(四) 利用 δ 函数序列分析周期工作过程的频谱

采用铣刀等多齿刀具切削时,切削过程的周期为 $T=60/(nz)$,式中 n 为刀具转速,z 为刀齿数。任一周期工作过程的频谱一般可用 Fourier 级数获得。但直接以多齿刀具进行切削试验,不易保证此周期工作过程的测试精度,因为刀具刃磨

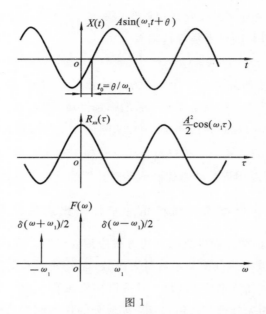

图 1

精度和机床主轴振摆等许多因素不易控制。所以,有的研究者以单齿刀具在相同条件下进行模拟切削试验[4],测出切削力等动态参数的变化规律 $f_z(t)$ ($0 < t < T$),再利用 δ 函数序列的一些性质重构多齿刀具切削过程的有关规律 $f(t)$,并求出其频谱 $F[f(t)]$。具体步骤如下。

第一步,将 $f(t)$ 分解为单齿切削规律的叠加:

$$f(t) = f_z(t) * \delta_T(t) = \sum_{n=-\infty}^{\infty} f_z(t - nT)$$

式中,$\delta_T(t) = \sum_{n=-\infty}^{\infty} \delta(t - nT)$,即为 δ 函数序列。

第二步,对 $\delta_T(t)$ 进行 Fourier 变换,有

$$F[\delta_T(t)] = \Delta_T(f) = \frac{1}{T} \sum_{n=-\infty}^{\infty} \delta\left(f - \frac{n}{T}\right)$$

而 $f_z(t)$ 的 Fourier 变换记为 $F_z(f)$,式中 f 为频率。

第三步,利用 Fourier 变换的卷积定理,有

$$F[f(t)] = F[f_z(t) * \delta_T(t)] = F_z(f) \cdot \Delta_T(f)$$
$$= F_z(f) \cdot \left[\frac{1}{T} \sum_{n=-\infty}^{\infty} \delta\left(f - \frac{n}{T}\right)\right]$$

分析过程如图 2 所示。可见,由单齿刀具模拟切削所得到的某动态参数的频谱 $F_z(f)$ 来重构多齿刀具实际切削过程的频谱,可归结为对 $F_z(f)$ 在频域内进"采

样",采样的频率间隔为原周期 T 的倒数的 2 倍。这样,用于分析周期函数频谱的 Fourier 变换就可以用于分析非周期函数频谱。

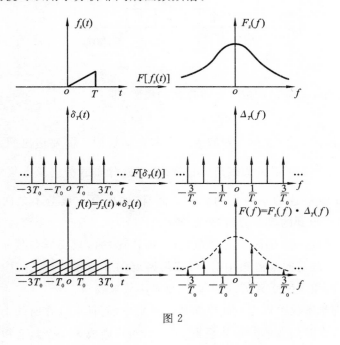

图 2

二

除了上述诸方面的具体应用以外,δ 函数引入机械系统的分析,还可有效地沟通了原先以为是互不相关的,甚至是相互对立的一些概念之间的内在联系。这对从事机械制造专业的人来说,有利于开拓思路,以便从更高的高度上来归纳与概括有关动态测试与分析的方法。

(一) 分布量与集中量

设作用在梁上分布载荷的幅值为 n,分布宽度为 $1/n$,如图 3 所示,因而作用在梁上的载荷总量为

$$\int_{-\frac{1}{2n}}^{\frac{1}{2n}} q_n(x)\,\mathrm{d}x = n\int_{-\frac{1}{2n}}^{\frac{1}{2n}} \mathrm{d}x = 1 \quad (单位力)$$

现令此分布载荷总量恒为 1 个单位力,缩短其分布宽度,增加其幅值,将分布载荷向原点集中。显然,当 $n\to\infty$ 时,$q_n(x)$ 转化为集中力的"分布",其总量为 1 单位力,作用在原点处。在这里,$q_n(x)$ 是普通函数,当 n 取值不同时,它就形成一个

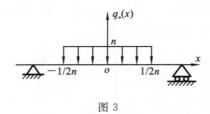

图 3

函数的序列,而 δ 函数可理解为这一序列的极限,即

$$\lim_{n\to\infty} q_n(x) = \delta(x)^*$$

这里将集中力的"分布",理解为分布载荷的极限。需要指出的是,只要保持 $\int_{-\infty}^{\infty} q_n(x)\mathrm{d}x=1$,而不必考虑载荷分布规律,当此分布载荷向原点集中时,均会形成一个以 $\delta(x)^*$ 描述的集中力的分布。而且,如果分布载荷不是向原点而是向 $x=a$ 点集中,则得到的集中力的"分布"为 $\delta(x-a)$。

同理,当考虑一个平面上的分布载荷(即比压)时,只要分布载荷总量为 1 单位力,则将分布载荷向某点集中时,可获得两维的 δ 函数。而空间中的集中力的"分布",如空间质点所受载荷,就要以三维 δ 函数来表示。

一个物理量不仅可以在空间里向某一点集中,而且可以在时间上向某一时刻集中,例如,力在时间上的分布关系为 $f(t)=\delta(t)$,则成为一个具有单位冲量的脉冲力。

由此可见,将分布函数取极限,其结果就得到了一个理想化的在空间或时间上集中的物理量的"分布函数",即 δ 函数。本来,集中物理量与分布物理量是相互对立的概念,对于它们的处理方法也不相同。但借助 δ 函数的概念,可以将集中物理量看成是一种特定的分布物理量,反过来,又可把分布物理量看做是一系列集中物理量的叠加,从而不仅在概念上,而且在处理方法上,有可能将这两者统一起来。下面以梁的挠度计算为例来证明这一点。梁的挠度曲线的微分方程为[5]

$$\frac{\mathrm{d}^4 y(x)}{\mathrm{d}x^4} = \frac{q(x)}{EI} \tag{8}$$

式中:$q(x)$ 为分布载荷;$y(x)$ 为挠度。

设梁的两端固定(见图4),有幅值为 p 的均布载荷和作用在 $x=a$ 处的集中载荷 f 作用在梁上,即

$$q(x) = pu(x) + f\delta(x-a) \tag{9}$$

---

* 严格地讲,δ 函数是在线性泛函数弱收敛意义下某些普通函数序列的极限。

式中，$u(x)$ 为单位阶跃函数。

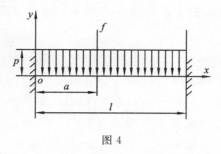

图 4

而边界条件为

$$y(0^+) = y'(0^+) = 0 \tag{10}$$
$$y(l) = y'(l) = 0 \tag{11}$$

采用 Laplace 变换，由式(8)~(11)可解得

$$y(x) = \frac{1}{2}y''(0^+)x^2 + \frac{1}{6}y'''(0^+)x^3 + \frac{px^4}{24EI} + \frac{f}{6EI}(x-a)^3 u(x-a)$$

式中，$y''(0^+)$ 和 $y'''(0^+)$ 可由上式据式(11)求出。值得注意的是，这里分布载荷 $pu(x)$ 与集中载荷 $f\delta(x-a)$ 是被同等地看待，且以同样的方式进入公式的。

(二) 连续量与间断量

本来，在 Newton-Leibniz 意义下，只有连续函数才有可能求导，但引入 δ 函数以后，函数在第一类间断点处也可求导。例如，单位阶跃函数 $u(x)$ 的导数即为 $u'(x)=\delta(x)$*。如将 $u(x)$ 理解为梁所受的切力，则 $u'(x)$ 就是分布载荷，而 $u'(x)=\delta(x)$ 正是材料力学中切力与分布载荷之间的关系。本来这一关系只对连续可微的切力分布才有意义，但引入 δ 函数后，它也可用于间断的切力，而且表明在切力的间断处载荷分布为 δ 函数，即存在一集中载荷。由此可见，δ 函数的引入将连续量与间断量沟通起来了。从这个意义上讲，对于具有第一类间断点的函数，至少在形式上可以像一般连续可微函数那样来对待和处理，例如，进行微分和积分等。

(三) 初始激励与过程激励

系统的初始条件称为初始激励，而系统的输入称为过程激励。以下证明具有

---

* 这里的导数是基于泛涵定义的广义导数。

δ 函数形式的过程激励,实际上等价于某一个初始激励。

设有某个多自由度的机械系统,如图 5 所示。对各个质块可列出运动微分方程

$$m_i\ddot{x}_i(t) - c_{i+1}\dot{x}_{i+1}(t) + (c_i + c_{i+1})\dot{x}_i(t) - c_i\dot{x}_{i-1}(t) - k_{i+1}x_{i+1}(t)$$
$$+ (k_i + k_{i+1})x_i(t) - k_i x_{i-1}(t) = f_i(t)$$
$$i = 1, 2, \cdots, n; x_n(t), x_{n+1}(t) \text{ 恒为 } 0, \tag{12}$$

设初始条件为零,在第 l 个质块上作用的力为 $f_l(t) = \delta(t)$,而其余质块上的作用力 $f(t_l) = 0, i \neq l$。在此条件下,式(12)的 Laplace 变换为

$$\begin{cases} m_l s^2 X_l(s) - c_{l+1} s X_{l+1}(s) + (c_l + c_{l+1}) s X_l(s) - c_l s X_{l-1}(s) \\ - k_{l+1} X_{l+1}(s) + (k_l + k_{l+1}) X_l(s) - k_l X_{l-1}(s) = 1 \\ m_i s^2 X_i(s) - c_{i+1} s X_{i+1}(s) + (c_i + c_{i+1}) s X_i(s) - c_i s X_{i-1}(s) \\ - k_{i+1} X_{i+1}(s) + (k_i + k_{i+1}) X_i(s) - k_i X_{i-1}(s) = 0, i \neq l \end{cases} \tag{13}$$

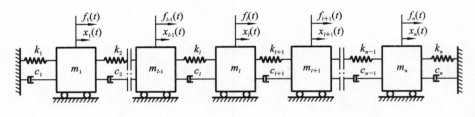

图 5

另一方面,又设整个系统不受任何外力作用,即 $f_i(t) = 0, i = 1, 2, \cdots, n$,但有一定的不全为零的初始条件 $\dot{x}_i(0)$ 与 $x_i(0), i = 1, 2, \cdots, n$。在此条件下,式(12)的 Laplace 变换为

$$m_i s^2 X_i(s) - c_{i+1} s X_{i+1}(s) + (c_i + c_{i+1}) s X_i(s) - c_i s X_{i-1}(s)$$
$$- k_{i+1} X_{i+1}(s) + (k_i + k_{i+1}) X_i(s) - k_i X_{i-1}(s) - m_i \dot{x}_i(0)$$
$$- m_i s x_i(0) + c_{i+1} x_{i+1}(0) - (c_i + c_{i+1}) x_i(0) + c_i x_{i-1}(0) = 0$$
$$i = 1, 2, \cdots, n \tag{14}$$

如果在以上两种情况下,系统具有相同的运动规律,即 $X_i(t)$ 或 $x_i(s)$ 相同,那么可由式(13)、(14)消去 $X_i(s)$ 等有关项,得出初始条件 $x_i(0)$ 和 $\dot{x}_i(0)$ 所应满足的方程为

$$\begin{cases} m_l \dot{x}_l(0) + m_l s x_l(0) - c_{l+1} x_{l+1}(0) + (c_l + c_{l+1}) x_l(0) - c_l x_{l-1}(0) = 1 \\ m_i \dot{x}_i(0) + m_i s x_i(0) - c_{i+1} x_{i+1}(0) + (c_i + c_{i+1}) x_i(0) - c_i x_{i-1}(0) = 1; i \neq l \end{cases}$$
$$\tag{15}$$

由于 s 的任意性,故可比较等式两边 s 的同次幂的系数,得出

$$\begin{cases} x_i(0) = 0, i = 1, \cdots, n \\ \dot{x}_i(0) = 0, i \neq l \\ \dot{x}_l(0) = \dfrac{1}{m_l} \end{cases} \quad (16)$$

这说明在机械系统中,对某一质块 m_l 上作用一个 δ 函数的力,其效果相当于 m_l 增加一个初速度 $\dot{x}_l(0) = 1/m_l$,而对其他初始条件无影响。事实上由式(16)有

$$\dot{x}_l(0) \cdot m_l = 1$$

表明 $\delta(t)$ 所代表的单位冲量,在作用的一瞬间转化成为质块 m_l 的单位动量,这符合冲量定理。

可以证明,当 $\delta(t)$ 所作用的质块 $m_l = 0$ 时,有

$$\begin{cases} x_l(0) = 1/(c_l + c_{l+1}) \\ \dot{x}_{l-1}(0) = c_l/m_{l-1}(c_l + c_{l+1}) \\ \dot{x}_{l+1}(0) = c_{l+1}/m_{l+1}(c_l + c_{l+1}) \end{cases} \quad (17)$$

这时 $\delta(t)$ 的作用点获得初位移 $x_l(0)$,而由 $\delta(t)$ 转化成的动量按阻尼系数 c_l 和 c_{l+1} 成正比地分配在距作用点最近的两个质块 m_{l-1} 及 m_{l+1} 上,使它们分别获得初速度 $\dot{x}_{l-1}(0)$ 及 $\dot{x}_{l+1}(0)$。由式(17),有

$$m_{l-1} \dot{x}_{l-1}(0) + m_{l+1} \dot{x}_{l+1}(0) = 1$$

表明这一结果符合冲量定理。

如果一个高阶系统的运动微分方程为

$$\begin{aligned} & a_0 x^{(n)}(t) + a_1 x^{(n-1)}(t) + \cdots + a_{n-1} \dot{x}(t) + a_n x(t) \\ & = b_0 f^{(n-1)}(t) + b_1 f^{(n-2)}(t) + \cdots + b_{n-2} \dot{f}(t) + b_{n-1} f(t) \end{aligned} \quad (18)$$

则以类似的方法可证明,当 $\delta(t)$ 的过程激励作用在该系统上面时,其等价的初始激励,即初始条件 $x^{(i)}(0) = c_i, i = 0, 1, \cdots, n-1$,满足以下代数方程组:

$$\begin{cases} b_0 = a_0 c_0 \\ b_1 = a_0 c_1 + a_1 c_0 \\ \vdots \\ b_{n-1} = a_0 c_{n-1} + a_1 c_{n-2} + \cdots + a_{n-1} c_0 \end{cases} \quad (19)$$

由此可解得:

$$\begin{cases} c_0 = x(0) = b_0/a_0 \\ c_1 = \dot{x}(0) = (b_1 - a_1 c_0)/a_0 \\ c_2 = \ddot{x}(0) = (b_2 - a_1 c_1 - a_2 c_0)/a_0 \\ \vdots \end{cases} \quad (20)$$

一般而言,在已知 $c_0, c_1, \cdots, c_i$ 的情况下,可递推出:

$$c_{i+1} = \frac{1}{a_0}(b_{i+1} - a_1 c_i - a_2 c_{i-1} - \cdots - a_{i+1} c_0)$$

例如,图 6 所示的三阶系统,其运动微分方程为

$$\frac{mc}{k_2}\dddot{x}(t) + m\ddot{x}(t) + \frac{c(k_1+k_2)}{k_2}\dot{x}(t) + k_1 x(t) = f(t) + \frac{c}{k_2}\dot{f}(t)$$

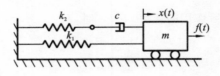

图 6

与式(18)比较,有 $a_0 = mc/k_2$, $a_1 = m$, $a_2 = c(k_1+k_2)/k_2$, $a_3 = k_1$, $b_0 = 0$, $b_1 = c/k_2$, $b_2 = 1$。按式(20)可算出 δ 函数作用在该系统上所产生的等价初始条件为 $c_0 = x(0) = 0$, $c_1 = \dot{x}(0) = 1/m$, 和 $c_2 = \ddot{x}(0) = 0$。对此问题由形如式(12)的微分方程组和式(16)也可得出同样的结果,这表明式(16)、(17)和式(20)实质上是一致的。

有趣的是,当式(18)左边只有 $f(t)$ 而无其导数项时,式(20)可大为简化,即

$$c_{n-1} = x^{(n-1)}(0) = b_{n-1}/a_0$$

而

$$c_0 = c_1 = \cdots = c_{n-2} = 0$$

总之,δ 函数对系统的作用,与一定的初始条件等价。或者说 δ 函数对于时刻 0^- 的初始条件为零的系统作用的结果,使系统在时刻 0^+ 产生一定的非零初始条件。这样,δ 函数又起到了沟通系统的过程激励与初始激励的作用。

按 δ 函数的"筛选性质",对任何形式的输入 $f(t)$,有

$$f(t) = \int_{-\infty}^{\infty} f(\tau)\delta(t-\tau)d\tau$$

这表明任一输入都可视为一系列前后相继的脉冲输入 $f(\tau)\Delta\tau \cdot \delta(t-\tau)$ 的叠加,而已经证明每一个脉冲输入又等价于 $t_0 = \tau$ 时一定的初始条件。由此可见,不仅是 δ 函数形式的过程激励,而且是任何形式的过程激励,都有可能转化为各个时刻的一系列初始激励,这样,又把系统在一定初态下的自由运动与外扰作用下的强迫运动联系起来了。

参 考 文 献

[1] 绪方胜彦. 现代控制工程(中译本). 科学出版社(1978).

[2] 井町勇. 机械振动学(中译本). 科学出版社(1979).

[3] 阳含和. 金属切削工艺过程工程控制论初步探讨. 西安交通大学学报, 第1期(1978).

[4] D. E. Gygax. *Dynamics of Single-Tooth Milling*. Annals of the CIRP, Vol. 281/1 (1979).

[5] Eginhard J. Muth. *Transform Methods*. Prentice—Hall, Inc. (1977).

[6] Leonard Meirovitch. *Elements of Vibration Analysis*. McGraw-Hill Book Company (1975).

[7] 邓必鑫. 关于 δ 函数. 长春光学精密机械学院学报, 创刊号(1978).

The Application of δ-Function in Machine-Building

Yang Shuzi　　Shi Hanmin

Abstract　Some typical applications of δ-function in machine-building are given with explanation. The significance of δ-function for the analysis of mechanical systems, the application of δ-function to establish the relations between periodic function spectrum analysis and non-periodic function spectrum analysis, distributed and concentrated quantities, continuous and discontinuous quantities as well as process excitement and initial excitement are discussed.

(原载《华中工学院学报(自然科学版)》1980年第8卷第4期)

A Study of the Static Stiffness of Machine Tool Spindles

Yang Shuzi

Abstract A thorough discussion about the radial stiffness of machine tool spindles is made. The concept of "influence factors" in mechanics is employed to study the effects of a number of factors on radial static stiffness, e. g. the structure of spindles and their parameters, especially the number of bearings (radial and thrust bearings) and their mounting positions. The graphical method has been used for determining the optimum span of a spindle mounted in two or multiple bearings. The direct and indirect influence factor methods for calculating the reaction of any bearing and the displacement of any position for an elastic beam possessing many elastic bearings have been developed (see Appendix 2) and used for analysing a spindle mounted in two and three bearings. Problems of uncoaxiality of the housing holes for a spindle possessing many elastic bearings have also been studied.

Introduction

Static stiffness of spindles is one of the important performances of machine tools. Hence it must be correctly defined and the structure and related parameters of spindles must be properly determined. The concept of "influence factors" in mechanics is made use of to discuss the definition of static stiffness (referring to radial static stiffness throughout this article) and to minimize the displacement at the spindle-nose Y_s in the selection of the structure and related parameters.

I. Discussion on the Definition of Static Stiffness of fhe Spindle

Stiffness of the spindle is defined as its ability to resist deflection under the action of cutting force. In general, "correlation stiffness" is used as the definition of static stiffness of the spindle[10, 12]. As is shown in Fig. 1, when a force exerted on the spindle-nose is P, the displacement at the spindle-nose in the same direction as that of P is Y_s^p and correlation stiffness is defined as $J_s^p = P/Y_s^p$.

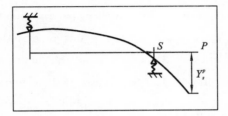

Fig. 1

The advantage of this definition is that the magnitude of the displacement of the spindle-nose which is just where the cutting forces are transmitted to the spindle is directly related to machining accuracy. And this definition, while dependent on the structure of the spindle, is independent of the manner in which the workpiece or tool is mounted, making it possible to make a comparison between spindles of different structures.

The reciprocal of stiffness $1/J_s^p$ is generally called softness. By the "influence factor" is meant the displacement produced at point i by a unit load at point k in an elastic system. If the load is a force and the displacement is a linear or angular one, then the influence factor is designated as α_{ik} or β_{ik}. If the load is a couple moment and the displacement is a linear or angular one, then the influence factor is designated as γ_{ik} or δ_{ik}. Obviously, if i and k are both located at s, then α_{ss} is $1/J_s^p$ and $Y_s^p = P\alpha_{ss}$.

However, during the operation of the spindle the displacement at the spindle-nose is caused not only by force P, produced by cutting force and exerted on the spindle-nose, but also by couple moment M, produced by cutting force and exerted on the spindle-nose, and by driving force Q, produced by cutting force and exerted on a certain point of the spindle. The displacement at the spindle-nose caused by M and Q are $Y_s^M = M\gamma_{ss}$ and $Y_Q^M = Q\alpha_{sQ}$ respectively, because now for γ_{ik}, i and k are both located at s, for α_{ik}, i is located at s and k is located at the point on which Q is exerted.

It can be seen clearly that α_{ss}, γ_{ss} and α_{sQ} all depend on the dimensions of the structure, the properties of a material, the bearing stiffness, etc. The result of defining the static stiffness of the spindle by only considering P while neglecting M and Q, and by making a study of α_{ss} while giving no attention to γ_{ss} and α_{sQ} is often an incomplete definition. This is because the structure and parameters selected only in accordance with α_{SS} may result in inappropriate structure and parameters under the action of M and Q, e. g. the inappropriateness of position of the driving elements may lead to too large a displacement at the spindle-nose.

However, during the operation of the spindle, the ratios of magnitudes of P, Q and M to each other vary with the variation in machining performed. Furthermore, the three are not on a plane, so the sum of displacements at the spindle-nose produced by them, i. e. the vector sum $Y_s = P\alpha_{ss} + M\gamma_{ss} + Q\alpha_{sQ}$ varies in magnitude and direction. The component of Y_s in the sensitive direction (i. e. the direction which exerts a decisive influence on machining accuracy) varies too. Thus, from the relationship between Y_s and P, M, Q, it is difficult to describe the static stiffness of the spindle by J_s^p alone. It will also be difficult to find a definite function between Y_s and α_{ss}, γ_{ss} and α_{sQ}. That is to say, it is generally impossible to choose a kind of spindle such that α_{ss}, γ_{ss} and α_{sQ} are determined by its structure and parameters will make Y_s minimum under any condition.

Under the action of P, M and Q, of course, Y_s can be made as small as possible. In order to do so, efforts must be made to minimize α_{ss}, γ_{ss} and α_{sQ} respectively as well as to make the components of Y_S^P, Y_S^M, and Y_S^Q partly offset each other in the sensitive direction. (This also depends on the selection of α_{ss}, γ_{ss} and α_{sQ}.) Briefly, the study of the static stiffness can be boiled down to the finding of the influence factors at the spindle-nose under different conditions and to make them minimal. This is the purpose of this article.

The so-called Maxwell's Reciprocal Theorem $\alpha_{ik} = \alpha_{ki}$, $\beta_{ik} = \gamma_{ki}$, $\delta_{ik} = \delta_{ki}$ are repeatedly made use of in this article.

As it is the author's intention to discuss shear in a future work, this topic is not included herein in spite of its significance with short spans between multiple bearings.

II. Stiffness of Spindles Mounted in Two Bearings

Spindles mounted in two bearings are roughly divided into four classes, namely: (1) those without driving force Q (i. e. the so-called unloaded spindles); (2) those with force Q exerted on the overhang; (3) those with force Q exerted on the protruding tail; and (4) those with force Q exerted on the span between two bearings.

Relevant formulae are found in Appendix 1.

For an unloaded spindle acted on only by force P, the influence factor at the spindle-nose is

$$\alpha_{ss} = \frac{a^3}{3EI_a} + \frac{la^2}{3EI} + \frac{1}{C_1}\left(1+\frac{a}{l}\right)^2 + \frac{1}{C_2}\left(\frac{a}{l}\right)^2, \tag{1}$$

where
  a: length of overhang,
  l: bearing span,
  E: elastic module of material,
  I_a: area moment of inertia of overhang,
  I: area moment of inertia of span between bearings,
  C_1, C_2: stiffness of front and rear bearing respectively.

Obviously, as shown in Fig. 2, under the action of a unit force, the first term of Eq. (1) is the deflection of the overhang beam at its nose; the second is the displacement at the spindle-nose produced by deflection of the simply supported beam; the third and the fourth are displacements at the spindle-nose produced by deflection of the front and rear bearings respectively. These four displacements are all in the same direction.

In order to increase the stiffness of the spindle, it is evident from Eq. (1) that the stiffness of the bearings, in particular that of the front bearing, must be as great as possible, the area moment of inertia must be as large as possible and the overhang length as short as possible while the bearing span must have an optimum value l_0. This means that if other parameters are the same but $l=l_0$, then α_{ss} is minimal.

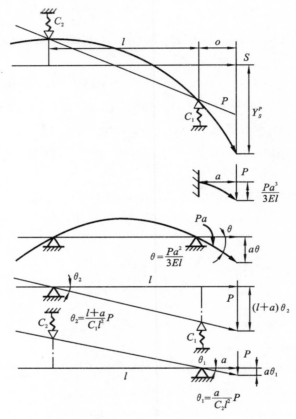

Fig. 2

Fig. 3 is designed according to $d\alpha_{ss}/dl=0$ to find l_0 when I, C_1, and C_2 are given. Here $K_1 = l/D$, $K_{c_1} = C_1/D$, $K_a = a/D$, $I = \pi D^4/64$ mm^{-4}, and K_{c_1} is calculated in such a way that the units of C_1 and D are 10 N $\cdot$ μm^{-1} and mm respectively.

For example, for the spindle of a lathe, if $C_1 = 1500$ N $\cdot$ μm^{-1}, $C_2 = 700$ N $\cdot$ μm^{-1}, $D = 88$ mm, $a = 118$ mm, then $C_1/C_2 = 2.14$, $K_a = 1.34$, $K_{c_1} = 1.71$. Hence from C_1/C_2 and K_{c_1} we can get a point in the first quadrant, from K_a and K_{c_1} we get a point in the third quadrant and from these points we get $K_1 = 2.65$

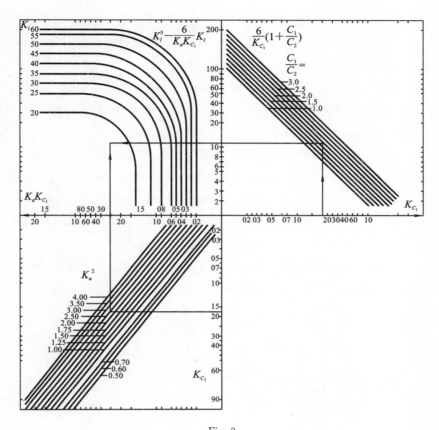

Fig. 3

in the second quadrant as shown in Fig. 3. Thus $l_0 = K_1 D = 233$ mm.

Compared with graphs of absolute values of parameters this graph of relative values is simpler, more practical and shows a wider range of parameter variation.

When the unloaded spindle is acted on only by M, (Fig. 4), the optimum value l_0^M can also be found from Fig. 3; if only from the third to the second quadrant the abscissa $K_a K_{c_1}$ is multiplied by 2. Obviously, $l_0^M < l_0$.

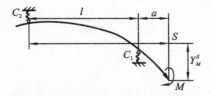

Fig. 4

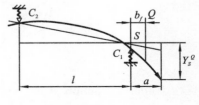

Fig. 5

When the overhang is acted on only by Q (Fig. 5), the optimum value l_0^Q can also be found from Fig. 3; if only from the third to the second quadrant the absissa $K_a K_{c_1}$ is multiplied by 2. Obviously, $l_0^M < l_0$.

When the overhang i, acted on only by Q(Fig. 5), the optimum value l_0^Q can also be found from Fig. 3; if only from the third to second quadrant $K_a K_{c_1}$ is multiplied by $b_f/(a + b_f)$. Because $b_f < a$, $l_0^Q > l_0$.

If we only consider force Q acting on the protruding tail (Fig. 6), we can see that the displacement at the spindle-nose produced by the deflection of the spindle and that produced by the deflection of the bearings are opposite in direction and that they can be offset by each other if the parameters are properly chosen. If l is equal to an optimum value l_0^Q, then $\alpha_Q = 0$. Now, l_0^Q can also be found from Fig. 3 if only $K_a K_{c_1}$ is multiplied by $[1+aC_1/(b_bC_2)]$. Obviously, when $l=l_0$, we can calculate the optimum value of b_b from Fig. 3.

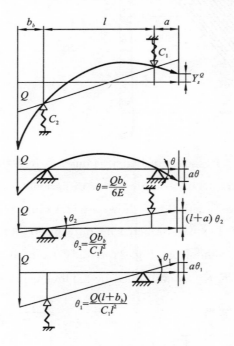

Fig. 6

Under the action of $P^{[2]}$, when $l=l_0$, the tail is parallel to the original axis, i.e. $\alpha_Q = a/(C_2 l)$ (Here α_Q is equal to the influence factor at the real bearing

a_{2s}.) Thus, we have $a_{sQ}=a_{Qs}=a/(C_2l)$ regardless of the position of Q. When $l<l_0$, the tail will tilt up, therefore, the greater b_b the greater a_{sQ} will be. When $l>l_0$, the reverse is true and the tail will intersect the original axis at a point, i. e. $a_{sQ}=0$. This point is called node j. When force Q is exerted on node j, $a_{sQ}=0$. The equation $a_{sQ}=0$ is just what has been used to find the optimum value of l or b_b.

If force Q is exerted on the span (Fig. 7), the displacements at the spindle-nose due to the deflection of the spindle and the rear bearing and that due to the deflection of the front bearing are opposite in direction, so they can be offset by each other. If the parameters are properly chosen, then $a_{sQ}=0$. However, when $l=l_0$, the optimum value of b_m can be chosen from Fig. 8, in which $q=l_0/a$, $K_b=b_m/D$. For example, if $C_1=1700$ N $\cdot$ μm^{-1}, $D=88$ mm and $l=l_0=233$ mm, then we can get $K_b=0.7$, thus $b_m=K_bD=62$ mm. When $l\neq l_0$, we can find K_b from a corresponding graph (see [5]).

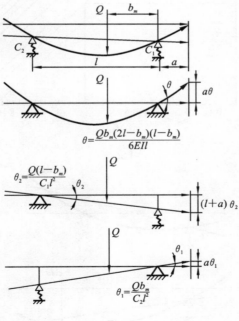

Fig. 7

It is clear that when choosing the parameters, the effects of **P**, **M** and **Q** must be considered as a whole so as to make Y_s^P, Y_s^M and Y_s^Q partly offset each other. This is even more important in the sensitive direction, e. g. the horizontal component of cutting force (P_y) and that of driving force (Q_H) exerted on a lathe spindle are generally opposite in direction (Fig. 9). If force Q is exerted in front of node j, then α_{ss} is opposite to α_{sQ}.

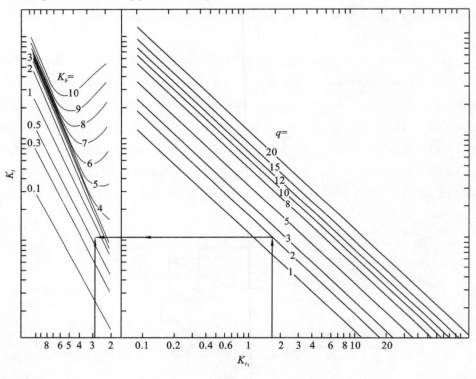

Fig. 8

Fig. 9

Ⅲ. Stiffness of Spindles Mounted in Multiple Bearings

It is noted that the addition of bearings has two effects. One is the effect of the force of reaction $\boldsymbol{R}$ due to addition of radial bearings on the static stiffness of the spindle, the other is that of the reaction moment $\boldsymbol{M_e}$ due to the addition of thrust bearings. If both the force of reaction and reaction moment are produced by a single bearing simultaneously, then this bearing can be regarded as a radial and a thrust bearing save that they are located in the same position.

The calculation of a beam mounted in many bearings involves the calculation of the reaction of bearings and the influence factor in an arbitrary position of the beam. By the application of the physical concept of influence factor we have proposed two calculating methods and derived corresponding calculating formulae (see Appendix 2). From them the following formulae and conclusions are obtained. The number "3" after the subscript of an influence factor will hereafter be understood as representing the number of bearings (not including the thrust bearing) as is customarily done.

(1) The spindle mounted in three bearings

Under the action of $\boldsymbol{P}$, we have

$$R_i^p = -\frac{\alpha_{is}}{S_i + \alpha_{ii}} \cdot P \qquad (2)$$

where

S_i: softness for the ith bearing, $S = 1/C_i$

R_i^p: the force of reaction for the ith bearing

$$\alpha_{ss.3} = \alpha_{ss} - \frac{\alpha_{is}}{S_i + \alpha_{ii}} \cdot \alpha_{si} \qquad (3)$$

where R_i^p, α_{ss}, α_{ii}, α_{is}, α_{si} are determined when there are only two bearings and no bearing i.

The effect of increasing the third radial bearing is $\alpha_{ss.3} < \alpha_{ss}$. Because $\alpha_{is} = \alpha_{si}$, S_i, α_{ii} and α_{ss} are positive, hence the two terms on the right-hand side of Eq. (3) are both positive and the difference between them is of course less than α_{ss}, $\alpha_{ss.3} = \alpha_{ss}$ only when $\alpha_{is} = 0$, i. e. the third bearing is mounted at node j.

Optimum span. In general, to make $\alpha_{ss.3}$ minimal, we have to find l and l_1 from $\partial \alpha_{ss.3}/\partial l = 0$ and $\partial \alpha_{ss.3}/\partial l_1 = 0$ where l_1 is the span between the front and middle bearing. Analysing Eq. (3) we know that if the middle bearing is replaced by the force of reaction R_2^p (Fig. 10) and $l = l_0$, or if the rear bearing is replaced by R_3^p and $l_1 = l_0$, then α_{ss} will be minimal. Because $\alpha_{is}^2/(S_i + \alpha_{ii})$ is always positive or zero (when $\alpha_{is} = 0$), $\alpha_{ss.3}$ remains small.

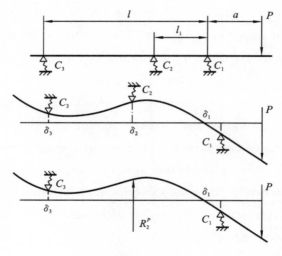

Fig. 10

We may take a spindle mounted in three bearings ($l_1 = 375$ mm) of a lathe as an example, whose $\alpha_{ss.3}$ is reduced by more than 36% as compared with α_{ss} of the spindle without the middle bearing. If $l_1 = l_0 = 233$ mm, then $\alpha_{ss.3}$ will be reduced by 42% as compared with α_{ss}. If $l_1 = 200 \sim 275$ mm, then $\alpha_{ss.3}$ will change only insignificantly as shown in Fig. 11, in which α_{22}^d is $S_2 + \alpha_{22}$. If the span between spindle bearings C_1 and C_2 is taken as $l_0 = 233$ mm and span l_1 between the third bearing C_3 and front bearing C_1 is varied, then it is clear from calculation that the third bearing has little desirable effect on the static stiffness. If the third bearing is mounted behind bearing C_2, then the effect on stiffness steadily decreases. If the span between C_1 and C_3 is over 500 mm, then there is scarcely any effect, as shown in Fig. 12.

Hence the evaluation of spindles mounted in three bearings under the action of **P** should be made from two angles.

A Study of the Static Stiffness of Machine Tool Spindles

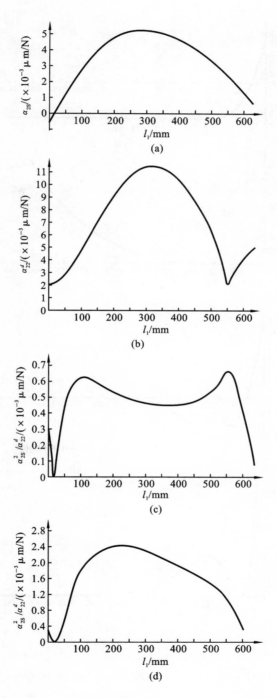

Fig. 11

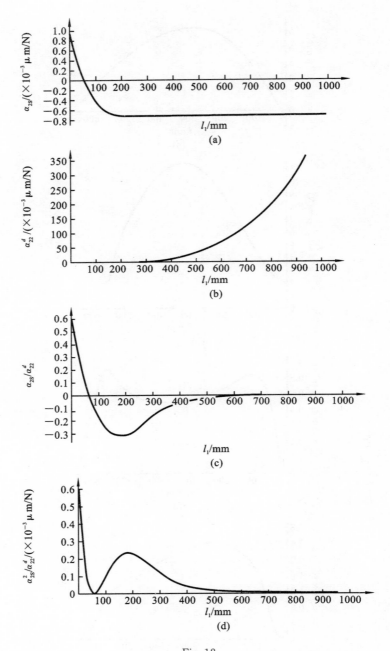

Fig. 12

① If the span between the front and the other bearing is l_0, then the effect of the third bearing on static stiffness of the spindle is insignificant, especially when the third bearing is a rear one.

② If l is much greater than l_0, then the effect of adding the middle bearing and making l_1, equal to l_0 which corresponds to the front and middle bearing is extremely significant.

Besides, the shorter the overhang, the larger the area moment of inertia and the greater the bearing stiffness, the smaller $\alpha_{ss.3}$ will be. It should be pointed out that the greater the stiffness of the bearings, the greater their reaction. And R_1^p is always opposite to P and greater than R_2^p and R_3^p [2, 7]. If all C_i's approach infinity and l_1 approaches zero, then $\alpha_{ss.3}$ approaches $a^3/(3EI_a)$, i.e. the spindle has become an overhang beam with a fixed end, and $\alpha_{ss.3}$ will be minimum.

Under the action of M and that of Q exerted on the overhang, we generally have $\gamma_{s.3} < \gamma_s$ and $\alpha_{sQ.3} < \alpha_{sQ}$ respectively (see Appendix 2).

If Q is exerted on the span or the tail, then compared with α_{sQ}, $\alpha_{sQ.3}$ may be greater, smaller, zero or of opposite sign. Hence, what should be done is either determine the position of the node by $\alpha_{sQ.3} = 0$ or find the position where the absolute value of $\alpha_{sQ.3}$ is minimal by graphical method.

(2) The spindle mounted in two bearings acted on by reaction moment

As mentioned earlier, if a bearing can produce force of reaction R and reaction moment M_c simultaneously, then the bearing can be regarded as a radial and a thrust bearing located in the same position. Thus, taking this into account a reaction moment is equivalent to adding a thrust bearing W.

Under the action of P (Fig. 13), we have

$$M_c^p = -\frac{\beta_{us}}{K_w + \delta_{uw}} \cdot P \qquad (4)$$

$$\alpha_{ss}^{M_c} = \alpha_{ss} - \frac{\beta_{us}}{K_w + \delta_{uw}} \cdot \gamma_{sw}$$

Where M_c^p is the reaction moment of the bearing W, α_{ss}, β_{us}, γ_{sw}, δ_{uw} are determined when bearing W is replaced by its reaction moment M_c and K_w is the angular softness of bearing W.

Effects of M_c. Because $\beta_{us} = \gamma_{sw}$, $\alpha_{ss}^{M_c} < \alpha_{ss}$. And when $\beta_{us} = 0$, $\alpha_{ss}^{M_c} = \alpha_{ss}$, i.e. the thrust bearing is mounted where the angular displacement is zero.

Optimal position of the thrust bearing. Let $\partial \alpha_{ss}^{M_c}/\partial e = 0$ or

$$\partial \left(\frac{\beta_{us}^2}{K_w + \delta_{uw}} \right)/\partial e = 0$$

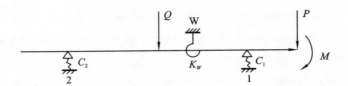

Fig. 13

then we can find the optimum value of e to minimize $\alpha_{ss}^{M_c}$. As the thrust bearing is generally mounted near a radial bearing, we must investigate $\beta_{us}^2/(K_w+\delta_{uw})$ near both ends of the interval$^{[0,1]}$ while e varies and the following conclusions are formed: ① when the bearing W is mounted between the front and rear bearing, it is advisable for W to be somewhat apart from the one or the other; ② when bearing W is mounted beyond either the front or the rear bearing, it is best for W to be close to the one or the other. Now let us compare the effect of bearing W mounted in the front bearing ($e=0$) with that when it is mounted in the rear bearing ($e=1$). It is known from the formulae of influence factor that $\beta_{1s}>\beta_{2s}$, $\delta_{11}=\delta_{22}$ and $K_1 \leqslant K_2$. Thus we have

$$\beta_{1s}^2/(K_1+\delta_{11}) > \beta_{2s}^2/(K_2+\delta_{22}), \quad \alpha_{ss}^{M_1} < \alpha_{ss}^{M_2}$$

This shows that the effect of $e=0$ is better than that of $e=l$. When $e=l=l_0$, $\beta_{2s}=0$, that is to say, bearing W mounted in the rear bearing has no effect on static stiffness. If we substitute formulae for β_{1s} and β_{2s} into the above formula, then we have

$$\left(\frac{\beta_{1s}^2}{K_1+\delta_{11}} - \frac{\beta_{2s}^2}{K_2+\delta_{22}}\right) \Big/ \frac{\beta_{1s}^2}{K_1+\delta_{11}} > 75\%$$

This is in conformity with experimental results given in Ref. [11] which shows that Y_s^P of the spindle of a lathe with a thrust bearing compared with that of this lathe spindle without the thrust bearing will be decreased by $25\% \sim 35\%$ when the thrust bearing is mounted in the front bearing and by 6% when it is mounted in the rear bearing. The relative difference between 25% and 35% and 6% is $76\% \sim 83\%$.

It is worthwhile pointing out that the force of reaction of the front and rear bearing due to reaction moment M_1 of the front bearing is opposite to that due to P. This is beneficial to bearing operation.

In general, M_1 is taken as ζ_1^P Pa. Some investigators often use $\zeta_1^P=0\sim0.35$

proposed in Ref. [8] without making any analysis. As shown in Appendix 2, we can say that when the angular softness of the thrust bearing is small enough, ζ_1^P can even be greater than 1. When $M_1 \zeta_1^P \to 1, C_1 \to \infty, \alpha_{ss}^{M_1} \to a^3/(3EI_a)$. Hence, when C_1 and $1/K_1$ are both large enough, $\alpha_{ss}^{M_1}$ will approach the influence factor at the unfixed end of an overhang beam, and is almost independent of l, C_2 and I.

Generally, under the action of M, $\gamma_{ss_c}^{M_c} < \gamma_{ss}$. It is also more desirable for bearing W to be located in the front than in the rear bearing. In $M_1 = \zeta_1^M \cdot M$, the maximum value of ζ_1^M can approach 1. If C_1 and $1/K_1$ are both large enough, then $\gamma_{ss}^{M_1} \to a^2/(2EI_a)$ (the influence factor at the unfixed end of an overhang beam).

Just as in the analysis of the spindle mounted in three bearings, the case of Q exerted on the overhang is similar to that of M. When Q is exerted on the span or the tail, either determine the node by $\alpha_{sQ}^{M_c} = 0$ or find the minimal absolute value of $\alpha_{sQ}^{M_c}$ graphically. It must be pointed out that if C_1 and $1/K_1$ are both large enough, then $\alpha_{sQ}^{M_c} \to 0$.

The results obtained for the spindle mounted in three bearings acted on by reaction moment M_c are similar to those discussed above.

IV. The Effect of Uncoaxiality of Bearing Housing Holes

For a spindle mounted in many bearings, when it is known that the ith bearing hole is deviated from the axis through the centers of two basic housing holes by Δ_i. we can find the bearing reaction of the spindle (see Appendix 3). For a spindle mounted in three bearings, we have

$$R_i^\Delta = -\Delta_i/(S_i + \alpha_{ii}) \quad \text{(when } M_c^\Delta \text{ is neglected)}$$

$$\begin{bmatrix} M_c^\Delta \\ R_i^\Delta \end{bmatrix} = -\begin{bmatrix} K_w + \delta_{wW} & \beta_{ui} \\ \gamma_{iw} & S_i + \alpha_{ii} \end{bmatrix}^{-1} \begin{bmatrix} 0 \\ \Delta_i \end{bmatrix} \quad \text{(when } M_c^\Delta \text{ is not neglected)}$$

It is clear that R_i^Δ and M_c^Δ are both proportional to Δ_i. If any C_i is equal to zero, then $R_i^\Delta = 0$, $M_c^\Delta = 0$. Thus, as long as the span between the front and one bearing is taken as l_0 under the action of P, and the diametral clearance of any other ith bearing is no less than $2\Delta_i$, the static stiffness of the spindle is great. And little will the static stiffness change when all other bearings are working under the action of P.

V. Conclusions

(1) Relevant structure and parameters must be properly chosen by considering the effects of P, M, and Q at the same time according to particular conditions.

(2) Shortening the overhang, increasing the area moment of inertia of the spindle and the stiffness of the front bearing are effective measures of enhancing the static stiffness of the spindle.

(3) Whether or not the static stiffness of a spindle can be improved by increasing the number of bearings depends on whether it is P or M or Q that is exerted on the spindle. For the case of P being exerted, it will be increased (at least not decreased); for the case of M and that of Q exerted on the overhang, it will in general be increased; while for the case of Q acting on the span or the tail it can be increased or decreased, or it may remain unchanged.

(4) The span must be properly chosen. For the case of P, as long as the span between the front and one bearing is chosen as an optimum span, the spindle will have considerable stiffness and the effect of the third bearing is insignificant. (Here the optimum span refers to the span of a spindle mounted in two bearings.) If l is much greater than the optimum span and a middle bearing is used with l_1 approximately equal to the optimum span, then the static stiffness will be greatly increased.

(5) It is much more desirable for the thrust bearing to be mounted in the from than in the rear bearing. If C_1 and $1/K_1$ are both large enough, then the influence factor of the spindle-nose approaches that of the overhang beam with a fixed end no matter whether the spindle is under the action of P, or M, or Q.

References

1 Machine Tool Teaching and Research Group of Huazhong Institute of Technology. Machine Tools(textbook) (1973).

2 Yang Shuzi. "An analysis and calculation of the static stiffness of machine tool spindles". J. Huazhong Inst. Technol. 6(1),(1978).

3 Yang Shuzi. "The effect of bearing reaction moment on static stiffness of machine tool spindles". J. Huazhong Inst. Technol. Suppl. to 7(1). (1979).

4 Yang Shuzi. "An analysis 6f some problems of uncoaxiaty of the housing holes in spindles mounted in many bearings". J. Huazhong Inst. Technol. 7(4), (1979).

5 Yang Shuzi. "An analysis and calculation of the static stiffness of the machine tool spindles mourned in two bearings". Machine Tools No. 3 (1979).

6 Yang Shuzi. "An analysis and discussion on the static stiffness of the machine tool spindles mourned in two bearings". Machine Tools No. 9(1979).

7 Dept. of Mechanics of Guangdong Teachers' College and Guangzhou (Canton) Machine Tool Center's Hydrostatic Techniques Division. "A calculation and test of the spindle mounted in many hydrostatic bearings". Machine Tool and Hydraulic Transmission No. 2 (1978).

8 D. N. Reishetov. "The calculation of machine tool elements". Mashgiz (1945).

9 V. L. K. Dondoshanski and V. E. Chireayev. "Forced vibration of machine tool". Mashgiz (1959).

10 J. KDie. "Bezongene Starrheit"—Die Bestimmung der Hauptspindd von Werkzeugmaschinen aufgrund einer Neucr Starrheitsbetrachtung". Maschinenmarkt 75(19) (1967).

11 Axial Kugeilagcr an Arbeitsspindeln. Maschinenmarkt 75(19) (1969).

12 H. Pitrovf and E. Wichf.. "Laufgute von Werkzengmaschinen-spindeln". Werkst. Betr. Wien. 12,Heft 8(1969).

13 H. Regah. "Dynamic behaviour of machine tool spindles mounted in three bearings". Machinery and Production Engineering 124(3193). (1974).

14 A. N. Bakovsk and Z. M. Levina. "Angular stiffness of thrust bearings of machine tool spindles and its influence on radial stiffness". Stankie Instrument No. 11 (1977).

15 R. Zpenkovic and V. Dukovski. "Die steifheit von Werkzeugma-schinenspindeln unter Einwirkung der Arbeitskraft". Werkst. Betr. Wien. 111(2). (1978).

Appendix 1

Below we shall derive formulae for spindles mounted in two bearings.

For an unloaded spindle acted on only by force P (Fig. 2), the displacement at the spindle-nose is

$$Y_s^P = P\left[\frac{a^3}{3EI_a} + \frac{la^2}{3EI} + \frac{1}{C_1}\left(1+\frac{a}{l}\right)^2 + \frac{1}{C_2}\left(\frac{a}{l}\right)^2\right] \tag{1}$$

Comparison with $Y_s^P = P\alpha_{ss}$ yields

$$\alpha_{ss} = \frac{a^3}{3EI_a} + \frac{la^2}{3EI} + \frac{1}{C_1}\left(1+\frac{a}{l}\right)^2 + \frac{1}{C_2}\left(\frac{a}{l}\right)^2 \tag{2}$$

Let $d\alpha_{ss}/dl=0$, and denote a/D by K_a, l/D by K_1, C_1/D by K_{c_1} and as $E=2\times10^5$ Nmm^{-2}, $I=\dfrac{\pi D^4}{64}$ mm, we have

$$K_1^3 - 6K_1/K_a K_{c_1} - \frac{6}{K_{c_1}}\left(1+\frac{C_1}{C_2}\right) = 0 \tag{3}$$

where D = average diameter of span between bearings.

According to Eq. (3), we have designed Fig. 3.

For an unloaded spindle which is acted on only by couple moment M (Fig. 4), the displacement at spindle-nose is

$$Y_s^M = M\left[\frac{a^2}{2EI_a} + \frac{la}{3EI} + \frac{1}{C_1}\cdot\frac{l+a}{l^2} + \frac{1}{C_2}\cdot\frac{a}{l^2}\right] \tag{4}$$

Comparison with $Y_s^M = M\gamma_{ss}$ gives

$$\gamma_{ss} = \frac{a^2}{2EI_a} + \frac{la}{3EI} + \frac{1}{C_1}\cdot\frac{l+a}{l^2} + \frac{1}{C_2}\cdot\frac{a}{l^2} \tag{5}$$

Similarly, let $d\gamma_{ss}/dl=0$, we have

$$K_1^3 - \frac{6K_1}{K_a K_{c_1}} - \frac{6}{K_{c_1}}\left(1+\frac{C_1}{C_2}\right) = 0 \tag{6}$$

Comparing Eq. (6) with Eq. (3), we find that only the coefficients in the second terms are different. So, if only the abscissa $K_a K_{c_1}$ is multiplied by 2, K_1 can be found in the same way as discussed previously.

For a spindle which is acted on only by Q exerted on the overhang (Fig. 5) the displacement at the spindle-nose is

$$Y_s^Q = Q\left[\frac{b_f^2}{2EI_a}\left(a-\frac{b_f}{3}\right) + \frac{lab_c}{3EI} + \frac{1}{C_1 l^2}(l+a)(l+b_f) + \frac{ab_f}{C_2 l^2}\right] \tag{7}$$

Comparison with $Y_s^Q = Q\alpha_{sQ}$ gives

$$\alpha_{sQ} = \frac{b_f^2}{2EI_a}\left(a-\frac{b_f}{3}\right) + \frac{lab_f}{3EI} + \frac{1}{C_1 l^2}(l+a)(l+b_f) + \frac{ab_f}{C_2 l^2} \tag{8}$$

Similarly, let $d\alpha_{sQ}/dl=0$. we have

$$K_1^3 - \frac{6K_1}{K_a K_{c_1}}\left(\frac{2b_f}{a+b_f}\right) - \frac{6}{K_{c_1}}\left(1+\frac{C_1}{C_2}\right) = 0 \tag{9}$$

Comparing Eq. (9) with Eq. (3), we find that Fig. 3 can also be used to find K_1 if only the $K_a K_{c_1}$ is multiplied by $2b_f/(a+b_f)$.

For a spindle which is acted on only by Q exerted on the tail (Fig. 6), the displacement at the spindle-nose is

$$Y_s^Q = Q\left[\frac{lab_b}{6EI} - \frac{b_b}{C_1 l^2}(l+a) + \frac{a}{C_2 l^2}(l+b_b)\right] \quad (10)$$

Thus,

$$\alpha_{sQ} = \frac{lab_b}{6EI} - \frac{b_b}{C_1 l^2}(l+a) + \frac{a}{C_2 l^2}(l+b_b) \quad (11)$$

Let $\alpha_{sQ}=0$, we have

$$K_1^3 - 6K_1\bigg/\left[K_a K_{c_1}\bigg/\left(1+\frac{aC_1}{b_b C_2}\right)\right] - \frac{6}{K_{c_1}}\left(1+\frac{C_1}{C_2}\right) = 0 \quad (12)$$

Obviously, we can also find K_1 from Fig. 3 if only $K_a K_{c_1}$ is multiplied by $1/[1+aC_1/(b_b C_2)]$.

For a spindle which is acted on only by Q exerted on the span (Fig. 7), we have

$$Y_s^Q = \left[\frac{ab_m}{6EIl}(2l-b_m)(l-b_m) - \frac{1}{C_1 l^2}(l+a)(l-b_m) + \frac{ab_m}{C_2 l^2}\right] \quad (13)$$

Thus,

$$\alpha_{sQ} = \frac{ab_m}{6EIl}(2l-b_m)(l-b_m) - \frac{1}{C_1 l^2}(l+a)(l-b_m) + \frac{ab_m}{C_2 l^2} \quad (14)$$

Let $\alpha_{sQ}=0$, $l=l_0$ and denote b_m/D by K_{b_m}, l_0/a by q, then according to Eq. (3), we have

$$K_{b_m}^3 - 3K_1 K_{b_m}^2 + 3K_1^2 K_{b_m} - \frac{6}{K_{c_1}}(q+1) = 0 \quad (15)$$

from which we can draw a graph as seen in Fig. 8.

Appendix 2

Common methods and formulae for calculating the bearing reaction and displacement in any position of an elastic beam mounted in many elastic bearings are derived as follows:

(1) The direct influence factor method

By this method, the beam mounted in many bearings is turned into one mounted in two bearings, and using the influence factors of the latter we can find the reaction of bearings and the influence factor in an arbitrary position of the former directly.

In Fig. (1), assume that the beam is mounted in $n+2$ bearings and load L is applied on it. A_i is a reaction to the beam from the ith bearing. Let the direction

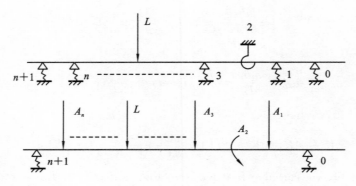

Fig. (1)

of L be positive (the assumed direction of A_i is shown in Fig. (1) and that of the displacement produced by L also be positive, then if the $1-n$ bearings are replaced by A_1-A_n ones (these n bearings are arbitrarily chosen), then the beam mounted in $n+2$ bearings is turned into one mounted in two bearings. The displacement f_i in the ith bearing position of the beam is

$$Lu_{iL} + A_1 u_{i1} + \cdots + A_i u_{i1} + \cdots A_n u_{in} = f_i \tag{16}$$

where u_{ik} is the influence factor, $k=L, 1, 2, \cdots, n$.

Let $S_i = 1/C_i$. As A_i and f_i are opposite in direction, we have

$$-A_i S_i = f_i \tag{17}$$

From Eq. (16) and (17), we have

$$A_1 u_{i1} + A_2 u_{i2} + \cdots + A_i (S_i + u_{ii}) + \cdots + A_n u_{in} = -Lu_{iL} \tag{18}$$

Rewriting Eq. (18) as a matrix equation, we have

$$\begin{bmatrix} S_1 + u_{11} & u_{12} & u_{13} & \cdots & u_{1n} \\ u_{21} & S_2 + u_{22} & u_{23} & \cdots & u_{2n} \\ \vdots & \vdots & \vdots & & \vdots \\ u_{n1} & u_{n2} & u_{n3} & \cdots & S_n + u_{nn} \end{bmatrix} \begin{bmatrix} A_1 \\ A_2 \\ \vdots \\ A_n \end{bmatrix} = -L \begin{bmatrix} u_{1L} \\ u_{2L} \\ \vdots \\ u_{nL} \end{bmatrix} \tag{19}$$

i. e.

$$[U_s] \cdot \{A\} = -L\{u_L\} \tag{20}$$

Thus,

$$\{A\} = -L[U_s]^{-1}\{u_L\} \tag{21}$$

Under the action of L the displacement x_g in an arbitrary position g of the beam is

$$x_g = Lu_{gL} + A_1 u_{g1} + A_2 u_{g2} + \cdots + A_n u_{gn} \tag{22}$$

i. e.

$$x_g = Lu_{gL} + [u_{g1} \quad u_{g2} \quad \cdots \quad u_{gn}] \begin{bmatrix} A_1 \\ A_2 \\ \vdots \\ A_n \end{bmatrix} \quad (23)$$

or

$$x_g = Lu_{gL} + \{u_g\}^T \cdot \{A\} \quad (24)$$

where $\{u_g\}^T$ transposed matrix of $\{u_g\}$.

From Eq. (21) and (24), we have

$$x_g = L[U_{gL} - \{u_g\}^T [v_s]^{-1} \cdot \{u_L\}] \quad (25)$$

Thus,

$$u_{gL_{n+2}} = u_{gL} - \{u_g\}^T [v_s]^{-1} \cdot \{u_L\} \quad (26)$$

Obviously, $u_{gL_{n+2}}$ the influence factor in an arbitrary position of a beam mounted in $n+2$ bearings. (Here, $n+2$ includes the number of thrust bearings.)

Relevant formulae for calculating influence factors of a beam mounted in two bearings can be found in textbooks on Mechanics of Materials.

(2) The indirect influence factor method

By this method, the number of bearings can be reduced step by step, that is to say, we can find the reaction of bearing and the influence factor in an arbitrary position of a beam with n bearings from the influence factors of a beam with $n-1$ bearings.

Fig. (2) shows a beam mounted in n bearings, relevant symbols and directions of load and displacement being the same as before. If the ith bearing is replaced by A_i, then the beam mounted in n bearings is changed into one mounted in $n-1$ bearings. The displacement in the ith bearing position of the beam is

$$Lu_{iL_{n-1}} + A_i u_{ii \cdot n-1} = f_i \quad (27)$$

$$-A_i S_i = f_i \quad (28)$$

From Eq. (27) and Eq. (28), we have

$$A_i = -\frac{u_{iL_{n-1}}}{S_i + u_{ii \cdot n-1}} L \quad (29)$$

Under the action of L, the displacement x_g in an arbitrary position g of the beam is

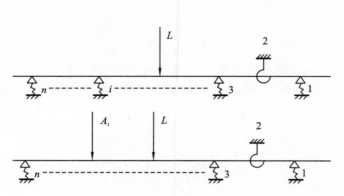

Fig. (2)

$$x_g = Lu_{gL\cdot n-1} + A_i u_{gi\cdot n-1} \tag{30}$$

$$x_g = L\left(u_{gL\cdot n-1} - \frac{u_{iL\cdot n-1}}{S_i + u_{ii\cdot n-1}} u_{gi\cdot n-1}\right) \tag{31}$$

$$u_{gL\cdot n} = u_{gL\cdot n-1} - \frac{u_{iL\cdot n-1}}{S_i + u_{ii\cdot n-1}} u_{gi\cdot n-1} \tag{32}$$

As it is, Eq. (21) and Eq. (26) can be regarded as expansions of Eq. (29) and Eq. (32) respectively.

From Eq. (32) we have formed a very important conclusion, namely, if position g of a spindle is the position where L is applied on the spindle and the direction of the displacement in position g is the same as that of L, then $u_{gL\cdot n-1} = u_{gi\cdot n-1}$ and the first and second term of Eq. (32) are positive, so $u_{gL\cdot n} < u_{gL\cdot n-1}$. This means that when a bearing is added to a spindle, in the direction of L the displacement in a position where L is applied on the spindle must be decreased (at least not increased), but not to zero.

(3) Special cases

For some special cases Eq. (21), Eq. (26) or Eq. (29). Eq. (32) can be written as follows:

(a) For a spindle mounted in three bearings.

① Under the action of P

$$R_i^P = -\frac{\alpha_{is}}{S_i + \alpha_{ii}} \cdot P \tag{33}$$

$$\alpha_{ss\cdot 3} = \alpha_{ss} - \frac{\alpha_{is}}{S_i + \alpha_{ii}} \cdot \alpha_{si} \tag{34}$$

② Under the action of M

$$R_i^M = -\frac{\gamma_{is}}{S_i + a_{ii}} \cdot M \tag{35}$$

$$\gamma_{ss\cdot 3} = \gamma_{ss} - \frac{\gamma_{is}}{S_i + a_{ii}} \cdot a_{si} \tag{36}$$

③ Under the action of Q

$$R_i^P = -\frac{a_{iQ}}{S_i + a_{ii}} \cdot Q \tag{37}$$

$$a_{sQ\cdot 3} = a_{sQ} - \frac{a_{iQ}}{S_i + a_{ii}} \cdot a_{si} \tag{38}$$

Now we shall discuss Eq. (36). It is known that node ($a_{js} = 0$) of the span under the action of P must be behind node ($\gamma_{js} = 0$) under the action of M, a_{js} and γ_{js} have different signs when between the two nodes and the same sign when beyond them. The addition of a radial bearing implies that I is generally large and a large l will make the two nodes approach each other and the front bearing at the same time. Apparently, it is quite unlikely that the radial bearing will be added between the two nodes. Hence, in general, $\gamma_{ss\cdot 3} < \gamma_{ss}$ Similarly, when Q is exerted on the overhang, $a_{ss\cdot 3} < a_{ss}$ is the general case.

(b) For a spindle mounted in two bearings acted on by reaction moment M_c.

① Under the action of P

$$M_c^P = -\frac{\beta_{us}}{K_w + \delta_{uw}} \cdot P \tag{39}$$

$$a_{ss}^{M_c} = a_{ss} - \frac{\beta_{us}}{K_w + \delta_{uw}} \cdot \gamma_{sw} \tag{40}$$

② Under the action of M

$$M_c^M = -\frac{\delta_{us}}{K_w + \delta_{uw}} \cdot M \tag{41}$$

$$\gamma_{ss}^{M_c} = \gamma_{ss} - \frac{\delta_{us}}{K_w + \delta_{uw}} \cdot \gamma_{sw} \tag{42}$$

③ Under the action of Q

$$M_c^Q = -\frac{\beta_{uQ}}{K_w + \delta_{uw}} \cdot Q \tag{43}$$

$$a_{ss}^{M_c} = a_{sQ} - \frac{\beta_{uQ}}{K_w + \delta_{uw}} \cdot \gamma_{sw} \tag{44}$$

(c) For a spindle mounted in two bearings acted on by $M_c = M_1$ or M_2 (i. e. bearing W is in the from or rear bearing).

① Under the action of P

$$M_1^P = \frac{1/3EI + 1/C_1l^2 + 1/C_2l^2 + 1/C_1la}{1/3EI + 1/C_1l^2 + 1/C_2l^2 + K} \cdot Pa = \zeta_1^P \cdot Pa \tag{45}$$

$$\alpha_{ss}^{M_1} = \left[\frac{a^3}{3EI_a} + \frac{la^2}{3EI} + \frac{l+a}{C_1l} + \frac{(l+a)a}{C_1l^2} + \frac{1}{C_2}\left(\frac{a}{l}\right)^2\right]$$

$$- \zeta_1^P\left[\frac{la^2}{3EI} + \frac{(l+a)a}{C_1l^2} + \frac{1}{C_2}\left(\frac{a}{l}\right)^2\right] \tag{46}$$

When $K_1 \leqslant 1/C_1la, \zeta_1^P \geqslant 1$

$$M_2^P = \frac{-1/6EI + 1/C_1l^2 + 1/C_2l^2 + 1/C_1la}{1/3EI + 1/C_1l^2 + 1/C_2l^2 + K_2} \cdot Pa = \zeta_2^P \cdot Pa \tag{47}$$

$$\alpha_{ss}^{M_2} = \left[\frac{a^3}{3EI_a} + \frac{la^2}{3EI} + \frac{l+a}{C_1l} + \frac{(l+a)a}{C_1l^2} + \frac{1}{C_2}\left(\frac{a}{l}\right)^2\right]$$

$$- \zeta_2^P\left[-\frac{la^2}{6EI} + \frac{(l+a)a}{C_1l^2} + \frac{1}{C_2}\left(\frac{a}{l}\right)^2\right] \tag{48}$$

Obviously, in general, $|\zeta_2^P|<1$; when $l=l_0, \zeta_2^P=0$.

② Under the action of M

$$M_1^M = \frac{1/3EI + 1/C_1l^2 + 1/C_2l^2}{1/3EI + 1/C_1l^2 + 1/C_2l^2 + K_1} \cdot M = \zeta_1^M \cdot Pa \tag{49}$$

$$\gamma_{ss}^{M_1} = \left[\frac{a^2}{2EI_a} + \frac{la}{3EI} + \frac{l+a}{C_1l^2} + \frac{a}{C_2l^2}\right] - \zeta_1^M\left[\frac{la}{3EI} + \frac{(l+a)}{C_1l^2} + \frac{a}{C_2l^2}\right] \tag{50}$$

$$M_2^M = \frac{-1/6EI + 1/C_1l^2 + 1/C_2l^2}{1/3EI + 1/C_1l^2 + 1/C_2l^2 + K_2} \cdot M = \zeta_2^M \cdot M \tag{51}$$

$$\gamma_{ss}^{M_2} = \left[\frac{a^2}{2EI_a} + \frac{la}{3EI} + \frac{l+a}{C_1l^2} + \frac{a}{C_2l^2}\right] - \zeta_2^M\left[-\frac{la}{6EI} + \frac{(l+a)}{C_1l^2} + \frac{a}{C_2l^2}\right] \tag{52}$$

Obviously, $|\zeta_1^M|<1, |\zeta_2^M|<1$.

(d) For a spindle mounted in three hearings acted on by reaction moment M_c.

① Under the action of P

$$\begin{bmatrix} M_c^P \\ R_1^P \end{bmatrix} = -P[U_s]^{-1}\begin{bmatrix} \beta_{us} \\ \alpha_{is} \end{bmatrix} \tag{53}$$

$$\alpha_{ss\zeta_3}^M = \alpha_{ss} - [\gamma_{sw}\alpha_{si}][U_s]^{-1}\begin{bmatrix} \beta_{us} \\ \alpha_{is} \end{bmatrix} \tag{54}$$

② Under the action of M

$$\begin{bmatrix} M_c^M \\ R_i^M \end{bmatrix} = -M[U_s]^{-1}\begin{bmatrix} \delta_{us} \\ \gamma_{is} \end{bmatrix} \tag{55}$$

$$\gamma_{ss}^{M_c} = \gamma_{ss} - [\gamma_{sw}\alpha_{si}][U_s]^{-1}\begin{bmatrix}\delta_{us}\\ \gamma_{is}\end{bmatrix} \qquad (56)$$

③ Under the action of Q

$$\begin{bmatrix}M_c^Q\\ R_1^Q\end{bmatrix} = -Q[U_s]^{-1}\begin{bmatrix}\beta_{uQ}\\ \alpha_{iQ}\end{bmatrix} \qquad (57)$$

$$\alpha_{sQ\cdot 3}^{M_c} = \alpha_{sQ} - [\gamma_{sw}\alpha_{si}][U_s]^{-1}\begin{bmatrix}\beta_{uQ}\\ \alpha_{iQ}\end{bmatrix} \qquad (58)$$

$$[U_s] = \begin{bmatrix}K_w + \delta_{uw} & \beta_{ui}\\ \gamma_{iw} & S_i + \alpha_{ii}\end{bmatrix} \qquad (59)$$

Appendix 3

As shown in Fig. (3), a beam is mounted in $n+2$ bearings (including thrust bearing). Let a force and a linear displacement in a given direction be positive and a couple moment and an angular displacement in a given direction also be positive. By the direct influence factor method, the displacement δ_i in the ith bearing position of the beam is

$$\Delta_i + R_1^\Delta \alpha_{i1} + R_2^\Delta \alpha_{i2} + \cdots + M_c^\Delta \gamma_{iw} + \cdots + R_i^\Delta \alpha_{ii} + \cdots + R_n^\Delta \alpha_{in} = \delta_i \qquad (60)$$
$$-R_i^\Delta S_i = \delta_i \qquad (61)$$

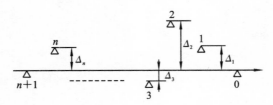

Fig. (3)

From Eq. (60) and Eq. (61), we have

$$R_1^\Delta \alpha_{i1} + R_2^\Delta \alpha_{i2} + \cdots + M_c^\Delta \gamma_{iw} + \cdots + R_i^\Delta (S_i + \alpha_{ii}) + \cdots + R_n^\Delta \alpha_{in} = -\Delta_i \qquad (62)$$

Similarly, the angular displacement θ at bearing W of the beam is

$$R_1^\Delta \beta_{i1} + R_2^\Delta \beta_{i2} + \cdots + M_c^\Delta \delta_{uw} + \cdots + R_i^\Delta \beta_{ui} + \cdots + R_n^\Delta \beta_{un} = \theta \qquad (63)$$
$$-M_c^\Delta K_w = \theta \qquad (64)$$

From Eq. (63) and Eq. (64), we have

$$R_1^\Delta \beta_{w1} + R_2^\Delta \beta_{w2} + \cdots + M_c^\Delta (k_w + \delta_{uw}) + \cdots + R_i^\Delta \beta_{ui} + \cdots + R_n^\Delta \beta_{un} = 0 \qquad (65)$$

Rewriting Eq. (62) and Eq. (65) as a matrix equation, we have

$$\begin{bmatrix} S_1+\alpha_{11} & \alpha_{12} & \cdots & \gamma_{1w} & \cdots & \alpha_{1n} \\ \alpha_{21} & S_2+\alpha_{22} & \cdots & \gamma_{2w} & \cdots & \alpha_{2n} \\ \vdots & & \cdots & \vdots & & \vdots \\ \beta_{w1} & \beta_{w2} & \cdots & K_w+\delta_{ww} & \cdots & \beta_{un} \\ \vdots & \vdots & & \vdots & & \vdots \\ \alpha_{n1} & \alpha_{n2} & \cdots & \alpha_{mw} & \cdots & S_n+\alpha_{m} \end{bmatrix} \begin{bmatrix} R_1^\Delta \\ R_2^\Delta \\ \vdots \\ M_c^\Delta \\ \vdots \\ R_n^\Delta \end{bmatrix} = - \begin{bmatrix} \Delta_1 \\ \Delta_2 \\ \vdots \\ 0 \\ \vdots \\ \Delta_n \end{bmatrix} \quad (66)$$

$$[U]\{A^\Delta\} = -\{\Delta\} \tag{67}$$

$$\{A^\Delta\} = -[U]^{-1}\{\Delta\} \tag{68}$$

(原载 International Journal of Machine Tool Design & Research, Vol. 21, Issue 1, 1981)

平稳时间序列的数学模型及其阶的确定的讨论

杨叔子

提 要 本文从系统分析角度讨论平稳时间序列的数学模型的物理意义,给出了模型的 Green 函数及逆函数的直接关系式;进而从应用角度提出确定模型形式及其阶的一种方法。

(一)

采用控制理论中识别系统建立数学模型的方法时,一般要求知道系统的输入与输出。但是,如同在其他领域中一样,在机械工程领域中的不少情况下,特别当所研究的系统是在工况下时,系统的输入往往是不可观测的,显然,这时一般无法直接应用控制理论的方法建立系统的数学模型,而时间序列分析的方法却为解决这类问题提供了途径。固然,就其实质而言,这两种方法建模是统一的*,但由于各自的发展历史的不同,两者又各具有本身的特点。

设 $\{x_t\}$ 为平稳的、正态分布的、零均值的时间序列,$\{a_t\}$ 为正态分布的白噪声,则此时可对 $\{x_t\}$ 拟合出 ARMA(n,m) 模型:

$$x_t - \phi_1 x_{t-1} - \cdots - \phi_n x_{t-n} = a_t - \theta_1 a_{t-1} - \cdots - \theta_m a_{t-m} \qquad (1)$$

式中,ϕ_i, θ_j 分别为自回归参数、滑动平均参数。

引入后移算子 B,则式(1)可改写为

$$(1 - \phi_1 B - \phi_2 B^2 - \cdots - \phi_n B^n) x_t = (1 - \theta_1 B - \theta_2 B^2 - \cdots - \theta_m B^m) a_t \qquad (2)$$

---

* 王治藩、杨叔子. 改进时间序列方法,建立过程模型. 全国机械工业自动化学会第二届年会宣读论文,1983 年.

或缩写为
$$\phi(B)x_t = \theta(B)a_t \tag{3}$$
从控制理论看来,显然 $\theta(B)/\phi(B)$ 是系统的传递函数。

$\phi(B)$ 表示系统与外界作用无关的本身固有特性(即系统本身的结构与参数),$\theta(B)$ 表示系统与外界作用的相互关系。自然,式(3)中 $\{a_t\}$ 为系统输入,x_t 为系统输出,如图 1(a)所示。

实际上,系统的输入不一定是白噪声 $\{a_t\}$,而是平稳的正态分布的零均值的时间序列 s_t,系统的传递函数为 $\theta_0(B)/\phi_0(B)$,如图 1(b)所示。由于 s_t 不可观测,故在模型输出与实际系统输出相同的基础上,建立了等价系统的模型,而以 a_t 为输入,如图 1(c)所示;其简化模型即为图 1(a)所示模型。由图 1 可知:
$$\phi(B) = \phi_0(B)\phi'_0(B) \tag{4}$$

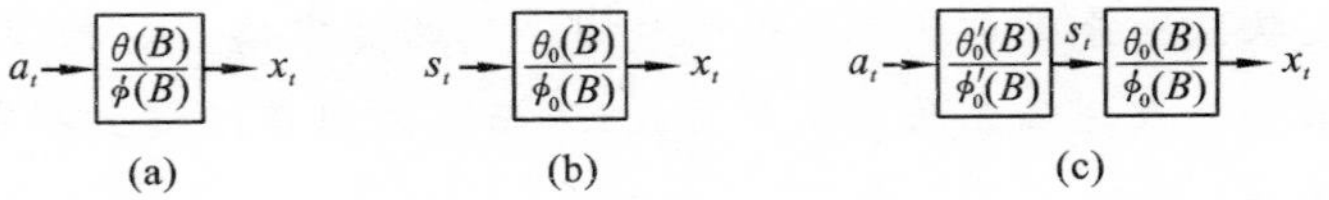

图 1

这表明等价系统的 $\phi(B)$ 包含了实际系统的 $\phi_0(B)$,而 $\phi(B)$ 中多包含的 $\phi'_0(B)$,则是由于 s_t 不可观测引起的。

对此,应指出两点。

(1) 对单输入单输出的线性系统而言,如图 1 所示,有 $\theta(B) = \theta_0(B)\theta'_0(B)$。但是,对多输入单输出的线性系统而言,$\theta(B)$ 中不一定包含 $\theta_0(B)$,而 $\phi(B)$ 中却一定包含 $\phi_0(B)$。

(2) 特别重要的是,时间序列通过具有惯性的物理系统后本身的相关性至少不会减弱,一般本身相关性较弱的时间序列通过此物理系统将变为本身相关性较强的时间序列,反之则不成立。因此,等价系统如以某一平稳时间序列 m_t 作为输入,而实际系统以平稳时间序列 s_t 作为输入,当 s_t 的相关性又较的 m_t 为弱时,等价系统的 $\phi(B)$ 将丧失实际系统的与外界无关的固有特性的一部分,这是不允许的。如图 2 所示,$\phi_0(B) = \phi_s(B)\phi_m(B)$,而 $\phi(B) = \phi_m(B)$,$\phi_s(B)$ 将丧失。这就是从系统分析角度看,为什么采用白噪声作为等价系统输入的根本原因。

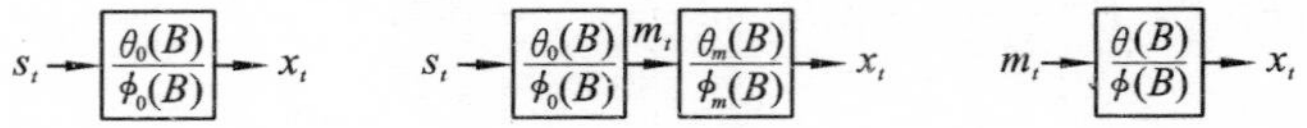

图 2

对 ARMA(n,m)模型而言，当 $m=0$ 时，则模型为纯自回归模型 AR(n)；当 $n=0$ 时，则模型为纯滑动平均模型 MA(m)。即后两者可视为前者的特例。实际上，就最小相位系统而言，对它所产生的平稳时间序列$\{x_t\}$可以拟合出 ARMA 模型，也可以拟合出 AR 模型或 MA 模型。当然，最适合的模型的参数将是最少的。从而利用这一模型进行计算，误差一般也是最小的。若所拟合的 AR 模型的阶数不是远比 ARMA 模型的阶数为大，则拟合 AR 模型远比拟合 ARMA 模型为易，利用 AR 模型进行预测与控制也比利用 ARMA 时简单[1-3]。

在拟合平稳时间序列模型时，存在如何确定模型的形式与阶数的问题。目前，问题并未获得满意的解决。本文进一步从应用角度在文献[3]的基础上提出一种确定模型形式与阶数的方法。在叙述这一方法之前，先给出模型的 Green 函数与逆函数的直接关系。

（二）

文献[2]给出了 Green 函数 G_i 与逆函数 I_j 的定义：

$$x_t = \sum_{j=0}^{\infty} G_j a_{t-j} \tag{5}$$

$$a_t = \sum_{j=0}^{\infty} (-I_j) x_{t-j} \tag{6}$$

依据此两式。可有图 3 与图 4 所示模型，两图从系统角度表明了 G_j 与 I_j 的含义。图中 δ_t 为 Kronecker δ 函数。

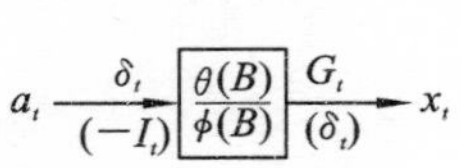

图 3

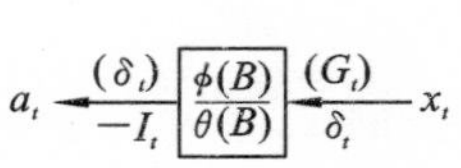

图 4

由式(6)有

$$a_t = \left(\sum_{j=0}^{\infty} -I_j B^j \right) x_t \tag{7}$$

将式(7)同 $a_t = \dfrac{\phi(B)}{\theta(B)} x_t$ 比较，得

$$\frac{\phi(B)}{\theta(B)} = \sum_{j=0}^{\infty} (-I_j) B^j \tag{8}$$

因 $G_t = \dfrac{\theta(B)}{\phi(B)} \delta_t$，有

$$\frac{\phi(B)}{\theta(B)}G_t = \delta_t \tag{9}$$

将式(8)代入,得 $\left(\sum_{j=0}^{\infty}-I_j B^j\right)G_t = \delta_t, \sum_{j=0}^{\infty}(-I_j)G_{t-j} = \delta_t$,从而有

$$I_t * G_t = -\delta_t \tag{10}$$

式(10)的物理意义是显然的:当系统"正"传时,如图 3 所示,G_t 是系统的单位脉冲响应函数,G_t 同任何输入的卷积将是此时的输出。今若输入 $-I_t$,则输出为 $-I_t * G_t$;而此时的输出,由图 3 可知为 δ_t,故得式(10)。当系统作"逆"传时,可作类似分析,亦得式(10)。

现将式(10)写成展开形式,则可求得用 Green 函数直接计算逆函数或用逆函数直接计算 Green 函数的递推算式:

$$\left.\begin{array}{ll} t=0, & I_0 G_0 = -1, \qquad\qquad 所以\ I_0 = -\dfrac{1}{G_0} = -1 \\ t=1, & I_0 G_1 + I_1 G_0 = 0, \qquad 所以\ I_1 = G_1 \\ t=2, & I_0 G_2 + I_1 G_0 + I_2 G_0 = 0, \\ \quad\vdots & \quad\vdots \\ t=j, & I_0 G_j + I_1 G_{j-1} + \cdots + I_j G_0 = 0, \end{array}\right\} \tag{11}$$

将上式写成矩阵形式

$$\begin{bmatrix} G_0 & & & & \\ G_1 & G_0 & & & \\ G_2 & G_1 & G_0 & & \\ G_3 & G_2 & G_1 & G_0 & \\ \vdots & \vdots & \vdots & \vdots & \end{bmatrix} \begin{bmatrix} I_0 \\ I_1 \\ I_2 \\ I_3 \\ \vdots \end{bmatrix} = \begin{bmatrix} -1 \\ 0 \\ 0 \\ 0 \\ \vdots \end{bmatrix} \tag{12}$$

或

$$\begin{bmatrix} I_0 & & & & \\ I_1 & I_0 & & & \\ I_2 & I_1 & I_0 & & \\ I_3 & I_2 & I_1 & I_0 & \\ \vdots & \vdots & \vdots & \vdots & \end{bmatrix} \begin{bmatrix} G_0 \\ G_1 \\ G_2 \\ G_3 \\ \vdots \end{bmatrix} = \begin{bmatrix} -1 \\ 0 \\ 0 \\ 0 \\ \vdots \end{bmatrix} \tag{13}$$

显然,对 AR(n) 模型而言,有

$$-I_0 = 1, I_1 = \phi_1, I_2 = \phi_2, \cdots, I_n = \phi_n; I_j = 0, j > n \tag{14}$$

对 MA(m) 模型而言,有

$$G_0 = 1, G_1 = -\theta_1, G_2 = -\theta_2, \cdots, G_m = -\theta_m; G = 0, j > m \tag{15}$$

（三）

现讨论如何确定模型形式及其阶数。

先对观测到的平稳时间序列 $\{x_t\}$ 拟合出适用的 AR(p) 模型

$$(1 - I_1 B - I_2 B^2 - \cdots - I_p B^p) x_t = a_t \tag{16}$$

再根据 Green 函数与逆函数的关系式求出 MA(q) 模型

$$(1 + G_1 B + G_2 B^2 + \cdots + G_q B^q) a_t = x_t \tag{17}$$

当然，由于所讨论的系统是稳定系统，G_j 是衰减的[2]，j 高至一定阶数 q 后，G_j 可略去，可获得适用的 MA(q) 模型。所确定的 MA(q) 模型是否适用，可用有关适用性检验准则对模型进行检验。

设所拟合出的 AR(p) 与 MA(q) 模型不是最适合的模型（即参数最少的模型），则所应拟合出的适用的模型为 ARMA(n,m)，且 $n+m \not> p, q$。由式(2)与式(16)可得算子恒等式

$$1 - \phi_1 B - \cdots - \phi_n B^n \equiv (1 - \theta_1 B - \cdots - \theta_m B^m)(1 - I_1 B - \cdots - I_p B^p) \tag{18}$$

比较 B 算子的同次幂项的系数，得

$$\left. \begin{aligned} \phi_1 &= \theta_1 + I_1 \\ \phi_2 &= \theta_2 - I_1 \theta_1 + I_2 \\ &\vdots \\ \phi_n &= \theta_n - I_1 \theta_{n-1} - I_2 \theta_{n-2} - \cdots - I_{n-1} \theta_1 + I_n \end{aligned} \right\} \tag{19}$$

$$0 = \theta_j - I_1 \theta_{j-1} - I_2 \theta_{j-2} - \cdots - I_{j-1} \theta_1 + I_j, \quad p \geqslant j > n \tag{20}$$

显然，当 $j > m$ 时，$\theta_j = 0$。

今取 $p = n' + m$，$n' > n$。而一般所用的 ARMA(n,m) 模型均为有理谱模型，$m \leqslant n$。故根据式(20)，得

$$\begin{bmatrix} I_{n'} & I_{n'-1} & \cdots & I_{n'-m+1} \\ I_{n'+1} & I_{n'} & \cdots & I_{n'-m+2} \\ \vdots & \vdots & & \vdots \\ I_{p-1} & I_{p-2} & \cdots & I_{n'} \end{bmatrix} \begin{bmatrix} \theta_1 \\ \theta_2 \\ \vdots \\ \theta_m \end{bmatrix} = \begin{bmatrix} I_{n'+1} \\ I_{n'+2} \\ \vdots \\ I_p \end{bmatrix} \tag{21}$$

由上式得

$$\theta_m = \frac{1}{\Delta_l} \begin{bmatrix} I_{n'} & I_{n'-1} & \cdots & I_{n'-m+2} & I_{n'+1} \\ I_{n'+1} & I_{n'} & \cdots & I_{n'-m+3} & I_{n'+2} \\ \vdots & \vdots & & \vdots & \vdots \\ I_{p-1} & I_{p-2} & \cdots & I_{p-m+1} & I_p \end{bmatrix} \tag{22}$$

式中，Δ_I 表示式(21)中矩阵的行列式。

按 $n'=p-2, p-3, \cdots$，逐步计算出 θ_m，即算出 $m=2$ 时的 θ_2，$m=3$ 时的 θ_3，$\cdots$。当算得 $m=k+1, \theta_{k+1} \approx 0$ 时，则取 $m=k, \theta_m=\theta_k, k<p/2$。

确定 m 后，按式(21)算出 $\theta_1, \theta_2, \cdots, \theta_m$，再按式(19)算出 $\phi_1, \phi_2, \cdots, \phi_n$。$n$ 值应根据 $\phi_{n+1} \approx 0$ 来确定，且 $n>m$。

这样可确定 ARMA 的阶数 n 与 m 及参数 ϕ_i 与 θ_j。但为使所确定的阶数较为可靠，再利用 Green 函数与 MA(m) 模型，仿上述做法，再次确定 n 与 m 及 ϕ_i 与 θ_j。

此即，据式(2)与式(17)又可得算子恒等式

$$(1-\phi_1 B-\cdots \phi_n B^n)(1+G_1 B-\cdots-G_q B^q) \equiv 1-\theta_1 B-\cdots-\theta_m B^m \quad (23)$$

比较 B 的同次幂项的系数，得

$$\left.\begin{aligned} \theta_1 &= \phi_1 - G_1 \\ \theta_2 &= \phi_2 + \phi_1 G_1 - G_2 \\ &\vdots \\ \theta_m &= \phi_m + \phi_{m-1} G_1 + \phi_{m-2} G_2 + \cdots + \phi_1 G_{m-1} + G_m \end{aligned}\right\} \quad (24)$$

$$0 = \phi_j + \phi_{j-1} G_1 + \phi_{j-2} G_2 + \cdots + \phi_1 G_{j-1} + G_j, \quad q \geqslant j > m \quad (25)$$

显然，当 $j>n$ 时，$\phi_j=0$。

同样，$q=n+m', m'>m$，故据式(25)可得与式(21)、(22)类似的算式；但应计及 $m \leqslant n$，G 的下标为 0 时，其值为 1，下标为负时，其值为 0。仿前，可算出 $\phi_1, \phi_2, \cdots, \phi_n$，再算出 $\theta_1, \theta_2, \cdots, \theta_m$。

根据两次计算出的阶数 n 与 m，恰当地选定 ARMA 模型的阶数；根据两次计算出的 ϕ_i 与 θ_j，恰当地选择参数初值，再对 ARMA 模型的参数进行非线性最小二乘估计；最后，对模型进行适用性检验，即使检验出此模型不适用，适用模型的阶数也不会比此模型的高多少。

文献[3]提出的方法中，取 $p=n+m$，这样的 ARMA(n,m) 模型的参数数目没有比相应的 AR(p) 模型的有所减少，这是没有根据的，也是不恰当的。另外，在此方法的论述中，还有些提法值得商榷。

当系统的输入、输出组成多维时间序列时，仍可采用如式(1)、(2)的模型形式，只不过 x_t 为多维的平稳时间序列，a_t 为同维的白噪声，ϕ_i 与 θ_j 为相应维数的系数矩阵而已。当然，此时模型阶数的确定与参数的估计就十分复杂了。

对一维的平稳时间序列而言，采用本文所提出的方法来确定最合适的(即参数数目最少的)模型的形式与阶数，计算工作量是较少的，特别是在模型阶数较高

时更是如此。同时，这一方法不仅不会丢失最合适的模型形式，而且还可获得 AR、MA 和 ARMA 三种形式的适用的模型。这就为选择便于应用的模型形式提供了条件。

参 考 文 献

[1] G. E. P. Box, G. M. Jenkins. *Time Series Analysis*. Forecasting and Control, Holden-Day, San Francisco (1976).

[2] S. M. Pandit, S. M. Wu. *Time Series and System Analysis with Applications*. John-Wiley&Sons (1983).

[3] M. D. Srinath, P. K. Rajasekaran. *An Introduction to Statistical Signal Processing with Applications*. John-Wiley&Sons (1979).

[4] G. C. Goodwin, R. L. Payne. *Dynamic System Identification, Experiment Design and Data Analysis*. Academic Press, New York (1977).

[5] 复旦大学，概率论（第三册，随机过程）．人民教育出版社（1981）．

A Mathematical Model and Its Order for Stationary Time Series

Yang Shuzi

Abstract From the point of view of system analysis, the physical meaning of a mathematical model for stationary time series is discussed and the direct relationship between the Green function and inverse function is given. A method for determining the model's form and its order is suggested.

（原载《华中工学院学报（自然科学版）》1983 年第 11 卷第 5 期）

时序建模与系统辨识

杨叔子　熊有伦　师汉民　王治藩

提　要　本文分析了时序方法的特点，并从系统角度对时序 ARMA 模型进行了分析，进而指出了 ARMA 模型确能反映系统与外界无关的固有特性；文中还研究了 ARMA 模型与系统辨识差分模型的关系，比较了这两种模型在建模方法上的特点；最后，指出了时序建模与系统辨识相结合这一趋势。

目前在采用时间序列分析方法（以下简称时序方法或时序分析）解决一类直接用控制理论中各种识别系统的方法所无法或难于解决的问题，建立有关系统的时序 ARMA 模型的同时，往往又从系统角度对此模型加以研究与分析，给予物理解释。这两者相互渗透与相互促进的趋向，在时序方法与系统辨识方法的发展与应用中，越来越明显。这是值得十分重视的。本文将在这些方面进行探讨，着重探讨时序 ARMA 模型与现代控制理论中系统辨识的差分模型之间的关系以及这两种模型的建模方法之间的关系。

一

时序方法实质上是对有序的随机的观测数据的处理方法。这种有序的随机数据一般称为时间序列或动态数据。这种数据的顺序与大小包含了数据结构的规律，反映了数据内部的相互关系，蕴含着有关的信息。

通常所谓的动态数据或随机数据处理方法其实也是一种时序方法，它对观测所得的有序的随机数据，建立非参数的数学模型，即自协方差函数（或自相关函数）与自谱函数（即功率谱密度 PSD），来揭示这些结构、关系与信息。设观测数据是一个平稳的、正态的、零均值的时间序列 $\{x_t\}$，$t=1,2,\cdots,N$，N 为观测数据数目或样本量,则观测数据的统计特性由其时域内的自协方差函数(此时亦为自相关函数)$r_k = E[x_t x_{t-k}]$或频域内的自谱函数$s_{xx}(\omega) = F[r_k]$所决定。

但是,由于所获得的观测数据 N 或样本长度 $N\Delta$(Δ 为采样时间间隔)是有限的,因此,不可能获得 r_k 与 $s_{xx}(\omega)$ 的真值,而只能通过某种数学方法获得它们的某种估值。通常所用的动态数据处理方法是直接利用观测数据来计算 r_k 与 $s_{xx}(\omega)$ 的估值,亦即直接利用观测数据建立自协方差函数与自谱函数这种非参数模型,再利用这种模型来揭示观测数据中所蕴含的有关信息。然而,这种方法有着固有的缺陷,即观测数据是加窗处理的,从而使估值偏离真值较远。谱线发生泄漏,频率分辨力低,弱信号被淹没。虽然目前提出了不少改进方法,但不能从根本上消除这些固有的缺陷[1],*。

现在所谓的时序方法,是利用观测所得的有序的随机数据,建立参数模型,即 ARMA 模型,来揭示观测数据所蕴涵的有关信息的方法。ARMA 模型即指差分方程

$$x_t - \sum_{i=1}^{n} \phi_i x_{t-i} = a_t - \sum_{j=1}^{m} \theta_j a_{t-j} \tag{1}$$

此处,式(1)左边的 n 阶差分多项式称为 n 阶自回归部分,式(1)右边的 m 价差分多项式称为 m 阶滑动平均部分,$\{x_t\}$ 为观测数据,$\{a_t\}$ 为残差,而且 $\{a_t\}$ 为正态白噪声,即 $a_t \sim \text{NID}(0; \sigma_a^2)$。

现在所要完成的工作不是直接通过观测数据寻求 r_k 与 $s_{xx}(\omega)$ 的估值,而是寻求 ARMA 模型的阶数 n、m 与参数 ϕ_i、θ_i。由于 ARMA 模型是一个动态模型,对 $\{x_t\}$ 具有外延性质;因此,模型一旦建立,这一外延性质就不受样本长度限制,即无加窗处理问题,从而免除了常用处理方法的固有缺陷。当然,模型的阶数 n、m 与参数 ϕ_i、θ_i 的估计精度是同样本长度密切相关的;在样本长度不足的情况下,同样也不能获得接近于理论上的模型。但是,实践表明,获得较好的 ARMA 模型比获得较好的传统谱(当然包括了 FFT)所需的观测数据为短。

正如上述,时序方法是处理动态数据的方法,ARMA 模型是描述动态数据结构,反映数据内部相互关系的数学模型;然而,从系统角度看来,这一模型也是描述一个系统的动力学特性的数学模型,a_t 为其输入,x_t 为其输出。其实,这是很自然的,因为观测数据势必来源于某一现象或过程,来源于某一广义系统;在这些数

---

* 王治藩,杨叔子. ARMA 谱估计及其应用. 全国第三次飞机结构动力学学术讨论会宣读论文,桂林阳朔,1983 年 11 月。

据所蕴涵的信息中,必然包含这一系统的固有特性,包含这一系统行为发展的趋势[2]。这就是说,还可以利用 ARMA 模型来了解系统的固有特性,监视系统目前的状态,预测数据未来的取值,对系统进行控制。

引入后移算子 B,式(1)可改写为

$$(1 - \sum_{i=1}^{n} \phi_i B^i) x_t = (1 - \sum_{j=1}^{m} \theta_j B^j) a_t \tag{2}$$

简写为 $\phi(B) x_t = \theta(B) a_t$,从而有

$$x_t = \frac{\theta(B)}{\phi(B)} a_t \tag{3}$$

显然,$\frac{\theta(B)}{\phi(B)}$ 是系统的传递函数。

从系统角度看来,从模型可得系统的动态特性如下:

(1) 单位脉冲响应函数 G_j

它就是时序中的 Green 函数,且有[3,4]

$$x_t = \sum_{j=0}^{\infty} G_j a_{t-j} \tag{4}$$

G_j 可从模型中算出;

(2) 频率特性 $H_x(\omega)$ [2]

$$H_x(\omega) = \frac{\theta(B)}{\phi(B)} \bigg|_{B=e^{-\tau \omega \Delta}} = \sum_{j=0}^{\infty} G_j B^j \bigg|_{B=e^{-\tau \omega \Delta}} \tag{5}$$

显然,$H_x(\omega)$ 就是 G_j 的 Fourier 变换。

从统计角度看来,从模型可得时序的统计特性如下:

(1) 自协方差函数 r_k [3,4]

$$r_k = E[x_t x_{t-k}] = \sigma_a^2 \sum_{j=0}^{\infty} G_j G_{j+k} \tag{6}$$

(2) 自谱函数 $s_{xx}(\omega)$ [2]

$$s_{xx}(\omega) = F[r_k] = \sigma_a^2 |H_x(\omega)|^2 \tag{7}$$

式中,$-\pi/\Delta < \omega < \pi/\Delta$。

显然,G_j,$H_x(\omega)$,r_k 与 $s_{xx}(\omega)$ 这四者在时域与频域内存在着严格的对应关系;而 r_k 与 $s_{xx}(\omega)$ 是从模型中而非直接从观测数据中算得的,故能较好地逼近它们的真值。总之,建立 ARMA 模型不仅可以更好地了解数据本身的统计特性,而且还可以了解产生数据的系统的动态特性。

二

将 ARMA 模型作为相应系统的数学模型,这就表明时序建模方法就是一种系统辨识方法。但是,同控制理论中的系统辨识方法相较,时序建模只是利用系统的输出,而系统的输入只作为白噪声(它的方差 σ_a^2 还有待于估计)处理。文献[5]的分析表明,ARMA 模型的自回归算子(即传递函数分母)$\phi(B)$ 一般是包含实际系统的传递函数分母 $\phi_0(B)$ 的,只要 $\phi_0(B)$ 与 $\theta'(B)$ 之间无相同因子相消即可;图 1 中,s_t 表示实际系统的未知的平稳时序输入,而 $\theta_0(B)/\phi_0(B)$ 是实际系统的传递函数,$\theta'(B)/\phi'(B)$ 是将白噪声 a_t 转变成 s_t 的成形滤波器。

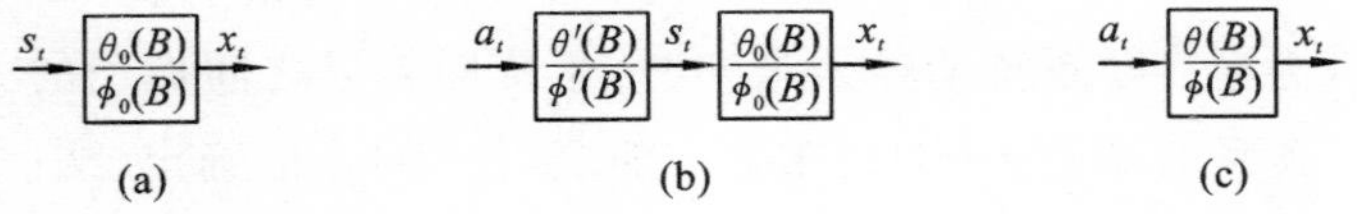

图 1

当实际系统在多个平稳时序输入作用下(见图 2)时,只要它是闭环系统,则不论系统在何处输入,何处输出,系统传递函数的分母都不变。而一般实际系统存在内反馈,事实上是闭环的。因此

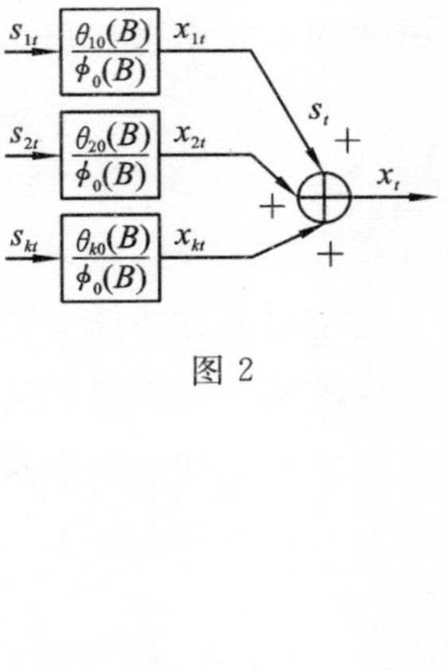

$$x_t = \sum_{i=1}^{k} x_{it} = \sum_{i=1}^{k} \frac{\theta_{i0}(B)}{\phi_0(B)} s_{it}$$

$$\phi_0(B) x_t = \sum_{i=1}^{k} \theta_{i0}(B) s_{it}$$

根据平稳时序的运算性质[6],可令

图 2

$$\sum_{i=1}^{k} \theta_{i0}(B) s_{it} = \theta_0(B) s_t$$

式中:s_t 为某一合适的平稳时序;$\theta_0(B)$ 为相应的算子。从而可将图 2 所示模型化为图 1(a)所示的单输入情况。当然,也可以将各 s_{it} 化为在相应白噪声 a_{it} 驱动下第 i 个成形滤波器的输出,而后再简化为单输入 a_t 作用时的情况。

当实际系统的输出受到某一平稳时序噪声 s_t 的干扰时,如图 3(a)所示,则 s_t 可视为在白噪声驱动下成形滤波器 $\theta'(B)/\phi'(B)$ 所产生的输出,如图 3(b)所示,此时有

$$x = y_t + s_t$$

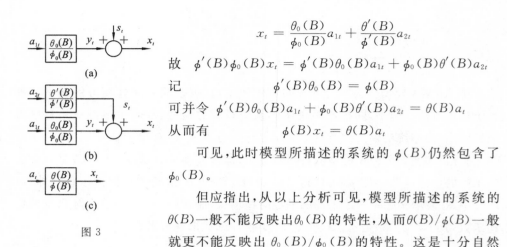

图 3

$$x_t = \frac{\theta_0(B)}{\phi_0(B)}a_{1t} + \frac{\theta'(B)}{\phi'(B)}a_{2t}$$

故　$\phi'(B)\phi_0(B)x_t = \phi'(B)\theta_0(B)a_{1t} + \phi_0(B)\theta'(B)a_{2t}$

记　　　　$\phi'(B)\theta_0(B) = \phi(B)$

可并令　$\phi'(B)\theta_0(B)a_{1t} + \phi_0(B)\theta'(B)a_{2t} = \theta(B)a_t$

从而有　　　　$\phi(B)x_t = \theta(B)a_t$

可见,此时模型所描述的系统的 $\phi(B)$ 仍然包含了 $\phi_0(B)$。

但应指出,从以上分析可见,模型所描述的系统的 $\theta(B)$ 一般不能反映出 $\theta_0(B)$ 的特性,从而 $\theta(B)/\phi(B)$ 一般就更不能反映出 $\theta_0(B)/\phi_0(B)$ 的特性。这是十分自然的。这正是因为当实际系统的输入未知时,系统同外界之间作用的关系就未知,从而反映这一关系的算子 $\theta_0(B)$ 也未知。加之, $\phi_0(B)$ 不能确定,从而系统的与输入有关的特性,即与 $\theta_0(B)/\phi_0(B)$ 有关的特性,例如,实际机械系统或机械结构的动刚度或动柔度等一般也是不知道的;然而,与 $\phi_0(B)$ 有关的固有特性,例如,实际系统的固有频率、阻尼比、谐振频率等一般是不会丧失的。

三

采用时序方法建立的 ARMA 模型同采用现代控制理论中系统辨识方法建立的差分模型一方面是相互一致的,另一方面又是相互补充的。

系统辨识的差分模型一般可写为[7,8]

$$y_t + \sum_{i=1}^{n} a_i y_{t-i} = \sum_{j=0}^{n} b_j u_{t-j} + \varepsilon_t + \sum_{k=1}^{n} c_k \varepsilon_{t-k} \tag{8}$$

式中:u_t 为输入;y_t 为输出;ε_t 为残差(为正态白噪声)。引用后移算子 B,上式可写为

$$a(B)y_t = b(B)u_t + c(B)\varepsilon_t \tag{9}$$

当 u_t 为不可观测的平稳、正态、零均值的时序时,则可将 u_t 视为在白噪声 ξ_t 驱动下某一成形滤波器 $b'(B)/a'(B)$ 的输出,从而有

$$a(B)y_t = b(B)\frac{b'(B)}{a'(B)}\xi_t + c(B)\varepsilon_t$$

即

$$a'(B)a(B)y_t = b'(B)b(B)\xi_t + a'(B)c(B)\varepsilon_t \tag{10}$$

记 $$a'(B)a(B) = \phi(B)$$
又可令 $$b'(B)b(B)\xi_t + a'(B)c(B)\varepsilon_t = \theta(B)a_t$$
从而有 $$\phi(B)y_t = \theta(B)a_t$$
这就是 ARMA 模型。

当 u_t 为不可观测的确定性函数时,系统相应的输出也是确定性函数,记为 d_t,则有
$$y_t = \frac{b(B)}{a(B)}u_t + \frac{c(B)}{a(B)}\varepsilon_t$$
故 $$y_t = d_t + x_t$$
式中,
$$d_t = \frac{b(B)}{a(B)}u_t \quad 或 \quad a(B)d_t = b(B)u_t \tag{11}$$

$$x_t = \frac{c(B)}{a(B)}\varepsilon_t \quad 或 \quad a(B)x_t = c(B)\varepsilon_t \tag{12}$$

显然,x_t 为一平稳、正态、零均值的时序,式(12)即 ARMA 模型,而式(11)与式(12)的综合就是时序方法对具有趋向性的非平稳时序$\{y_t\}$(即所谓自协方差平稳的时序)所建立的模型[2,4]。

当 u_t 为具有趋向性的非平稳时序而又不可观测时,情况就是上述两种情况的综合。

由上可知,当式(8)或式(9)中的 u_t 为不可观测时,系统辨识的差分模型就化为时序方法的模型。

当 u_t 为可观测的平稳、正态、零均值的时序时,式(8)或式(9)可化为一个二维的 ARMA 模型。因为此时,y_t 与 u_t 可组成二维向量 $\boldsymbol{x}_t$,从而

$$\begin{bmatrix} 1 & -b_0 \\ 0 & 1 \end{bmatrix}\begin{bmatrix} y_t \\ u_t \end{bmatrix} + \begin{bmatrix} a_1 & -b_1 \\ 0 & g_1 \end{bmatrix}\begin{bmatrix} y_{t-1} \\ u_{t-1} \end{bmatrix} + \cdots + \begin{bmatrix} a_n & -b_n \\ 0 & g_n \end{bmatrix}\begin{bmatrix} y_{t-n} \\ u_{t-n} \end{bmatrix}$$
$$= \begin{bmatrix} 1 & 0 \\ 0 & 1 \end{bmatrix}\begin{bmatrix} \varepsilon_t \\ \eta_t \end{bmatrix} + \begin{bmatrix} c_1 & 0 \\ 0 & d_1 \end{bmatrix}\begin{bmatrix} \varepsilon_{t-1} \\ \eta_{t-1} \end{bmatrix} + \cdots + \begin{bmatrix} c_n & 0 \\ 0 & d_n \end{bmatrix}\begin{bmatrix} \varepsilon_{t-n} \\ \eta_{t-n} \end{bmatrix} \tag{13}$$

上式就是一个二维的 $\phi(B)x_t = \theta(B)a_t$,将它展开后有两个标量式;第一式就是式(8),第二式就是一维的 ARMA 模型,$g(B)u_t = d(B)\eta_t$,它将 u_t 化为在白噪声 η_t 驱动下成形滤波器 $d(B)/g(B)$ 的输出。当然,$d(B)$ 的阶数不一定达到 n,有关参数可以为零。

当 u_t 为可观测的确定性函数或可观测的具有趋向性的非平稳时序时,易于参照上述各种情况变换为确定性模型与二维 ARMA 模型的综合。

当式(8)本身为向量模型时,则可变换为更高维的 ARMA 模型。

显然,ARMA 模型比控制理论中系统辨识的差分模型包括的范围更为广泛。今考察以下四种情况:

(1) 系统的输入无法观测到(例如飞机、车辆、机床等在工作情况下);

(2) 对一个系统观测到多个时序(例如大气中的降雨量、气温、气压;又如切削过程中的切削力、切削温度、刀具磨损),而它们之间的因果关系不清楚或不完全清楚;

(3) 系统受到噪声的干扰太大,以致按控制理论中系统辨识方法所施加的有用信号都完全被淹没;

(4) 系统本身的边界或系统本身是什么也不清楚(例如,以市场商品价格作为输出,以磨床砂轮表面的起伏作为输出,那么什么是相应系统的边界或相应系统的本身这是讲不清楚的)。

处理上述情况时,是不可能直接应用控制理论中的系统辨识的方法建模的,也不可能获得系统本身的固有特性,因为这时都缺乏明确的或完全的输入与输出的因果关系,而控制理论中的频率特性、状态模型、差分模型等系统辨识方法正是建立在输入与输出的因果关系完全清楚与输入、输出等价的基础之上的。但是,在采用时序方法建模时,就不存在什么困难,因为它只是建立在输出等价的基础之上的。采用时序方法,可将所观测到的时序(也包括输入)作为系统的一维或多维的输出,而将模型所描述的等价系统视为在与输出同维白噪声驱动下产生这一输出的。

ARMA 模型之所以包含的内容与适用的范围来得广泛,是因为它从统计角度来揭示各时序内部的统计关系与各时序之间的统计关系,而控制理论中系统辨识的差分模型则是从系统角度来揭示各时序之间的因果关系的。显然,对所观测的多个时序而言,统计关系可以包括因果关系,但因果关系却不能包括统计关系。毫无疑问,能获得时序之间的因果关系,则可获得系统更多的特性,特别是可获得系统的与外界有关系的特性,这当然是所希望的。

四

对有序的随机的观测数据的处理,目前的时序方法之所以优越于常用的传统的动态数据处理方法,在于前者能建立一个对观测数据具有外延性质的参数模型。这就是说,只有这一参数模型的阶数 n、m 与参数 ϕ_i、θ_i 能有良好的估值时,其优越性方能显示。它们估值是否优良,不仅同观测数据的样本量密切相关,而且也同估计方法紧密相连。在时序建模方法中,模型参数的估计往往起着特别重要的作用。

对于 ARMA 模型与它的两个特例 AR 模型与 MA 模型,固然可建立一个统一的非线性最小二乘估计方法[4],但它们还各有其特有的参数估计方法[1,6],并且这些估计方法还在不断发展之中。特别值得指出的是 ARMA 模型的参数估计。由于 ARMA 模型的无条件回归是非线性的,因此,参数估计十分麻烦。但是,进一步对系统辨识的差分模型式(8)或式(9)作分析,就可将残差 ε_t 视为不可观测的平稳输入,u_t 为可观测的输入,而 y_t 为 ε_t 与 u_t 共同作用时的输出;那么,当式(8)或(9)中可观测的输入 u_t 为零时,就获得了 ARMA 模型这一特殊情况。从而,将系统辨识差分模型的参数估计的算法与公式中的 $b_i(i=1,2,\cdots,n)$ 全部取为零,则这些算法与公式(例如 LS、GLS、ML 以及它们的递推形式 RLS、RGLS、RML 等[7,8])除极个别外可用于 ARMA 模型的参数估计。我们在建立车床主轴轴心随机漂移的 ARMA 模型时,采用 GLS 法,建模速度比采用非线性最小二乘法的为快*。

另外,现代控制理论中的 Kalman 滤波理论与参数实时估计方法则是采用时序参数模型进行在线预测与监控往往必须用到的。只要令

$$\boldsymbol{\phi}^T = [-\phi_1, -\phi_2, \cdots, -\phi_n, \theta_1, \theta_2, \cdots, \theta_m]^T$$
$$\boldsymbol{\psi}^T = [x_{t-1}, x_{t-2}, \cdots, x_{t-n}, a_{t-1}, a_{t-2}, \cdots, a_{t-m}]^T$$

就可写出最简单的状态方程

$$\begin{aligned}\phi_{t-1} &= \phi_t \\ x_t &= \psi_t \phi_t + a_t\end{aligned} \tag{14}$$

从而就可利用 Kalman 滤波公式,不断根据新的观测数据,修正模型参数*。

但是,由于时序方法是数理统计的一个分支,因此,在模型适用性方面(核心问题是定阶问题)发展了一系列的检验准则,其中不少准则已为现代控制理论中系统辨识所采用,而现代控制理论中系统辨识在模型适用性检验方面还没有提出系统的理论与检验准则。

五

时序建模同控制理论中系统辨识相结合,时序分析同系统分析相结合,使时

---

* 王治藩,杨叔子. 改进时间序列方法,建立过程模型. 全国机械工业自动化学会第二届年会宣读论文,大连,1983 年 10 月.

序分析能吸取控制理论的长处，使其内容更加丰富与完善，同样，也将使控制理论中系统辨识领域得到扩大，方法更为完善，从而使控制理论进一步得到发展与应用。例如，可以将现代控制理论中系统辨识的随机状态模型中输入与噪声分别视为可观测输入与不可观测输入，则当输入未知时，可观测输入为零，模型即成时序模型；同时，对状态模型的参数的估计方法可相应地予以应用。当然，如同讨论差分模型与 ARMA 模型的关系一样，只要将这种时序模型维数增高，就又可包括随机状态模型。这样，时序建模又可按状态模型的建模方法进行。这既是为时序建模也是为控制理论中状态模型辨识增添了新的内容。

1962 年 Zadeh 给系统辨识下的定义是："系统辨识是在输入与输出的基础上，从一类系统中确定一个与所测系统的等价系统。"[9] 这一定义只能作为对系统辨识在当时情况下的一种描述。目前，系统辨识的发展已远远超出这一描述的范围。系统辨识不仅与控制理论而且也与数理统计互相渗透，互相结合，时序建模已成为现代系统辨识中的一个不可分割的部分。

参 考 文 献

[1] S. M. Kay and S. L. Marple. PIEEE, Vol. 69, No. 5, pp. 1380~1419(1981).

[2] 杨叔子. 动态数据的系统处理. 机械工程，1983 年第 5 期~1984 年第 3 期.

[3] G. E. P. Box, G. Jenkins. *Time Series Analysis, Forecasting and Control*. Hoiden-Day San Francisco (1976).

[4] S. M. Pandit, S. M. Wu. *Time Series and System Analysis with Applications*. John Wiley & Sons(J983).

[5] 杨叔子. 平稳时间序列的数学模型及其阶的确定的讨论. 华中工学院学报，第 11 卷，第 5 期，第 9~14 页(1983).

[6] 安鸿志，陈兆国，杜金观，潘一民. 时间序列的分析与应用. 科学出版社(1983).

[7] G. C. Goodwin, R. L. Payne. *Dynamic System Identification, Experiment Design and Data Analysis*. Academic Press, New York(1977).

[8] 韩光文，辨识与参数估计. 国防工业出版社(1980).

[9] L. A. Zadeh. Proc. IRE, Vol. 50：856~865 (1962).

Time Series Modelling and System Identification

Yang Shuzi　Xiong Youlun
Shi Hanmin　Wang Zhifan

Abstract　The ARMA model in time series is analyzed from the system point of view and it is pointed out that the ARMA model does reflect the system's inherent characteristics of being independent of external factors. The relationship between the ARMA model in time series and the difference model in system identification is discussed and the features of their modeling methods are compared.

（原载《华中工学院学报（自然科学版）》1984年第12卷第6期）

金属切削过程颤振预兆的特性分析[*]

杨叔子　刘经燕　师汉民　梅志坚

提　要　本文从模式识别理论与切削颤振特性出发,通过大量试验研究,讨论了几种特征信号的选择、模式向量的获得与特征主分量的抽取,提出了判别函数的构造问题,以便能在线及时判别颤振预兆的出现与否,为对金属切削过程颤振进行监控提供了前提条件。

符　号:

σ_p——切削深度

N——采样点数;机床主轴转速(r/min)

$S(f)$——信号的自功率谱

x_t——采样数据

γ_0——刀具前角

κ——主偏角

σ_{ay}^2——刀架 y 向振动加速度信号的方差

σ_{az}^2——刀架 z 向振动加速度信号的方差

$\sigma_{ay^2}^2$——刀架 y 向振动加速度信号二阶自回归模型的残差方差

$\sigma_{az^2}^2$——刀架 z 向振动加速度信号二阶自回归模型的残差方差

$\sigma_{az^n}^2$——尾顶尖 z 向振动加速度信号 n 阶自回归模型的残差方差

$\sigma_{Fz^n}^2$——z 向切削力信号 n 阶自回归模型的残差方差

f——频率(Hz);进给率(mm/r)

n——自回归模型阶次

t——时间

α_0——刀具后角

Δ——采样间隔

κ'——负偏角

一、概　　述

金属切削过程颤振的监控是提高加工质量、提高生产效率、发展机床适应控

---

[*]　中国科学院科学基金赞助课题。

制,特别是发展柔性生产系统中不可回避的一个重要问题。显然,进行这样主动的监控,关键或前提条件在于能在线识别出颤振即将发生的预兆。对切削过程而言,产生颤振的机理是复杂的。多年来不少专家学者在切削颤振机理方面开展了不少研究工作。特别是近十几年来,由于计算机技术的迅速发展,信息理论与控制理论的广泛应用,不少学者置颤振机理的研究于次要地位,而直接从切削过程的有关信号中,考察有关物理量(或数学量)的变化,提取同颤振有关的信息,寻求颤振即将发生的预兆,以判断颤振是否即将发生[1-5],并采取相应的控制措施。但是这些成果有其局限性,例如:对车削过程而言,S. Braun 考察刀架振动加速度信号中颤振频率附近的相频特性变化情况[1],显然,这需要预先知道颤振频率;T. L. Subramanian 等人考察刀架振动信号的幅值变化情况[3],但信号幅值受干扰影响较大;K. F. Eman 等人考察尾顶尖振动加速度信号时序模型的阻尼率的变化情况[3],但实际上难以获得可靠的阻尼率,而且计算繁杂,预测下一步的阻尼率也无必要;Y. Miyoshi 等人考察了他们定义的均方频率的变化情况[5],这颇有特色,但均方频率没有考虑信号的总能量变化而存在一些缺陷。然而他们的工作还是富有启发性的。

实际上,对切削颤振预兆的识别问题,即切削颤振的早期诊断问题,是一个模式识别问题,而且只是识别两种不同的模式的问题,一是有颤振预兆的模式,另一是无颤振预兆的模式。对于颤振已发生的模式,不采用监控技术,人们也可识别。对颤振的监控而言,这种模式是不允许出现的。

二、试　验

试验是在一台最大加工直径为 500 mm 的西德 VDF 车床上进行的。所选择的试验条件如表 1 所示。

表 1　试验条件

试验	检测信号	试验条件			
		刀具几何角度	切削用量	工件直径 /mm	工件长度 /mm
1	尾架顶尖 Z 向振动加速度	$\kappa=85°$　$\kappa'=8°$　$\alpha_0=8°$　$\gamma_0=13$	$N=560$ r/min　$\alpha_p=0.5$ mm　$f=0.2$ mm/r	48.0	900

续表

试验	检测信号	试验条件			
		刀具几何角度	切削用量	工件直径 /mm	工件长度 /mm
2	刀架 z 向振动加速度 刀架 y 向振动加速度	$\kappa=90°$ $\kappa'=6°$ $\alpha_0=12.8°$ $\gamma_0=12.4$	$N=542$ r/min $\alpha_p=0.25$ mm $i=0.1$ mm/r	38.3	900
3	工件振动位移	$\kappa=90°$ $\kappa'=2.6°$ $\alpha_0=9°$ $\gamma_0=0°$	$N=560$ r/min $\alpha_p=0.5$ mm $f=0.1$ mm/r	44.5	980
4	Z 向切削力	$\kappa=90°$ $\kappa'=6°$ $\alpha_0=9.2°$ $\gamma_0=8.1°$	$N=614$ r/min $\alpha_p=0.5$ mm $f=0.2$ mm/r	44.2	900

[注] 刀具材料 YT15　工件材料 45 钢

试验原理如图 1 所示，在离三爪卡盘 170 mm 处，在床身上支持有涡流式位移传感器，以测量工件与传感器间的相对振动位移。在四方刀架上装夹有八角环测力仪，以测量切削力。在尾架顶尖的垂直方向、四方刀架的垂直方向与水平横向装有 B&K 加速度计，以测量该处振动的绝对速度。

图 2(a)、(b)、(c)、(d)、(e)所示的分别表示在表 1 试验条件下某次试验的尾顶尖 z 向振动加速度 a_z、刀架 z 向振动加速度 a_z、刀架 y 向振动加速度 a_y、工件相对振动位移 d_y 和 z 向切削力 F_z 的时域信号图，图中表示出无颤振、将振与已振的发展过程。显然，无颤振与已颤振的时域信号有明显差别，后者的振动周期性明显，振幅增长极大，这充分表明具有颤振频率的周期振动（即极限环）的存在。这无疑给人以启发，能否从颤振即将发生之前的信号中，寻找出与幅值变化、频率变化有关的最为敏感的模式向量、主分量及判别函数，获得与颤振即将发生的预兆有关的尽可能准确的信息。

为了获得这些信息，必须对这些信号在以合适的采样间隔 Δ 采样后，将采样数据加以处理。目前，对随机数据的处理一般有两大类方法：一是以 FFT 为基础的谱分析与相关分析；另一是时序分析[6,7]。在此，从模式识别角度出发，应予指出的是，时序模型的参数凝聚了也反映了时域采样数据中所包含的信息[8]。显然，采用时序分析方法，建立时序模型，为将模式识别用于工况监视与故障诊断开

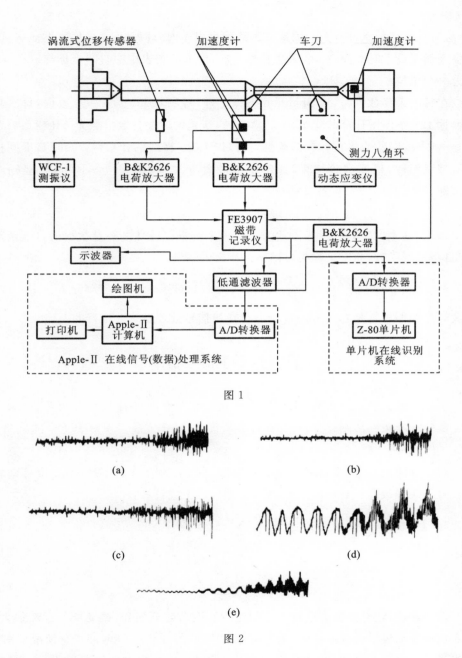

图 1

图 2

拓了广阔领域。

  大量试验资料表明,与整个切削过程有关的信号频率结构的变化主要在 1000 Hz 以下,故以 1000 Hz 作为截止频率对时域信号进行低通滤波后再进行采样计算。为研究颤振从无到有的发展过程及其特征,又考虑到颤振一般在 1~3 s 内产生(这与具体切削条件有关),因此在远离颤振发生时每隔 3.6 s 采集数据一次,而在临近颤振发生时每隔 0.9 s 采集数据一次,以下的图表中的时间坐标均以 3.6 s 作为一个单位。

在每次采集数据时,按 Shannon 采样定理,每隔 0.5 ms 采一个数据,每次采集数据 128 个用于计算与建模,为计算迅速,所建的时序参数模型为 AR 模型。在建立 AR 模型时,均将采样数据减去均值,以 Burg 算法估计参数,以 FPE 准则检验模型适用性,根据试验结果,规定模型最大阶次为 10,建立 AR(n)模型($n\leqslant 10$),即

$$x_t - \varphi_1 x_{t-1} - \varphi_2 x_{t-2} - \cdots - \varphi_n x_{t-n} = a_t \tag{1}$$

式中,a_t 为残差,其均值为零,方差为 σ_a^2。模型的重要特性之一是数据的自谱函数(即自功率谱函数,此时又称为自回归谱或最大熵谱)。

$$s(f) = \sigma_a^2 \Big/ \Big| 1 - \sum_{k=1}^{n} \varphi_k e^{-j2\pi/k\Delta} \Big|^2, \quad -1/(2\Delta) < f < 1/(2\Delta) \tag{2}$$

有关时序模型计算是在 Apple-Ⅱ 微型机在线信号处理系统上进行的[8]。

三、时域特征分析

所进行的试验研究表明,采用振动加速度信号作为特征信号较为适宜,因为它受试验环境条件限制较少,传感元件装置简单,便于安装。现对表 1 中的试验 2 的 a_z 进行特征分析。

对一个随机信号(采样数据)x_t 一般以二阶矩的方差 σ_x^2 来描述其位移动特性。对于采样数据 $x_t(t=1,2,\cdots,N)$,方差可按下式计算:

$$\sigma_x^2 = \frac{1}{N} \sum_{t=1}^{N} (x_t - \bar{x})^2 \tag{3}$$

式中

$$\bar{x} = \frac{1}{N} \sum_{t=1}^{N} x_i \tag{4}$$

正如前述,随着颤振的孕育、形成与发展,振动逐渐加强,振动能量逐渐加大,这表现为振动信号的幅值加大,表 1 试验 2 中 a_z 的方差 σ_{az}^2 随时间变化情况如图 3 实线所示。由图 3 可知,在颤振即将发生的瞬间,信号的方差迅速增大。一旦颤振发生后,信号方差有所下降,这表明振幅逐渐趋于稳定,但方差仍远大于无颤振时信号的方差。

显然,按式(4),需在采样完毕后,才能计算 $\bar{x}$,然后再求方差。为提高计算速

度,我们采用下式计算:

$$\sigma_x^2 = \frac{1}{N}\sum_{t=1}^{N} x_t^2 - \left(\frac{1}{N}\sum_{t=1}^{N} x_t\right)^2 \tag{5}$$

$$\begin{cases} s_1 = x_1, & s_2 = x_1^2, & t = 1 \\ s_1 \Leftarrow s_1 + x_t, & s_2 \Leftarrow s_2 + x_t^2, & t = 2, 3, \cdots, N \end{cases} \tag{6}$$

由式(6),当采第 t_1 个样本($2 \leqslant t_1 \leqslant N$)时,由于前面 $t_1 - 1$ 个样本的 s_1, s_2 的已算好,此时可利用采样间隔中的时间计算 $s_1 \Leftarrow s_1 + x_{t_1}, s_2 = s_2 + x_{t_1}^2$。这样当采样完毕,即 $t_1 = N$ 时,$\sum_{t=1}^{N} x_t$ 和 $\sum_{t=1}^{N} x_t^2$ 都已分别计算好,然后只要按照式(5)就可以计算 σ_x^2。显然,这可大大提高计算速度。

但是,用方差作为特征量可能引起误判,因为外界干扰使得振动加速度幅值加大时,方差也会增大。

现考察所建立 AR 模型的残差方差。图 3 虚线表示试验所得的 a_z 的二阶自回归模型的残差方差 $\sigma_{a_z}^2$ 随时间变化情况。显然,在颤振即将发生时,残差方差 $\sigma_{a_z}^2$ 迅速增大。这一现象的实质可作如下理解:如果将采样数据 x_t 视为模型所描述的等价系统的输出,则残差 a_t 为系统的输入。x_t 的 σ_x^2 方差增大的原因有二:一是系统的阻尼减小;二是输入 a_t 的方差 σ_a^2 加强。但是在所有试验中,阻尼率的变化并无一定规律,并不如文献[4]中所述,颤振将发生时 AR(2)模型的阻尼率 ξ 接近于零,而是仍然保持较大的值,有时还大于 0.5。因此,σ_x^2 的增大主要与 σ_a^2 的增大有关,它们之间的关系是 $\sigma_x^2 = \sigma_a^2 \sum_{j=0}^{\infty} G_j$。综上所述,AR 模型残差方差作为特征量是

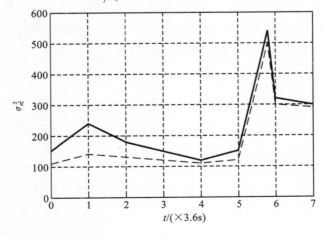

图 3

能反映出切削颤振的发生,而且是同振动信号的能量密切相联的。

再考察 AR 模型的参数特性。在计算中我们发现,随着颤振的即将发生,φ_{21} 的值是增大的,而且对 a_z 而言,φ_{21} 的值还由负变正。值得注意的是,φ_{21} 大小变化规律与采样数据的一步自相关函数 ρ_1 的大小变化规律相同。其 ρ_1 定义为

$$\rho_1 = \sum_{t=2}^{N} x_t x_{t-1} / \sum_{t=2}^{N} x_{t-1}^2 \tag{7}$$

图 4 实线和虚线分别表示 a_z 信号的 φ_{21} 与 ρ_1 的变化情况。

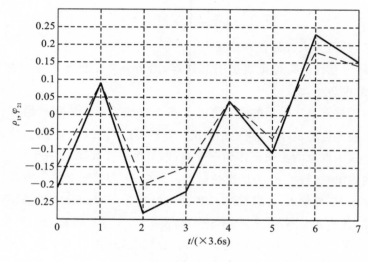

图 4

我们在另文[9]中将讨论 ρ_1 的大小变化(在合适的采样间隔条件下,ρ_1 内的正负性变化)可反映切削过程颤振预兆的出现。在这里,研究 φ_{21} 与 ρ_1 变化相同的原因。为此考察所采用的 Burg 算法中参数的递推公式:

$$\varphi_{11} = 2 \sum_{t=2}^{N} x_t x_{t-1} / \left(\sum_{t=2}^{N} x_t^2 + \sum_{t=2}^{N} x_{t-1}^2 \right) \tag{8}$$

$$\varphi_{21} = \varphi_{11}(1 - \varphi_{22}) \tag{9}$$

$$\varphi_{22} = \frac{2 \sum_{t=3}^{N} (x_t - \varphi_{11} x_{t-1})(x_{t-2} - \varphi_{11} x_{t-1})}{\sum_{t=3}^{N} (x_t - \varphi_{11} x_{t-1})^2 + \sum_{t=3}^{N} (x_{t-2} - \varphi_{11} x_{t-1})^2} \tag{10}$$

对于式(8),当 N 较大时显然有

$$\varphi_{11} = \rho_1 \tag{11}$$

将式(10)展开,且分子分母同除 $N-1$,则有

$$\varphi_{22} = M/D \tag{12}$$

式中

$$M = \frac{1}{N-1}\sum_{t=3}^{N}x_t x_{t-2} - \frac{\varphi_{11}}{N-1}\sum_{t=3}^{N}x_{t-1}x_{t-2} - \frac{\varphi_{11}}{N-1}\sum_{t=3}^{N}x_t x_{t-1} - \frac{\varphi_{11}^2}{N-1}\sum_{t=3}^{N}x_{t-1}^2$$

$$D = \frac{1}{2}\Big[\frac{1}{N-1}\sum_{t=3}^{N}x_t^2 - \frac{2\varphi_{11}}{N-1}\sum_{t=3}^{N}x_t x_{t-1} + \frac{\varphi_{11}^2}{N-1}\sum_{t=3}^{N}x_{t-1}^2$$

$$+ \frac{1}{N-1}\sum_{t=3}^{N}x_{t-1}^2 - \frac{2\varphi_{11}}{N-1}\sum_{t=3}^{N}x_t x_{t-1} + \frac{\varphi_{11}^2}{N-1}\sum_{t=3}^{N}x_t^2\Big]$$

当 N 较大时，且考虑到式(11)，有

$$\varphi_{21} = \rho_1 \frac{1-\rho_2^2}{1-\rho_1^2} \tag{13}$$

又 $|\rho_2|<1$，故由式(13)知，φ_{21} 的正负性与 ρ_1 相同，φ_{21} 的大小主要由 ρ_1 决定，当 ρ_1 增大时，φ_{21} 将增大，当 ρ_1 减小时，φ_{21} 将减小；比较图 4 中实线与虚线，可清楚地看到这一点。

在试验中，还证明了模型特性之一的 Green 函数 G_i（单位脉冲响应函数）衰减程度的变化也是颤振即将发生的预兆，故 G_i 也可作为模式向量，但计算较繁。

四、频域特征分析

在频域中考察，颤振预兆是极为明显的。自回归谱或自回归谱峰频率就是采样数据经建模与求模型特性这两个变换所得到的一个频域中的模式向量。

图 5～图 7 所示分别是 $a_{z'}$、a_z、a_y 的自回归谱阵。

对于振动加速度信号，所考察的频率范围为 0～1000 Hz，这因为在切削过程中，振动加速度信号的背景噪声的高频成分居多，例如各种冲击所导致的振动等。在平稳切削过程中，切削过程产生的振动较小，振幅较小，故自回归谱的主频带（亦即峰值频率构成的模式向量的主分量）在高频段；而当颤振在孕育与形成时，由图 5～图 7 可知，频振从无到有，其能量从小到大，切削系统振动加强，主峰频率逐渐向低频段方向移动到颤振频率附近，这时切削系统失稳。

比较图 5 至图 7 可知，尾顶尖处的 $a_{z'}$ 信号较刀架处 a_z 与 a_y 信号更为敏感。同时受到切削过程中各种影响较小；在颤振发生时，$a_{z'}$ 信号高频成分受到抑制而接近于消失，但 a_z 与 a_y 的高频成分依然存在。因此，就频域而言，采用尾顶尖垂直方向振动加速度信号来提取模式向量中的主分量——主峰频率是更为合适的。

对 d_x 与 F_z 进行自回归谱阵分析也可得知，当颤振将要发生时，其信号的主频

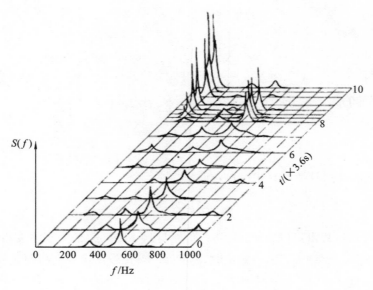

图 5

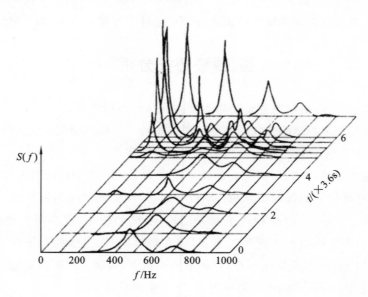

图 6

带向颤振频率移动。

为加强主频带改变时所呈现的差异,定量描述这一现象,提高对颤振预兆判别的准确性,采用均方频率是较好的[5],其定义如下:

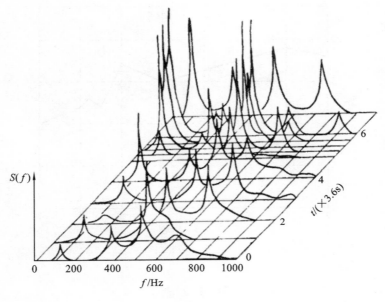

图 7

$$FH = \sum_{i=1}^{H} f_i^2 S(f_i) \Big/ \Big[c \sum_{i=1}^{H} S(f_i) \Big] \tag{14}$$

式中，$f_1, f_2, \cdots, f_H$ 为所感兴趣的频率范围的频率；$S(f_i)$ 为相应的 f_i 的功率谱值；c 为一常数，使 FH 变为无量纲的量，c 一般取颤振频率的二次方。此判别函数的特点是，信号的功率谱 $S(f)$ 作为频率二次方 f_i^2 的权函数，故当主频带在高频段时，FH 显著增大，而当主频带在低频段时，FH 显著减小。

今取 $c = 110^2 = 12\,100$，$f_H = 600$ Hz，根据所求的自回归谱，按表 1 所示条件，用式(14)计算 FH。

在试验 2 条件下，由 a_z 与 a_y 信号所得的 FH 变化情况如图 8 所示。显然，FH 值的变化规律是明显的，当颤振即将发生时，FH 值迅速下降。

然而采用 FH 作为判别函数有其缺陷，因为 FH 只是考察了信号能量在频域中的分布情况，而没有考察信号总能量本身的变化情况。如果信号主频带在低频段，但其能量却不很大，这时 FH 就会导致误判。

综前所述，采用信号方差、自回归模型残差方差、φ_{21}、ρ_1、主频带位置与 FH 作为模式向量的主分量或判别函数，来判别颤振是否即将发生，都有缺陷；显然，我们既应考虑主峰频率大小的变化，又应考虑主峰值（能量）大小的变化。为此，我们提出了一新的综合判别函数，以判别颤振是否即将发生，这在另文中论及[9]。

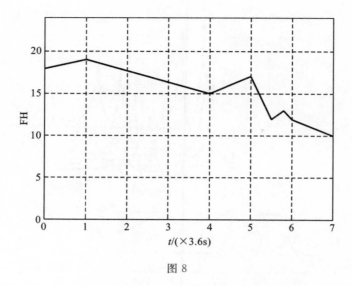

图 8

参 考 文 献

[1] S. Braun. Annales of the CIRP, Vol. 24, No. 1, pp. 315~320 (1975).

[2] T. Sata, T. Iramura. Annales of the CIRP, Vol. 24 No. 1, pp. 309~314 (1975).

[3] T. L. Subramanian, M. F. Devies, S. M. Wu. Journal of Engineering for industry ASMB Series B, Vol. 98, pp. 1209~1214 (1976).

[4] K. F, Eman, S. M. Wu. ASME Trans Series B, Vol. 102, PP. 315~321 (1980).

[5] Y. Miyoshi, H. Nakazawa. Proceedings of the 5th International Conference on Production Engineering Tokyo, pp. 528~533 (1984).

[6] S. M. Kay, S. L. Maple. Proc IEEE, Vol. 69, pp. 1380~1419 (1981).

[7] 杨叔子,王治藩.《机械工程》增刊,第 1~17 页(1984).

[8] 杨叔子等. 华中工学院学报,第十二卷,第 6 期,第 93~100 页(1984).

[9] 梅志坚等. 华中工学院学报,第十三卷,第 5 期,第 87~95 页(1985).

Chatter Omen During Metal Cutting

Yang Shuzi Liu Jingyan

Shi Hanmin Mei zhijian

Abstract Based on the pattern recognition theory and a large amount of experimental data, the choice of characteristic signals, the acquisition of pattern vectors and the extraction of principal components are discussed. The method of constructing the decision function is suggested for a timely on-line recognition of chatter omen to make it possible to monitor and control chatter during metal cutting.

（原载《华中工学院学报（自然科学版）》1985 年第 13 卷第 5 期）

机械设备诊断学的探讨

杨叔子　师汉民　熊有伦　王治藩

提　要　本文提出了机械设备诊断学的目的、任务、内容与诊断过程,给出了按信号测取方式、信息提取方式与状态诊断方式对设备诊断方法进行分类,指出了目前分类中的一些混乱情况,并对设备故障情况的分析作了论述。文中还指出了在设备诊断中专家系统应用的广阔前景。

关键词:设备,故障,诊断学,诊断,分类,机械设备诊断学,工况,征兆,状态

随着科学技术与生产的发展,机械设备的工作性能越来越好,工作强度越来越大,自动化程度与生产效率越来越高,同时,设备的结构也越来越复杂,各部分的关联也越来越密切,往往某处稍有故障,不仅会直接或间接地造成巨大的经济损失,而且会危及人身安全,造成极为严重的后果。因此,随着测试技术与计算机技术的发展,机械设备(包括结构)诊断技术应运而生,日益发展[1-8]。不仅在生产实践中发挥越来越大的作用,而且已有关于机械设备诊断学方面的专著出版。现在的问题是:机械设备诊断学的目的、任务、内容、体系是什么?设备诊断方法应如何分类?设备诊断学的重点又何在?由于设备诊断学还正在形成与发展,这些问题有的并未解决,有的解决得不很完善,人们还未能较好地将机械设备诊断技术及实践加以概括,提到应有的理论的高度,并将理论与实践较为完美地结合起来。本文拟就有关方面作些研究与探讨,以期有助于机械设备诊断学的建立与机械设备诊断技术的发展。

一、机械设备诊断学的基本内容

就机械设备诊断技术的起源与发展考察,机械设备诊断学的目的应是"保证

可靠地有效地发挥机械设备的功能"。这里包含了三点：一是保证设备无故障，运行可靠；二是要"物尽其用"，保证设备发挥其最大效益；三是要保证设备如将有故障或已发生故障，能及时而正确地诊断出来，加以维修，以减少维修时间，提高维修质量，应使重要的设备能按设备状态进行维修（即视情维修或预知维修），改革目前的按时维修的维修体制。应当指出，设备诊断技术应为设备维修服务，应是设备维修技术的重要内容；然而，它绝不仅限于维修方面，而正如其目的的第一、二点所述，它还应保证设备能处于最佳的运行状态，这意味着它还应为设备的设计、制造与运行服务。还应指出，所谓故障应是指设备的功能低于规定的水平，即设备的有关状态处于低功能的状态；显然，故障不等于失效，更不等于损坏，失效与损坏属于严重的故障。

同机械设备诊断学的目的相应，机械设备诊断学最根本的任务就是通过对设备的观测信号来识别机械设备的状态，在一定程度上也可以说，机械设备诊断学就是机械设备状态识别学。概括讲来，如同对人体的诊断，一是预防与保健，二是看病与处置一样，对机械设备的诊断，一是防患于未然，早期诊断，二是诊断故障，分析情况，采取措施。

具体讲来，机械设备诊断学至少包括如下五个方面的内容：

(1) 正确地根据设备性质与工况，选择与测取设备有关状态的特征信号；

(2) 正确地从特征信号中提取设备有关状态的有用信息（征兆）；

(3) 正确地根据设备的征兆，识别设备的有关状态（状态诊断），这里包含识别设备的有关状态将有异常（故障早期诊断）与已有异常（故障诊断）；

(4) 正确地根据设备的征兆与状态，进一步分析设备的有关状态及其发展趋势（状态分析），这里包含当设备有故障时，分析故障位置、类型、性质、原因与趋势等；

(5) 正确地根据设备的状态及其趋势，作出决策，干预设备及其工作过程，所谓干预包括控制、自诊治、调整、维修、继续监视等措施。

在此应指出，征兆既用于由外表现象推断内部状态，这时可称为症候；又用于由现在状态推断未来状态，这时可称为预兆。

一般的工况监视实际上是状态监视，故障是设备的一种状态，因此，工况监视是故障诊断的基础，故障诊断只不过是一种特殊的工况监视与分析。设备诊断的过程也可以说是工况的监视、分析与干预的过程。

由机械设备诊断学的内容可将设备诊断过程概括为图1所示的框图。

显然，从设备所测取的信号应包含同设备有关状态的有关信息，所以称这种

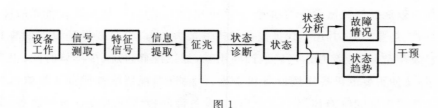

图 1

信号为特征信号。例如,要诊断飞机机翼这种结构有无裂纹,而竟去测取它的温度,其势必无法判断裂纹这一故障的有无。

但是,一般还难于直接从特征信号判明设备有关状态的情况,查出故障的有无。例如,结构的振动信号包含了结构有无裂纹的信息,然而一般难于直接从振动信号做出有无裂纹的结论;这还需要振动理论、信号分析理论、控制理论等等提供的理论与方法,加上裂纹诊断的实践提供的经验,提取有用的信息(征兆),而且这种征兆应对状态的变化最为敏感,方有可能判明设备有关状态的情况。例如,理论分析与试验研究表明,从振动信号算出的结构固有频率这一征兆,对结构有无裂纹并不敏感,不易判明裂纹是否发生,而结构的频率特性(有时称频率响应函数)却对裂纹是否发生存在十分敏感的频带[*]。

一旦获得征兆,就可采用多种的模式识别理论与方法,将征兆加以处理,构成判据,以进行状态识别与分类,即状态诊断。显然,状态诊断这一步是机械设备诊断的重点所在,也是设备诊断技术成败之所系;当然,这绝不意味着诊断技术的成败只取决于状态诊断这一步。

当状态为无故障时,还可采用 Kalman 滤波、时序模型等方法,进一步分析状态的发展趋势,预计未来情况。当状态为有某种故障时,则可进一步采用故障树分析、模式识别、信号分析等方法分析故障所在、类型、性质、原因与趋势等。在这一基础上,就可进行决策分析,做出决策,干预设备及其工作,以保证设备能可靠地、有效地发挥其功能,达到诊断目的。

根据模式识别理论来考察机械设备诊断的过程,可将特征信号称为初始模式,征兆称为最终模式,状态称为状态模式,状态趋势称为模式趋向。整个诊断过程主要是去伪存真、除粗取精、由表及里、由近及远的模式识别过程。信号测取时除噪技术的应用,就是去伪存真;由初始模式获得最终模式,就是除粗取精,有时还得经过中间模式,再经变换与降维处理,获得最终模式;而由最终模式获知状态模式,就是由表及里;实际上,有时事先已建立了同状态模式对应的基准模式,由

---

[*] 王谓季. 结构故障的振动诊断与监测研究. 南京航空学院博士学位论文,1986 年 2 月.

最终模式获知的只是基准模式；至于由最终模式与状态模式了解模式趋势，就是由近及远了。例如，采用时序分析这一统计分析方法从特征信号中提取信息，得到的是参数模型；由模型的参数或有关特性可构成模式向量，有时此向量维数嫌高，只能作为中间模式，经过变换与降维，提取主特征量，才形成最终模式；此时，在进行状态识别与分类时，只能根据统计试验得到与不同状态一一对应的基准模式，将待检的最终模式归为某一模式[4]。因此，更为复杂的设备诊断过程如图2所示。

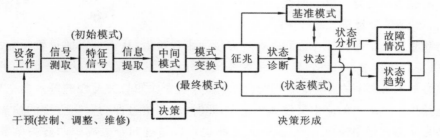

图 2

在此应指出，由上述分析可知，机械设备诊断学是建立在以实践为基础之上的融合多学科的一门新兴学科。

二、机械设备诊断方法的分类

我们认为，机械设备诊断学中关于诊断方法的分类问题是一个极为重要的问题；它不仅是建立一个学科所必需的条件，给人以明确的科学体系，而且更能启发人们改进已有的方法，寻求新的方法。目前的分类方法有的是不完善、不够科学的，有的在概念上是混乱的。

显然，从机械设备诊断的内容与过程来考察，我们认为，将诊断方法按信号测取方式、信息提取方式、状态诊断方式与状态分析方式加以分类为宜。

1. 按信号测取方式分类

按信号测取方式，诊断方法可分为用人的感觉器官测试与用测试装置测试，而人们感兴趣的是后者。采用后者，又可分为采用物理方法测取信号与采用化学方法测取信号，如振动与噪声、力、温度、电、磁、光、超声、声发射、各种射线和气污染、油污染、锈蚀等。其中，振动与噪声信号应用极为广泛，不仅运动的机械设备一定具有振动与噪声信号,而且静止的机械设备也可受激励而产生振动*,[5]。

2. 按信号提取方式分类

按照信息提取方式,诊断方法可分为两大类。一类是直接法,此时特征信号即征兆,例如当测定油耗率与废气烟度来直接诊断柴油机有无故障时,油耗率与废气烟度既是特征信号又是征兆;一类是间接法,此时特征信号并非征兆,而需从特征信号中提取征兆。

间接法又可分为以下两类。

(1) 函数分析法

此时特征信号与征兆之间存在着定量的函数关系,从而可以通过数学分析方法,由特征信号求得征兆。例如,最为典型的分析方法就是采用状态空间分析(包括线性的与非线性的,确定性的与随机性的,渐变性的与突发性的等),而在化工等连续生产过程中的应用又较为普遍[6];此时,特征信号即为输出方程的输出,征兆即为状态方程与输出方程的状态,征兆也可以是这两个方程中的有关参数(包括物理参数、特性参数等),输入可根据实际情况加以确定。这样,就可建立设备的状态空间模型,再采用各种 Kalman 滤波,直接预测下一步征兆[6,7]。对前述的结构有无裂纹的诊断,建立结构的振动方程,进行结构的模态分析,计算结构的固有频率、阻尼、振型与频率特性等,亦属此类。

(2) 统计分析法

此时特征信号与征兆之间存在着统计关系。此法又分以下两种。

① 非参数模型法。它就是目前广泛采用的传统的统计分析法,亦即一般信号分析理论中所阐述的相关分析、周期图分析以及以它们为基础的有关统计分析。此时,还有各种快速算法,FFT 就是典型代表。这一方法的特点就是从特征信号(经过采样)直接计算出统计特性,这些统计特性就是特征信号的非参数模型,有关的统计特性可取为征兆。显然,根据统计特性的性质,又可分为时域法、频域法、倒频域法以及某些特殊方法。其中,频域法又是应用最为广泛的,往往又称为频谱分析技术,采用频域法进行诊断往往是十分有效的。还应指出,倒频域法的发展与应用值得格外重视,因为倒频域法与频域法相比,在一些方面具有不可比拟的优越性。

② 参数模型法。它是根据测试所得的信号(经过采样)建立起差分方程形式

---

* 杨叔子,师汉民,熊有伦,王治藩. 机械设备诊断技术与振动工程. 中国振动工程学会振动工程在国民经济建设中的作用专题讨论会,1986.10.31—11.4,杭州.

的参数模型,再用模型的参数或用由模型计算出的信号统计特性、系统固有特性或其他特性作为征兆。目前已经兴起的参数模型方法之一就是时序模型的诊断方法[4]。采用时序模型,同样可以获得信号在时域、频域与倒频域中的有关统计特性。在此应指出,人们一般忽视了现代控制理论系统辨识的差分方程,这一参数模型的参数或有关特性,也可作为获得征兆之用。其实,如能获知系统的输入或能对系统施加输入,则此种差分方程比时序模型可提供更为完备的信息[8]。与非参数模型法相比,参数模型法的一大优点是在计算信号统计特性时可以没有对采样数据的加窗影响,另一大优点是几乎能将蕴涵在采样数据中的全部信息凝聚在少数的几个模型参数之中。

3. 按状态诊断方式分类

按照状态诊断方式,诊断方法也可分为两大类。一类是直接法,此时征兆即状态,例如,采用状态空间分析进行系统的状态反馈控制时,由输入、输出(特征信号)计算出的状态,既是系统的征兆,也是系统的状态;一类是间接法,此时征兆并非状态,而须由征兆诊断出状态来。

间接法又可分以下五类。

(1) 对比诊断法

这是目前广为应用的方法。此法是事先通过计算分析、试验研究、统计归纳等方式,确定同各有关状态一一对应的征兆(即基准模式)。因此,在获得设备工作时的征兆后,由人将此征兆直接同基准模式对比,立即可确定设备的状态。这一方法最典型的应用,就是求取设备在有关状态下的作为基准模式的振动信号自功率谱图与主峰频率。当将待诊断状态的自功率谱图与主峰频率同基准模式对比时,就可诊断出设备相应的状态。如果只诊断有无故障两种状态,则只要有"无故障"或"有故障"的基准模式及其允许变动范围即可,这就是一般所谓的简易诊断关键所在。例如,通过理论计算与试验研究,可以证明,对旋转机械而言,转子不平衡引起的主峰频率为转子工作频率 f,轴线不对中频率为 $2f$(还有轴向振动的 f),轴的径向裂纹频率为 $2f$(但无轴向振动的 f),滑动轴承的油膜涡动频率为 $(0.42\sim 0.46)f$,非线性因素有时会引起亚谐波频率等。应指出,采用振动信号,应用传统的非参数模型分析法(特别是频域法)加上对比诊断法,就是目前在机械设备诊断中应用最广泛的方法*,[1],[5]。

---

* 曾正明编辑. 实用设备诊断技术. 第一汽车厂设备管理协会(1985).

(2) 函数诊断法

在征兆与状态之间,如果存在定量的函数关系,则在获得征兆后即可算出相应的状态。其实,上述信息提取方式中的函数分析法往往同状态诊断法中的函数诊断法紧密联系在一起。例如,结构的裂纹的位置、尺寸、形状(即设备的状态)同结构的刚度(即作为设备的征兆)有定量的数学关系,而结构的刚度又同结构的振动信号(即作为设备的特征信号)有定量的数学关系,这就是能够通过振动信号来诊断结构有无裂纹的理论依据。自然,当征兆与状态之间的数学关系极为复杂时,可通过试验研究找出它们的对应关系,进而可采用对比诊断法。

(3) 逻辑诊断法

在征兆与状态之间,如果存在逻辑关系,则可通过征兆以推理方式诊断出设备的状态。此法又可分为以下两种。

① 物理逻辑诊断法。它根据征兆与状态之间的物理关系,进行推理、诊断。例如,润滑油污染分析即一典型方法,此时通过光谱、铁谱、磁塞或磁棒方式,分析设备润滑油中所含的金属微粒的情况,而这些金属微粒是从设备有关运动部分互相摩擦、产生磨损而得来的。这些微粒的情况就是征兆。根据设备有关零件的材料与成分,就可由微粒情况推断出设备的磨损零件与零件磨损情况。又如,从柴油机的油耗率与废气烟度(征兆)可推断出柴油机有无故障。

② 数理逻辑诊断法。它根据征兆与状态之间的数理逻辑关系(即决策布尔函数),在获得征兆(征兆布尔函数)后,按照规定的逻辑运算规则,即可求出状态(状态布尔函数)。这种方法只局限于有无故障的诊断,但是,往往很有效[3]。

(4) 统计诊断法

它就是一般的模式识别书刊中所说的统计模式识别法。它将所得到的征兆构成一模式向量(最终模式),并按下述方法之一将此模式向量划归某一基准模式。一种方法是用距离函数形式来检查待检的模式向量同各基准模式向量在模式空间相距的远近,将待检模式划归相距最近的基准模式。另一种方法是用分区方式,通过一定的判据,将待检模式划入相应的基准模式区域;而 Bayes 分类器等就是这种分区划入方式,此时可使错划(错误诊断)所造成的相应损失为最小。例如,在船舶发动机的故障诊断的研究中,也采用 Bayes 分类器作为诊断方法之一*。这种统计诊断往往同参数模型分析法联合使用。

---

* Proceedings. 4th International Symposium on Technical Diagnostics. Kupari-Dubrovnik, Yugoslavia, Oct. ,1986.

(5) 模糊诊断法

它是一种颇有前途的诊断方法,其特点有二:第一,它采用多因素诊断;第二,它计及人利用模糊逻辑而能精确识别事物这一特性。对前者而言,因为一种状态可以引起多种征兆,而一种征兆也可在不同程度上反映多种状态。因此,要正确诊断出设备有无某些状态,就必须尽可能利用多种征兆进行综合诊断。对于后者,可采用模糊数学方法,对于每一状态与每一征兆均应有相应的隶属度,对多个状态与多个征兆则应有隶属度模糊向量,这两个向量可用以大量实践为基础所获得的模糊矩阵来联系。这样,一旦获得征兆的隶属度模糊向量后,就可通过此向量与模糊矩阵,求出状态的隶属度模糊向量,进而根据状态的隶属度模糊向量各元素的大小就可诊断出设备状态的情况。这一方法在我国对大型汽轮机组振动的工况监视与故障诊断系统的研究中,已获得成功的应用*。

在诊断方法的分类中,最后要指出以下两点。第一点,在一些情况下,特征信号就是状态,目前采用的许多无损检测与无损探伤技术,如超声波、光全息、光导纤维、各种射线等,还有目视检查,得到的特征信号就是设备的相应的状态。这十分类似于医学上的外科与放射科。这样获得的信息称为一次信息,前面所讲的通过信息提取、状态诊断所获得的信息称为多次信息。采用一次信息进行诊断当然是更为准确的,但往往受到许多条件的限制而无法或难于应用。第二点,由于机械设备及其工况往往是很复杂的,在进行设备诊断时,往往同时采用多种信号、多种信息提取方式与多种状态诊断方式,甚至采用本文最后所述的专家系统,以期尽可能获得全面而准确的诊断结果。

三、故障及其分析

机械设备诊断学的重点不在于研究故障的本身,而在于研究状态识别的方法,即故障诊断的方法。然而,故障本身的类型与性质是极为重要的,不同的故障往往决定了不同的诊断方法,正如不同的疾病往往决定了不同的诊断方法一样。如果对故障的情况毫无所知,则诊断是难于甚至无法进行的。可见,对故障也必须分类。按故障的类型,大体可分为结构型故障(如裂纹、磨损、腐蚀、不平衡、不对中等)与参数型故障(如流体涡动、共振、配合松紧不当、过热等)。按故障的性质,则可以从多方面分类。例如:按危险的程度分,有危险性的与非危险性的;按

---

* Proceedings of CSMDT'86 Conference. 沈阳,1986 年 6 月.

发生的速度分,有突发性的、随机性的与渐变性的;按影响的程度分,有全局性的与局部性的;按持续的时间分,有持续性的与临时性的;按产生的原因分,有先天性的、劣化性的与滥用性的;按信号的来源分,有设备本身的与监控系统的。显然,人们所注意的是危险性的、突发性的、全局性的、持续性的故障,因为它们造成的损失往往是灾难性的,而且有时也难于预防。对于这些故障,早期诊断尤为重要。

一旦判明设备有故障后,往往还应进一步查清故障情况,即进行故障分析,分析故障的位置、类型、性质、原因乃至趋势,以便采用相应的决策,对设备及其工作进行干预。故障树是一种成功的故障分析法。它根据设备的结构情况,绘出同故障有关的环节之间的具有逻辑关系的故障树,用于搜索与查清发生故障的可能原因与概率,查清设备的薄弱环节。这样,不但便于故障一旦发生,立即查明故障的原因与情况,而且还可用于改进设备的设计,以防患于未然。当然,有的故障原因可称为第二级、第三级、第四级故障原因。例如,滚动轴承烧坏是故障,其原因可能是缺乏润滑油,而缺油的原因又可能是油管堵塞,而油管堵塞的原因又可能是滤油器失效,则它们可分别称为第一、二、三、四级故障。由于故障分析可能就是进行第二级故障诊断、第三级故障诊断等,因此,前述的许多状态诊断方法此时也可应用。

在此,还应指出两点。第一,一般还将设备诊断分为简易诊断与精细诊断。所谓的简易诊断是指只能判明设备状态是正常(无故障)还是异常(有故障)的状态诊断,特别是指其中的对比诊断;所谓的精细诊断是指不仅能判明状态有无异常,而且在有异常时还能判明是什么样的异常?性质如何?原因何在?趋势怎样?这就可能不仅要求进行能判明设备多种状态诊断,而且甚至还包括了状态分析(主要是故障分析)。第二,在判明某一状态时,可采用与之相应的基准模式。但是,例如,与正常状态相应的可能有多种基准模式,与某种异常状态相应的也可能有多种基准模式,而人们又往往不能穷尽这些基准模式。因此,一旦以相应于正常状态的若干基准模式来进行状态诊断,认为出现了非这些基准模式的模式时,就表明设备状态异常,这就可能出现"虚判";反之,一旦以相应于某种异常状态的若干基准模式来进行状态诊断时,与上类似,也可能出现"漏判"。显然,在进行状态分析时也可能会出现类似的情况。因此,采用多因素的状态诊断法与多因素的状态分析法,都是十分有价值的。

在此,应特别提出的是专家系统[*],[9-12]。专家系统是人工智能应用中新发展

---

* Program and Abstracts MFPG,40th Meeting,April,Gaithersburg,Maryland,U.S.A.,1985.

的最活跃的分支。它用于设备诊断时,不仅包括从信号测取到状态分析,而且还包括了决策形成与干预。专家系统将专家的宝贵的经验、智慧与思想方法同计算机的巨大存储、运算与分析能力相结合,从而对设备诊断而言,具有十分有效的诊断与干预能力,这对于重要的设备与生产过程是极为值得重视的。应该说,设备诊断领域也是专家系统应用的新领域,因此,从理论上与实践上,专家系统在机械设备诊断中的应用,前景是极其光明而广阔的,意义是极为重大而深远的。

参 考 文 献

[1] R. A. Collacoit 著,孙维东等译.机械故障的诊断与情况监视.机械工业出版社(1983).

[2] 塩见弘.故障解析と诊断.日科技连出版社(1977).

[3] 屈梁生.机械故障诊断学.上海科技出版社(1986).

[4] 杨叔子.时序模型的诊断方法.机械工程,第2、3期(总62、63期)合刊,第5～14页(1986).

[5] R. A. Collacotl. *Vibration Monitoring and Diagnosist*. Chapman & Hall, London (1979).

[6] 叶银思等.动态系统的故障检测与诊断方法.信息与控制,第6期,第27-34页(1985).

[7] 秋月影雄.设备诊断技术と安全.设计と制御,第24卷,第4期,第301-306页(19S5).

[8] 杨叔子等.时序模型与系统辨识.华中工学院学报,第12卷,第6期,第85-92页(1984).

[9] 熊范纶等.计算机专家咨询系统及其建立.信息与控制,第1期,第33-38页(1986).

[10] P. P. Bonissone, H. E. Johnson. *DELTAS An Expert System for Diesel Electric Locomotive Repair*. AD-P003 941, pp. 397～413(1983).

[11] M. S. Fox. *Techniques for Sensor-Based Diagnosis*, 5th IJCAI, pp. 158-163(1983).

[12] 出海滋,木口高志.异常诊断のための情报处理.计测と制御,第25卷,第10期,第871-878页(1986).

On Mechanical Equipment Diagnostics

Yang Shuzi　　Shi Hanmin
Xiong Youlun　　Wang Zhifan

Abstract　The necessity of developing mechanical equipment diagnostics is discussed; The purpose, task, content, and dignostic process are suggested with the flowchart given; It is proposed that the method of mechanical equipment diagnosis be classified according to how the signal is measured, how the symptom is extracted and the state diagnosed. Some confusions in the existing classification of diagnosis methods are pointed out.

Keywords: Equipment, Fault, Diagnostics, Diagnosis, Classification, Mechanical equipment diagnostics, Working conditions, Symptom, State.

（原载《华中工学院学报（自然科学版）》1987 年第 15 卷第 2 期）

灰色预测和时序预测的探讨

吴 雅 杨叔子 陶建华

提 要 本文讨论了灰色模型,特别是 GM(1,1)模型的特点和适用范围,并将 GM(1,1)模型和时序 AR(n)模型结合起来(称为组合模型),对我国轻工业产量发展指数等三个项目分别进行了组合模型预测。结果表明,在一般 GM 模型中引入 AR 模型可显著提高预测的准确度;在非平稳时序建模中引入 GM 模型,可作为提取趋势项的另一种方法。文中还从预测的角度将灰色模型和时序模型进行了比较和分析,对"灰"的物理概念进行了初步探讨。

关键词:灰色模型,时序模型,灰色预测,时序预测,组合模型预测,惯性系统

灰色预测和时序模型预测都已在社会、经济和工程领域中得到了成功的应用[1,2]*。灰色系统理论是新近提出和发展起来的,而时序方法本是一种数据处理方法,那么,从对系统的预测(或预报)这一角度来看,此两者之间是否具有某些联系?如何将此两者结合起来?都是很有意义的问题。本文将首先讨论灰色模型和时序模型结合的问题,然后给出预测实例,在此基础上进而讨论上述问题。

一、灰色模型和时序模型

1. 灰色模型

由于各种灰色模型 GM 的前提条件和建模原理都是一致的,此处仅以最基本

---

* 杨叔子,吴雅,丁洪,梅志坚,等. 动态数据的时间序列分析. 中国振动工程学会成立大会暨第一次代表大会专题报告,1987 年 4 月.

的灰色模型 GM(1,1)为例对灰色模型进行分析(后述灰色模型即泛指所有的 GM 模型,以区别于 GM(1,1)模型)。

按灰色理论[1,2],设原始离散数据序列 $x^{(0)} = \{x_1^{(0)}, x_2^{(0)}, \cdots, x_N^{(0)}\}$,对 $x^{(0)}$ 进行一次累加生成处理(记为 AGO),即可得到一个生成序列 $x^{(1)} = \{x_1^{(1)}, x_2^{(1)}, \cdots, x_N^{(1)}\}$,对此生成序列可建立一个一阶微分方程,记为 GM(1,1),其形式为:

$$\frac{dx^{(1)}}{dt} + ax^{(1)} = u \tag{1}$$

GM(1,1)的解为:

$$\hat{x}_{t+1}^{(1)} = \left(x_1^{(0)} - \frac{u}{a}\right)e^{-at} + \frac{u}{a} \tag{2}$$

式(2)即为灰色预测公式。对 $\hat{x}_{t+1}^{(1)}$ 求导(或作累减还原处理 IAGO)可得原始数据的预测公式为:

$$\hat{x}_{t+1}^{(0)} = (-ax_1^{(0)} + u)e^{-at} \tag{3}$$

分析以上叙述,灰色模型具有以下特点。

(1) 灰色建模的前提是原始数据序列 $x^{(0)}$ 为"光滑的离散函数"。从信号分析的角度来看,一般可将信号(即观测数据)分为确定性信号和随机性信号,确定性信号可用明确的数学关系式来描述,而随机信号则不能。显然,所谓"光滑的离散函数",实际上是要求原始数据序列是确定性的(即具有确定性的趋势),而且还可用一般的初等函数来表达。由此可见,灰色模型适用于分析确定性信号,而不适用于分析随机性信号。

(2) 灰色建模的特点是:对原始序列 $x^{(0)}$ 进行 AGO,得到生成序列$\{x^{(1)}\}$,而用灰色模型描述 $x^{(1)}$,从而间接地描述 $x^{(0)}$。应该指出,AGO 实质上是一种数据预处理的方法。这是因为观测数据 $x^{(0)}$ 总是不可避免地含有随机干扰的成分,那么,对 $x^{(0)}$ 的"光滑"要求显然是一种理想状况,而对 $x^{(0)}$ 进行 AGO 可带来两点明显的好处:第一,可使 $x^{(0)}$ 中的随机干扰成分在通过 AGO 后得到减弱或消除,即使 $x^{(0)}$ 不满足"光滑"条件,也能保证 $x^{(1)}$ 满足光滑条件,而且由式(1)显见,GM(1,1)是直接描述 $x^{(1)}$ 的,也无须对 $x^{(0)}$ 提出如此苛刻的要求(实际上灰色建模中的光滑度检验就是针对 $x^{(1)}$ 进行的)。第二,可使 $x^{(0)}$ 中所蕴涵的确定性信息在通过 AGO 后得到加强,即使 $x^{(1)}$ 成为单调增长且增长速度很快的数据序列,从而保证 $x^{(1)}$ 能用"函数"式(而且是指数函数式,见后述)来表达。因此实用中的灰色模型并不一定严格要求 $x^{(0)}$ 是"光滑的离散函数",而允许 $x^{(0)}$ 带有一定的随机性,这样,在一定程度上拓宽了灰色模型的实用范围,使得灰色模型得以在实际中成功地应用。我们认为,灰色建模中的 AGO 方法是值得数据处理方法借鉴的。

(3) 式(2)表明,GM(1,1)模型描述了一个随时间 t 按指数规律单调增长或衰减的过程。当然,由于上述分析特点(2),此处所指出的"单调"是指从 $x^{(0)}$ 的发展趋势上看的单调性,而并不排斥由于随机干扰所引起的局部波动。

综上所述,我们认为,一方面由于 GM(1,1)是一个微分方程模型,它要求原始序列 $x^{(0)}$ 具有光滑性;但另一方面由于 AGO 过程,灰色模型本身是生成序列 $x^{(1)}$ 的模型,因而它又允许 $x^{(0)}$ 具有相当的随机性,所以,GM(1,1)模型描述了数据序列中所蕴涵的、确定性的指数函数规律,这就是 GM(1,1)模型的适用范围。

2. 时序 ARMA 模型

时间序列方法是利用观测所得的有序的随机数据 $\{x_t\}$(时间序列)建立差分方程形式的参数模型,其中最常用的是对平稳、正态、零均值的时序 $\{x_t\}$ 建立的 ARMA 模型[3]:

$$x_t = \sum_{i=1}^{n} \phi_i x_{t-i} + a_t - \sum_{j=1}^{m} \theta_j a_{t-j} \quad (a_t \sim \text{NID}(0, \sigma_a^0)) \tag{4}$$

当式(4)中各 $\theta_j = 0$ 时,为其特例——AR 模型。时序方法中常用 AR 模型进行预测,其预测公式为:

$$\hat{x}_t(l) = \begin{cases} \sum_{i=1}^{l-1} \phi_i \hat{x}_t(l-i) + \sum_{i=l}^{n} \phi_i x_{t+l-i} & (l=1,2,\cdots; l \leqslant n) \\ \sum_{i=1}^{n} \phi_i \hat{x}_t(l-i) & (l > n) \end{cases} \tag{5}$$

将灰色模型与 ARMA 模型相比较,可作如下分析。

(1) 按照时间序列的定义,一组有序的观测数据即称为一个时间序列,显然,灰色建模所需的"原始离散数据序列" $x^{(0)}$ 就是一个时间序列,那么仅从这一角度来看,灰色模型也是一种时序模型,因它也是基于观测时序建立的一种数学模型。

(2) 按照灰色的概念,部分信息确定、部分信息不确定的系统称为灰色系统,描述灰色系统的数学模型为灰色模型,而时序模型本身仅以系统的输出 $\{x_t\}$ 为基础,并不需了解系统的原理结构和输入情况,那么仅从这个角度来看,时序模型又是一种灰色模型,因它是某一仅知道输出的系统的数学模型。

(3) ARMA 模型要求 $\{x_t\}$ 是平稳、正态的时序,此即要求原始数据是平稳的正态随机过程的一个样本。因而,与灰色模型相反,ARMA 模型适用于分析平稳的随机性信号。

(4) 因为 $\{x_t\}$ 是随机性信号,故无法用任何函数形式来描述 $\{x_t\}$ 随时间 t 变动的规律。而时序方法采用随机过程自身的历史值(确定性的) $x_{t-i}(i=1,2,\cdots,n)$ 的线性组合来描述随机过程自身的方法（由式（4）可见），能从杂乱无章的随机过程中提取大量的有规律的信息，直至模型残差 a_t 为白噪声。因而，ARMA 模型以系统的历史行为 x_{t-i} 来描述系统的现在行为 x_t，即用 x_{t-i} 而不直接通过时间 t 来描述系统的"动"。灰色模型则与之相反，它直接反映系统行为随时间 t 变动的规律。显然，灰色模型与 ARMA 乃至所有时序模型的这种根本差别来源于处理确定性信号和随机性信号的两种根本不同的思想方法。

综上所述，我们认为，虽然灰色模型和时序模型都可相互包容，但它们却存在着本质上的差异，GM(1,1)适于描述具有确定性指数函数规律的过程，ARMA 模型则适用于分析平稳的随机性信号，因此 GM(1,1)直接反映系统随时间 t 变动的指数规律，ARMA 模型则反映系统本身的历史行为与现在行为间的线性规律。

3. 组合模型

在许多实际问题中，观测得到的时间序列 $\{x_t\}$ 既不是"光滑"的，也不是"平稳"的，而可能是这两种情况的组合，这在时序方法中称为具有趋向性的时间序列。从我国轻工业产量发展指数的情况看，其具有稳定增长的趋势，但同时存在着随机波动。在这种情况下，$\{x_t\}$ 中含有两种成分：一是随时间 t 增长的确定性部分（记为 d_t），二是随机性部分（记为 y_t），其组合模型为：

$$x_t = d_t + y_t \tag{6}$$

按时序方法[8]，式(6)中的 d_t 称为趋势项，它是 $\{x_t\}$ 中的非平稳部分，可用多元回归方法来提取；y_t 是 x_t 中经提取了 d_t 后所剩下的平稳时序，可用 AR 模型来描述。根据上述对灰色模型的分析，我们可用 GM(1,1)模型来提取趋势项，再与 AR(n)模型组合以形成组合模型：

$$x_t = (-ax_1 + u)\mathrm{e}^{-a(t-1)} + \sum_{i=1}^{n} \phi_i y_{t-i} + a_i \tag{7}$$

式中，第一项称为 GM(1,1)部分；第二、三项称为 AR(n)部分。组合模型的预测公式为：

$$\hat{x}_t(l) = (-ax_1 + u)\mathrm{e}^{-a(t+l-1)} + \hat{y}_t(l) \quad (l = 1, 2, \cdots) \tag{8}$$

式中，$\hat{y}_t(l)$ 按式(5)计算。

对组合模型式(7)，可作两方面的理解，从时序方法的角度来看，式(7)是一个具有指数趋势项 d_t 的时序模型，只不过 d_t 具有 GM(1,1)的形式；从灰色理论的角度来看，式(7)是一个加入了摆动分量的 GM(1,1)模型，AR(n)部分即是摆动分量。因而，仅对观测数据的描述而论，灰色建模也是一种对动态数据的处理方法，它可与任何数据处理方法组合使用，以达到所期的目的。

二、组合模型预测

1. 组合模型预测

利用组合模型,对我国 1981—1990 年的轻工业产量发展指数、重工业产量发展指数和全国总人口数进行了预测,预测曲线分别如图 1～图 3 所示。图中曲线 ① 为实际值,② 为长数据 GM 预测,③ 为短数据 GM 预测,④ 为组合模型预测。组合模型的前四步预测精度见表 1。

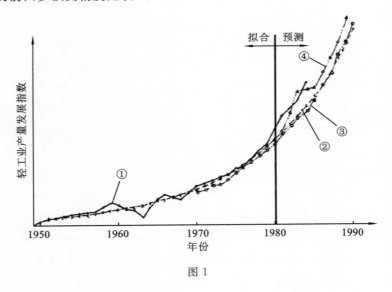

图 1

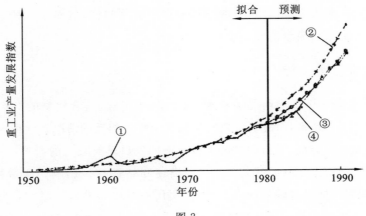

图 2

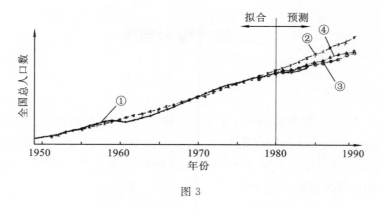

图 3

表 1 组合模型预测结果

项 目		轻 工 业	重 工 业	总人口数
预测准确度/(%)	1981 年	84.21	82.90	99.08
	1982 年	97.58	93.51	99.14
	1983 年	94.21	99.81	99.93
	1984 年	93.39	98.01	99.35
	平均	92.35	93.56	99.38
平均预测误差/(%)	长数据 GM(1,1)	14.80	33.16	3.81
	短数据 GM(1,1)	8.61	16.25	0.23
	组合模型	7.65	6.44	0.62

需要指出,由于短数据 GM(1,1)模型的预测准确度高于长数据模型的,我们对组合模型式(7)中的 GM(1,1)部分使用短数据模型,但由于短数据 GM(1,1)模型的残差序列 y_t 太短(只有几个数据),不适于时序建模,故对于 AR(n)部分,使用长数据 GM(1,1)模型的残差序列建模。

2. 预报结果分析

(1) 比较表 1 中平均预报误差项,对于轻、重工业,组合模型的平均预报误差均低于 GM(1,1)预报,而总人口数的预报则出现了相反的情况。这种现象反映了两方面的问题:一方面,随机因素的干扰对前两项目具有较大的影响(由图 1、图 2 亦可见),应该使用 AR(n)模型来反映、描述和估计这一影响。在这种情况下,AR(n)模型对预测具有较强的修正能力,实用中可以将 GM(1,1)与 AR(n)组合使用,以进一步提高灰色预测的准确度;另一方面,对于总人口数,当 GM(1,1)预测已达到了足够高的精度(如此时,平均预测误差仅为 0.23%,平均预测准确度高达 99.77%)时,使用 AR(n) 修正可能反而适得其反。

(2) 由式(3)和式(7)可见,GM(1,1)预测公式是显式,未来时刻 $l(l=1,2,\cdots)$ 的预值 $\hat{x}^{(0)}_{t+l}$ 只取决于时刻 $t+l$,而与时刻 $t+l$ 以前的预测值无关,因而灰色预测不受累计预测误差的影响,适宜于作长期预测。因而,对于组合模型预报,其长期预测依赖于 GM(1,1),AR(n) 模型只适于在 GM(1,1) 预测的基础上作短期修正。

(3) 值得提出的是,由于 AGO 过程的使用,用 GM(1,1) 建模提取指数趋势项来取代常用的多元回归方法,具有计算简单、速度高的优点。同时,还可考虑采用 GM(2,1) 模型来提取非平稳时序的周期性趋势项。这种采用 GM 建模提取趋势项的办法可作为建立非平稳时序模型的另一条途径。

三、关于"灰"的讨论

下面,根据 GM(1,1) 模型和前面的预测情况对"灰"的物理概念作初步的探讨。

(1) GM(1,1) 模型描述了一个惯性系统。将 GM(1,1) 模型式(1)与一阶微分方程

$$\frac{\mathrm{d}x}{\mathrm{d}t} + ax = u \tag{9}$$

比较,可见两者完全一致。对式(9)进行 Laplace 变换并移项,有 $x(s)=\frac{1}{s+a}u(s)$,显然,这是一个以 u 为输入、x 为输出的惯性环节。故从系统的角度来看,GM(1,1) 模型描述了一个惯性系统,只不过它又将 u 作为待估计的模型参数而已。而大多数的社会、经济系统都具有很大的惯性(惯性系统)[3](如本文所研究的三个项目均属此列),这也正是 GM(1,1) 模型成功地用于社会、经济问题的根本原因。

(2) 对灰色模型而言,所谓"灰",是指系统因素、模型结构和模型阶数已知,而模型参数未知。一般的灰色模型 GM(n,h),n 代表模型阶数,h 代表变量个数。当我们试图对给定数据序列建立某一特定的 GM(n,h) 时,实质上就已经设定了系统因素(h 个),模型结构(某一特定的函数式)和模型阶数(n 阶),而建模的真正目的只是估计出模型参数。如 GM(1,1) 是一个一阶微分方程,那么,在对给定数据序列 $x^{(0)}$ 建立 GM(1,1) 模型之前,实质上我们就已经清楚了(更确切地说是已经设定了)下面的问题——

① 系统因素:系统行为 x 与时间 t 有关;

② 模型结构：指数函数规律；

③ 模型阶数：一阶，惯性环节。

而建模过程只是估计出了模型参数 a 和 u。如 GM(2,1)模型是一个二阶微分方程[1,2]，那么在建 GM(2,1)模型之前，我们也清楚了(设定了)——

① 系统因素：系统行为 x 与时间 t 有关；

② 模型结构：谐波函数规律；

③ 模型阶数：二阶，振荡环节。

而所谓建模只是估计出了模型参数 a_1、a_2 和 u。

从这个角度来看，所有用各种理论建立数学模型的过程，都已蕴涵了"灰"的概念：建模之前，首先考察数据序列的特性，然后确定采用哪种数学模型(如：按时序方法，采用时序模型；按系统辨识方法，采用差分模型；按控制理论，采用状态模型；等等)，一旦确定了模型方案，实质上就确定了(清楚了)模型结构(甚至模型阶数)，但尚不清楚模型参数。

(3) 比较"灰"的程度，时序模型和系统辨识中的差分模型所描述的系统均比 GM 模型描述的系统更"灰"。如同灰色模型一样，当对给定的时间序列 $\{x_t\}$ 建立 ARMA(n,m) 模型式(4)或按系统辨识理论利用输入、输出时序 $\{u_i\}$、$\{y_i\}$ 建立差分模型[4]

$$y_t + \sum_{i=1}^{n} a_i y_{t-i} = \sum_{i=0}^{n} b_i u_{t-i} + \varepsilon_t + \sum_{i=1}^{n} c_i \varepsilon_{t-i} \tag{10}$$

时，实质上就已经设定了(清楚了)模型结构(均为线性关系)。但是 ARMA 模型式(4)和差分模型式(10)均是采用"由低阶至高阶、逐步升阶，逐次检验"的建模方案，其阶数 n,m 与模型参数($\phi_i, \theta_j, a_i, b_i, c_i$)都是经过建模过程以后才能得到的，即是说，在建模之前，其阶数是不清楚的，而 GM 模型的阶数是清楚的，从这一角度看，模型式(4)、式(10)所描述的系统比 GM 模型描述的系统更"灰"。

本文得到了陈绵云副教授、南京工学院钟秉林老师的帮助，特此致谢。

参 考 文 献

[1] 邓聚龙. 灰色系统·社会·经济. 国防工业出版社(1985).

[2] 邓聚龙. 灰色控制系统. 华中工学院出版社(1985).

[3] S. M. Pandit, S. M. Wu. *Time Series and System Analysis with Applications*. John Wiley and Sone(1983).

[4] G. C. Goodwin, R. L. Payne. *Dynamic System Identification, Experiment Design and Data Analysis*. Academic Press, New York(1977).

On Forecasts by Grey Model and Time Series Model

Wu Ya   Yang Shuzi   Tao Jianhua

Abstract   The specific features and the scope of application of Grey Model (GM), especially GM (1,1), are discussed. GM (1,1) is combined with time series AR model for forecasting purposes. The combined model has been used in the prediction of three projects including the production growth rate of China's light industry. The results show that the accuracy by the combined model is better than that by GM(1,1). It is suggested that GM (1,1) model be used for cases with an exponential trend in a non-stationary time series instead of the regressive method so far used, and that GM (2,1) be used in the case of a harmonic trend.

A comparison between Grey Model and time series model for forecast shows that:

1. GM(1,1) is good for exponential case and ARMA model for cases with stationary data; one can be explained by the other.

2. GM (1,1) is an expression for an inertial system and that is why GM(1,1) has been found successful in social and economic prediction.

3. "Grey" here means that system factors, model structure and order are known while model parameters have to be estimated.

4. The systems represented by either ARMA or difference model are even "greyer" than those represented by GM because the order of the former is yet to be found in modeling and the order of the latter is already known.

Keywords: Grey model, Time series model, Grey forecast, Time series forecast, Combined model forecast, Inertia system.

Quantitative Wire Rope Inspection

Wang Yangsheng   Shi Hanmin   Yang Shuzi

Abstract   This paper describes a complete system for inspecting wire ropes automatically. Obstacles to the development of such a system are analyzed in detail, followed by a description of the hardware and software of the system. The hardware uses 12 Hall effect sensors arranged in a circle around the wire rope. Signal analysis is based on a pattern recognition approach. Selection of the feature vector and the use of a trainable non-linear classifier are described.

Keywords: Hall effect, Pattern recognition, Spatial frequency spectrum, Wire rope, Automatic inspection

Wire ropes are extensively used in many industries and are often safety-critical components. In many cases failure of a wire rope could lead to expensive damage to equipment or even to loss of life. To prevent such failures, current methods of wire rope inspection are usually carried out manually and in some industrial practices wire ropes are often replaced at regular intervals without testing. Yet research has shown that over 2% of ropes still in service had lost over 30% of their normal strength, whereas over 70% of ropes removed from service had little or no strength loss[1].

The problem of inspecting wire ropes automatically has attracted a considerable body of research[2]. However, many of the instruments developed as a result of this work, while a significant advance over previous methods, are unable to determine the exact number of broken wires in a short segment of rope. Most wire ropes in service suffer from broken strands, but there is little strength loss unless these are concentrated in a short segment and so the ropes can still be used safely. As a wire rope nears the end of its working life, broken wires tend to congregate within short segments, resulting in a significant strength loss. Hence a factor governing the serviceability of wire ropes is the number of broken wires in a given segment. To be effective, an automatic wire rope inspection system must be able to determine the exact number of broken wires per segment.

Major Obstacles to Automatic Wire Rope Inspection

Automatic wire rope inspection is highly desirable for economic reasons but is difficult to achieve reliably. The major difficulties are as follows:

The Structure of Wire Rope is Complex

A typical cross-section of wire rope is shown in Fig. 1. Wire rope has a complex internal structure consisting of a large number of steel wires.

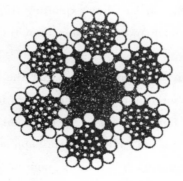

Fig. 1 Typical cross-section of wire rope

The Inspection Environment is Difficult

The length of wire ropes in service is often several hundred meters. In the working environment ropes often become smeared with lubricating grease or coated with mud. These make the sensor design and the analysis of the resulting signal more difficult.

Problems Resulting from the Sensor's Basic Principle of Operation

Many methods of signal detection have been tried in wire rope inspection, including the use of acoustic emission[3], eddy currents[4], radioactivity and electromagnetism[5]. Among these methods, the electromagnetic approach using the Hall effect seems to offer the best prospects and is the method adopted in this work.

The Hall effect[6] is based on the relationship
$$V = KI \times B \tag{1}$$
where V is the Hall potential, K is a constant, I is the control electric current and B is the magnetic fiux density. In practice the current I is held constant and so V is proportional to B.

When the rope is magnetically saturated and contains no broken wires, the leakage flux will mainly be distributed along the rope surface. However, if there are broken wires, a strong magnetic flux, called stray flux, will be developed in the vicinity of the gap in each broken wire. The combination of the leakage flux and stray flux increases the magnetic induction density. Generally, as the number of broken wires increases, the magnetic induction density and the range of influence of the resulting stray magnetic field increase correspondingly. The system described in this paper is based on exploiting this relationship. However, the practical situation is much more complex than the simple model described above.

The main problem with this approach is that the relationship between the detected signal and the underlying physical quantity of interest is a complex one. The magnetic induction density at any given point is a vector sum with contributions from many sources. Hence in this application, if there are many broken wires in a small segment, B will be the sum of the stray fluxes from each gap. The same number of broken wires in a different pattern will probably generate a different B.

When a wire is broken, the length and shape of the gap are uncertain, so the magnitude and direction of B are uncertain also.

Usually a rope is composed of six twisted strands, which can be thought of as a rope wave on the surface of the rope. The resulting B from a gap in a strand will be dependent on the position of the gap within the "rope wave".

There will be vibration and deviation between the rope and a sensor during wire rope inspection, because a wire rope is long and flexible, and this also influences B.

In addition, B will also be influenced by many types of leakage magnetic field and disturbing magnetic field. In summary, B is influenced by many sources and its value is uncertain. The change in B determines the change in V. So, even if the number of broken wires is the same, V changes randomly.

The Speed of a Sensor Relative to a Rope During Inspection is Variable

Generally, when the number of broken wires concentrated in a small segment is increased, the range of influence of the stray magnetic field will be increased also. This provides some important information. Unfortunately, because of variable inspection speed, which compresses or extends the detected signal in the time coordinate axis, it is difficult to use these range data to provide useful information.

Weak Output Signal

The output Hall potential is in the range $0 \sim 200$ mV and highly susceptible to disturbances from many sources, especially in the harsh conditions of inspecting wire ropes in the field.

Inspection System

Twelve Hall effect sensors are placedat regular intervals on a circle concentric with the center of the wire rope. These are protected from the influence of external magnetic flux by a shield. The complete inspection system comprises a sensor assembly, signal preprocessing equipment, a computer and various other items as shown in Fig. 2. The signal preprocessing module amplifies and improves the signal/noise ratio of the raw sensor output. The computer is mainly used to process the digitized sensor data but is also used to control the system, display sensor data and output analysis results. The use of the pulse generator will be discussed in the next section.

A typical signal waveform output by the system is shown in Fig. 3 where the abscissa represents distance along the wire rope. Adjacent signals in the figure come from adjacent Hall effect devices on the sensor assembly. The parts of Fig. 3 circled and labeled a-f show abnormal signal variations caused by broken strands in the wire rope. The cyclic signal variation elsewhere in the figure results from the normal wire rope wave. Fig. 3 shows clearly the advantages gained by using multiple Hall sensors arranged in a concentric circle and being sampled concurrently. Firstly, the resolution of the sensor is increased. For

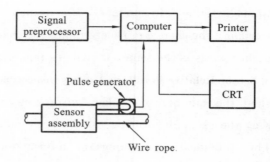

Fig. 2 Schematic of complete inspection system

example, the parts of the signal labeled a, d and f come from three separate areas of broken wires; if only a single sensor was used, it would not be possible to make this discrimination. Secondly, the cyclic effect of the rope wave can be eliminated using two adjacent signals that are 180 out of phase.

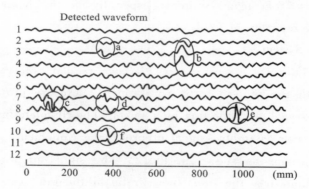

Fig. 3 Typical signal waveform output by the system

Spatial Signal and Smoothing

Some sources of interference can be eliminated by the signal preprocessing module shown in Fig. 2. However, additional methods must be used to reduce other perturbations and to compensate for errors introduced.

Eliminating the Effect of Variable Inspection Speed

Sensor output signals are sampled at regular intervals of length D along the rope by connecting a pulse generator to the drive wheel of the sensor assembly. When the drive wheel rotates through a certain angle, the pulse generator produces a pulse which in turn triggers the signal-sampling operation. Hence the sampling is related to distance along the wire rather than to travel speed.

Signal Smoothing

Theoretically the sampling span D is constant, but in fact D will vary occasionally as a result of the drive wheel skidding rather than rolling along the rope. This results in a signal delay. Other noise sources exist, generating both Gaussian-type noise and noise impulses. These are reduced by a combination of linear and non-linear filtering[7]. The non-linear component is a median filter described by

$$y(m) = \mathrm{median}[x(m), x(m+1), x(m+2)] \tag{2}$$

The linear component is a filter described by

$$s(m) = \sum_{n=-x}^{x} y(n) h(m-n) \tag{3}$$

$$h(n) = \begin{cases} 1/4, & (n = 0) \\ 1/2, & (n = 1) \\ 1/4, & (n = 2) \\ 0, & (\text{otherwise}) \end{cases} \tag{4}$$

The quality of signal after smoothing is distinctly improved.

A pattern Recognition Approach to Signal Analysis

As a result of the sensor design and signal processing described above, the computer is provided with a set of 12 reasonably well behaved signals. The technique used to analyse these signals is based on pattern recognition. In the next section, selection of the feature vector is described. Then, since the class boundaries in feature space are not well defined, a trainable non-linear classifier is used to generate the system output.

Selecting a Feature Vector

In the classical pattern recognition approach, a key step is selection of the feature vector. In this work we use a feature vector with three components.

The times of differential value exceeding threshold (DO)

It is immediately obvious that the absolute difference between two consecutive signal samples from the same sensor is greater in abnormal signal areas than in normal signal areas, as in Fig. 3, a-f.

If a threshold which is large relative to the normal absolute difference is used, the abnormal areas can be identified.

The feature of DO located at l meters from some reference inspection position can be described as

$$DO_l = \sum_{i=1}^{12} \sum_{m=-\infty}^{+\infty} T[x_i(m)]W(1-m) \tag{5}$$

$T[x(m)]$ is a non-linear transform of the signal $x(m)$:

$$T[x(m)] = C[x(m+1) - x(m)] \tag{6}$$

$C(u)$ is defined as a threshold function:

$$C(u) = \begin{cases} 1 & |u| > t \\ 0 & \text{otherwise} \end{cases} \tag{7}$$

where t is the preset threshold. In Eq. (5) above $W(n)$ is a window function:

$$W(n) = \begin{cases} 1 & -N_j < n < 0 \\ 0 & \text{otherwise} \end{cases} \tag{8}$$

where N_j is a window width which can be adapted to change with the change of range of abnormal signal. The window function can be used to separate abnormal signal areas from the signal.

The use of i ranging from 1 to 12, sums the 12 channel signals; this can compensate for some errors resulting from vibration or deviation between the censor and rope to some degree.

Short-distance energy (ENG)

As positions on the rope corresponding to broken wires, it can be seen from Fig. 3 that signal amplitudes increase. This feature can be selected with the short-distance energy:

$$ENG = \sum_{i=1}^{12} \sum_{m=0}^{N-1} x_i^2(m) \tag{9}$$

where N is the number of samples. Again i is the Hall effect sensor number.

The sum of the spatial spectrum (SP)

The spatial spectrum is a discrete Fourier transform of the signal

$$X(k) = \sum_{m=0}^{N-1} x(m) \exp\left(-j \frac{2\pi}{N} km\right) \tag{10}$$

Both the amplitude of individual frequency components and the number of significant components increase with an increasing number of broken wires. This feature can be selected with the sum of spatial frequency components:

$$SP = \sum_{i=1}^{12}\sum_{k=0}^{N-1} X(k) \qquad (11)$$

where N and i are the same as in Eq. (9). Grouping DO, ENG and SP, we get the feature vector Y:

$$Y^T = (DO, ENG, SP) = (y_1, y_2, y_3) \qquad (12)$$

Training the Non-linear Classifier

Since the set of sensor signals is influenced by many sources of noise and interference, it is difficult to partition the feature space. For this reason, a trainable statistical classifier rather than a deterministic classifier is used:

$$d_i(y) = p(w_i/y), \quad i = 0,1,2,\cdots,m \qquad (13)$$

In this case y is the feature vector and i is the number of broken wires; if $d_i(y) > d_j(y)$, $\forall j \neq i$, then $y \ni w_i$. The $d_i(y)$ is a criterion function. The $p(w_i/y)$ is a probability density when $y \ni w_i$.

To evaluate $p(w_i/y)$ immediately is difficult. A polynomial can be used to approximate $p(w_i/y)$; this has the advantage of being easy to compare. We assume

$$p(w_i/y) = \sum_{j=1}^{k} w_{ij}\phi_j(y) = W_i^T \Phi \qquad (14)$$

$$W_i^T = (w_{i1}, w_{i2}, \cdots, w_{ik}) \quad \Phi^T = (\phi_1(y), \phi_2(y), \cdots, \phi_k(y))$$

W_i is a weight factor which can be recursively evaluated with the Robins-Monro algorithm; Φ is a basic function vector; ϕ_j is a basic function which is orthogonal in the pattern definition domain. Hermite polynomial functions are orthogonal with the recursive definition

$$H_0(y) = 1 \quad H_1(y) = 2y$$
$$H_{l+1}(y) - 2yH_l(y) + 2lH_{l-1}(y) = 0 \qquad (15)$$

If Y is an n-dimensional vector, first a group of Hermite functions for every element of Y is computed, then a basic function $\phi_j(y)$ is a product of Hermite functions.

During wire rope inspection, first a feature vector is computed, then a criterion function is computed; finally we can get the exact number of broken wires according to Eq. (13).

If only one feature element, DO, is selected, the computing time will be reduced significantly. A statistical analysis with one feature, DO, revealed that the accuracy of judging the number of broken wires is over 73 %. If a feature vector given by Eq. (12) is used, the accuracy of judgment will increase significantly. In addition, the present method can also locate the position of broken wires as well as determining their exact number.

Conclusions

It is very desirable to be able to inspect wire ropes automatically for both safety and economic reasons. Previous attempts to design such systems suffered from not being able to give a quantitative measure of the number of broken wires. The system described in this paper, based on the use of multiple Hall effect sensors and signal analysis by pattern recognition techniques, has proven to be successful. After successful laboratory trials, a prototype system is currently in operation in a coal mine. Work on a full production version will begin shortly.

Acknowledgement

We wish to thank Dr Bob Beattie of Meta Machines Ltd, Oxford, UK, for modifying the paper carefully and for providing many good suggestions. Mr. Tony Pugh, also of Meta Machines Ltd, is thanked for drawing the figures.

References

1 Rice, R. C. and Jentgen, R. L. *Statistical analysis of wire rope*. Contract J0215012 Battelle Columbus Laboratories, US Bureau of Mines (February 1983).

2 Wall, T. F. *Electromagnetic testing for mechanical flaws in steel wire rope*. J InstElecEng 67 (1929) pp. 899-911.

3 Casey, N. F. and Taylor, J. L. *The evaluation of wire ropes by acoustic emission techniques*, Brit J NDT 27 6 (November 1985) pp. 351-356.

4 Bundy, S. A. H. *Eddy current testing of wire and bar*, Wire Industry 48 576 (December 1981) pp. 887-888.

5 Wait, J. R. *Review of electromagnetic methods in nondestructive testing of wire ropes*. Proc IEEE 67 (June 1979) pp. 892-903.

6 Hall, E. H. *On a new action of the magnet on electric current*. Am J Math 2 (1879) p. 287.

7 Jukey, J. W. *Nonlinear (nonsuperposable) method for smoothing data*. Congress Record, 1974 EASCON (1974) p. 673.

Authors

The authors are in the Department of Mechanical Engineering, Huazhong University of Science and Technology, Wuhan, People's Republic of China. Mr. Wang, as a senior academic visitor, is currently doing some research work in the area of machine vision at Meta Machines Ltd, Oxford, UK.

(原载 NDT International, Volume 21, Issue 5, October 1988)

钢丝绳断丝定量检测的原理与实现

王阳生　师汉民　杨叔子　李劲松　叶兆国

韩连生　刘连顺

提　要　本文论述了当前钢丝绳定量探伤中所存在的一些主要困难。为了克服这些困难，本文设计了以计算机和传感器为核心的检测系统，排除了因检测中速度不均衡所造成的影响，消除了某些噪声的干扰；提出了以差分超限数为核心的特征矢量，并最终利用模式识别技术实现了对钢丝绳断丝的定量检测。在实验室及矿井现场进行的大量试验表明，该系统能以不小于95％的可靠性，保证集中断丝数的检测精度高于±1根。

关键词：霍尔效应，非线性平滑，特征矢量，差分超限数，模式识别

　　钢丝绳作为一种极其重要的承重构件，被广泛使用。它的状态直接关系到生命和财产的安全。为了防止意外断绳事故的发生，至今，仍主要采取定期"手摸""眼观"等办法来检查钢丝绳的状态，或采用强制性定期更换的办法，以保证安全。这些方法劳动强度大或带有较大的盲目性，既浪费人力物力，又难以杜绝事故。据文献[1]介绍，在对8000多个实验室和现场记录数据进行统计分析后发现：一方面大约2％的钢丝绳强度损失30％，处于非常危险的状态，但仍在使用；另一方面大约70％的钢丝绳强度损失很少，甚至没有损失，却被强制更换下来。

　　上述事实很早就引起了人们的关注[2]，并竞相研制钢丝绳探伤仪。至今已推出多种产品，其中有代表性的是英国研制的LMA型[3]、加拿大研制的磁图式[4]、日本研制的MRT-07型[5]以及我国生产的TGS*型探伤仪。这些产品比靠人体

---

*　TGS-46.5型钢丝绳探伤器使用说明书，营口仪器厂，1966。

感官进行检查,无疑是一个很大进步,但其主要不足是不能分辨集中在一起的断丝的准确数目。钢丝绳在使用中,往往存在断丝,分散的断丝所引起的强度损失不大,一般不影响正常使用。但是在使用的后期,断丝有一种趋于集中的现象,也就是说在某些捻距(类似于螺纹的导程)中出现较多的断丝,这种集中的断丝,严重削弱钢丝绳的承载能力。按照安全规程的规定,一个捻距中的断丝数是决定钢丝绳是否需要更换的主要指标。因而钢丝绳探伤的最关键的任务,是应对绳上集中出现的断丝数目作出定量判断的措施。

一、实现钢丝绳定量探伤的主要障碍

钢丝绳定量探伤是重要的,然而又是非常困难的,主要表现在如下几个方面。

1. 钢丝绳结构复杂、工作环境恶劣

图1所示的是一种典型的钢丝绳横截面,由图可见,钢丝绳是一种由许多细钢丝捻制在一起的复杂集合体。使用中钢丝绳往往长达几百米,甚至数千米。另外钢丝绳主要用于野外,环境往往非常恶劣,其表面又常涂满润滑油脂,粘满尘垢。这些情况都为传感器的设计和信号的分析与识别带来了较大的困难。

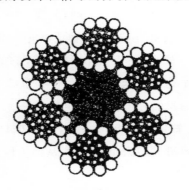

图1 一种典型的铜丝绳横截面

2. 由检测原理所引入的问题

在钢丝绳探伤中,至今已尝试过多种检测方法,例如,声发射检测[6]、涡流检测[7]、视觉检测[3]、放射线检测[8],以及电磁检测等[9]。在诸种检测方法中,人们公认电磁法较好。本文所用传感器也采用了以霍尔片为敏感元件的电磁检测传感器。该方法基于霍尔效应原理[10],即

$$V_H = KI \times B \tag{1}$$

式中：V_H 为霍尔电势；K 为比例常数；I 为控制电流；B 为磁感应强度。传感器调试好后，I 为常数，因而 V_H 的变化主要受到 B 的影响。

当钢丝绳被磁化（见图2）而绳上没有断丝时，漏磁场主要沿绳表面分布。当绳上出现断丝后，在断口处出现扩散磁场，扩散磁场与漏磁场叠加的结果，使磁感应强度 B 增加。一般来说，随着断丝数的增加，磁感应强度和扩散磁场的影响范围也相应增加。这正是钢丝绳断丝定量探伤的根据，但具体实现起来，却相当复杂。

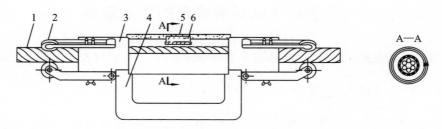

图 2　检测传感器

1—钢丝绳；2—导轮；3—极靴；4—永久磁铁；5—屏蔽罩；6—霍尔片

从电磁理论上看，电磁探伤主要是利用了 B 与物体状态之间的关系，然而绳上某一位置的 B 与该位置上的断丝状态之间，并无一一对应的关系，而是一种积分对应关系，也就是说，在一定范围内，各个物理点的状态都会影响某一位置上的 B 值。就钢丝绳探伤而言，如果在一定范围内集中了多根断丝，那么在任一位置检测到的检测信号 V_H 就是这些断丝各自产生的扩散磁场影响的叠加。因此，要想由检测到的 V_H 去判断断丝的分布及数目，并非易事。

从钢丝绳的损伤情况来看，钢丝断口的大小以及断口处钢丝的挠屈形状是变化不定的，这些变化将影响到 B 的大小和方向，从而给检测的结果带来随机分散，造成判断误差。

钢丝绳的表面并非准确的圆柱面，而存在绳股，如果断口在绳股的不同位置（股峰或股谷），则也会影响到检测信号的大小和方向。

由于钢丝绳为柔性体，且很长，因而在检测中，霍尔片和绳之间将难免发生相对位移和振动，这些也会使检测信号发生波动。

另外，B 还将受到各种外界干扰磁场的影响。总之，B 受到的影响是多方面的和不确定的，因而在断丝数目一定的情况下，检测信号仍会有随机的变化。只有排除这些变化的影响，才有可能准确判断断丝数目。

3. 检测速度的不均衡

一般来说，当集中断丝区中的断丝数增加时，扩散磁场的影响范围也将发生变化，这将为定量探伤提供重要信息。但实际检测中，钢丝绳相对于传感器的移动速度往往不能保证均衡，致使信号在时间轴方向发生拉伸、压缩的变化，这将使扩散磁场影响范围所提供的信息受到严重畸变。

4. 弱信号的处理

传感器输出的霍尔电势在 $200\mu V$ 左右，在恶劣的使用环境下，该电势将极易受到各种干扰源的影响。

二、传感器和检测系统

为实现定量探伤而设计的传感器如图 2 所示。图中，永久磁铁为励磁器，屏蔽罩是为了减小漏磁场的影响而设置的，12 个霍尔片在圆周方向等距布置。传感器加上计算机并辅以其他装置形成检测系统，如图 3 所示。图中模拟信号处理装置主要用于将传感器测到的弱信号进行放大、屏蔽和消除各种噪声的干扰；等距脉冲发生器的作用详后；计算机用于对数字断丝信号进行分析处理并借助于显示器和打印机输出图形或数字结果。

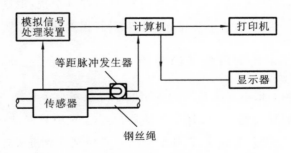

图 3　检测系统

图 3 所示系统在一次检测中输出的信号波形如图 4 所示。图中横坐标表示相对于绳上某一点的沿轴向的距离，图中 a,b,c,d,e,f 等信号异常区对应于钢丝绳上的断丝，而其余部分则是由于钢丝绳的绳股所造成的股波信号。结合图 4 和图 2 可以看出，本文所述传感器的主要特点是，12 个霍尔片环形布置，单独输出。其优点是：①有利于提高检测局部损伤的分辨率，例如从图 4 中可分辨出在 a,d,f 处

存在三处断丝,它们分布在同一横截面上,但处于不同的周向位置,可是,如果单路输出,将只能发现一处断丝;②它可提高对钢丝绳锈蚀段断丝的分辨能力,当绳上出现锈蚀点时,信号将存在锈蚀噪声,单路输出时,整圆周上的锈蚀噪声的合成有可能淹没较小的断口信号;③12路信号单独输出,使计算机处理更加灵活,例如,可利用相邻两路股波信号相位相差180°的特点作相加处理,以消除股波的影响。

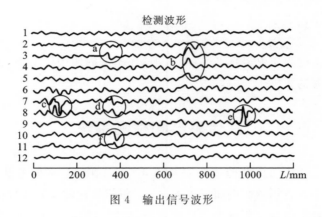

图 4　输出信号波形

三、空域信号及其平滑

利用图3所示的模拟信号处理装置可排除一部分干扰,但这是很不够的,还有必要采用如下一些措施,进一步排除干扰和进行误差补偿。

1. 消除速度不均衡所造成的影响

消除速度影响的根本方法是使用等空间采样技术。参见图3,等距脉冲发生器安装在传感器的导向轮上,只有在传感器与绳之间相对移动一个确定的间距时,脉冲发生器才发出一个脉冲,以外部中断方式控制计算机采样,因而采样信号仅仅是空间的函数,这类信号可称为空域信号 $x(m)$,空域信号与时间和速度无关,因而从根本上排除了速度的干扰。

2. 信号的平滑

由图3可见,等距脉冲发生器安装在导向轮上,只有在轮与绳纯滚动的条件下,才是理想的等距采样,轮发生滑动,采样点就会偏离正常位置,在信号上则出现"延迟"现象。另外,信号由于受到各种干扰影响,还可能出现"野点"及一些其他噪声,表现在信号上则是显得粗糙而不光滑。例如,图 5(a)所示信号。对于上述情况,可用滑动中值平滑器和线性平滑器组合成的一种非线性平滑器[11],对信号进行平滑处理。中值平滑器的输出为

$$y(m) = \text{Median}[x(m), x(m+1), x(m+2)] \quad (2)$$

线性平滑器可选用汉宁滤波器,即

$$S(m) = \sum_{n=-\infty}^{\infty} y(n)h(m-n) \quad (3)$$

$$h(n) = \begin{cases} \frac{1}{4}, & n = 0 \\ \frac{1}{2}, & n = 1 \\ \frac{1}{4}, & n = 2 \\ 0, & \text{其余} \end{cases} \quad (4)$$

信号经平滑后得结果如图 5(b)所示。可见信号质量得到明显改善。

(a)　　　　　　　　　　　　　　(b)

图 5　信号平滑结果

四、检测信号的定量解释

在上述工作的基础上,可利用计算机模式识别技术,以一定的精度从检测信号中判断断丝的根数。

1. 特征量的选择

在模式识别方法中,特征量选择非常重要,能否选取合适的特征量往往成为能否解决问题的关键。

(1) 差分超限数 $D0$

这是我们提出的一个新的特征量,用于识别与分析局部异常信号。如对图 4 所示信号进行等空间采样,并计算相邻两采样点信号幅值之间的差分绝对值,将不难发现,在信号局部变化异常区 a,b,c,d,e,f 中,差分绝对值一般远大于非异常

区的差分绝对值。如果预先设定一个门限,据之即可分离各个局部变化异常区,将分离出来的局部变化异常区中的差分绝对值超过门限值的采样点的个数累加起来,即得特征量——差分超限数 $D0$。

距检测基准点 L 处的断丝集中处的 $D0$ 值可用如下公式描述:

$$D0_L = \sum_{i=1}^{12} \sum_{m=-\infty}^{+\infty} T[x_i(m)]W(L-m) \tag{5}$$

式中 $T[x_i(m)]$ 表示对待分析信号作非线性变换,变换式为

$$T[x_i(m)] = c[x(m+1) - x(m)], \quad m = (-\infty, \infty) \tag{6}$$

$c(u)$ 为所定义的门限函数,即

$$c(u) = \begin{cases} 1, & |u| > t \\ 0, & \text{其余} \end{cases} \tag{7}$$

t 为预设定的门限。式(5)中 $W(n)$ 为窗函数,即

$$W(n) = \begin{cases} 1, & -N_i \leqslant n \leqslant 0 \\ 0, & 0 \end{cases} \tag{8}$$

式中,N_i 为窗宽,其大小能依信号局部变化异常区的范围变化而变化。该窗函数的功用是将信号的局部变化异常区分离出来。

式(5)中 i 从 1 变化到 12 表示对 12 种信号进行求和,这在一定程度上补偿了钢丝绳在传感器中径向与周向随机晃动所造成的误差。

(2) 短距能量 E

由图 4 可见,存在断丝处,信号幅值增大,波形发生明显变化。该特点可用一个捻距中信号的能量这一特征量表示,即

$$E = \sum_{i=1}^{12} \sum_{m=0}^{N-1} x_i^2(m) \tag{9}$$

式中: N 为一个捻距中的采样点数; i 对应霍尔片号数。

(3) 空间频谱幅值之和 SP_s

对检测信号作离散 Fourier 变换,可得空间频谱 $X(K)$,即

$$X(K) = \sum_{m=0}^{N-1} x(m) e^{-j\frac{2\pi}{N}Km} \tag{10}$$

式中,N 为采样点数。通过频谱分析发现,如果不考虑股波产生的谱峰,那么随着断丝数的增加,频谱幅值加大,频带加宽。利用此特点可得特征量——空间频谱幅值之和 SP_s,即

$$SP_s = \sum_{i=1}^{12} \sum_{k=0}^{N-1} |X(K)| \tag{11}$$

式中：$|X(K)|$ 为频谱的幅值；N,i 定义同式(9)。

将 $D0、E、SP_s$ 组合起来则得三维特征矢量 Y，即

$$Y^T = (D0, E, SP_s) \tag{12}$$

2. 非线性分类器的训练

检测信号受到多方面的随机干扰，使模式样本没有严格的归属关系。对于这种情况，可依概率密度进行分类。即如果令

$$d_i(Y) = p(\omega_i/Y), \quad i = 1, 2, \cdots, m \tag{13}$$

且有

$$d_i(Y) > d_j(Y), \quad \forall j \neq i \tag{14}$$

则

$$Y \in \omega_i \tag{15}$$

式中：$d_i(Y)$ 为判别函数；$p(\omega_i/Y)$ 表示当特征矢量取值为 Y 时，状态属于叫 ω_i 类的概率密度。

可用一多项式来逼近 $p(\omega_i/Y)$，即令

$$p(\omega_i/Y) = \sum_{j=1}^{k} \omega_{ij} \phi_j(Y) = \omega_i^T \boldsymbol{\Phi} \tag{16}$$

$$\omega_i = \begin{bmatrix} \omega_{i1} \\ \omega_{i2} \\ \vdots \\ \omega_{ik} \end{bmatrix}, \quad \boldsymbol{\Phi} = \begin{bmatrix} \phi_1(Y) \\ \phi_2(Y) \\ \vdots \\ \phi_k(Y) \end{bmatrix}$$

式中：ω_i 为权矢量，其值可利用 Robbins-Monro 算法递推估计；$\boldsymbol{\Phi}$ 为基函数矢量；$\phi_i(Y)$ 为基函数，它需要在模式定义域中正交。因为 Hermite 多项式为正交函数，在 Y 为 n 维矢量时，可用递推关系式

$$\left. \begin{aligned} H_0(y) &= 1 \\ H_1(y) &= 2y \\ H_{L+1}(y) &- 2yH_L(y) + 2LH_{L-1}(y) = 0 \end{aligned} \right\} \tag{17}$$

先求出 Hermite 函数集，在函数集中取出几个函数，并代入 Y 的各个分量 $y_1, y_2, \cdots, y_n$，然后将它们相乘，即可求出基函数 $\phi_i(Y)$。

本文求出的一个基函数矢量和利用一组样本求得的权矢量分别为

$$\boldsymbol{\Phi}^T = (1, 2y_3, 2y_2, 2y_1, 4y_2y_3, 8y_1y_2y_3)$$

$$\omega_1^T = (0.369, -0.059, -0.009, -0.079, -0.025, -0.013)$$

$$\omega_2^T = (0.290, 0.077, 0.055, 0.065, 0.013, 0.001)$$

$$\omega_3^T = (0.233, 0.133, 0.066, 0.181, 0.030, 0.015)$$

实际检测中,先求出特征矢量,将特征矢量代入式(16)则可求出判别函数,依式(15)则可最终判断断丝的准确根数。

特征矢量中,如果只取一个分量 D_0,计算量将大为减少并可实现在线检测。另外,本文所述方法,除了可判断断丝数以外,还可准确指示断丝的位置,以及断丝最多的危险捻距的位置。

五、实验与试用效果

该系统研制成功以后,在实验室对 $\phi=28$ mm,6×19 交捻钢丝绳及 $\phi=28$ mm,6×30 的顺捻钢丝绳进行了 100 余次检测试验,其结果是:准确判定每一个集中断丝处的断丝根数的概率为 73%,误差为 ±1 根的为 26%,±2 根的概率仅为 1%。断丝位置的误差均值为零(即系统误差为零),标准差为 13.82 mm,即 95.44% 的置信区间为 ±27.64 mm。以上仅为使用 D_0 值这一特征量,进行在线实时判别的结果(即钢丝绳通过传感器以后,立即显示结果)。试验表明,如同时采用前述三种特征量,组成模式向量,进行模式识别,并根据对一条钢丝绳长期监测的累积资料进行判断,其判别精度还可大为提高。以上检测精度已经由正式的技术鉴定确认[*]。该系统在通过实验室技术鉴定以后,又于 1987 年 10 月 25 日至 1988 年 1 月 26 日在某矿井进行了 300 余次现场试验,其结果为:准确判断集中断丝根数的概率达 68%,±1 根的为 27%。以上均按 D_0 值进行在线实时判别的结果。

参 考 文 献

[1] Rice,R. C. & Jentgen,R. L.. Contract J0215012,Battcllc Columbus Laboratories,US Bureau of Mines,February,1983.

[2] Wall,T. F.. J. Inst. Elec. Eng. ,67(1929),899—991.

[3] Herbert,R. Wcischclel. Operation and Maintenance Manual of LMA-250y 1985.

[4] Underbakke,L. D,& Haynes, H. H. Test anl Eraluatton of the Magnograph Unit——A Nondestructive WireRope Testery T\ NO. iV-1639, Naval Engineering Laboratory, Port Hueneme CA,1982.

[5] Operation Mannal for Mitsui Steel Wire Rope Tester,Mittsui Muke Machinery CO. ,

---

[*] 煤炭工业部司局文件,(87)煤技字第 183 号,关于印发《钢丝绳断丝定量分析与识别的研究学术评定证书》的函。

LTD,Tokyo,Japan,1978.

[6] Casey,N. F. &Taylor,J. L.. Brittsh Journal of NDT,Nov. 1985,351—356.

[7] Bandy,S. A. H. *Eddy Current testing of Wire and Bar*. Wire Industry,Dec. 1981, 887—888.

[8] 小林秀男,各国のワィャロヘプヲヌタヘじつぃて,と探鉱与保安,17(1971),3:130—143.

[9] Wait,J. R.. Proc. of the IEEE,67(1979),6:892—903.

[10] Hall,E. H.. Amer. J. Math. ,2(1879),287.

[11] Tukey,J. W.. Congress Record,1974 EASCON,1974,673.

（原载《中国科学（A 辑 数学 物理学 天文学 技术科学）》1989 年第 9 期）

复杂系统诊断问题的研究[*]

丁 洪 杨叔子 桂修文

提 要 本文基于复杂系统的定义和特点,详细地讨论了复杂系统诊断问题中的概念体系,并在此基础上,根据复杂系统的层次性和故障产生与传播的机理,提出了复杂系统诊断问题的层次因果模型。该模型的中心思想是将复杂系统诊断问题分解成有限个层内诊断与层间诊断的循环交迭求解过程。

关键词:复杂系统,诊断模型,知识,层次性,概念体系,专家系统

1 引 言

有关复杂系统诊断问题,虽然已有些文章论及[1-4],但作者认为至少还存在以下几个方面的问题:①哪些系统属于复杂系统?复杂系统的明确定义是什么?②在基于知识的复杂系统的诊断问题中,求解应包含哪些方面的知识?领域专家是如何运用和控制这些知识进行诊断和推理的?③复杂系统的故障产生与传播机理是什么?它对诊断求解有何作用?以上这些问题都是建造一个有效的、实用的复杂系统的诊断系统和进行基于知识的诊断理论的研究所必须努力解决的问题。

诊断概念和知识是复杂系统诊断的基础。在已有的诊断系统中,都试图用故障、征兆以及由故障到征兆的有向因果关系(即"浅知识")作为全部的诊断知识来描述该诊断问题,即使在基于诊断对象的结构、功能与行为等"深知识"诊断中,关于故障产生与传播机理对诊断问题求解的作用的认识也是不充分的,甚至是贫乏

---

[*] 国家教育科学基金资助项目。

的。在深知识的定义和表示方面仍存在问题。一些研究对象,尤其是复杂系统的本体知识的不易描述和获取是目前"深知识"表示方面的研究和应用难以深入的一个重要原因。另一方面,合适的诊断模型和诊断策略是诊断问题得以有效求解的关键,它们是目前诊断问题求解中的主要研究课题。文献[4]～[6]分别提出了反演绎推理和概率计算相结合的概率因果模型、基于深知识的模型和集成符号推理与模糊测量的推理模型。然而,这些模型在复杂系统的诊断问题求解中的应用都受到一定的限制。例如,复杂系统的深知识是不易表示和获取的;复杂系统中部件级的故障与系统级的观测征兆之间的因果关系知识也是不易获取的。因此,如何有效地综合运用浅知识与深知识仍是一个重要的研究课题。

2 复杂系统诊断问题的概念体系

目前,在诊断问题的研究中,有关诊断对象、诊断知识以及诊断问题的性质等方面的概念是十分含糊的,甚至存在相互矛盾的情况;另一方面,虽然已提出一些诊断方法和诊断模型,但大多是针对特定领域和特殊情况的,而又没有考虑诊断对象与诊断问题的性质,因而在实际中很难扩大应用。所有这些问题都表明有必要建立一个包含诊断对象性质的诊断问题的一般概念体系。

虽然,实际中的诊断对象之间可能存在很大差异,然而,由系统分解理论,根据诊断对象的结构特点,可将它们分成简单系统、复合系统和复杂系统三类。相应地,诊断问题也可分为简单系统诊断、复合系统诊断和复杂系统诊断三类。本文重点讨论复杂系统的诊断问题,但为了讨论清楚,下面将从简单系统开始,讨论诊断问题的概念体系。

定义 1 一个系统是简单系统的充分必要条件是:

a. 在结构上,该系统直接由一个或多个基本的物理或电子等元器件组成,并且各元器件之间的连接关系是确定的,可描述的;

b. 在功能上,该系统的输出与输入之间存在严格的定量或逻辑关系,并可进行描述。

讨论 1 基本的物理或电子等元器件是组成系统的基本单元,例如:电阻、电容、数字触发器、逻辑门、螺钉、杆、圆柱体等等。事实上,它们中的一些本身就是简单系统,各自完成一定的功能。即便是由多个单元组合起来的系统,每个单元之间的连接以及系统本身的输出与输入之间都存在着可描述的严格的定量或逻辑关系,例如数字乘法器、理想状态的连杆组件等。

定义 2 复合系统是由一些在功能与结构上各异的简单系统通过可描述的严格的定量或逻辑关系组合起来完成特定功能的多层次系统。

讨论 2 复合系统与简单系统相比,存在两点根本的区别:第一,复合系统是一个多层次系统,即由系统分解的观点,复合系统至少应由系统级、子系统级和部件级三个层次组成;第二,在复合系统中,同一层次的各子系统之间无论是在结构上,还是在功能上都存在着一定的差异。与简单系统相类似,复合系统中的任一层次级无论是在结构上,还是在功能上都是可描述的。例如:电子计算机以及一般的由纯数字电路组成的仪器等。

定义 3 一个系统是复杂系统的充分必要条件是:

a. 根据系统分解的观点,该系统是一个多层次系统;

b. 该系统中,同一层次的各子系统之间无论是在结构上,还是在功能上都存在着一定的差异;

c. 同一层次的各子系统以及不同层次的各子系统之间虽然可能相连,但在结构与功能上大多无严格的逻辑与定量关系。

讨论 3 由上述定义 1～3 可看到,在结构上,简单系统⊂复合系统⊂复杂系统。对一台汽油发动机可进行如图 1 所示的结构分解,显然它是一多层次系统。虽然曲轴-连杆子系统可以分解成活塞组件、连杆组件和曲轴组件等子系统,但是,由于油膜和间隙的存在(这是这些组件正常工作所必需的),这些组件之间并不存在逻辑关系或易于描述的定量关系。因此我们说汽油发动机是一个复杂系统。

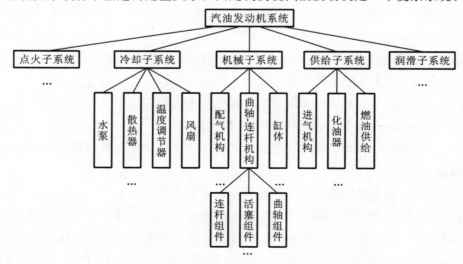

图 1　汽油发动机系统的分解

目前已有的许多诊断方法与模型可用来解决简单系统和复合系统的诊断问题。实际上,这些模型和方法的提出和讨论始终都是以简单系统和复合系统为研究对象的[7,8]。文献[7]虽然对复杂系统的诊断问题作了初步探讨,但到目前为止,不但没有一个有效实用的诊断模型,而且该诊断问题中的许多概念问题还没有得到很好的解决。

定义 4　指定层次的子系统的故障是指该子系统的功能失调,即它不能在规定的时间内和条件下完成预定的功能。

讨论 4　复杂系统是一个多层次系统,所以该系统的故障也具有层次性。系统级故障对应于系统级功能失调或破坏,一定层次的子系统级故障对应该子系统级的功能失调或破坏。部件级的故障为该复杂系统的最低层故障。

定义 5　指定层次子系统的征兆是指该子系统的有关输出。

讨论 5　显然,征兆也具有层次性,不同层次的子系统对应有不同层次的征兆,并且下一层次子系统的输出是上一层次子系统的输入。根据这个定义,可以看到任一子系统的征兆包括两部分:异常征兆和正常征兆。前者对应于该子系统的异常输出,后者对应其正常输出。一个子系统的异常征兆是该子系统本身的故障和(或)其他与之相连的子系统的故障所引起的系统响应。它是我们识别这些故障的重要知识,但不是唯一的诊断知识。Peng[4],De Kleer[8]等人在他们各自提出的诊断模型中均未考虑正常征兆的知识,其实这一诊断知识常常会大大加速诊断推理过程。

定义 6　指定层次子系统产生故障的原因是引起该子系统功能失调或破坏的因素。它包含以下三个方面的内容:

a. 来自与该子系统相连的下一层次子系统的异常征兆;

b. 来自与该子系统相连的同一层次其他子系统的异常征兆;

c. 若该层次为系统的最低层次,即部件级,则还包括部件本身的失效。

讨论 6　故障是针对系统而言的,它与征兆在概念上是有着本质区别的。根据系统分析的观点,对于指定层次的某子系统来说,该子系统的一种故障对应于系统固有特性的一种所不允许的劣化,这种劣化通常是由外界作用(输入)而引起的。我们知道,当外界输入在规定的时间和条件范围内时,系统的固有特性是不随外界输入变化而发生变化的,这是进行系统分析的基本假设,但当外界输入超过该范围时,该子系统就会产生某种物理或化学的变化,即固有特性产生劣化,不能完成预期的功能,这就是一种故障。此时,即使外界输入回到规定的范围内,该子系统也将保持这种故障特性。

指定层次的某子系统出现了故障,即使子系统的输入正常,也必然会产生异常输出,即产生异常征兆。如图 2 所示,其中 S^{ij} 表示第 i 层的第 j 个子系统;X^{ij} 是该子系统的输出;输入 X^{ik} 是来自同一层次的相联第 k 个子系统的输出;输入 $X^{(i+1)j}$ 是来自第 $(i+1)$ 层次的第 j 个子系统的输出。显然,当 S^{ij} 出现故障时,即使 X^{ik} 和 $X^{(i+1)j}$ 正常,X^{ij} 也是异常的。反过来,如果已知 X^{ij} 为异常征兆,则它可能是由于子系统 S^{ij} 出现故障而引起的;也可能是由于 X^{ik} 和/或 $X^{(i+1)j}$ 的异常而引起的,而此时系统 S^{ij} 并无故障。根据上述定义,可看到当 S^{ij} 为非部件级子系统时,作为 S^{ik} 子系统和 $S^{(i+1)j}$ 子系统的征兆 X^{ik} 和 $X^{(i+1)j}$ 是 S^{ij} 子系统产生故障的原因。此时 X^{ik} 和 $X^{(i+1)j}$ 又是 S^{ik} 和 $S^{(i+1)j}$ 子系统的异常征兆。所以,一个异常征兆即可能是一有故障子系统的输出,也可能是引起另一个子系统产生故障的原因,对于复杂系统的最低层次,即部件级子系统来说,它产生故障的原因主要是定义 6 中的后两条因素,其中部件本身的失效包括超过使用周期、工作环境影响以及受力作用等因素,这些都与系统的设计有一定关系。

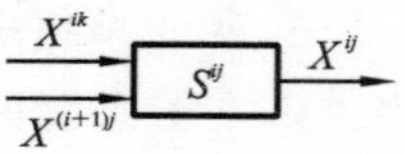

图 2 故障、征兆、原因及系统四者之间的关系

定义 7 指定层次某子系统的工作状态是指该子系统工作过程中的行为。

讨论 7 当该子系统出现某种故障时,在任一输入的情况下,该子系统都将输出某些异常征兆,这就是该子系统的一种故障行为,或称异常工作状态。当该子系统无故障时,可能输出正常征兆,也可能输出某些异常征兆,这些都属于子系统的正常工作状态。状态知识正是表征了系统或子系统工作状态的知识,它是复杂系统诊断中的重要知识之一[8]。

定义 8 复杂系统的诊断问题是指基于一定的诊断模型,利用各种诊断知识和合适的诊断方法,由给定的系统级和其他层次级的征兆集合找出引起这些征兆出现的指定层次各子系统的故障集合。

3 复杂系统诊断问题的求解

3.1 层内诊断与层间诊断

根据定义 8,我们看到复杂系统的诊断过程是一个逐层搜索过程,虽然指定层次的深度不同,诊断问题的难度也不同,但实质上都可分解成两类诊断子问题:层内诊断和层间诊断。层内诊断是指由已知的征兆集合到产生该征兆集合的该层中各子系统故障集合的识别;而层间诊断则是由已知上一层的故障集合,识别引起该故障集合的原因集合,即输入到上一层子系统级的该层中各子系统的征兆集合,如图 3 所示。这样,复杂系统诊断问题的求解就是这两类诊断子问题的循环求解过程,并以层内诊断作为该过程的开始和结束。

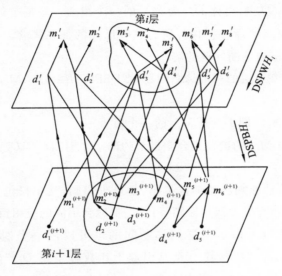

图 3　层内诊断与层间诊断

3.2　展次因果诊断模型

第 i 层的层内诊断问题可用一个四元式 $DSPWH_i = \langle D_i, M_i, C_i, M_i^* \rangle$ 来描述,式中: $D_i = \{d_1^i, d_2^i, \cdots, d_N^i\}$ 表示第 i 各子系统所有故障的非空有限集合; $M_i = \{m_1^i, m_2^i, \cdots, m_N^i\}$ 表示第 i 层中故障集含 D_i 所引起的所有征兆的非空有限集合; C_i 是描述 D_i 到 M_i 之间的因果关系集合,即 $C_i = \{\langle d_j^i, m_l^i \rangle | d_j^i \in M_i, m_l^i \in M_i\}$,其中 $\langle d_j^i, m_l^i \rangle$ 表示故障 d_j^i 的产生可能会引起征兆 m_l^i,并且 m_l^i 可能是异常的,也可能是正常的。如果 m_l^i 是一正常征兆,则 $\langle d_j^i, m_l^i \rangle \in C_i$ 表示故障 d_j^i 的产生在一定的程度上会阻止征兆 m_l^i 变为异常。这样如果已知或观测得征兆 m_l^i 是异常的,则这一事实也将反过来在一定程度上否定故障 d_j^i 的产生。这种否定关系其实是一种很有用的诊断知识。M_i^* 为属于 M_i 的一个已知征兆集合,它可通过测试或通过其他中间假设和结论推理获得。用于层内诊断的这些知识集合可表达如下:

$$M(d_j^i) = \{m_l^i | \langle d_j^i, m_l^i \rangle \in C_i\}, \quad \forall d_j^i \in D_i$$

$$D(m_j^i) = \{d_l^i | \langle d_j^i, m_l^i \rangle \in C_i\}, \quad \forall m_j^i \in M_i$$

$$M(D_i^*) = \bigcup_{d_j^* \in D_i^*} M(d_j^i)$$

式中:$D_i^* \subseteq D_i$ 是一假设的故障子集;$D(M_i^*) = \bigcup_{m_j^i \in M_i^*} D(m_j^i)$;$M^*(d_j^i) = M(d_j^i) \cap M_i^*$;$M^*(D_i^*) = M(D_i^*) \cap M_i^*$。

如果 $M_i^* \subseteq M(D_i^*)$,则称 D_i^* 为 M_i^* 的一个覆盖。根据节约覆盖理论[4],该故障子集 D_i^* 必须能解释(或引起)所有这些已知的征兆集合。基于上述这些概念和知识,可给出层内诊断问题的解。

定义 9 对于第 i 层的层内诊断问题,即 $DSPWH_i = \langle D_i, M_i, C_i, M_i^* \rangle$,当且仅当

a. $M_i^* \subseteq M(D_i^*)$,即 D_i^* 是 M_i^* 的一个覆盖;

b. D_i^* 是所有 M_i^* 的覆盖集合中最节约的一个时,$D_i^* \subseteq D_i$ 是 $DSPWH_i$ 的一个解。

一些不同的节约原则已相继被提出,如最少原则,其中仅包含为了覆盖所需的最少元素,此外还有极大似然原则等,它们都适合不同的应用情况。

从第 i 层到第 $(i+1)$ 层的层间诊断问题也可以用一个四元式来描述,即 $DSPBH_i = \langle R_i, D_i, A_i, D_i^* \rangle$,其中 R_i 是引起上述故障子集 D_i 产生的原因的非空有限集合,事实上,该原因集合也正是第 $(i+1)$ 层次的征兆集合 M_{i+1},即 $R_i = M_{i+1} = \{m_1^{i+1}, m_2^{i+1}, \cdots, m_K^{i+1}\}$,$D_i$ 的含义与上述层内诊断相同;A_i 则是描述 M_{i+1} 到 D_i 之间的因果关系集合,即 $A_i = \{\langle m_l^{i+1}, d_j^i \rangle | m_l^{i+1} \in M_{i+1}, d_j^i \in D_i\}$,其中 $\langle m_l^{i+1}, d_j^i \rangle \in A_i$ 表示 m_l^{i+1} 可能是 d_j^i 产生的原因,或 m_l^{i+1} 的存在可能支持 d_j^i 产生这一假设。当 m_l^{i+1} 为正常征兆时,则因果关系 $\langle m_l^{i+1}, d_j^i \rangle \in A_i$ 表示故障 d_j^i 的产生可能支持征兆 m_l^{i+1} 是正常的这一假设;D_i^* 是上述定义的第 i 层的层内诊断问题的解。这样,层间诊断问题的知识集合可表述如下:

$$D(m_l^{i+1}) = \{d_j^i | \langle m_l^{i+1}, d_j^i \rangle \in A_i\}, \quad \forall m_l^{i+1} \in M_{i+1}$$

$$R(d_j^i) = \{m_l^{i+1} | \langle m_l^{i+1}, d_j^i \rangle \in A_i\}, \quad \forall d_j^i \in D_i$$

$$D(M_{d+1}^*) = \bigcup_{m_l^{i+1} \in M_{i+1}^*} D(m_l^{i+1})$$

式中:$M_{i+1}^* \subseteq M_{i+1}$ 是一个假设的征兆子集;$R(D_i^*) = \bigcup_{d_j^* \in D_i^*} R(d_j^i)$;$D^*(m_l^{i+1}) = D(m_l^{i+1}) \cap D_i^*$;$D^*(m_{i+1}^*) = D(m_{i+1}^*) \cap D_i^*$。

如果 $D_i^* \subseteq D(M_{i+1}^*)$,则 M_{i+1}^* 也被定义为 D_i^* 的一个覆盖。基于上述的基本

概念和诊断知识,层间诊断问题的解可定义如下:

定义 10 对于从第 i 层次到第 $(i+1)$ 层次的层间诊断问题 $DSPBH_i = \langle R_i, D_i, A_i, D_i^* \rangle$,当且仅当

  a. $D_i^* \subseteq D(M_{i+1}^*)$;

  b. M_{i+1}^* 是所有 D_i^* 的覆盖中节约的一个时,$M_{i+1}^* \subseteq M_{i+1}$ 是 $DSPBH_i$ 的一个解。

在定义 9 和定义 10 的基础上,本文进一步定义复杂系统诊断问题 $DPCS = \langle D, M, C, R, M_i^* \rangle$ 的解如下:

对于复杂系统诊断问题 $DPCS = \langle D, M, C, R, M_i^* \rangle$,第 i 层次中故障的有限子集 D_i^* 作为该诊断问题的解可以通过 i 次层内诊断和 $(i-1)$ 次层间诊断的循环交迭求解过程获得。

4 结　　语

诊断技术发展至今已经历了三个发展阶段。在第一阶段,诊断结果在很大程度上取决于领域专家的感官和专业经验。传感器技术、动态测试技术以及信号分析技术的发展使诊断技术进入了第二阶段,并且在工程中得到广泛的应用。近年来,为了满足复杂系统的诊断要求,随着计算机技术的发展及人工智能技术尤其是专家系统在诊断问题求解中的应用,诊断技术发展正进入它的第三阶段——智能化阶段。在这一阶段,人类领域专家的知识将得到充分的重视,诊断问题的研究正致力于模拟领域专家的推理过程、控制和运用各种诊断知识的能力以及诊断问题的实质等以解决问题。可以说这些问题才刚刚被人们所重视,远远没有得到很好的解决。本文讨论的复杂系统诊断问题的概念体系和提出的层次因果模型正是对该研究问题的一点探讨。该模型的主要思想是根据复杂系统及其故障的层次性和特点,将复杂系统诊断问题分解成多个层内诊断和层间诊断的循环交迭求解过程。

实践证明,仅采用单一的诊断知识(如深知识、浅知识等)来解决复杂系统的诊断问题是不现实的,事实上领域专家在诊断过程中通常要用到各种知识,包括常识。值得指出的是,状态知识在复杂系统诊断问题中有着重要的应用价值。虽然诊断技术进入了它的智能化阶段,专家经验和知识得到重视,但我们仍毫无理由忽视现代测试技术、信号分析、模式识别等技术在诊断问题求解中的应用。事实上,目前常用的各种诊断模型都在一定程度上依赖于动态测试的结果,从这些

测试数据还可以提取出反映复杂系统或子系统工作状态的信息,即状态知识。在复杂系统的诊断过程中,状态知识能为推理过程提供一些有价值的中间假设,如一定层次的某正常征兆 m_{ln}^i 和异常征兆 m_{la}^i 的假设等,从而加速诊断过程。

  从诊断技术的发展来看,基于知识的诊断问题应是与智能设计问题(设计型专家系统)密切相连的。人类领域专家的经验和知识中有很大一部分是关于诊断对象的结构、工作原理、功能和行为方面的,对于复杂系统来说,这方面的知识尤为宝贵,因为它不易获得。因此,如果在复杂系统的设计阶段就考虑上述知识的表示和获取,将其提取出来并显式化,则毫无疑问会大大增加诊断结果的可靠性和加速诊断过程。

参考文献

[1] Milne R. *Strategies for Diagnosis*. IEEE Trans. Syst., Man., Cybern., 1987, SMC-17(3):333-339.

[2] Fink P K, Lush J C. *Expert Systems and Diagnostic Expertise in the Mechanical and Electrical Domains*. JEEE Trans. Syst., Man., Cybern., 1987, SMC-17(3):340-349.

[3] De Kleer J, Willians I. C. *Diagnosing Multiple Faults*. Artificial Intelligence, 1987, 32:91-130.

[4] Peng Y, Reggia J A. *A Probabilitic Causal Model for Diagnostic Problem Solving—Part one and Fart two*. /EEJE Trans., Syst., Man., Cybern., 1987, SMC-17(2,3):369-379.

[5] Kuipers B. *Qualitative Simulation as Causal Explanation*. IEEE Trans., Syst. 9 Manti Cybern., 1987, SMC-17(3):C2-4.

[6] Yager R R. *An Explanatory Model in Expert Sy zizm*. Int. J. of Man-Machine Studies, 1087, 23:539-519.

[7] 杨叔子,等. 机诚设备诊断学的探讨, 华中工学院学报, 1987, 15(2):1-8.

[8] Ding, H, Gui X W, Yang S Z. *State Ktioyvledge Representaticn Techniques*. Proc. of ICESEA'89, 1989.

On Diagnosis of Complex System

Ding Hong Yang Shuzi Gui Xiuwen

Abstract The objects to be diagnosed can be divided according to structure and function into the Simple, Composite and Complex Systems. Most of the diagnostic models and techniques in current use are developed for simple and/or composite systems while little progress has been made so far for complex systems (DPCS) in wide use in practice. Based on the definitions and specific features of complex systems, a complete list of conceptual systems of DPCS is proposed and discussed. A hierarchical causal model for the solution to DPCS is presented according to the hierarchy and the mechanism of fault occurrence and propagation. The basic idea of the model is to solve the problem with circulative and alternative procedures by dividing it into a finite number of sub-problem diagnoses within the hierarchy and those between the hierarchies. A brief account is made of the ciosc relationship between the knowledge-based diagnostic problem-solving and the intelligent design.

Keywords: Complex system, Diagnostic model, Knowledge, Hierarchy, Concepual system, Expert system

（原载《华中理工大学学报(自然科学版)》1989 年第 17 卷第 4 期(增刊)）

Plant Condition Recognition—
A Time Series Model Approach

Zheng Xiaojun   Yang Shuzi   Wu Shouxian

Abstract This paper presents a method of recognizing the conditions of a plant with time series models. In the method, the task of recognition is achieved in two stages. The first stage is condition monitoring, the goal of which is to quickly find abnormal conditions, and second stage is condition identifying, the goal of which is to identify the classification of the abnormal conditions in order to find the causes and locations of the abnormal condition (e.g. a system fault). The whole system is configured with a microcomputer and some general instruments. Results from recognizing the conditions in vibration signals of an electrical motor by using the system are shown.

Keywords: Pattern recognition, Statistical pattern recognition, Time series model, Testing, Quality, Industrial plant, Condition recognition

1. Introduction

In recent years applications of pattern recognition techniques to fault diagnosis techniques for plants have been studied more and more frequently. Before this paper several scientists have studied application of the pattern recognition techniques based on time series analysis methods to plant diagnosis problems. In 1980 Tobin[1] used the DDS method (Dynamic, Data System, e.g. ARMA model method) to diagnose a motor's vibration faults. The method can monitor the motors condition by means of checking a residue square sum of an ARMA model and identify causes and locations of the motor's faults by ARMA power spectrums. On the basis of it a method applying a normalized residue square sum to monitor condition of a plant was also presented. However, the most valuable researches are those performed separately by Nakamizo[3] and Gersch[5]. They apply AR modeling methods to plant diagnosis independently, where AR model coefficients which are features of the plant's conditions, reference features which are features of the standard normal condition of the plant, and distance measures are used together to achieve a diagnosis task.

In this paper a new pattern recognition method based on time series models for plant diagnosis will be presented which also consists of two stages: the first stage is condition monitoring and the second one is abnormal condition analyzing. An experimental system and an example study for diagnosing an electrical motor are introduced.

2. Condition Recognition Principle

As a general pattern recognition method, the method based on time series model is also divided into two algorithms: a training algorithm and a recognizing one. Each new object, at first, is processed by the training algorithm, the goal of which is to provide the microcomputer with the information of plant's normal condition. Then the recognizing algorithm is preformed to carry out the recognizing task of plant's conditions.

Fig. 1 is a principle diagram of the training algorithm. The first step is to compute separately a group of features from a group of training samples, and the second one is to use the correspondent reference feature exacting algorithm to get a group of reference features, and the third one is to compute the threshold of decision functions.

Fig. 2 is a principle diagram of the recognizing algorithm which consists of two steps: the first step, called on-line monitoring, is a standard pattern recognition process achieved automatically by computer, and the second one, called abnormal condition analyzing, is achieved with the information provided by the computer, for example, power spectrum, power cepstrum, and so on.

There are many key theoretical issues that are yet unsolved, such as feature extracting, reference feature selecting, condition monitoring, and abnormal condition analyzing. The following are the viewpoints and methods used by authors.

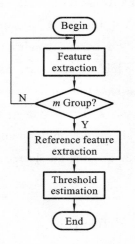

Fig. 1 Diagram of training algorithm   Fig. 2 Diagram of recognizing algorithm

2.1 Feature Extraction

Generally, the features of a dynamic signal are the pattern-type characteristics which can be acquired from three different domains, e. g. time domain, frequency domain, and correlation domain (including cepstrum domain). The features in frequency domain are often used as a recognizing feature. Advantages of these features are their high precision and properties of representing directly the statistic properties of a dynamic signal. But their available dimension is so high that it is very difficult to get them. In recent years many features in time domain, such as AR model coefficients, are also used. Advantages of the features are their small calculation amount and clear physical meaning. But they cannot represent directly the statistic properties of the dynamic signal and therefore have low recognizing accuracy. In order to solve this problem a new kind of feature —AR cepstrum coefficients—is introduced, which is a kind of feature in the cepstrum domain and can represent the statistic properties of the dynamic signal more effectively than that in the frequency domain. Besides this, a new feature extracting method for it has been proposed, which can reduce a lot the amount of calculation necessary for extracting feature.

The fundamental procedure of the feature extracting method proposed here is: first, using an AR coefficient estimating algorithm, for example, the one in the literature [5], to get a group of AR coefficients and a residue square sum of the signal; then using the following method to get a group of AR cepstrum coefficients.

Based on the concept of complex cepstrum, the cepstrum coefficient is defined as a group of sample values of a complex cepstrum function. However, complex cepstrum is defined as an inverse Fourier transformation series of the log series which is a Fourier transformation series of the original signal.

In the literature [6] authors detailed the knowledge on this aspect and developed an algorithm of achieving directly the AR cepstrum coefficients from a group of AR coefficients and a residues square sum. That is, if the group of AR coefficients and residue square sum were $a_{p,1}, a_{p,2}, \cdots, a_{p,p}$ and σ^2 separately, then we would have had group of cepstrum coefficients

$$\begin{cases} C_{p,0} = \log \sigma \\ C_{p,1} = -a_{p,1} \\ C_{p,t} = -\left[\sum_{k=1}^{t-1}\left(1-\frac{k}{t}\right) \cdot a_{p,k} \cdot C_{(t-k)} + a_{p,t}\right], & 1 < t \leqslant p \\ C_{p,t} = -\sum_{k=1}^{p}\left(1-\frac{k}{t}\right) \cdot a_{p,k} \cdot C_{(t-k)}, & t > p \end{cases} \quad (1)$$

In fact the former $p+1$ cepstrum coefficients have included the whole information included in the group of AR coefficients. As obtained from σ^2, $C_{p,0}$ value is not reliable. So another group of features should be employed, that is $C = (C_{p,1}, C_{p,2}, \cdots, C_{p,p})$ which can be called as P-order AR cepstrum coefficients according to the order number of the correspondent AR coefficients, P. Because of the low dimension of the features, a feature selecting process in general pattern recognition is not necessary for them.

2.2 Reference Feature Selection

During on-line monitoring only the reference features for normal condition need be estimated because the task of the process is to distinguish normal condition from abnormal one. The following are the reference feature selecting methods adopted.

(A) *A reference feature selecting method for simple systems (e.g. systems which have constant model coefficients).* According to the results of literature [6] features of the training samples from simple systems, $A_i (i = 1, 2, \cdots, r)$, can be considered as a group of statistic samples with a multidimension normal distribution so their reference features should be

$$\begin{cases} A^r = \dfrac{1}{m} \sum_{k=1}^{m} A_k \\ W = \dfrac{1}{m} \sum_{k=1}^{m} A_k^\mathrm{T} A_k - A^{r\mathrm{T}} A^r \end{cases} \quad (2)$$

where $A_k = (a_{p,1}^k, a_{p,2}^k, \cdots, a_{p,p}^k)$ is a feature vector of the kth group of training samples, $A^r = (a_{p,1}^r, a_{p,2}^r, \cdots, a_{p,p}^r)$ is a reference feature vector, and W a covariance matrix of the feature vector.

(B) *A reference feature selecting method for complex systems (e.g. systems which have random model coefficients).* In the literature [7] a reasonable reference feature selecting method has been presented. In real-world, the real-model coefficient vectors of most systems are not constant but random with some distribution, usually normal distribution. The real feature vectors coming from different groups of training samples are unequal but all submit the normal distribution, so it is not reasonable to average them with the same weight when estimating the reference features. The result presented in literature [7] was that the weights in reference features estimating process should be determined according to some real-world conditions and so the methods used should be

$$\begin{cases} A^r = \dfrac{\left\{\sum\limits_{k=1}^{m}\dfrac{A_k}{\sigma_k^2}\right\}}{\left\{\sum\limits_{k=1}^{m}\dfrac{1}{\sigma_k^2}\right\}} \\ W = \dfrac{1}{m}\sum\limits_{k=1}^{m}\dfrac{(A_k-A^r)^T(A_k-A^r)}{\sigma_k^2} \end{cases} \quad (3)$$

where σ_k^2 is the residues variance of the kth group of training samples.

2.3 Condition Monitoring

The basis of condition monitoring is: if the condition of a system is normal, then the feature vector of its output signals must be very close to the reference feature vector, so the value of distance measured that represents the distance between them must be smaller than the predetermined decision threshold. Therefore, in each time computing the feature vector, and the value of the distance measure between the feature vector and the reference feature vector, and then comparing the measure value with the predetermined threshold, are the whole work of checking the system.

Obviously the key issue of condition monitoring is the selection of a distance measure. The schemes adopted in this paper are the following.

(A) *For a simple system the scheme employs a Mahalanobis measure*:

$$D(A^t, A^r) = (A^t - A^r) W^{-1} (A^t - A^r)^T \quad (4)$$

where A^r, W are the reference feature, and A^t is the feature vector of the samples to be tested.

(B) *For a complex system the reference feature vector is not a constant but a random one which submits a multidimension distribution* with an average value A^r and a covariance matrix W. For the situation Wald presented a distance measure[8], that is

$$D(A^t, A^r) = (A^t - A^r) C^{-1} (A^t - A^r)^T \quad (5)$$

where C is general covariance matrix defined as

$$C = W + \frac{R'^{-1}}{N}[1, A^t] R\ [1, A^t]^T$$

where R is the correlation matrix of the samples to be tested, R' is a submatrix of R, and N is the number of samples to be tested.

The first part of C is an item with relation to the reference feature, and the last part is the item with relation to the samples to be tested. In Eq. (4) and Eq. (5) W has different meaning: e. g. W in Eq. (4) comes from the errors of statistics samples and W in Eq. (5) from the randomized of real model of the system. So the reference features must be obtained by means of different methods.

2.4 Abnormal Condition Analysis

The methods used for analyzing the abnormal conditions are mainly the general signal processing methods such as normalized AR spectrum analysis, AR cepstrum analysis, and so on.

2.4.1 Normalized AR Spectrum Analysis

The definition of AR spectrum is

$$P(f) = \frac{\sigma^2 \cdot \Delta t}{\left|1 + \sum_{k=1}^{p} a_{p,k} \exp(-j2\pi fk \cdot \Delta t)\right|^2} \tag{6}$$

where σ^2 is residue variance, and Δt is the sampling interval.

Because of its variety a normalized AR spectrum with the normalized frequency and amplitude is often used to take a place of it:

$$S(f) = -10\log\left\{\left|1 + \sum_{k=1}^{p} a_{p,k} \exp(-j2\pi fk)\right|^2\right\} \tag{7}$$

where the unit of $S(f)$ is dB, and f represent the ratio of real frequency and sampling frequency.

2.4.2 Peak Frequency Analysis of Power Spectrums

In previous $P(f)$ let $z = \exp(j2ft)$, then

$$P(f) = \frac{\sigma^2 \cdot \Delta t}{A_{(z)} A(1/z)}$$

where $A(z)$ is an auto-regressive inverse filter. Therefore the peak frequency of the power spectrum can be obtained by the following procedures: first solving the equation $A(z) = 0$ and getting its roots $Z_k = a_k + jb_k$, then selecting the root with an amplitude close to 1 in the upper half part of the complex plane as the object to compute the peak frequency, and then computing the angle ϕ_k ($0 \leqslant \phi_k < \pi$) in the Z-plane with

$$\phi_k = \begin{cases} \arctan\left(\dfrac{b_k}{a_k}\right), & a_k > 0, b_k > 0 \\ \arctan\left(\dfrac{b_k}{a_k}\right) + \pi, & a_k < 0, b_k > 0 \end{cases} \quad (8)$$

and lastly transforming it into the required frequency.

$$F_k = \dfrac{\phi_k}{2\pi \Delta t} \quad (9)$$

2.4.3 AR Cepstrum Analysis

For the minimum phase signal, we can use the special relation between cepstrum coefficients and power cepstrum to get the latter. According to the result of literature [6]:

$$\begin{cases} C(t) = 0, & t < 0 \\ C(t) = 2C_{p,0}, & t = 0 \\ C(t) = C_{p,t}, & t > 0 \end{cases} \quad (10)$$

Therefore the algorithm for getting directly power cepstrum from AR coefficients and residues variance is the following:

$$\begin{cases} C(0) = \log \sigma^2 \\ C(1) = -a_{p,1} \\ C(n) = -\sum_{k=1}^{n-1}\left(1 - \dfrac{k}{n}\right) \cdot a_{p,k} \cdot C(n-k) - a_{p,n}, & 1 < n \leqslant p \\ C(n) = -\sum_{k=1}^{n-1}\left(1 - \dfrac{k}{n}\right) \cdot a_{p,k} \cdot C(n-k), & n > p \end{cases} \quad (11)$$

In the situation requiring amplitude cepstrum, it is enough to get an absolute value of the previous value.

3. A Recognition System

The system consists of a microcomputer and some general instruments, and can be divided into three parts: a measuring subsystem, a data acquiring subsystem, and a data processing subsystem. The function of the measuring subsystem is to get the vibration signal of the diagnosed object (it must be accompanied with some fault information). The speed measuring instruments are equipped for measuring the momentary rotational speed so that the effects of variable rotational speeds can be put away. The data acquiring subsystem consists of a NOBUS microcomputer and a 12-bit A/D, D/A data sampling board. The data processing subsystem is realized mainly by a software system. The main programs of the software are divided into five modules, e. g. data sampling, feature extracting, system training, condition monitoring, and abnormal condition analyzing. Except for the data sampling module, which is programed in Macro Assemble language, modules are all programed in Pascal language. The software system can achieve a series of functions from data sampling, data preprocessing to outputting the results.

4. Sample Results

4.1 Introduction

In an experiment, a small electrical motor with three phases is used as the diagnosed object. In it, two faults (e. g. bearing faults and fan unbalancing) are produced artificially.

Because the vibration signal of an electrical motor has a wide frequency range a scheme that monitors two or more frequency bands should be employed: the first one, the sampling frequency of which is 200 Hz, is used to monitor the abnormal condition of the fan; the second one, the sampling frequency of which is 2 kHz, is used to monitor the abnormal condition of the bearing.

4.2 System Training

Before working the system needs to be trained, the goal of which is to get the reference feature of the motor's normal condition and the decision threshold. For the frequency bands 25 groups of training samples are first used to compute the feature vectors (AR cepstrum coefficient vectors) and residue variances. Next the correspondent reference features are computed by means of separate reference feature extraction methods for a complex or simple system. Lastly the decision thresholds are computed.

4.3 Condition Monitoring

In a frequency band the distance measure values between the features of the sampled to be tested and the correspondent reference features are computed, and then compared with the correspondent decision threshold. On the basis of this comparison a decision that whether or not the motor is normal is made. The results of an experiment, in which the diagnosed object is considered as a simple system and the frequency band monitored is 0 to 200 Hz, are shown in Tab. 1. In the table the conditions with numbers 1~10 are normal, and that with 11~13, the fan faults. It can be seen from the table that the results obtained by the monitoring system are the same as the real conditions.

Tab. 1 Condition monitoring experiment (I), $f=200$ Hz, simple systems, fault detection threshold is 18.50

	Test number												
	1	2	3	4	5	6	7	8	9	10	11	12	13
Condition	N	N	N	N	N	N	N	N	N	N	F	F	F
Measure value	13.42	13.98	12.99	10.04	11.23	13.69	9.98	13.2	13.6	8.46	19000	35.5	186.3
Test results	N	N	N	N	N	N	N	N	N	N	F	F	F

The results of another experiment, in which the diagnosed object is considered as a complex system and the frequency band monitored is 200 Hz to 2 kHz, are shown in Tab. 2. In the table the conditions with numbers 1~14 are normal and that with 15~20, the bearing faults. It can be seen from the table that the results obtained by the monitoring system are also the same as the real conditions.

Tab. 2 Condition monitoring experiment (II), $f=2$ kHz, complex systems, fault detection threshold is 42.55

	Test number																			
	1	2	3	4	5	6	7	8	9	10	11	12	13	14	15	16	17	18	19	20
Condition	N	N	N	N	N	N	N	N	N	N	N	N	N	N	F	F	F	F	F	F
Measure value	22.8	13.08	15.40	28.64	14.53	13.4	7.43	37.7	14.9	30.46	8.44	4.57	8.23	9.92	61	73.27	7300	939	2166	276
Test results	N	N	N	N	N	N	N	N	N	N	N	N	N	N	F	F	F	F	F	F

4.4 Abnormal Condition Analyzing

According to the AR coefficients of the abnormal condition signal stored during condition monitoring the AR spectrum analyzing, AR cepstrum analyzing and spectrum peak frequency solving can be performed. During the process the AR spectrum and AR cepstrum can be provided for comparing.

Two spectrum diagrams of normal condition and fan faults of the motor are given, where the analyzing frequency band is $0 \sim 200$ Hz, the rotational speed of the motor is 1384 r/min (e. g. the rotational frequency of the motor is $f_i = n/60 = 23.07$ Hz). Comparing the normal AR spectrum diagram with the abnormal one, it is obvious that there exists an additional spectrum peak of 23 Hz in the spectrum diagram of fan faults, e. g. the vibration component caused by the unbalancing of the spin parts.

Two spectrum diagrams of normal condition and bear faults of the motor are given (Fig. 3 and Fig. 4), where the analyzing frequency band is 200 Hz to 2 kHz, the rotational speed of the motor is 1240 r/min, so the theoretical fault frequency should be the rotational frequency of the motor $f_i = n/60 = 20.67$ Hz, the fault frequency of the bear outer $f = 2.5764 f_i = 53.25$ Hz, the fault frequency of the bear inner $f = 4.4236 f_i = 91.42$ Hz, the fault frequency of the rolling ball $f = 3.5256 f_i = 72.86$ Hz.

Comparing the normal AR spectrum diagram with the abnormal one, it can be found that the 53.25 Hz spectrum component increases obviously, however, the 91.4 Hz spectrum component does not appear. This fact verifies further the results that the faults of bear outer can be found by means of power spectrum analysis, but the faults of bear inner cannot. Comparing the normal AR cepstrum diagram with the abnormal one it can be found that there exist two additional spectral lines in the cepstrum diagram of fault condition, e. g. 6.9 ms (144.92 Hz) and 13.8 ms. (72.46 Hz). 144.92 Hz is oviously the double frequency of 72.46, so it is sure that the faults of rolling balls, which can produce a frequency modulating signal with basic frequency 72.86 Hz, produce it.

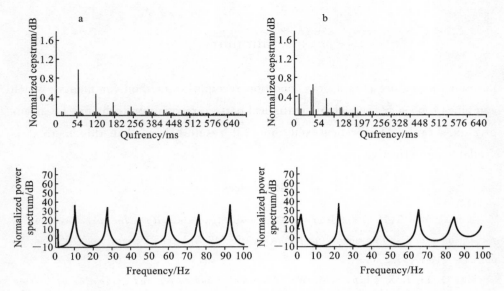

Fig. 3 Abnormal condition analysis with frequency range 0~200 Hz
(a) AR cepstrum and spectrum figure in normal condition;
(b) AR cepstrum and spectrum figure in abnormal condition

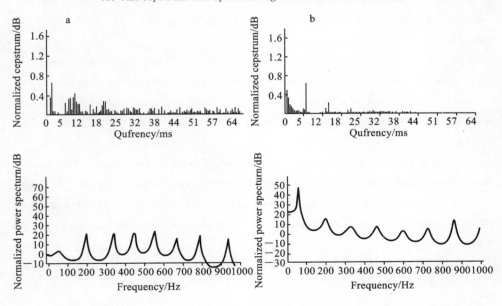

Fig. 4 Abnormal condition analysis with frequency range 200 Hz~2 kHz
(a) AR cepstrum and spectrum figure in normal condition;
(b) AR cepstrum and spectrum in abnormal condition

5 Summary

This paper presents a new condition recognition method for plants, which combines time series model and statistical pattern recognition technique by means of a new style. Theoretical and practical results show that this method is available and effective.

References

1 S. M. Wu, T. H. Tobin Jr and C. M. Chow. *Signature analysis for mechanical systems for dynamic data system monitoring technique*. AS ME J. Mech. Design, Vol. 102, No. 2, pp. 217-221 (1980).

2 C. Q. Li, K. X. Chen, S, M. Wu. *A scheme of monitoring and diagnosing of bearing malfunction by microprocessor*, in: Proceedings of 2rd International Conference on Computer Engineerings Vol. 2, pp. 215-218 (1982).

3 S. O. T. Soeda and T. Nakamizo. *A method of predicting failure or life for stochastic systems by using autoregressive models*. Intemat. J. System Sci. ,Vol. 11, No. 10, pp. 1177-1188 (1980).

4 W. Gersch. *Nearest neighbor rule classification of sta-tionary and nonstationary time series*, in: D. F. Findley, ed Applied Time Series Analysis //, Academic Press, London, 1981.

5 Zheng Xiaojun and Wu Shouxian. *A method for AR spectrum analysis by using both symmetric matrix facotrization method*. J. of Math. Stat. Appl. Prob. Vol. 2, No. 1, pp. 15-24 (1987).

6 Zheng Xiaojun and Wu Shouxian. *A method for AR cepstrum analysis by using directive coefficients conversion method*. J. Math. Stat. Appl Prob. Vol. 1, No. 1, pp. 93-101 (1986).

7 Zheng Xiajun and Wu Shouxian. *Reference feature extracting in classiHcation problems with AR process of random coefficients*. Chinese J. Appl Prob. Stat. Vol. 1, No. 1, pp. 73-75 (1985).

8 H. B. Mann and A. Wald. *On the statistical treatment of linear stochastic difference equation*. Econometrica, Vol. 11, No. 3 and 4, pp. 173-220 (1943).

Authors

Zheng Xiaojun was born in the People's Republic of China in 1962. He received his BSc degree in industrial engineering and MSc degree in measurement and instrument engineering from Changsha Institute of Technology in 1982 and 1985, respectively, and his PhD degree in applied artificial intelligence from Huazhong University of Science and Technology in 1988. In 1985 he was employed as an assistant professor in information engineering with Changsha Institute of Technology. Now he is looking for an opportunity to do research as a postdoctoral research associate. His current research interests include knowledge-based diagnostic reasoning, expert system, knowledge acquisition and learning, and time series analysis. He has written some 40 papers on these subjects.

Yang Shuzi was born in the People's Republic of China in 1933. He graduated from Mechanical Engineering Department, Huazhong University of Science and Technology in 1956. During 1981 to 1982 he was a senior visiting scholar at University of Wisconsin-Madison, U. S. A. In 1956 he joined Huazhong University of Science and Technology, where he is currently a professor of computer application. His research interests include application of signal processing, time series analysis, and artificial intelligence in industrial engineering. He has written 4 books and more than 100 papers on those subjects.

Wu Shouxian was born in the People's Republic of China in 1927, and died in 1987. He graduated from Mechanical Engineering Department, Dalian University of Science and Technology in 1952. In 1987 he was an associate professor of measurement and instruments. His research domain included signal processing, time series analysis, and dynamic measurement. He wrote more than ten papers on those subjects.

(原载 Computers in Industry, Volume 11, Issue 4, February 1989)

Space-domain Feature-based Automated Quantitative Determination of Localized Faults in Wire Ropes

Li Jingsong   Yang Shuzi   Lu Wenxiang   Wang Yangsheng

Abstract　At present, one of the most difficult problems in wire rope inspection is quantitative determination of localized faults, such as a series of broken wires or corrosion pitting. This paper presents a practical method using feature-based analysis for quantitative interpretation of test signals with respect to the types and sizes of defects present in wire rope. An improved Hall-effect sensor used for detecting the defects is first described. Next, the analytical procedure for test signals obtained from the sensor, using space-domain features readily discernible by automated data-processing techniques, is presented. Finally, a prototype instrument for in-service wire rope inspection is introduced. The laboratory and field results show that, using the instrument, not only the types and sizes but also the locations of the defects can be determined quantitatively and automatically, and if the defect's severity exceeds the retirement condition set by safety code the instrument will alarm automatically.

Keywords: Automatic inspection, Hall-effect sensor, Localized faults, Magnetic testing, Pattern recognition, Quantitative inspection, Space-domain features, Wire ropes

Introduction

Wire ropes are used extensively in many industries and life-sustaining situations. In the great majority of cases, failure of a wire rope could lead to serious damage to equipment or even to loss of life. For this reason, the quantitative assessment of wire rope is of particular importance.

In-service wire rope inspection can be divided into two cases. The first case is for loss of metallic cross-sectional area (LMA), a deterioration mode including distributed defects such as external and internal corrosion or abrasion. The second case is for localized faults (LFs), a deterioration mode including, primarily, broken wires and corrosion pitting. In recent years, the art of wire rope inspection has progressed rapidly. Test data from currently available electromagnetic instruments can provide accurate quantitative estimates of safety-related wire rope LMA parameters. However, quantitative estimates of LFs by these instruments are, in many cases, far from satisfactory[1,2].

Quantitative determination of LFs is difficult because the structure of wire rope is complex and the inspection environment is adverse. In addition, the inspection technique is less well developed. To solve the problem, more-accurate test instrumentation and better data-processing techniques for test signals are necessary.

Reliable and accurate defect detection is an essential component of wire rope inspection. Many nondestructive testing (NDT) procedures have been proposed and tried. At present, the magnetic testing method using DC magnetization of the rope in the longitudinal direction and measurement of flux leakage caused by defects appears to be the most suitable. Some electromagnetic flaw detectors that could find LFs in wire rope qualitatively were developed in recent years.[1-6] Traditionally, the detectors used inductive coils as transducers of the leakage flux into electric signals. However, the amplitude and waveform characteristics of output signals determined by using an inductive sensor may be affected by the inspecting speed of the transducer relative to the rope during inspection. This leads to difficulties in quantitative signal interpretation. In contrast to the inductive sensor, a Hall-effect sensor provides a velocity-independent signal.[7-9] Hence, a Hall-effect sensor rather than inductive coils should be used in the detector for quantitative wire rope inspection. High quantitative resolving power of the flaw detector is also important in quantitative wire rope inspection.

However, the separate LFs in the rope cannot be distinguished from the output signals by most of the currently available flaw detectors, especially when many LFs—such as a series of broken wires or corrosion pitting — occur in a small range of rope.

To improve the performance of these detectors and to overcome the problems associated with the quantitative resolving power, reliability, and stability of the test signal, then to determine the LFs quantitatively, an improved electromagnetic flaw detector was designed and built.

Quantitative signal interpretation is another important component of the inspection process. In many industrial applications, wire rope inspections depend heavily on visual interpretation of graphically displayed data by a human operator.[1,2] Thus, an experienced inspector, with a thorough understanding of both wire ropes and NDT procedures, is necessary. Despite all this, the results of interpretation are inaccurate because of the subjectivity, fallibility, and variability of humans. Moreover, the interpretation of data from densely occurring defects, especially when different types of defects are superimposed, is very difficult for a human operator. To make on-line automated quantitative determination of LFs possible, a practical method using feature-based analysis for quantitative interpretation of test signals with respect to the type and size of the defects is presented. This method uses space-domain features readily discernible by automated data-processing techniques. The application of feature-based pattern-recognition techniques led to the identification of various space-domain quantities from which the safety-related wire rope parameters of LFs can be determined.

Improvement of Electromagnetic Flaw Detector

The following work focused on some solutions that should be considered in the improvement of a flaw detector for overcoming the problems associated with the quantitative resolving power and reliability and stability of the test signals.

In the axial plane of the rope, the leakage flux B caused by a defect can be resolved into orthogonal components B_r (perpendicular to the rope axis) and B_t

(parallel to the rope axis) (Fig. 1(a)). The radial component B_r of the magnetic flux density can be measured when the normal direction n of the Hall-effect plates is parallel to the direction of B_r; the tangential component B_t can be measured when the normal direction n is parallel to B_t. It was found experimentally that the actual signal-to-noise ratio (SNR) of test signals during wire rope inspection was smaller when B_t was measured than when B_r was measured because of the influence of stray magnetic flux around the rope (Fig. 1 (b)). Therefore, in the sensor, the Hall-effect plates should be arranged in the mode in which B_r can be accurately measured.

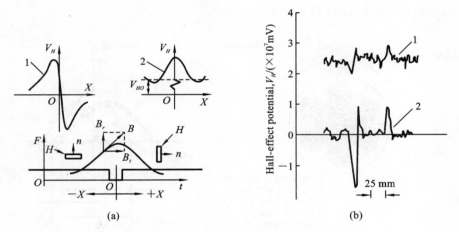

Fig. 1 Solution and measurement of leakage flux B
(a) Standard signal of 1, B_r; 2, B_t
(b) Actual chart of 1, B_r; 2, B_t. V_H=Hall-effect potential; H=Hall-effect plate.

The Hall-effect plates used as detecting elements should be small so that the individual defects distributed along the longitudinal direction of the rope can be distinguished. However, small Hall-effect plates lead limitation of the inspection range of a single plate in the cross-sectional area of the rope. The inspection range of a single plate can be described by its covering angle θ (Fig. 2(a)), which may be defined as an angle in the area of which the leakage flux caused by the detectable LFs can be efficiently measured by the single plate. To simplify the analysis, we took single broken wires as samples of detectable LFs. Fig. 2(b) shows three relation curves between θ of a Hall-effect plate and the SNR of its output signal. The data were obtained from experiments using a single Hall-effect plate to measure the leakage flux caused by three models of single broken wire. By analogy, the same curves can be obtained with models of corrosion pitting. Assuming a fixed value SRN0, we can get a minimum covering angle θ_{min} from the relation curves between θ and SNR. θ_{min} can be considered as the maximum inspection range of a single Hall-effect plate for the detectable LFs in the cross-sectional area of the rope sample. By considering Fig. 2(b), we obtain

$$\theta_{min} = \min\{\theta_1, \theta_2, \theta_3\} \tag{1}$$

As shown in Fig. 2(a), one can see that to ensure inspection of the whole wire rope, several Hall-effect plates are required.

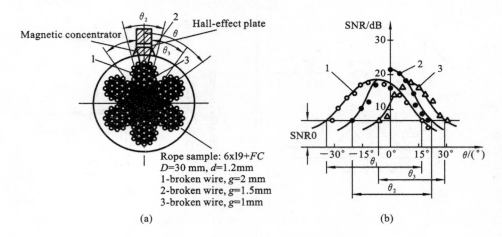

Fig. 2 Detection of covering angle

(a) Cross-sectional area of rope; D=rope diameter, d=wire diameter

(b) Relationship between θ and SNR. g=size of break gap.

If the plates are arranged uniformly around the rope, the minimum total number N_{min} of required plates must satisfy

$$N_{min} \times \theta_{min} > 360° \tag{2}$$

Usually, the value of θ_{min} will vary as the change of diameter of the tested rope. To keep a fixed value of θ_{min}, magnetic concentrators may be used in the sensor (Fig. 2(a)). However, for densely occurring LFs, the flux leakages B_r of neighboring LFs superimpose on each other, causing gradual cancellation of the signal in the magnetic concentrators.[10]

To sidestep the difficulty, several small-size concentrators may be used, one for each Hall-effect plate. Each plate uses an independent output channel; thus multichannel test signals can be obtained (Fig. 3). The Hall-effect plates placed at regular intervals in a circle concentric with the center of the rope function as several units of a defect scanner. In this way, the test signals from separate LFs, distributed not only along the longitudinal direction of the rope but also in the cross-sectional area of the rope, can be distinguished. Hence, the quantitative resolving power of the test signal from the sensor can be improved.

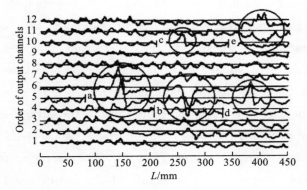

Fig. 3 Strip-chart recording with the flaw detector developed for this study

The periodic signals shown in Fig. 3 are referred to as intrinsic noise, caused by the complex structure of wire rope. These structure-related signals cause serious problems and always make test signals very noisy.[1] It has been found in theory and in practice that the periodicity of the intrinsic noise relates mainly to the number of strands of rope. For a p-strand wire rope, the difference between the wave crest and the wave hollow of intrinsic noise is $180°/p$ in the circumference of the cross-sectional area of the rope. In other words, if two Hall-effect plates are arranged so that the central angle between which is $180°/p$, the phase difference between the intrinsic noises from the two plates is 180 degrees. This is shown clearly by the strip chart of Fig. 3, obtained by using a twelve-channel scanner to inspect a six-strand rope. Based on this, when the Hall-effect plates are arranged uniformly surrounding the rope, and the total number N of plates satisfies the relation

$$N = i \times p \tag{3}$$

where i is an even number and p is the number of strands of the rope, the intrinsic noise can be decreased significantly by a superposition method for output signals from an adjacent output channel. In practical application, the superposition method is carried out by a microcomputer, as mentioned in the next section.

The Hall-effect plates are usually fixed in an annular sensor assembly and protected from the influence of external stray magnetic flux by a shield. Because the sensor is subjected to frequent bouncing and vibration as well as the diameter change of the rope during inspection, differences in inspection clearance between the rope's surface and the annular sensor assembly are inevitable, thus producing a great deal of noise. On the other hand, theoretical calculations revealed that the output voltage of a Hall-effect plate was proportional to the distance S from the plate to the flaw in the rope.[11] Therefore, the sensor assembly must be held constantly and quite close to the surface of wire rope for optimum response to LFs. It was for this purpose that a mechanical centering structure was adopted in the sensor (Fig. 4). The operating principle of the mechanical self-adapting centering structure of the sensor can be described as follows: The Hall-effect plates are fixed in several probes of the sensor, and these probes can slide in the radial direction of the tested rope during inspection. The flexible connecting link allows the change of inspection clearance S to be offset automatically. Thus S remains nearly constant and the reliability and stability of the test signals can be improved.

The output signals from Hall-effect plates are weak and highly susceptible to disturbances from many sources, especially in the harsh conditions of inspecting wire ropes in the field. To further improve the signals, a multichannel analog signal preprocessor was built for the sensor.

On the basis of the above-mentioned results, an electromagnetic flaw detector, consisting of an exciter, a Hall-effect sensor, and an analog signal preprocessor, was developed.

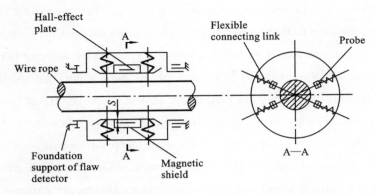

Fig. 4　Mechanical self-adapting centering structure of probes of the improved sensor

Quantitative Interpretation of Test Signals

A typical signal waveform output by the developed flaw detector is shown in Fig. 3. We can see that LFs cause abnormal signal variations, as shown in the parts of Fig. 3 circled and labeled a~e. In 1987, Kalwa and Piekarski[10] reported experimental work concerning the output signals when a Hall-effect sensor was used to detect multiple fractures of wires and corrosion patches in models. The results demonstrated that each kind of damage in wire ropes has its own signal pattern. Further research has shown that not only the amplitude but also the waveshape of the output signal received may vary with the type of defect present. To some degree, the defects can be inferred quantitatively by the operator's interpretation of the recorded signal pattern. However, for on-site field inspection, large quantities of material must be inspected and the inspection data must be analyzed accurately, often at considerable speeds. For this reason, an automated quantitative characterization scheme for the test signals becomes desirable. It was for this purpose that, a computer-aided quantitative signal-processing technique was developed.

To meet the needs for a computer to process output signals from the detector, analog-to-digital conversion is necessary. For the multichannel test signals output by the flaw detector, a multichannel analog-to-digital converter (ADC) must be used; thus the waveform of the sampled data is a multichannel waveform (Fig. 3). In addition, because the leakage flux caused by LFs in the rope is time-invariant, if the sampling for the test signals is in the time domain, the shape of the sampled data will be strongly affected by velocity perturbation during inspection, even though magnitude of the data is not affected because of the velocity-independence of the Hall-effect sensor. This is very unfavorable for quantitative analysis by shape of the test signal curves. Therefore, a sampling controller was built and connected to the flaw detector to control the ADC to sample the output signal in the space domain. The sampling controller is composed of a control wheel and a pulse generator. During wire rope inspection, when the control wheel rotates along the rope through a certain interval of length Δs of the flaw detector relative to the tested rope, the pulse generator produces a pulse that in turn triggers the signal-sampling operation; hence, the space-domain sample data of the test signals can be obtained by analog-to-digital conversion. The spatially equidistant sampled data are not affected by inspecting speed and reflect the real state of the leakage flux caused by LFs in the tested rope.

Before further data analysis, an additional data preprocessing method must be used to reduce other perturbations and to compensate for errors. The errors of sampling interval Δs as a result of the control wheel skidding rather than rolling along the rope can be compensated by a combination of linear and nonlinear filtering,[12] and the random noise from other sources may be reduced by a special noise-reducing method.

After the stages mentioned above, the feature-based technique is applied to the quantitative interpretation of the sampled data. In the pattern-recognition approach, the selection of a key feature vector is very important. In 1988, Wang et al.[12] outlined a pattern-recognition approach using the times of differential value exceeding threshold as the key feature for quantitative determination of broken wires in rope. Further research revealed that features involving information on waveform distortion of the test signals were more sensitive to LFs; therefore, an improved pattern-recognition approach to signal analysis, in which the waveform distortion factor (WDF) feature was selected as the key feature, was presented.

Feature Selection and Pattern-Recognition Approach

Because the test signals from detector are sampled at spatially equidistant intervals, the features selected from the digitized signal are space-domain features. If the sampled data of the test signals are given by $\{x_i(m)\}$, where the subscript i is the order of the signal output channels and m is the serial number of the sampled data, then the *WDF* feature can be described as

$$WDF_j = \sum_{i=1}^{N} \sum_{m=0}^{\infty} D[x_i(m)] \cdot W_j(n_j - m) \qquad (4)$$

where N is the total number of signal output channels, the subscript j indicates the jth waveform distortion area in the test signals caused by LFs in the rope, $D[x_i(m)]$ is the differential value exceeding the variable-threshold function, and n is the first data serial number in the jth waveform distortion area.

When the calculus of forward first difference to sampled data $\{x_i(m)\}$ is expressed as

$$\Delta x_i(m) = x_i(m+1) - x_i(m) \qquad (5)$$

Then $D[x_i(m)]$ can be given by

$$D[x_i(m)] = \begin{cases} \mathrm{int}\left(\frac{|\Delta x_i(m)|}{t}\right), & |\Delta x_i(m)| \geqslant 1 \\ 0, & \text{otherwise} \end{cases} \qquad (6)$$

where int $(\ldots)$ is an integer function and $t(t>0)$ is a preset threshold value.

The function of Eq. (5) is to increase the shape-change information of waveform distortion of the test signals and to decrease the low-frequency noise involved in the sampled data. By use of Eq. (6), the useful information related to the waveform distortion range and magnitude can be gleaned from the sampled data. The preset threshold value t is variable according to the types and specifications of the tested wire rope and can be determined experimentally.

Generally, there are many separated LFs in an in-service wire rope; these cause many waveform distortion ranges in test signals. In Eq. (4), $W(\cdots)$ is defined as an adaptive-window function and is used to dissociate the waveform distortion ranges from the test signals. Assuming L_j is the jth width of the window, the function will have the form

$$W_j(k) = \begin{cases} 1, & -L_j \leqslant k \leqslant 0 \\ 0, & \text{otherwise} \end{cases} \tag{7}$$

After the separated waveform distortion ranges are dissociated, the use of i ranging from 1 to N sums the N-channel signals; this can decrease the influence of the intrinsic noise.

To obtain a more exact solution, the auxiliary information may be supplemented. Analysis of the data revealed the short-range energy (ENG) feature and the sum of the spatial spectrum (SP) feature of the test signals to be the available features for quantitative determination of LFs present in the rope.

The ENG feature may be defined as

$$ENG_j = \sum_{i=1}^{N} \sum_{m=0}^{\infty} x_i^2(m) \cdot W_j(n_j - m) \tag{8}$$

where $W(\cdots)$ is defined as in Eq. (5).

The SP feature can be obtained in the space-frequency domain. The spatial spectrum is a discrete Fourier transform of the sampled data

$$X_{ji}(k) = \sum_{m=0}^{M_j - 1} x_i(m) \cdot \exp\left\{-j \frac{2\pi}{M_j} km\right\} \tag{9}$$

Where M_j is the total number of the data in the jth waveform distortion area. Both the amplitude of individual frequency components and the number of significant components increase or decrease with the difference of the defects. Thus, this feature can be selected with the sum of spatial frequency components

$$SP_j = \sum_{i=1}^{N} \sum_{k=0}^{M_j - 1} |X_{ji}(k)| \tag{10}$$

Grouping WDF, ENG, and SP, we get a feature vector $\mathbf{Y}$. The feature vector of the jth waveform distortion range obtained from the sampled data can be written as

$$Y_j = (WDF_j, ENG_j, SP_j)^T = (y_{j1}, y_{j2}, y_{j3})^T \tag{11}$$

On the basis of the algorithm, a set of characteristic feature vectors will be obtained during wire rope inspection. With this set of feature vectors, the types and sizes of the defects may be determined quantitatively by a classifier. Because the class boundaries in feature space are not well defined (test signals are influenced by many sources of noise and interference), a trainable nonlinear classifier was used for classifying the feature vector.[12]

Before classifying the feature vectors, the nonlinear classifier must be trained by a set of feature vectors from the training samples. Such data are referred to as prototypes and belong to various pattern classes. The pattern classes may be determined on the dominant criterion for flaw detectability as well as the type and size of the localized defects (for example, the number of broken wires). When different types of defects are superimposed, other pattern classes may be adopted.

Program Structure of Quantitative Signal Processing

Computer software for automated quantitative determination of LFs was developed. Basically, the software consists of three parts. With the help of the first part of the software, the inspector inputs original data via the keyboard. The original data consist of the parameters of the rope to be tested, the length of rope to be inspected, and the retirement conditions corresponding to the tested rope. Additionally, the software, upon user request, can retrieve values of earlier calculations that have been stored on disk for use in additional calculations (for instance, the parameters of the classifier). After these, the control section begins to control the *ADC* to work.

The second part of the software is used to analyze the sampled data. The feature-based approach is carried out in several steps. First, a feature vector is computed, then a criterion function is computed, and finally the results that can be used to properly assess the degree of rope degradation, such as the exact number of broken wires densely occurring in a small range of the tested rope, can be determined.

Remarkably, from the definition of the *WDF* feature, one can see that the algorithm to obtain the feature is very simple and requires less computing time. In practice, most calculation of the algorithm can be carried out between the sampling intervals. On the other hand, it has been found that the many kinds of defects present in the rope can be judged directly according to the *WDF* feature.

This makes an on-line quantitative wire rope inspection method using only one feature possible. A subprogram for this purpose was developed, and its flowchart is shown in Fig. 5.

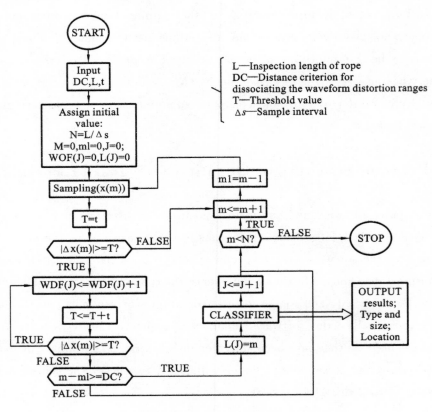

Fig. 5 Flowchart of subprogram

Also, according to the pulse given by the sampling controller, the locations of defects in wire rope can be determined. Furthermore, a subprogram that can control the computer alarm was programmed. Thus when the defect's severity, such as the maximum number of broken wires per unit of the tested rope length, exceeds the retirement condition set by safety code, the computer can alarm automatically.

The last part of the software is used for output of the results. Results of inspection can be expressed in the form of numerical values displayed on the computer screen and printed for a hard copy.

Results and Discussion

On the basis of the techniques introduced above, a prototype in-service inspection instrument was set up by combining the electromagnetic flaw detector with a computer working on the aforementioned software and various other items. Its structural block diagram is shown in Fig. 6.

A variety of tests for evaluating the performance of the above-mentioned inspection instrument have been performed.

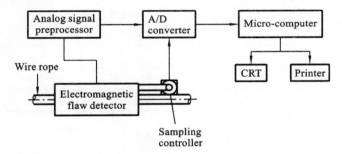

Fig. 6 Schematic of the improved inspection instrument

The laboratory and field results show that, using the instrument, not only the type and size but also the location of a defect can be determined quantitatively and automatically, whether it occurs alone or in a cluster. Tab. 1 shows the important results from this work for the purpose of judging the exact number of broken wires in the rope. The results were obtained by 600 laboratory records using the prototype instrument to inspect a wire rope sample in which three kinds of representative models of densely occurring broken wires were simulated (the first model had a mode of a single broken wire, the second had a mode of three broken wires, and the third had four broken wires). In the work, the sample was a 6×19 —$\phi 30$ mm fiber core rope. During inspection, only one feature, WDF, was used. Statistical analysis of the results revealed that the average accuracy (AA) of judging the number of broken wires was over 80%. In the same work, the location parameters of the models were obtained; the location error calculated was less than ± 25 mm within 100 m length of the rope sample.

The same series of experiments was conducted by using the feature vector described in Eq. (11). The results of the work showed that the accuracy of judgment would increase significantly. In practical applications, the one-feature method can be used for regular on-line safety inspections of wire ropes. The feature vector method can be used for further inspection of the critical area of the rope, such as the area where the most broken wires are concentrated.

Tab. 1 Test Results from Laboratory Experiments

Model	Number of broken wires	Number of times of experiments	Results of judging the number of broken wires in rope sample			
			No error	± 1 broken wire	± 2 broken wires	$> \pm 2$ broken wires
1	1	200	190	10	0	0
2	3	200	164	28	8	0
3	4	200	133	55	9	3
SUM		600	487	93	17	3

Average accuracy of judging the broken wires in rope sample (no error)

$$AA_1 = (487 \div 600) \times 100\% = 81.2\%$$

Average accuracy of judging the broken wires when ± 1 broken wires of judgment error are permissible

$$AA_2 = [(487 + 93) \div 600] \times 100\% = 97\%$$

After successful laboratory trials, the prototype inspection instrument has been in in-service operation in several mine hoists. The results from on-site field applications are heartening, as in the following two examples.

The first example was from a wolfram (tungsten) mine in the south of China, where the instrument was used on a pit rope for four months. Although the tested wire rope was covered with grease and there were many man-made interferences from the work environment, the inspection was performed on-line reliably and stably. The results given by the instrument were much more accurate than the results determined by visual inspection by a human inspector, and the accuracy of judging the number of broken wires by statistical analysis of the results was 85%.

The second example was from another wolfram mine, where a wire rope over 300 m long was inspected by the instrument. As soon as the first inspection was completed, the instrument alarmed and showed that there was a dangerous part in the rope. From the record given by the instrument, the dangerous part of the rope was easily found; the maximum number of broken wires within the rope lay length was 13. According to the China Mine Safety Code, that greatly exceeded the retirement condition, for wire ropes. Despite the fact that the rope was in very heavy use it was removed from service at once. Thus a rope failure was avoided.

Conclusion

It has been demonstrated that the above-mentioned inspection instrument is very useful in quantitative in-service inspection of localized faults in rope. The improved flaw detector can offer multichannel output signal having high reliability and stability and good quantitative resolution. The WDF feature is sensitive to the abnormal signal variation caused by LFs in the rope and can be selected as the key feature of the feature-based pattern-recognition approach to signal analysis in wire rope inspection. The space-domain feature-based method has made on-line automatic quantitative determination of LFs technically and practically feasible. The instrument can remedy the shortcomings of visual wire rope inspection by a human operator, and when the defect's severity reaches or exceeds a certain limit value set by the safety code, the instrument will alarm automatically.

References

1 Weischedel, H. R.. "The Inspection of Wire Ropes In-Service: A Critical Review". Materials Evaluation, Vol. 43, No. 13, Dec. 1985, pp 1592-1605.

2 Weischedel, H. R.. "Quantitative In-Service Inspection of Wire Ropes". Materials Evaluation, Vol. 46, No. 4, Mar 1988,. pp 430-437.

3 Babel, H.. "Destructive and Nondestructive Test Methods for the Delermination of the Life Expectancy of Wire Ropes, Part 11"(in German). Drant, Vol. 30, No. 4,1070 pp 354-359.

4 Taylor, j. L., and N. F. Casey. "The Acoustic Emission of Steel Wire Ropes". Wire Industry, Vol. 52. No. 601, 1984.

5 Stachursky. "Non-Destructive Testing and Inspection of Wire Ropes". International

Seilbahn-Rundschau, No. 6, 1984, pp 330-331.

6 Fuiinaka Y., K. Hanasaki, K. Tsukada. "Electromagnetic Inspection Equipment for Parallel Wire Strand Ropes" (in Japanese). Journal of the Mining and Metallurgical Institute of Japan, Vol. 102, No. 1185, 1986, pp 783-788.

7 Swider, W.. "Magnetic Test Method for Steel Wire Rope". British Journal of Non-Destructive Testing, Vol. 25, No. 2, Mar. 1983, pp 72-74.

8 Kokado J. I., Y. Fujinaka, K. Morita. "Velocity Independent Electromagnetic Inspection of Steel Wire Ropes" (in Japanese). Journal of the Mining and Metallurgical Institute of Japan, Vol. 94, No. 1081, 1978, pp 157-162.

9 Piao C. F., Y. Fujinaka, K. Hanasaki, K. Tsukada. "Electro-magnetic Inspection of Steel Wire Ropes with Hall Element Detector" (in Japanese). Journal of the Mining and Metallurgical Institute of Japan, Vol. 100, No. 1155, 1984, pp 411-415.

10 Kalwa E., and K. Piekarski. "Qualitative and Quantitative Determination of Densely Occurring Defects in Wire Ropes by Magnetic Testing Method". Materials Evaluation, Vol. 46, No* 6, May 1988, pp 767-770.

11 Liu K. M., J. S. Li, W. X. Lu, S. Z. Yang. "On the Electromagnetic NDT Principle for Wire Rope and Detection Signal Preprocessor" (in Chinese). Strength and Environment, No. 3, June 1988, pp 17-22.

12 Wang Y. S., H. M. Shi, S. Z. Yang. "Quantitative Wire Ropes Inspection". NDT International, Vol. 21, No. 5, Oct. 1988, pp 337-340.

Authors

Li Jingsong received his B. Sc. in mechanical engineering from Chengdu University of Science and Technology, Sichuan, PRC, in 1982. After graduation, he was employed as an assistant engineer at the Tool Research Institute of the State Commission of Machinery Industry, PRC. Since 1985, he has been a graduate research student at Huazhong University of Science and Technology, Hubei, PRC. He received his M. Sc. in measurement and instrument engineering in 1988 and is currently working toward a Ph. D. His current research interests include quantitative NDE, digital signal processing, and pattern recognition in industrial engineering.

Yang Shuzi graduated from the Mechanical Engineering Dept., Huazhong University of Science and Technology, Hubei, PRC, in 1956. During 1981 to 1982, he was a senior visiting scholar at the University of Wisconsin-Madison, WI, USA. In 1956, he joined Huazhong University of Science and Technology, where he is currently a professor of computer application.

His research interests include application and development of NDE methods for machinery monitoring and diagnosis, signal processing, and artificial intelligence in industrial engineering. He has written four books and more than 200 papers on these subjects.

Lu Wenxiang graduated from Beijing Specialized School of Mechanical Engineering, Beijing, PRC, in 1955 and from the Mechanical Engineering Dept., Huazhong University of Science and Technology, Hubei, PRC, in 1962. After graduation, he joined Huazhong University of Science and Technology, where he is currently an associate professor of mechanical engineering. His research interests include measurement and instrumentation, machinery monitoring and diagnosis, and signal processing.

Wang Yangsheng received his B. Sc. in mechanical engineering, and M. Sc. and Ph. D. in measurement and instrument engineering in 1982, 1985, and 1989, respectively, from Huazhong University of Science and Technology, Hubei, PRC. During 1988, he was a senior academic visitor at Meta Machines Ltd., Oxford, UK. His current research interests include digital signal processing, pattern recognition, and machine vision.

(原载 Material Evalution, Volume 48, Number 3, March 1990)

基于深知识的多故障两步诊断推理*

郑小军　杨叔子　师汉民

提　要　基于知识的诊断推理研究是当今应用人工智能中比较活跃的一个课题。本文具体地研究了两步多故障诊断推理方面的问题，提出了同时性诊断方式中候选产生的新算法，解决了序贯诊断方式中的最佳测量点选择问题和测量结果的归结问题，最后还提出了组合同时性诊断方式和序贯诊断方式的综合诊断策略。

关键词：基于知识系统，故障诊断，深知识，推理

一、引　言

　　基于深知识的诊断推理就是指基于结构、性能和功能的诊断方法，其早期的工作主要限于单故障诊断[1,2,4,5]，只是近年来才开始研究多故障的诊断问题[3,7]。J. de Kleer[3]运用 ATMS 对多故障的诊断进行了探讨，提出了一系列有关多故障诊断的概念，用于诊断推理机 GDE 和比较一般性的诊断解的求解法。然而，他对候选产生（candidate generation）和测量结果的归结（measurement resolution）等问题并未提出任何具体的算法。在 J. de Kleer 的工作[3]和 Genesereth 的工作[2]的基础上，R. Reiter 以一阶谓词逻辑为基础，提出了一种形式化的第一定律诊断方法，并给出了其具体的诊断算法。然而 Reiter 的方法并未能继承 J. de Kleer 的方法中将诊断算法分解为领域相关和领域无关两部分的思想，且未能给出其序贯诊断形式。

　　本文将提出多故障诊断的一种新的形式化方案。诊断工作仍由旧领域有关

---

*　作者郑小军，工学博士，现从事人工智能应用研究工作，杨叔子、师汉民均为教授、博士生导师，现从事智能应用研究工作。

的冲突识别(conflict recognition)和领域无关的候选产生两部分组成。冲突识别采用 J. de Kleer 的广义约束传播技术,而候选产生则采用我们提出的新方法。

二、问题的形式化

定义 1 一个系统是一个二元组 $S = \langle SP, SD \rangle$,此处①$SP$(Sub-Parts)是一个组成该系统的各子系统的有限集合;②SD(System Description)是描述这些子系统相互之间关系的一阶语句的集合。

定义 2 一个系统的一次测量是一组一阶语句的有限集合,用 OBS 表示。它反映该系统的某些可测量点在特定情况下的值。而系统的性能则是指系统所有可测量点在系统的各种正常工作情况下正常值组的集合,用 BEH 表示。

因此,当系统无故障时,任作一次测量,均应有 $OBS \in BEH$。通常,BEH 无法显式表示,仅当需要时通过推理程序直接对特定值作出预测。

定义 3 一个征兆是指任意一次测量 $OBS \overline{\in} BEH$。

定义 4 系统测量 $OBS \overline{\in} BEH$ 的一个冲突集是一个假设集合 $\{SS_1, SS_2, \cdots, SS_K\} \subseteq SP$,它能使 $SD \cup OBS \cup \{SS_1, SS_2, \cdots, SS_k\}$ 为不相容的,但由于 $OBS \overline{\in} BEH$,至少有一个元素 $SS_i \in \{SS_1, SS_2, \cdots, SS_k\}$ 是故障的。一个冲突是最小的,当且仅当它没有任何子集也是冲突集。

定义 5 设 $C_i (i = 1, 2, \cdots, k)$ 为集合,$\{C_1, C_2, \cdots, C_k\}$ 为任意集合组,若集合 H_e 满足

① $H_c \cap C_i \neq \Phi (1 \leqslant i \leqslant k)$;② $H_c \subseteq \bigcup_{i=1}^{k} C_i$,

则称集合 H_c 为集合组 $\{C_1, C_2, \cdots, C_k\}$ 的 hitting 集。一个 hitting 集被称为最小 hitting 集,当且仅当它没有任何子集也是 hitting 集。

例 1 如图 1 所示的是 Davis[2]、De Kleer[3] 和 Reiter[6] 引用过的一个例子,这是一个有 5 个元素的装置,其中 M_1、M_2、M_3 为乘法器,A_1 和 A_2 为加法器,则该系统的一次测量 OBS 应为 $\{in1(M_1) = 3, in2(M_1) = 2, in1(M_2) = 3, in2(M_2) = 2, in1(M_3) = 3, in2(M_1) = 2, out(A_1) = 10, out(A_2) = 12\}$ 可以注意到,在这次测量中输入值与输出值是不相容的。在该组输入下,通过预测得到的输出值应该是 $out(A_1) = 10, out(A_2) = 12$,因此有 $OBS \overline{\in}$

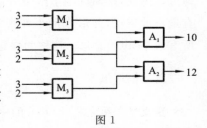

图 1

BEH,则该次测量表明一个征兆存在。

同样,该征兆将引出若干个冲突集合。根据 J. de Kleer 的方法来推断[3],这些冲突集是 $\langle A_1, M_1, M_2 \rangle$ 和 $\langle A_1, A_2, M_1, M_3 \rangle$ 以及它们的所有超集,而最小冲突则恰恰就是 $\langle A_1, M_1, M_2 \rangle$ 和 $\langle A_1, A_2, M_1, M_3 \rangle$ 本身。

三、同时性诊断问题求解

1. 系统的诊断解和最小诊断解

定义 6 系统测量 $OBS \in BEH$ 的一个诊断解是一个假设集合 $\Delta \subseteq SP$,它能使

$$SD \cup OBS \cup \{\overline{C}_1 \wedge \cdots \wedge \overline{C}_k | C_i \in \Delta\} \cup \{C_{k+1} \wedge \cdots \wedge C_n | C_j \in SP - \Delta\}$$

为相容的。一个诊断解被称为最小诊断解,当且仅当它没有任何子集也是诊断解。

事实上,可能的诊断解包括一个由最小诊断解以及其所有可能的超集组成的解空间中所有元素。该解空间可用子集超集网络表示,因此只需求出所有的最小诊断解即可得到所有的可能诊断解。

例 1(续) 对于该例,目前为止诊断解为由 $\{A_1\}, \{M_1\}, \{A_2, M_2\}, \{M_2, M_3\}$ 以及它们的所有超集组成解空间,而最小诊断解则恰恰是 $\{A_1\}, \{M_1\}, \{A_2, M_2\}$ 和 $\{M_2, M_3\}$ 本身。从诊断解任选一个,比如说 $\{A_2, M_2\}$,我们可以看到,如果 $\{A_2, M_2\}$ 产生故障,则该装置的异常输出情况就完全可以解释了。

定理 1 当且仅当 $\Delta \subseteq SP$ 是系统测量 $OBS \in BEH$ 的最小冲突集合的一个最小 hitting 集时,$\Delta \subseteq SP$ 是该系统测量 $OBS \in BEH$ 的一个最小诊断解。

证明略。

由此可见,求出所有的诊断解仅需求出所有最小冲突集的所有最小 hitting 集即可。因此,诊断问题求解可分两步来完成:①求出所有测量 $OBS \in BEH$ 的所有最小冲突集,并形成最小冲突集组。这一步叫冲突识别;②求出该冲突集组的所有最小 hitting 集。这称为候选产生。

2. 冲突识别

定义 7 系统测量 OBS 的一个环境是指当 $OBS \in BEH$ 时,存在一个 $ENV \subseteq SP$ 使得 $SD \cup OBS \cup ENV$ 为相容的,该 ENV 即是测量 OBS 的一个环境。一

个 OBS 的环境称为最小环境,当且仅当它没有任何子集也是 OBS 的环境。

冲突集的识别可以通过以下方式来进行:选择一个环境,测试它是否与目前已作的测量相容,若相容则说明该系统是正常的,若不相容则这个环境是一个冲突。这一过程要求一个推理策略 $C(OBS,ENV)$,它能在已知目前为止的测量 OBS 和一个环境 ENV 的情况下,决定是否其组合 $SD \cup OBS \cup ENV$ 是相容的。函数 $C(OBS,ENV)$ 是与领域有关的,对于数字电路诊断领域而言,已有几种方法可供使用,如 Genesereth 在 DART 系统中使用的疑点 计算方法[1],Davis、de Kleer 等使用的约束传播技术和广义约束传播技术[3,4]。

为了识别最小的不相容环境组(同时,最小冲突集),我们应当从空环境开始搜索,逐步从老环境向它们的超集作宽度优先移动。对每个环境,都应用 $C(OBS,ENV)$ 去决定是否 ENV 是一个冲突,在评价新环境之前,所有老环境的子集都需要先被评价。当找到一个不相容的环境时,该环境成为冲突,且它的超集不再需要评价。

识别所有最小冲突集的算法 MCRA,略。

3. 候选产生

候选产生是与领域无关的,这也就是说,只要提供最小冲突集组,就完全能够不受领域的限制而求出其所有最小诊断解,因而求出整个解空间。以下本文提出一种求任意集合组的所有最小 hitting 集的方法。

定义 8 设任意一集合组 $AR = (AR_1, AR_2, \cdots, AR_n)$,其中 $AR_i \subseteq (S_1, S_2, \cdots, A_m)(i = 1, 2, \cdots, n)$ 为不一定同基的集合,则其模式集为

$$P_S = \{Ps_1, Ps_2, \cdots, Ps_m\}$$

$$Ps_i = \{AR_k / S_i \in AR_k, k = 1, 2, \cdots, n\} \quad (j = 1, 2, \cdots, m)$$

例 2 设 $AR = (AR_1, AR_2, \cdots, AR_7)$,其中 $AR_1 = (S_2, S_4, S_5)$

$AR_2 = (S_1, S_2, S_3)$, $AR_3 = (S_1, S_3, S_5)$, $AR_4 = (S_2, S_4, S_6)$

$AR_5 = (S_2, S_4)$, $AR_6 = (S_2, S_3, S_5)$, $AR_7 = (S_1, S_6)$

因此 $SP = (S_1, S_2, \cdots, S_6)$。则其对应的模式集为

$$Ps_1 = (AR_2, AR_3, AR_7)$$

$$Ps_2 = (AR_1, AR_2, AR_4, AR_5, AR_6), \quad Ps_3 = (AR_2, AR_3, AR_6)$$

$$Ps_4 = (AR_1, AR_4, AR_5), \quad Ps_5 = (AR_1, AR_3, AR_6)$$

$$Ps_6 = (AR_4, AR_7), \quad P_S = (Ps_1, \cdots, Ps_6)$$

定理 2 对于任意集合组 AR,若取一集合 $h \subseteq (S_1, S_2, \cdots, S_m)$,有 $\bigcup_{S_i \in k} Ps_i =$

AR,则 h 为集合组 AR 的一个 hitting 集。

证明略。

事实上,由定理 1 可知,我们关心的是最小 hitting 集,为了求出任意集合组的所有最小 hitting 集,我们仍旧需要在由 SP 的幂集组成的超集子集网格图中由空集向全集方向作宽度优先搜索,当找到一个 hitting 集时,从网格图中删除其所有的超集,并继续搜索下去,直至全部剩余节点均被搜索完毕为止。

对此我们有以下识别所有最小 hitting 集的算法,略。

例 2(续) 由候选产生算法可求得任意集合组 AR 的最小 hitting 集为
$\{S_1, S_2\}, \{S_1, S_3, S_4\}, \{S_3, S_4, S_6\}, \{S_2, S_3, S_6\}, \{S_2, S_5, S_6\}, \{S_1, S_4, S_5\}$

4. 求解所有可能的解

一般诊断问题中的所有可能解的求解可由以下两步来完成:(1) 使用算法 MCRA 从所有测量中识别出所有最小冲突集,并形成最小冲突集组;(2) 使用算法 MHGA 从该最小冲突集组中识别出所有最小 hitting 集,它们即为所有的最小诊断解;而这些最小诊断解的所有超集之和便构成诊断解空间。

四、序贯诊断问题求解

序贯诊断问题的求解由两个步骤组成:一是最佳测量点的选择;二是测量结果的归结。

1. 最佳测量点的选择

在序贯诊断问题中,为了精炼候选假设,需要进行进一步测量。本节提供能选择下一个最好区分候选假设,即能为区分候选假设提供最大信息量的测量方法。

假定每次测量的损耗是等量的,则求最佳测量的目的是以最少的测量次数求出实际的候选假设。在此我们修改了 J. de Kleer 的一步预测最小熵法,这是一种次优的求解方法,它在每步测量时均选择能使系统的信息量最多的那个测量点,即能使系统熵最小的测量点。由于是序贯诊断,公式仅考虑一步测量的结果。

从前面的叙述,我们可以得知,进行一次测量必将减小候选空间。设测量 x_i 时,有:

(1) $x_i = v_{ik}$,则留下的候选空间为 R_{ik};

(2) $x_i \neq v_{ik}$,则必须从 R_{ik} 中删去的空间为 S_{ik};

(3) 对 x_i 不预测任何固定值的候选假设空间为 v_i。

这三个集合有以下关系:
$$R_{ik} = S_{ik} + v_i$$

对 x_i 实施测量,可能有 $m+1$ 种结果,即
$$x_i = v_{ij} \quad (j=1,2,\cdots,m) \quad \text{和} \quad x_i \neq v_{ij} \quad (j=1,2,\cdots,m)$$

即 x_i 不等于预测值中的任何一个,因此测量 x_i 后希望的熵为
$$H_S(x_i) = \sum_{k=1}^{m} P(x_i = v_{ik}) \cdot H(s \mid x_i = v_{ik}) + P(v_i) H_{v_i}$$

式中
$$H(s \mid x_i = v_{ik}) = -\sum_{C_l \in S_{ik}} P(C_l \mid x_i = v_{ik}) \log P(C_l \mid x_i = v_{ik})$$

如果 x_i 测量有一个未预料的值时的期望熵(即除 v_i 以外所有的候选假设均被删除)。考虑一次测量 $x_i = v_{ik}$,则一个候选假设 C_l 为实际诊断解的条件概率为[7]

$$P(C_l \mid x_i = v_{ik}) = \begin{cases} 0, & \text{如果 } C_l \in R_{ik} \\ \dfrac{P(C_l)}{P(x_i = v_{ik})}, & \text{如果 } C_l \in S_{ik} \\ \dfrac{P(C_l)}{P(v_i)}, & \text{如果 } C_l \in v_i \end{cases}$$

因此,$H_S(x_i) = -\sum_{k=1}^{m} \sum_{C_l \in S_{ik}} P(C_l) \log \dfrac{P(C_l)}{P(x_i = v_{ik})} - \sum_{C_l \in v_i} P(C_l) \log \dfrac{P(C_l)}{P(v_i)}$

所以最佳测量点即是使 $H_S(x_i)$ 最小的 x_i。

2. 测置结果的归结

在序贯诊断中,每次测量后需要对以前的结果进行修正,以便将历次测量与该次测量结合在一起来考虑,求得新的诊断解,这个过程我们称之为测量结果的归结。

继续考虑任意集合组 AR 的 hitting 集求解。当任意集合组增加新的元素 $AR_1', AR_2', \cdots, AR_m'$ 时,对 hitting 集的求解有以下定理。

定理3 设有两个任意集合组 AR、AR'。

$AR = (AR_1, AR_2, \cdots, AR_n)$ 且其最小 hitting 集组为 $H = (h_1, h_2, \cdots, h_k)$

$AR' = (AR_1', AR_2', \cdots, AR_m')$ 且其最小 hitting 集组为 $H' = (h_1', h_2', \cdots, h_l')$

则任意集合组 $AR \cup AR' = (AR_1, AR_2, \cdots, AR_n, AR_1', AR_2', \cdots, AR_m')$ 的最小

hitting 集组为由 $h''_i = h_j \bigcup h'_p (j = 1, 2, \cdots, k; p = 1, 2, \cdots, l; i = 1, 2, \cdots, k+l)$ 组成的集合组的最小集合。

证明略。

例 2(续) 设现进行一次测量，得到一组附加任意集合 $AR_8 = (S_3, S_4)$，$AR_9 = (S_1)$，求其新老任意集合的最小 hitting 集。

附加集合 AR_8, AR_9 的最小 hitting 集为 $\{S_1, S_5\}, \{S_1, S_4\}$，因此其新老任意集合的最小 hitting 集应从以下集合组中产生：

$$\{S_1, S_2, S_3\}, \quad \{S_1, S_2, S_3, S_6\}, \quad \{S_1, S_2, S_3, S_5, S_6\}$$
$$\{S_1, S_3, S_4\}, \quad \{S_1, S_3, S_4, S_5\}, \quad \{S_1, S_3, S_4, S_6\}$$
$$\{S_1, S_2, S_4\}, \quad \{S_1, S_2, S_4, S_6\}, \quad \{S_1, S_2, S_4, S_5, S_6\}$$
$$\{S_1, S_3, S_4\}, \quad \{S_1, S_4, S_5\}, \quad \{S_1, S_3, S_4, S_6\}$$

通过比较可得出其最小 hitting 集为

$$\{S_1, S_2, S_3\}, \quad \{S_1, S_2, S_4\}, \quad \{S_1, S_3, S_4\}, \quad \{S_1, S_4, S_5\}$$

五、综合诊断策略

在实际诊断中，常常是联合使用以上两种诊断方式的。事实上这是因为同时性诊断方式要求一次测量所有的可能测量点，而序贯诊断方式在诊断的初始阶段需要相当多的计算资源，因此单独使用任何一种方式来进行诊断问题求解都是不现实的。唯一可行的办法是采用这两种方式相结合的综合诊断策略，即将诊断过程分为以下两个阶段。①封闭求解阶段：从所有可能的测点中选约一半最易测试且最易出现征兆的测点作为实际测量点，根据测量结果求出其所有的最小诊断解。②开放求解阶段：利用一步最小熵法选择下一个最佳测量点来进行测试，并将其结果与原最小诊断解进行测量结果归结；然后判断系统的熵是否小于某个门限，如是则退出，以此时的最小诊断解作为最终的最小诊断解；否则继续进行开放求解阶段。

在基于深知识的诊断推理研究中，本文着重研究了两步诊断推理方面的某些问题，给出了新的诊断问题形式化方案，提出了同时性诊断方式中候选产生的新算法，解决了序贯诊断方法中的最佳测量点选择问题和测量结果的归结问题，最后还提出了组合同时性诊断方式和序贯诊断方式的综合诊断策略。

参 考 文 献

[1] R. Davis et al. *Diagnosis based on description of structure and function*. Proc. of

AAAI—82,1982.

[2] R. Davis. *Diagnostic reasoning based structure and behavior*. Artificial Intelligence, 24 (1984).

[3] J. De Kleer and B. C. Williams. *Diagnosing multiple faults*. Artificial Intelligence, 32 (1987).

[4] M. R. Genesereth. *Diagnosis using hierarchical design models*. Proc. of AAAI-82, 1982.

[5] M. R. Genesereth. *The use of design descriptions in automated diagnosis*. Standford Heuristic Programming Project Memo HPP-81-20, 1984.

[6] R. Reiter. *A theory of diagnosis from first principles*. Artificial Intelligence, 32 (1987).

[7] 郑小军. 基于知识的诊断推理——理论与系统, 华中理工大学博士论文, 1988, 7.

Deep Knowledge-based Two-step Diagnostic Reasoning for Multiple Faults

Zheng Xiaojun Yang Shuzi Shi Hanmin

Abstract Some concrete issues on the deep knowledge-based two-step diagnostic reasoning are studied and a new algorithm for candidate generation in the concurrent diagnostic method is presented. In the sequential diagnostic method the problem of how to select the best measurement and resolute the measuring results are solved. A compound diagnostic strategy integrating the concurrent diagnostic method with the sequential one is presented.

Keywords: Knowledge-based system, Fault diagnosis, Deep knowledge, Reasoning

(原载《计算机学报》1991 年第 3 期)

机床切削系统的强迫再生颤振与极限环

吴 雅　杨叔子　柯石求　李维国

提　要　本文首先对强迫再生颤振的机理进行了研究,指出这是一种由强迫激励激起的机床再生颤振失稳模态的共振,提出了强迫再生颤振的判别方法,并用于第二汽车制造厂 MX-4 车床振动类型的判别。本文还提出了对稳态颤振下的机床切削系统建立门限自回归模型的原理,揭示了极限环与稳态颤振之间的关系,提出稳态颤振下的机床切削系统是一种轨道稳定的非线性系统,并用于 MX-4 车床颤振的研究。

关键词:机床切削系统,强迫再生颤振,非线性系统,极限环,门限自回归模型

1　强迫再生颤振的机理

　　机床切削过程中的强迫再生颤振是一种混合型的颤振,是由强迫振动频率与机床切削系统的再生型颤振频率相接近时所产生的一种极其剧烈的振动。特别是对于机床空运转振动所激起的强迫再生颤振,机床上各运动部件(如齿轮、轴承等)在运转中的不平衡、破损、存在缺陷等,都会形成一种强迫激励作用到机床上,而由于机床的运动部件较多,这就使强迫激励的频率分布较密,从而很有可能导致机床切削系统的再生颤振频率与某一强迫激励的频率相接近,产生空运转强迫激励下的再生颤振。然而,现有的理论对这种类型的颤振研究不多,其原因在于目前对机床颤振的研究大多是在实验室人为地产生某种单一类型的颤振。星铁太郎首次发现这种混合型颤振并进行了阐述[1]。当时的研究是基于再生颤振下切削过程的频率特性的实部为常数,即在复平面上,切削过程的动柔度曲线是一条平行于虚轴的直线。对切削过程动态特性的研究表明[2],切削过程同时具有刚度和阻尼的特性,在复平面上,其动柔度曲线为具有某一斜率的直线,则由此导出

的机床再生颤振频率将高于以往得出的再生颤振频率。应该说,近年来的研究更符合实际。下面,对强迫再生颤振的机理进行研究。

1.1 强迫再生颤振的数学模型

在再生效应下,如果机床切削系统可化为图 1 所示形式[3],且受到频率为 ω_0、力幅为 P_0 的谐波激励,则其运动微分方程为

$$m\ddot{x}(t) + c\dot{x}(t) + kx(t) = -K[x(t) - x(t-\tau)] - C[\dot{x}(t) - \dot{x}(t-\tau)] + P_0\cos\omega_0 t \tag{1}$$

式中:m,c,k 分别是机床结构在再生效应下失稳模态的模态质量、模态阻尼、模态刚度;K、C 分别是等效切削刚度和等效切削阻尼;τ 是相邻两次切削振动波纹的滞后时间;$-K[x(t)-x(t-\tau)]$ 和 $-C[\dot{x}(t)-\dot{x}(t-\tau)]$ 分别是切削厚度和切入率的再生效应引起的动态切削力[3,2]。稳态颤振时,可只考虑 $x(t)$ 中频率为 ω 的基波[4],令 $x(t) = a\cos\omega t$,$\beta = \omega\tau$,则 $x(t-\tau) = a\cos(\omega t - \beta)$。将黏性阻尼作为结构阻尼处理,$c = h/\omega, C = H/\omega$,则可将式(1)写为

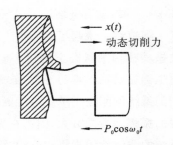

图 1 强迫再生颤振的物理模型

$$m\ddot{x}(t) + (\gamma/\omega)\dot{x}(t) + k_s(t) = P_0\cos\omega_0 t \tag{2}$$

式中,
$$\begin{cases} \gamma = h + K\sin\beta + H(1-\cos\beta) \\ k_s = k - H\sin\beta + K(1-\cos\beta) \end{cases} \tag{3}$$

式(2)即为机床切削系统(由机床结构和切削过程两者共同组成)强迫再生颤振的数学模型,γ、k_s 分别为系统的等效结构阻尼系数和等效刚度。系统的固有频率即再生颤振频率为

$$\omega_s = \sqrt{k_s/m} \tag{4}$$

作为式(2)的特例,当等号右边等于零时,即为单纯再生颤振的数学模型,且当 $\gamma = 0$ 时,机床切削系统失稳,产生再生颤振。

根据结构阻尼系统的共振条件[5],当谐波激励频率 ω_0 等于系统固有频率 ω_n 时,系统产生共振,由于式(2)表示再生颤振,故本文称为强迫再生颤振,其振幅 $X = P_0/\gamma$。而在非切削状态下,机床结构的失稳模态在 $P_0\cos\omega_0 t$ 激励下的运动微分方程为

$$m\ddot{x}_0(t) + (h/\omega)\dot{x}(t) + kx_0(t) = P_0\cos\omega_0 t$$

失稳模态固有频率 $\omega'_n = \sqrt{k/m}$。当 $\omega_0 = \omega'_n$ 时，产生共振，共振振幅 $X_0 = P_0/h$。从而可得强迫再生颤振对非切削状态下的共振的放大比

$$\eta = X/X_0 = h/\gamma \tag{5}$$

分析式(5)、式(3)可见，强迫再生颤振的放大比由机床再生颤振失稳模态的阻尼 h、等效切削刚度 K、切削阻尼 H 和相位差 β 所共同决定，而由于恒有 $1-\cos\beta \geqslant 0$，表示切削正阻尼的作用总是减小 η。而 $\pi < \beta < 2\pi$ 时对应于颤振发生[4]，此时有 $K\sin\beta < 0$，表示有再生颤振时，切削刚度的作用是增大 η。不仅如此，即使是在无再生颤振情况下，机床切削系统的阻尼 γ 虽然尚未减小至零，但已经存在 $\gamma < h$ 时，就有 $\eta > 1$，产生强迫再生颤振。从物理意义上来看，这种颤振相当于某一强迫激励激起了机床再生颤振失稳模态的共振，而对机床非切削状态下的强迫振动在切削状态下进行了放大。显然，机床切削系统的阻尼 γ 愈小、愈接近于再生颤振的临界点($\gamma = 0$)，放大比 η 愈大，强迫再生颤振愈强烈；当 $\gamma = 0$ 时，机床切削系统刚刚处于单纯再生颤振的临界点，就有 $\eta = \infty$。当然，实际强迫再生颤振的振幅不会无穷大，因为在振幅增大后，小振幅假设下的线性理论将不再成立，系统的非线性因素将起作用，而将振幅稳定在某一程度上。

1.2 强迫再生颤振的复平面表示

在切削加工状态下，机床切削系统是由机床结构与切削过程构成的闭环系统。可以证明，机床切削系统的特征方程为：$1 - G(s)/H(s) = 0$，其中，$G(s)$ 是机床结构失稳模态的传递函数；$1/H(s)$ 是切削过程的传递函数，以 $s = j\omega$ 代入，可以证明[2]

$$1/G(\omega) = k - m\omega^2 + j\omega c \tag{6}$$
$$1/H(\omega) = (1-\cos\beta)[K + \omega C\mathrm{ctg}(\beta/2) + j(\omega C - K\mathrm{ctg}(\beta/2))] \tag{7}$$

在复平面上，$H(\omega)$ 是一条斜率为 $K/(\omega c)$ 的直线，其与实轴、虚轴的交点分别为 $(-1/(2K), 0)$ 与 $(0, 1/(2\omega c))$，如图 2 所示。若只考虑单纯再生颤振，当 $H(\omega)$ 与 $G(\omega)$ 不相交时，$H(\omega)$ 如图 2 中实线所示，机床切削系统不发生再生颤振；当 $H(\omega)$ 与 $G(\omega)$ 相切时，如图 2 中虚线 $H(\omega)$ 所示，机床切削系统处于再生颤振的临界点 A，再生颤振频率为 ω。此时，在动态切削力 F 的作用下，机床结构的振动为 $F \cdot \overrightarrow{OA}$。然而，当强迫激励频率 $\omega_0 = \omega_n$ 时，振动 $F \cdot \overrightarrow{OA}$ 上就叠加一个同频率的振

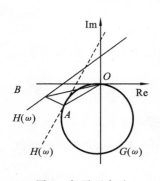

图 2 复平面表示

动 x_0。当 $x_0 = F \cdot \overrightarrow{AB}$ 时，此两振动的叠加（向量相加）就相当于将本来不相交的 $H(\omega)$ 和 $G(\omega)$（对应于无再生颤振）在再生颤振频率 ω_n 处加以连接，从而激起了强迫再生颤振。强迫再生颤振的振幅为 $F \cdot \overrightarrow{OA} + F \cdot \overrightarrow{AB} = F \cdot \overrightarrow{OB}$，它引起强迫振动的放大比为

$$\eta = |F \cdot \overrightarrow{OB}|/|x_0| = |\overrightarrow{OB}|/|\overrightarrow{AB}| \tag{8}$$

式(8)成立的条件为

$$|x_0|/F = |\overrightarrow{AB}| \tag{9}$$

对于给定的机床结构和切削过程，$G(\omega)$ 和 $H(\omega)$ 是确定的，点 A、B 由频率 $\omega_n = \omega_0$ 确定，则向量 $\overrightarrow{AB}$ 是唯一确定的，这表明，当强迫激励频率 ω_0 接近于机床切削系统再生颤振频率 ω_n 时，不论强迫振动 x_0 的大小如何，只要它存在，它就决定了动态切削力 F 的幅值和相位，使得 F 与 x_0 的幅值比和相位差能自动地调节而满足式(9)，形成强迫再生颤振。显然由式(8)可见，$H(\omega)$ 与 $G(\omega)$ 相距愈近（$\overrightarrow{AB}$ 愈小），η 愈大，极限情况是 $|\overrightarrow{AB}| = 0$，则 $\eta = \infty$，这一结果与前述分析是一致的，即：机床切削系统自身的稳定性愈低，表现为式(3)中 γ 值愈小，图 2 中 $G(\omega)$ 与 $H(\omega)$ 相距愈近，强迫再生颤振就愈剧烈，它将机床非切削状态下的强迫振动放大得愈大。

需要指出，上述分析是在颤振频率 ω_n 处进行的，可以证明，这是机床切削系统对于强迫振动放大效应的特例，上述分析还适用于机床一般类型的振动而不仅限于颤振。

2 强迫再生颤振的判别

由上分析可知，对机床切削系统来说，强迫再生颤振较单纯再生颤振更危险，它可以在尚未构成再生颤振条件时就导致严重的后果。当然，对这种颤振的防治措施也是较简单的，只需移动强迫振源的频率或拆除振源即可。因此，问题的关键在于对强迫再生颤振的判别。由于强迫再生颤振兼有强迫振动与再生自激振动的特点，因此，很难将其区分出来。星铁太郎曾提出过简单的判别方法[1]，根据我们对第二汽车制造厂 MX-4 曲轴连杆颈车床颤振类型的判别实践，对强迫再生颤振可按以下步骤判别。

① 按现有方法判别是否为再生颤振，因为强迫再生颤振是由再生效应而放大了的振动。

② 在机床空运转状态和切削状态下测取机床振动信号，进行谱分析，检查空

运转状态下强迫振动的频率是否与切削状态下的振动频率相接近,若两者接近,就有可能构成强迫再生颤振。

③ 对整个切削过程的振动信号(最好是振动加速度信号)进行谱阵分析,强迫再生颤振频率不随加工时间的推移而改变,而单纯再生颤振频率可能随加工时间的推移而改变。这一判别与目前许多研究结果是一致的。因为在单纯再生颤振的建立、发展过程中,振动能量会发生转移,大多数情况是颤振频率由高频向低频移动,而由于强迫激励的频率总是不变的,所以,强迫再生颤振的频率不会随加工时间的推移而改变,甚至可能"吸引"其他频率成分的振动能量进一步加入至强迫再生颤振。采用振动加速度信号的优点在于,剧烈颤振的位移信号谱图一般只有一个主峰,其谱阵很难反映颤振频率的移动,而加速度信号谱图对高频振动描述清楚,其谱阵能够清楚地反映颤振频率的移动情况。

以对 MX-4 车床的判别为例。MX-4 车床是第二汽车制造厂引进的关键设备之一,在车削状态下,该车床发生了剧烈的振动,致使连杆颈端面跳动值高达 1.8 mm,径向跳动值高达 0.4 mm,切削噪声高达 105 dB(A)以上。该车床为宽刀刃、横进给加工方式,加工出的连杆颈外圆表面振纹清晰可见,符合再生颤振原理,根据振纹算得振动频率为 527.5~563.0 Hz,初步判断为再生颤振。在机床空运转状态下测取刀架上的振动加速度信号,其谱图如图 3 所示。又在切削状态下测取振动加速度信号,经 8 次平均后的平均谱图如图 4 所示。由图 3 可见,机床具有一个频率为 532.5 Hz 左右的强迫振源(此谱图的频率分辨力为 2.5 Hz)。其引起的机床空运转振动很小,似乎可忽略不计。但由图 4 可见,在切削状态下振动加速度信号的第一主峰频率为 530 Hz(此谱图的频率分辨力为 5 Hz),经定标和平均处理后,算得在 530 Hz 时,切削状态的振动加速度较空运转振动加速度放大了约 196 倍,即 $\eta \approx 196$。由于放大比极大,初步判断为强迫再生颤振。最后对一个工作循环的振动加速度进行谱阵处理,图 5 示出了由 60 张谱图构成的谱阵。由图 5 可见,在整个工作循环中,频率为 530 Hz 的谱峰始终存在且位置不变,而频率为 705 Hz 的谱峰逐渐向 530 Hz 移动,特别是在第 40 张谱图后,移动速度加快,至第

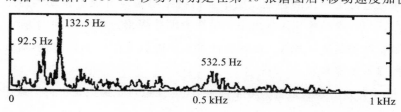

图 3　MX-4 车床空运转振动加速度谱

48张谱图加工结束时,已移至 600 Hz 频率处。因此,最后判断 MX-4 车床为强迫再生颤振。

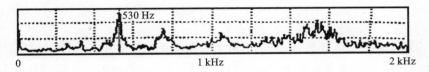

图 4　MX-4 车床切削状态振动加速度平均谱

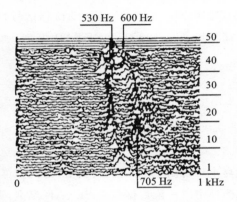

图 5　MX-4 车床振动加速度谱阵

3　机床切削系统的极限环

前面已述,当大振幅时,线性理论不再成立,应将机床切削系统作为非线性系统处理。稳态颤振下的机床切削系统是一个稳定的非线性系统,且这种稳定一般是轨道稳定的。轨道稳定性通常表现为非线性系统的极限环。极限环代表了非线性系统的一种稳定状态,在这一状态下,非线性系统保持周期性的平衡运动。因此,可以采用时序分析中的 SETAR 模型及其相应的极限环来描述机床切削系统的稳态颤振,并根据极限环的周期和大小来描述稳态颤振的频率和振幅。SETAR 模型的一般形式为

$$x_l = \varphi_0^{(j)} + \sum_{i=1}^{n} \varphi_i^{(j)} x_{t-i} + a_t^{(j)} \quad (r_{j-1} < x_{t-d} \leqslant r_j, j=1,2,\cdots,l) \tag{10}$$

记为 SETAR $(l;d;n_1,n_2,\cdots,n_l)$,其意义参见文献[5]。一般常采用两个门限段($l=2$)的模型,即 SETAR $(2;d;n_1,n_2)$ 其形式为

$$x_t = \begin{cases} \varphi_0^{(1)} + \sum_{i=1}^{n_1} \varphi_i^{(1)} x_{t-i} + a_t^{(1)} & (x_{t-d} \leqslant r) \\ \varphi_0^{(2)} + \sum_{i=1}^{n_2} \varphi_i^{(2)} x_{t-i} + a_t^{(2)} & (x_{t-d} > r) \end{cases} \quad (11)$$

特别是对于稳态颤振,机床切削系统的振动位移信号是平稳随机过程,采用 $t=2$ 的 SETAR 模型是适宜的。在式(11)中 $\{x_t\}$ 是稳态颤振时的振动位移采样时间序列。

测取 MX-4 车床刀架进刀抗力方向上的振动位移信号,经离散采样后得振动位移时序 $\{x_t\}$,按数据长度 $N=128$ 建立 SETAR 模型,典型的模型为 SETAR $(2;2,4,3)$,

$x_t =$
$$\begin{cases} -4.792 + 0.225 x_{t-1} + 0.037 x_{t-2} + 0.308 x_{t-3} + 0.826 x_{t-4} + a_t^{(1)} & (x_{t-2} \leqslant 122.62) \\ 18.150 + 0.248 x_{t-1} - 0.997 x_{t-2} + 0.275 x_{t-3} + a_t^{(2)} & (x_{t-2} > 122.62) \end{cases}$$

根据模型进行外推计算不难得出 $\{x_t\}$ 的预测波形,由模型算得的极限环如图6(a)所示。取另一次加工循环中的一段数据建立 SETAR 模型,算得极限环如图6(b)所示。分析所得的极限环,其具有以下几个特点。

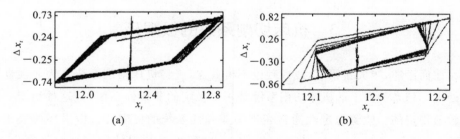

图 6 极限环

① 极限环的周期(频率)描述了机床切削系统稳态颤振的主频率。由图 6 所示的极限环以及所得的所有极限环,其周期为 $4/f_s$,f_s 为 $\{x_t\}$ 的采样频率。根据 f_s 算得频率为 530 Hz。将该段信号送至 CF-940 信号处理机上进行谱分析,所得的 16 次平均谱如图 7 所示,可见振动位移主峰频率为 530 Hz,与极限环的频率以及前述由振纹数算得的频率完全一致,这一频率就是 MX-4 车床的强迫再生颤振频率。

② 极限环的大小描述了机床切削系统稳态颤振的振幅和振动速度。极限环的横坐标是振动位移,纵坐标是振动速度,因而,极限环的最大、最小横坐标之差

即为稳态颤振振幅的双峰值,极限环的最大、最小纵坐标之差即为稳态颤振速度的双峰值。例如,图 6(a)所示极限环的横坐标排列为{12.1312,12.7588,12.4960,11.8552},纵坐标的排列为{0.27598,0.62759,−0.26288,−0.64071},则振幅双峰值为 0.9036,振动速度双峰值为 1.2683,经定标处理后即可算得振幅和振动速度的实际值。

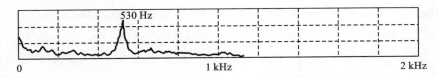

图 7　MX-4 车床切削状态振动位移平均谱

③ $\{x_t\}$ 的预测波形描述了所测部件处的颤振时域波形。例如,本文测取 MX-4 刀架处的振动位移信号,则 $\{x_t\}$ 的预测波形即为刀架处的颤振波形。如果测取刀架与工件之间在敏感方向上的相对位移信号,则 $\{x_t\}$ 的预测波形还可以代表工件表面的振痕。当然,这种描述(包括上述②)只对稳态颤振中的周期振动而言,不包括稳态颤振中的随机成分。

④ 极限环描述了机床切削系统的稳定性质。图 6 所示的极限环表示了两种不同的逼近情况:图 6(a)所示的是初始位置在环内,从环内逼近极限环;图 6(b)所示的是初始位置在环外,从环外逼近极限环的情况。这种现象充分说明稳态颤振下的机床切削系统是一种轨道稳定的非线性系统。

第二汽车制造厂闻耀祖、蒋其昂高工,徐善祥、张启林工程师参加了本文试验与研究工作,特致以衷心的感谢。

参 考 文 献

[1] 星铁太郎. 机械加工颤振现象——解析与对策. 工业调查会,1977.

[2] 杨绪光,陈继武. 切削过程动力特性和机床工作稳定性分析. 机床振动与噪声,1984,1:17～29.

[3] Tobias S A. *Machine Tool Vibration*. London BLACKIE Glasgow,1965.

[4] Shi Hanmin, Tobias S A. *Theory of Finite Amplitude Machine Tool Instability*. J. of MTDR,1984,24(1):45～69.

[5] 吴雅,杨叔子,师汉民. 门限自回归模型与非线性系统的极限环. 华中理工大学学报,1988,16(3):35～41.

Forced Regenerative Chatter and Limit Cycle of Machine Tool Cutting System

Wu Ya   Yang Shuzi   Ke Shiqiu   Li Weiguo

Abstract The mechanism of forced regenerative chatter is studied. It is pointed out that the chatter is a kind of resonance of regenerative instable mode caused by a forced excitation. A method for judging the chatter is developed and it has been used in the judgement of the vibrating mode of a type MX-4 lathe. A threshold auto-regressive model (SETAR) in time series analysis is worked out and applied to the representation of the steady-state chatter in machine tools. The relationship between a limit cycle and the steady chatter is given. It is suggested that the machine tool cutting system under steady chatter is a nonlinear one in the sense of orbital stability. This has also been applied to the investigation of chatter in the type MX-4 lathe.

Keywords: Machine tool cutting system, Forced regenerative chatter, Nonlinear system, Limit cycle, Threshold auto-regressive model

机械设备诊断学的再探讨*

杨叔子　丁　洪　史铁林　钟毓宁

提　要　本文进一步探讨了机械设备诊断学的基本体系与基本概念。从系统论的观点出发，研究了机械设备的系统特性与系统分类；深化了故障、征兆、特征信号的概念；提出了故障传播过程的实质与故障的特性（层次性、相关性、延时性、不确定性）；论述了"超层诊断"。文中从知识推理这一角度出发，进一步讨论了机械设备诊断学的目的、任务、诊断内容与诊断过程。

关键词：机械设备诊断，设备诊断学，系统，故障特性，知识推理，模式识别，诊断方法

随着现代机械设备与结构的大量应用，它们的工况监视与故障诊断技术也在迅猛发展，成为现代科技热点之一。考察这一技术的现状，显然可见：这一技术实质上还是诊断经验的堆积与有关学科方法的运用以及这两者的混合，缺乏自身的学科体系，从而阻碍了这一技术更为迅速、深入、可靠、有效的发展。为了摆脱这一现状，就必须建立这一技术自身的、以诊断实践为基础的、与多种学科的相关内容相互融合的、并且独立发展的学科体系，以指导机械设备（包括机械结构，下同）的工况监视与故障诊断的实际工作。

本专辑的若干论文正是为此而努力的。本文是文献[1]的继续与发展，主要是研究了机械设备的系统特性，深化了特征信号、征兆、故障的概念，提出了故障传播过程的实质与故障的特性；同时，还从知识角度，进一步讨论了文献[1]中所研究的若干问题。至于从宏观上来研究诊断策略，本专辑文献[2]中特别作了初步探索。

---

* 国家自然科学基金资助项目。

1 关于机械设备诊断的若干概念

为了研究机械设备的诊断问题,必须深刻了解与掌握机械设备及其故障的特点,必须明确与统一有关的基本概念与定义。

从系统论的观点考虑,机械设备也是一个系统,也是由元素按一定的规律聚合而成的,可以认为是由"元素"加上元素之间的"联系"而构成的。当然,系统的"元素"可以是子系统,子系统的"元素"还可以是更深层次的子系统,如此类推,直至元素是物理元件为止。显然,系统是有层次的。

系统的基本性质(或状态)取决于元素的性质(或状态)与联系的性质(或状态),而系统的行为(输出)则取决于系统的基本性质与系统同外界的关系(输入、客观环境的作用)。现将系统中的"元素"与"元素间的联系"这一总体称为系统的"构造",而系统的行为中人们所需的(即设计中所要求实现的)部分称为系统的"功能"。据此,虽然工程中的机械设备种类繁多,但基于其"构造"与"功能",可分为三类——

① 简单系统:在构造上,此系统由若干物理元件组成,元件间的联系是确定的;在功能上,系统的输出与输入之间存在着由构造所决定的定量的或逻辑的因果关系。

② 复合系统:在构造上,此系统由多个简单系统作为元素组合而成,这种组合可以是多层次的,层次之间的联系都是确定的,因而,在功能上,复合系统的特点和简单系统的完全相同。

③ 复杂系统:在构造上,此系统由多个子系统作为元素组合而成,这种组合是多层次的,在子系统内、层次之间的联系至少不完全是确定的;在功能上,系统的输出与输入之间存在着由构造所决定的一般并非严格的定量的或逻辑的因果关系。

显然,机械设备是复杂系统。因为这类系统的输出一般表现为模拟量,而对相同的机械设备而言,它们相同的机械元件的本身几何特性(尺寸、形状、表面形貌……)与物理特性(硬度、强度……)不可能完全一样,相同的联系(压力、间隙、介质状况……)也不可能完全一样,因此,即使在完全相同的输入(工作环境)下,相同的机械设备的状态与输出(行为)也就难于完全一样,并非确定。

所谓系统的故障,是指系统的构造处于不正常状态(劣化状态),并可导致系统相应的功能失调,即导致系统相应的行为(输出)超过允许范围。这种劣化状态称为故障状态。

　　显然,原级系统有故障,或者说,原级系统的构造处于劣化状态,必定是它的相应的元素、联系处于劣化状态;某级子系统有故障,或者说,某级子系统的构造处于劣化状态,必定是此子系统的相应的元素、联系处于劣化状态。由此可知,若上一级系统的元素有故障,则此故障必源于其下一级相应的元素、联系;但是,若上一级系统的联系有故障,则此故障并不一定源于其下一级的元素、联系。

　　系统的元素、联系处于劣化状态,原因有二:a. 其工作环境变化非正常,即系统的输入超过允许的范围;b. 在其正常工作环境下,元素、联系的状态由量变发展到质变而劣化。当然,也可以是这两者的联合作用。

　　所谓设备的故障诊断,是指在一定工作环境下查明导致系统某种功能失调的、所指定层次的子系统或联系的劣化状态(故障状态)。显然,故障诊断的实质就是对状态的识别。必须强调,要诊断故障,识别状态,一定要同系统的层次相关联,若不指定诊断所应达到的层次(即子系统级别),则故障诊断的概念是不清楚的,故障诊断的内容是不确定的。如果所指的层次是最高的层次,则此层次为原级系统;如果所指的层次是最低的层次,则此层次即为物理元件。

　　所谓系统的特征信号,是指系统的某部分行为(输出),而这部分是同所关心的系统的功能紧密相联的。对原级系统而言,所关心的功能往往是特征信号的一部分。系统无故障、有故障时的输出分别称为正常的、异常的输出;相应的,有故障、无故障时的特征信号分别称为正常的、异常的特征信号。显然,特征信号必然包含了系统中相应的元素、联系的有关状态的信息。因此,如何选取包含有关状态信息量最多的特征信号,成为机械设备诊断学中重要内容之一。

　　所谓系统的征兆,是指对特征信号加以处理而提取的、直接用于诊断故障的信息。显然,这种处理是去粗取精、"提炼"信息的过程。当然,特征信号本身有时也可以作为征兆。因此,如何提取最有效地用于诊断的征兆,也是机械设备诊断学中重要内容之一。

　　元素间的"联系"分为两类:"功能联系"与"非功能联系"。所谓"功能联系",是指起着将有关元素相联而构成系统这一作用的联系;所谓"非功能联系"即为不属于上述的联系。例如,在机床切削系统中,以工件系统与刀具系统各作为子系统之一,它们之间的联系是切削条件,这一联系是功能联系。在机械设备的结构中,轴与孔的联系是配合,配合是功能联系。至于机械设备在工作中,由于工作元件或工作副产生的振动、热量等,甚至会波及设备各有关元素、联系,由这种振动、热量等产生的联系是非功能联系。以前所讲的联系均为功能联系,以后如不特别指明的联系一般都指功能联系。

应指出,功能联系一般表现为"参数联系"。例如,机床切削系统如切削条件不当,则将产生强烈的颤振;改变切削条件(例如,调整切削速度、进给量,改变刀具前角、后角等等),可以抑制颤振,甚至消除颤振这一故障。这里,是"参数"起作用。这类由于功能联系中参数不当而产生的故障,称为参数型故障,往往可通过调整有关参数而加以消除。至于系统中由于物理元件损坏(例如,齿轮的轮齿折断,轴的危险裂纹等)而导致的故障,则必须修理或更换元件,方可消除。这类由于系统的物理元件损坏而产生的故障称为物态型故障。

所谓故障的传播过程,是指系统中异常输出与异常特征信号的传播过程。系统中某元素或同它有关的联系处于故障状态后,不论此元素的输入如何,其输出必然异常;而这种异常输出又必将通过"功能联系"与"非功能联系"作为同它相关的元素的异常输入,从而可能导致这些元素、联系的状态劣化,由此可能进一步激发上一层次系统的故障,直至原级系统的故障。还应指出,由于异常输出众多,传递途径众多,因而故障传播途径众多,传播同时进行,甚至还可交互影响。

2 机械设备故障的特性

在对系统、故障与故障传播的分析基础上,可以看出,机械设备的故障具有如下特性。

2.1 层次性

这是系统故障的最基本的特性。如上所述,上一层次系统的故障必源于其子系统、子系统间联系的故障。这是故障的"纵向性"。该特性为机械设备这一复杂系统的诊断提供了一个有效的策略与实用的模型,即层次诊断策略与层次诊断模型,将多层系统化为多个两层系统,逐步求解,直至所指定的层次为止。

2.2 相关性

当某一元素、联系产生故障,同它相关的元素、联系的状态也可能劣化。这是故障的"横向性"。这一特性带来了同一层次系统中多个故障并存的现实。众所周知,多故障的同时诊断导致对故障能否准确诊断这一十分困难的诊断问题,这也是机械设备诊断中的一个关键问题。

2.3 延时性

系统的故障是要传播的,从而系统、元素、联系的故障会有一定的发生与发展的过程,即由量变到质变的过程。这是故障的"时间性"。从实用角度上看,这是故障的最重要的特性。因为这一特性深刻表明,故障可以预测,可以早期诊断,诊断工作可以达到"防患于未然"的水平。显然,只要在相应的输出、特征信号、征兆的变化尚未超过允许范围之前,不仅可测量与分析出这些变化,而且还能获得这些变化的规律,则由此可能作出有关系统、元素、联系的目前状态、状态趋势与未来状态的判断。故障的"纵向性"、"横向性"与"时间性"是故障的三个"坐标轴",构成故障的"空间"。

2.4 不确定性

这是复杂系统故障的一个重要特性,它给诊断工作带来了巨大的困难,它是目前诊断理论与实践中的一个研究热点。这一特性是由下面三个方面的因素所决定的。

① 系统的元素特性与联系特性的不确定性。如前所述,在机械设备这类复杂系统中,对相同的系统而言,最低层次的元素(即物理元件)的特性以及这些元素间的联系的特性一般是不可能完全确定的,各层次的元素特性及元素间的联系特性当然也是不可能完全确定的,从而在一定的工作环境中与一定期限内,系统、元素、联系的状态或行为(输出)是否在劣化,劣化是否超过允许范围及其超过程度,一般也是不可能完全确定的。这就从本质上决定了故障发生的不确定性。

② 故障检测与分析装置特性的不确定性。由于故障检测与分析装置一般也属于复杂系统,从而其特性也不可能完全确定,这决定了故障检测与分析结果的不确定性,但一般均应限制在允许范围之内。

③ 系统、元素、联系的状态描叙方法与工作环境的不确定性。这点导致了故障程度描述的不确定性。

显然,a 与 c 两点最为重要。前者常用发生概率来表示,后者常用模糊隶属度来表示,至于 b 点,目前尚未有一种较为有效的统一的表示方法。

在此应着重提出"超层诊断"这一重要概念与方法。前已指明,下一层次系统、联系的故障不仅导致其上一层次系统的故障,甚至还可导致更高层次系统的乃至原级系统的故障,同时,可导致更高层次系统的乃至原级系统的直接同此故障有关的征兆。征兆的这一特性可称为超层性。只要对故障层次性与征兆超层性两者均有清楚了解，就完全可能从高层次系统所表现出的征兆出发，超越中间层次，直接诊断低层次系统的故障；甚至还可能从原级系统所表现出的有关征兆，直接诊断出物理元件或其间联系的故障。当然，这实际上是以承认系统层次性与故障层次性为前提的。这就是"超层诊断"。

3 机械设备诊断学的基本内容

从机械设备诊断技术的起源、发展与现状考察，从设备的系统特性与故障的系统特性考虑，机械设备诊断学的目的应是：在一定的工作环境中与一定的工作期限内，保证机械设备可靠地有效地在允许的状态下实现其功能。同文献[1]所提出的目的相比，本文主要的发展是：

① 指明了"一定的工作环境"，即在一般情况下，明确排除了由于系统的输入超过了允许的范围而使系统产生故障的情况；又在强干扰的条件下，则还需诊断出哪些工作条件被破坏；

② 指明了"一定的工作期限"，即一方面排除了设备过分超期服役而引起的故障，另一方面更为重要的是，明确了基于故障延时性的故障预测与早期诊断这一诊断学内容。

当然，本文的提法继承了文献[1]提法中所包含的内容：保证设备工作不仅可靠，而且有效；不仅着眼于设备维修，而且也应着眼于设备的设计与制造，只不过提法更为确切而已。

同机械设备诊断学的目的相应，机械设备诊断学最根本的任务，就是通过对设备观测所获得的信息来识别机械设备的有关状态。

文献[1]的论述是基于采用传感器测取设备状态的有关特征信号与采用数值计算（数据处理）来分析特征信号、识别设备状态的。显然，文献[1]提出的将"模式"概念加以扩展，则诊断过程即为模式识别过程这一论述，是以数值计算为主要基础的。然而，在工程（包括故障诊断）中，有相当一部分问题并不能采用数值计算来解决，而需要采用知识推理（符号推理）来解决。但是，可认为数值计算是一种特殊的知识推理，这样，模式识别的数值计算过程可扩展为相应的知识推理过程，基于数值计算的模式识别只不过是一种特殊的知识推理。因此，仍可沿用文献[1]的观点，即设备诊断过程实质上是模式识别过程。

这里，"知识"这一概念已经扩展，它不仅包括由符号体现的知识，而且包括由数据体现的知识；"推理"不仅包括非计算形式的推理，而且包括计算形式的推理。

由此出发,还应将"知识"这一概念进一步扩展:在人工智能领域中,"知识"包括浅知识(纯经验)与深知识,前者是表明因果关系完全不清楚的知识,后者是表明因果关系完全清楚的知识。然而,从知识的深度出发,存在大量介于这两者之间的中间知识。"知识"由浅而深,浅知识、中间知识、深知识并无截然的界限,而且表明因果关系在不同程度上清楚(或不清楚)的中间知识无疑是更多的。所谓"分析知识"、"状态知识"[2,3]即属于此类。例如,通过对随机数据的统计分析而获得的数学关系或结果就是这类知识。当然,各层知识有的可用符号关系表示,有的可用数值关系表示;不过,浅知识一般用符号关系表示,成为经验形式。

综上所述,我们将文献[1]所表述的机械设备诊断的内容与过程更准确地表述如下。

机械设备诊断的内容是,在尽可能正确地掌握与依据设备性质与工作环境的条件下——

① 采用合适的特征信号及相应的观测方式(包括合适的传感装置、人的感官、设备调试),在设备合适的部位,测取同设备有关状态的特征信号。

② 采用合适的征兆提取方法与装置,从特征信号中提取设备有关状态的征兆。

③ 采用合适的状态识别方法与装置,依据征兆进行推理而识别出设备的有关状态。显然,有关状态包括正常的与不正常的有关状态,不正常状态时,即为故障诊断。

④ 采用合适的状态趋势分析方法与装置,依据征兆与状态进行推理而识别出有关状态的发展趋势,这里包括故障的早期诊断与预测。

⑤ 采用合适的决策形成方法与装置,从有关状态及其趋势形成正确的干预决策:或者深入系统的下一层次,继续诊断;或者已达指定的系统层次,作出调整、控制、自诊治、维修等某类决策。

根据机械设备诊断的内容,显然,机械设备诊断的步骤与过程如图1所示。由图可知,诊断过程的每一步骤(阶段)都是同上述有关诊断内容相应的。

在设备诊断中,毫无疑问,应要求在每一步骤(阶段)花费尽可能少,有关状态的信息获取(或利用)尽可能多,结果尽可能好。然而,各诊断阶段是彼此相互联系与相互影响的,各局部最优并不能保证全局最优,因此,还必须从全局出发,深刻考虑整个诊断过程,从宏观上制定尽可能好的诊断策略。有关诊断策略问题见文献[2]。

关于设备的工况监视,可认为它是设备诊断的基础。这里有两层含义:第一,

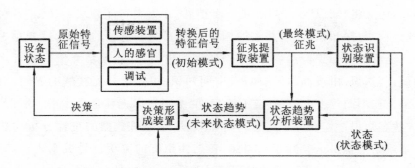

图 1　机械设备诊断过程

对某一层次子系统进行正常的与异常的两种状态的识别,即两种状态的监视或诊断;第二,为更深层次子系统的故障诊断提供依据与基础,即决定是否需要进行更深层次子系统的诊断与为此诊断提供必要的素材。

应该强调指出,机械设备诊断方法的分类问题,一直是一个极为重要的研究课题。文献[1]提出了十分关键的一点,即诊断方法的分类也必须基于同一概念或同一基础。据此,文献[1]提出了基于诊断过程的相同阶段的分类方法,这对改善当时在诊断方法分类问题上的不完善、不妥当、乃至在概念上发生混乱的情况是有益的。然而,文献[1]在基于诊断过程的相同阶段进行分类时,大致以数值计算作为基础,而未充分计及知识推理,因此,就难以全面、系统、深刻地反映诊断方法的本质与全貌。显然,在基于诊断过程的相同阶段进行分类时,必须以本文所扩展了的知识推理作为基础;由于状态识别这一阶段是设备诊断的重点所在,则更尖锐地显示出知识推理的重大作用。在以知识推理作为基础分类时,例如,必须考虑知识深度、推理方式、不确定性描述方式等等,这样的分类将另文讨论。

参 考 文 献

[1] 杨叔子,师汉民,熊有伦,等.机械设备诊断学的探讨.华中工学院学报,1987,15(2):1~9.

[2] 史铁林,杨叔子,师汉民,等.机械设备诊断策略的若干问题探讨.华中理工大学学报,1991,19(增刊(II)):9~14.

[3] Ding Hong, Gui Xiuwen, Yang Shuzi. *The State Knowledge Based Diagnosis for Engines*. Proc. of the 12th Biennial ASME Conf. on Mechanical Vibration and Noise, Montreal. Canada, 1989.

More on Diagnostics for Mechanical Devices

Yang Shuzi   Ding Hong   Shi Tielin   Zhong Yuning

Abstract The basic content and conceptual system of mechanical device diagnostics are studied. The system characteristics and system classification of a mechanical device are studied from the point of view of the system theory. The concepts of fault, symptom, and characteristic signal are further developed. The characteristics of fault, that is, the hierarchy, correlativity, time-delay and uncertainty as well as the nature of fault propagation are proposed. The "overstepped hierarchical" diagnosis is described. The aim, task, content, and process of diagnostics for mechanical devices are also discussed using knowledgeable reasoning.

Keywords: Mechanical device diagnosis, Device diagnostics, System, Fault characteristics, Knowledgeable reasoning, Pattern recognition, Diagnostic method

(原载《华中理工大学学报(自然科学版)》1991 年 8 月第 19 卷(增刊))

机械设备诊断策略的若干问题探讨[*]

史铁林　杨叔子　师汉民　钟毓宁

提　要　本文根据机械设备诊断问题的特点,深入讨论了机械设备诊断问题的概念体系,研究了机械设备诊断问题的知识策略和求解策略,提出了分析知识和基于知识的诊断环境等重要概念,并对其在诊断问题中的应用进行了探讨。

关键词:诊断策略,诊断问题求解,专家系统,概念体系,故障诊断

1　引　言

诊断策略是对诊断过程与诊断方法的宏观研究,它包含两方面的内容:其一是诊断问题的知识策略;其二是诊断问题的求解策略。前者从知识的角度研究求解诊断问题所用的知识及其组织、表达及获取等问题;后者从推理的角度研究诊断方法与问题求解方法。目前尽管在诊断策略的研究方面做了许多工作[1-3],但这些工作并未涉及诊断问题的概念体系,而诊断策略同诊断问题的概念体系是密切相关的。机械设备诊断问题的概念体系有其特殊性,是研究机械设备诊断问题的基础。文献[4]对此进行了深入的研究,本文将在此基础上对其进行深化,并以此为基础研究机械设备的诊断策略。

机械设备特别是大型机械设备是一类复杂设备,其复杂性至少表现在以下几个方面。

① 它是一个多层次系统,一般至少具有系统级、子系统级、部(组)件级及零件级四个层次。

② 对这样一类复杂设备,由于其结构异常复杂,加之输入、输出不明确,因而

---

[*] 国家自然科学基金资助课题。

无论是定量还是定性都难以用比较完备而比较准确的模型对其结构、功能以及状态等进行有效的表达。

③ 机械设备的故障及产生故障的原因有时是模糊不清的。一方面设备从正常状态到故障状态可能是一个连续量变的过程，而不存在明显的质变点；另一方面，一个故障可能是多种因素综合作用的结果，要比简单的因果对应关系复杂得多。此外，还存在多故障并存的困难情况。

以上复杂性大大增加了机械设备诊断问题的难度，使机械设备的诊断策略同简单系统或复合系统的诊断问题的诊断策略有很大的不同。本文针对机械设备诊断问题的特点，在文献[4]的基础上，研究了机械设备诊断问题的知识策略和求解策略，提出了分析知识和基于知识的诊断环境等重要概念。

2 机械设备诊断问题的概念体系

文献[4]从系统论的观点对机械设备诊断问题的概念体系进行了深入的研究，本文从诊断策略的角度出发，对此进行深化，并着重阐明这些基本概念的相互关系。

文献[4]对故障给出了明确的定义。从其产生的因果关系上来分，它可分为两类：一类是原发性故障，即故障源；一类是引发性故障，即这类故障是由其它故障引发的，当原发性故障消失时，这类故障也自然消失。当然，引发性故障也有可能成为一个新的故障源，这种情况当原发性故障使引发性故障相应的子系统或联系的固有特性发生劣化时就会出现。对于原发性故障，从其产生的性质来分，也可分为两类：一类是物态性故障；一类是参数性故障。物态性故障是不可恢复的，除非修理或替换零部件，而参数性故障可以通过调整工作参数而消除或避开。此外，故障还是一个相对的概念，同所给定的约束条件有关。在某一约束条件下，它可能属于故障状态，在另一约束条件下，又可能属于正常状。

系统的所有行为称为系统的输出。文献[4]明确定义了系统的功能、特征信号以及征兆等基本概念。可以用图 1 来直观地表示系统的输出、功能和特征信号之间的相互关系。如图 1 所示，当 $D \cap B \neq \emptyset$ 时，则意味着系统发生了故障，否则，系统处于正常状态。

引发指定层次的子系统产生故障的原因可以分为以下几类：

a. 同该子系统相关的下一层次的子系统发生故障；

b. 同该子系统相关的同一层次的子系统发生故障；

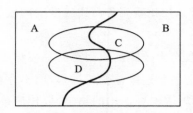

图 1　系统输出、功能与特征信号三者关系示意图
A—正常输出　B—异常输出　C—特征信号　D—系统功能

c. 该子系统同其他相关子系统之间的联系或该子系统内部子系统之间的联系失调；

d. 由于外部因素，使得系统的工作条件被破坏。

在一般情况下认为工作条件是满足的，d 故障原因主要是针对强干扰条件下的故障诊断而言的，因为在工作条件恶劣的环境下，不能保证工作条件常常是满足的，而诊断的目的之一，也包括诊断出哪些工作条件被破坏。

文献[4]明确阐述了故障诊断的概念和任务，本文根据故障的传播机理将其进一步具体化，定义诊断问题的诊断任务为：

对给定的征兆集合，运用诊断知识和合适的问题求解策略，诊断出系统的原发性故障源集合，这个故障源集合应能解释系统所出现的各种异常输出；如果指定了所要诊断到的层次，则亦称包含原发性故障源的该指定层次的子系统为一个故障源，这时，作为诊断解故障源集合应由指定层次的故障源组成。

3　机械设备诊断问题的知识策略

知识策略是知识系统的核心与关键，对诊断问题而言，其知识策略包括：a. 求解一个诊断问题，要用到哪些领域和哪种层次的知识，这是知识策略的核心，这在很大程度上决定了诊断策略的其他方面；b. 求解诊断问题的元知识，这是关于知识的知识，它在更高层次上反映了人类求解诊断问题的思路和方法(a 和 b 统称为诊断知识)；c. 所用知识的组织、表达与协调；d. 知识的获取与自学习；e. 知识的质量与可用性。

以上五个方面是知识策略的基本内容，下面将对某些内容进行较深入的阐述。

关于什么是知识，这是一个哲学问题，至今仍没有一种统一的定义，这也不是本文的研究内容。就诊断领域而言，我们认为，诊断知识是人类专家对诊断对象的一种认识。诊断知识的浅与深同这种认识水平有关。这种认识从宏观到微观逐层深入,因而也就形成了不同层次的诊断知识,如经验知识、因果知识以至结构知识等。求解诊断问题时,究竟要用到哪些领域和哪种层次的知识,这同具体的诊断对象有关。机械设备是一类复杂系统,求解这类诊断问题所采用的诊断知识有以下特点:a. 所使用的知识范围广、层次多,除使用经验、功能以及结构知识外,还广泛使用一种对诊断对象进行间接描述的知识,本文称其为分析知识;b. 由于机械设备属于复杂系统,因而其结构、功能以至所有输出都具有不确定性,这种不确定性必然导致领域知识和诊断信息的不确定性;c. 领域知识和诊断信息的不完备性与不一致性。领域知识的不完备性和不一致性同人类专家对诊断对象的认识水平及知识的获取有关,而诊断信息的不完备性和不一致性同特征信号的获取手段有关,一般说来,不可能完全准确地获得诊断所需的全部信息,然而机械设备诊断就是要求在这种诊断信息不确定、不完备、甚至不一致的情况下得出较准确的诊断结果,这是机械设备诊断最重要的特征之一。

　　求解诊断问题的元知识是更高层次的诊断知识,是人类专家形成的规划性知识。一般说来,它是同领域无关的,它在宏观上对诊断问题的求解进行总体规划,将复杂的诊断问题分解为简单的诊断子问题,并选择合适的求解方法和所需的领域知识。

　　诊断知识的表达,同所使用的知识层次及诊断模型有关。对于元知识,一般采用规则来表达,这些规则称为元规则。对于领域知识,由于机械设备诊断采用多种层次的知识,因而宜采用规则、框架和过程的混合表达技术,这种混合表达技术使得不同层次的诊断知识可以协调使用。值得指出的是,诊断知识不仅仅是符号形式的,也有数值形式的。这类诊断方法的知识表达不具有显示的符号表达特征,而只是由若干层节点及节点之间的联系所构成。知识就包含于这些节点的参数和联系的权重值之中。这是一种分布式知识表达技术,它模拟人脑神经元存储知识的方式,对此本文不进行深入的探讨。

　　就机械设备诊断知识而言,无论采取何种表达技术,其共同的特征是都具有层次结构的知识组织形式,这是由机械设备及其故障的层次性所决定的。层次结构的知识组织形式不仅有效地降低了求解问题的复杂性,同时也增加了知识组织的模块性,有利于知识库进一步完善与扩充。

　　知识的获取与自学习是知识策略的一个重要方面,也是任何知识系统的瓶颈问题。知识的获取应满足以下基本要求:a. 可表达性,即所获取的诊断知识应能有效地予以表达;b. 获取的方便性,即能用尽可能简单、方便的式进行获取;c. 一致性,即应使获取的知识不发生矛盾或应提供解决这种矛盾的方法。对机械设备诊断问题来说,知识的获取还有一个重要方面,那就是特征信号选择规则与征兆提取规则的获取。一般说来,我们总是选取包含有关状态信息量最多的系统输出作为特征信号,同时从特征信号里提取对系统状态最敏感的征兆作为诊断信息,这是两条基本的规则。

知识的自学习是使诊断知识不断完善的一种手段,而知识的质量与可用性同诊断系统的可靠性有密切的关系,本文对此不做进一步讨论。

使用分析知识是机械设备诊断问题知识策略的重要特点。前文述及[1],征兆是对特征信号加以处理而提取的特征信息,是对特征信号的一种抽象描述。通过信号处理技术、统计模式识别技术等数值分析手段可从系统特征信号中获得各种征兆,这些征兆从不同角度间接地反映了系统的固有特性,因而据此可进一步诊断出系统所处的状态以及引起系统处于故障状态的原因,这就是分析知识的含意。分析知识既不同于高层次的经验知识,又不同于低层次的结构、功能知识,它是人类专家对诊断对象的一种间接描述,而这种描述是通过数学抽象来实现的。以回转机械的诊断问题为例,一般是先测取其振动、位移等特征信号,再对所测取的特征信号进行诸如 Fourier 谱、AR 谱以及轴心轨迹等分析,获取系统的征兆,最后,用这些征兆进行诊断推理。此例展示了分析知识最一般的使用方式。分析知识一般包含以下内容:a. 所需测取的特征信号与测取方法;b. 所要提取的征兆与提取方式(一般为信号处理方法);c. 分析结果;d. 分析结果所支持的故障假设与原因假设;e. 可信度。这些内容构成了分析知识的基本框架。分析知识架起了数值计算与符号推理相结合的桥梁,是机械设备诊断最重要的知识之一。

4 机械设备诊断问题的求解策略

当设备处于故障状态时,必然引起一系列的特征信号发生变化,根据故障传播机理[4],这种变化是逐级传播的。诊断过程是按故障传播的反方向由上向下进行的。由故障的传播机理和诊断任务的定义,可以得到求解机械设备诊断问题的一般过程为:a. 特征信号的获取与征兆的提取;b. 根据所提取的征兆,提出系统所处故障状态的假设集合;c. 对这一集合中的每一个假设分别进行验证,若为真,则继续验证下一个假设,若为假,则转回 b;d. 对确认的故障状态假设,进一步提出故障原因的假设集合;e. 对故障原因假设集合中的每一个原因假设进行验证,若为真,则继续验证下一个故障原因,若为假,则返回 d;f. 对确认的故障原因,转到下一层次子系统的诊断过程。诊断过程即是这样一个以层次方式进行的循环搜索过程，直到所形成的故障原因假设均是指定层次的原发性故障为止。这一诊断过程是文献[4]中所提出的机械设备诊断过程的具体体现，是一种宏观求解策略，要有效地实现机械设备的故障诊断，还要从微观角度考虑诊断过程中的每一步。步骤 b 将诊断问题分解为若干子问题；步骤 c 针对各个不同的子问题，可以选择各自合适的推理方法进行求解。步骤 d 在下一层次寻找引起本层次故障的原因；步骤 e 与 f 完成本层次的诊断而转入下一层次。

机械设备诊断问题的求解策略有以下特殊性：第一，要用到多种推理方式，并且各种推理方式应能协调使用；第二，对机械设备的诊断问题，单纯的符号推理或数值计算都不能有效地予以解决，必须实现符号推理与数值计算的有机结合，更进一步，还应实现模拟人类逻辑思维的人工智能技术与模拟人类形象思维的神经网络技术的有机结合；第三，在机械设备诊断问题里，诊断知识与诊断信息都存在不确定性，因此诊断系统应含有各种相应的不确定性处理方法；第四，不仅所用到的知识是多领域、多层次的，而且所用到的诊断信息是多方面的，如符号、数值及图像等。针对机械设备诊断问题的特殊性，我们提出了基于知识的诊断环境的概念。这一环境包含了求解诊断问题所需要的知识、方法和手段，其结构如图 2 所示。这一诊断环境由元级系统、推理系统、知识库和数值计算系统组成。推理系统、知识库和数值计算统一在元级系统的管理之下。元级系统像一般的智能系统一样，也有它自己的知识库和推理机。它管理和控制这些程序和知识的选择、操作及其相互之间的通信，并执行不同程序间数据的相互转换。推理系统包含多种推理机制，在元级系统的管理调度之下，针对不同的诊断子问题，激活最合适的推理机制，求解当前的诊断子问题。知识库包含求解诊断问题所需要的各类知识，如结构知识、经验知识以及分析知识等。各种推理机制在元级系统的控制下，可激活与选择所需的知识源进行推理。数值计算系统包含求解诊断问题所需的各种数值计算程序，如各种信号处理程序、征兆提取程序等。为简明计，图 2 中的第三层只表示出各系统中所含的若干组成部分。

诊断环境中的各系统既可采用一种统一的语言，如 C、C++，又可采用各自更为合适语言，如推理系统可采用 Prolog 或 Lisp，数值计算可采用 C,Pascal 以至 Fortran 等。之所以把这样一种结构称之为环境，而不是系统，原因正是由于这一环境提供了设备诊断所需要的各种知识、方法及手段，其概念要比一般的诊断系统的概念更广泛。"置身"于这样一个环境中，如同拥有一批专家一样，可选择最有效的途径求解所面临的诊断问题，正是基于此，称其为基于知识的诊断环境。

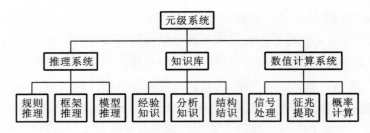

图 2 基于知识的诊断环境

参考文献

[1] Milne R. *Strategies for Diagnosis*. IEEE Trans. Syst. on Man and Cybern,1987,SMC-17(3):333~339.

[2] De Kleer J,Willians B C. *Diagnosing Multiple Faults*. Artificial Intelligence,1987,32:91~130.

[3] Peng Y,Reggia J A. *A Probabilistic Causal Model for Diagnostic Pyoblem Solving*. IEEE Trans. on Man and Cybern,1987,SMC-17(2):146~162,SMC-17(3):395~406.

[4] 杨叔子,丁洪,史铁林等.机械设备诊断学的再探讨.华中理工大学学报,1991,19(增刊(Ⅱ)):1~7.

[5] 丁洪,杨叔子,桂修文.复杂系统诊断问题的研究.华中理工大学学报,1989,17(4):9~16.

On Diagnostic Strategies for Mechanical Devices

Shi Tielin Yang Shuzi Shi Hanmin Zhong Yuning

Abstract By taking into account its characteristics the conceptual system for mechanical device diagnostic problems are treated in detail and the knowledge strategies and solving strategies are studied. Two important concepts,the analytic knowledge and knowledge based diagnostic environment are proposed. Their application in diagnostic problems is discussed.

Keywords:Diagnostic strategy,Diagnostic problem solving,Expert system,Conceptual system,Fault diagnosis

(原载《华中理工大学学报(自然科学版)》1991 年 8 月第 19 卷(增刊))

Forced Regenerative Chatter and its Control Strategies in Machine Tools

Wu Ya   Ke Shiqiu   Yang Shuzi
Li Weiguo   Xu Shanxiang   Jiang Qiang

Abstract   The mechanism of forced regenerative chatter in machine tools is studied in this paper. It is pointed out that this kind of chatter is a resonance of the chatter regenerative instable mode caused by a forced excitation. A method for judging the chatter is developed, which is based on the 3-dimensional spectrum of chatter control are developed: one is called the strategy of shifting the main mode of chatter aimed to increase the stiffness or damping of a machine tool to make the higher-order mode of the system into the main mode of chatter so that the forced regenerative chatter is restrained, and the other is called the strategy of removing forced excitation aimed to remove forced excitation directly or to change the speed of spindle to avoid the frequency of forced excitation so that the chatter is eliminated. The research results in this paper have been applied to a type MX-4 crank lathe.

Keywords: Forced regenerative chatter, Machining system, Mechanism, Judgement, Control

0 Introduction

Forced regenerative chatter in machining is a mixed chatter and an extremely violent vibration generated when the frequency of forced vibration is close to that of regenerative chatter in machine tools. However, little has been reported on this kind of chatter since most ongoing investigations are made in laboratories with chatter artificially generated as a pure one. Tetsutaro Hoshi (星铁太郎) published his discovery of forced regenerative chatter and the results of overcoming it in 1977 in his famous book[1]. No similar investigation results have been published since then. A further study on the mechanism of the forced regenerative chatter is performed in this paper, a method of judgement is then developed and two control strategies are proposed, which have been applied to an MX-4 lather in the Second Automobile Works of China.

1 The Mechanism of Forced Regenerative Chatter

1.1 A Mathematical Model for Forced Regenerative Chatter

Under the effect of regeneration in orthogonal cutting[3,4], if a harmonic force with a frequency ω_0 and amplitude P_0 excites to the cutting tool and workpiece, the differential equation of the vibration is

$$m\ddot{x}(t) + c\dot{x} + kx(t) = -K[x(t) - x(t-\tau)] - C[\dot{x}(t) - \dot{x}(t-\tau)] + P_0 \cos \omega_0 t \tag{1}$$

where m, c and k are the modal mass, damping and stiffness of the instable mode of machine tool structure under the effect of regeneration (i.e., the main mode of regenerative chatter); K and C represent equivalent cutting stiffness and damping; τ is the lag time for vibrational wave during two (adjacent) cuttings. When vibration $x(t)$ is rather small, linearization treatment can be made and only the fundamental wave with frequency ω in $x(t)$ is considered. Let $x(t) = a\cos\omega t$. $x(t-\tau) = a \cos(\omega t - \beta)$, $(\beta = \omega\tau)$, and treating the damping as a structural damping, that is $c = h/\omega, C = H/\omega$, then Eq. (1) can be written as

$$m\ddot{x}(t) + \frac{1}{\omega}[h + K\sin\beta + H(1 - \cos\beta)]\dot{x}(t)$$
$$+ [k - H\sin\beta + K(1 - \cos\beta)]x(t) = P_0 \cos \omega_0 t \tag{2}$$

The above equation is the differential equation model for the forced regenerative chatter of the machine tool cutting system (hereinafter the system for short, which is composed of both machine tool structure and cutting process). As a special case of the above equation, when the right side of the equation is equal to zero, we have pure regenerative chatter. The natural frequency of the system, or the regenerative chatter frequency, is

$$\omega_n = \sqrt{[k - H\sin\beta + K(1 - \cos\beta)]/m} \quad (3)$$

The equivalent structure damping coefficient of the system is

$$\gamma = h + K\sin\beta + H(1 - \cos\beta) \quad (4)$$

According to the resonance conditions for the structural damping system[5], when the excited frequency of the harmonic force is expressed as $\omega_0 = \omega_n$, the system will lead to resonant. As regenerative chatter is represented by Eq. (2), the resonance is called forced regenerative chatter, whose amplitude is

$$X = P_0/\gamma = P_0/[h + K\sin\beta + H(1 - \cos\beta)] \quad (5)$$

whereas is idle running, $K=0, C=0$, the modal frequency of the machine tool structure is $\omega'_n = \sqrt{k/m}$. At $\omega = \omega'_n$ resonance occurs, the amplitude of which is

$$X_0 = P_0/h \quad (6)$$

Then from Eq. (5) and Eq. (6), we have the amplification ratio of the forced regenerative chatter to the resonance in idle running as

$$\eta = X/X_0 = h/\gamma = h/[h + K\sin\beta + H(1 - \cos\beta)] \quad (7)$$

An analysis of Eq. (7) shows that the amplification ratio is jointly determined by the instable modal damping h, cutting stiffness K, cutting damping H and phase difference β. Moreover we have $\eta > 1$ when

$$\beta < 2\text{arctg}(-K/H) \quad (8)$$

from the condition $\gamma < h$. It is shown that, provided the value of β satisfies Eq. (8), the forced regenerative chatter will occur even in a stable state. From the physical point of view, such a kind of chatter is equivalent to the phenomenon that the resonance of the instable mode for the regenerative chatter in a machine tool is induced by a forced excitation while the vibration of a machine tool in a non-cutting state is amplified in cutting process. Evidently, the smaller the γ, and the closer it is to the critical point of regenerative chatter ($\gamma=0$), the greater the η value of and the more violent the vibration. At $\gamma=0$, there will be $\eta=\infty$. Of course, the amplitude of an actual forced regenerative chatter will not be infinity. This is because when the amplitude is increased, the nonlinear factors of the system will take effect so that the amplitude will be fixed at a certain level.

1.2 The Complex Plane Representation of the Forced Regenerative Chatter

It can be proved that the characteristic equation for a machine tool cutting system in the cutting state is $1 - G(s)/H(s) = 0$, where $G(s)$ represents the transfer function of the instable mode for a machine tool structure; $1/H(s)$ is the equivalent transfer function of a cutting process. The specific forms of the both are[4]

$$G(\omega) = 1/(k - m\omega^2 + jwc) \tag{9}$$

$$1/H(\omega) = (1 - \cos\beta)[(K + C\omega \text{ctg}\beta/2 + j(C\omega - K\text{ctg}\beta/2)] \tag{10}$$

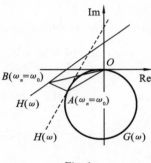

Fig. 1

which are represented in complex plane as shown in Fig. 1. If only pure regenerative chatter is considered, when the two the curves intersect as shown by the dotted line $H(\omega)$ in the figure, the system is at the critical point A of the regenerative chatter, the frequency of which is ω_n; when the two curves do not intersect as shown by the solid line in the figure, no regenerative chatter occurs in the machine tool cutting system. Now, under the action of the dynamic cutting force F, the vibration of the machine tool structure is $\boldsymbol{F} \cdot \overrightarrow{OA}$. However, when a forced excitation frequency is $\omega_0 = \omega_n$, a forced vibration $\boldsymbol{x}_0$ with an identical frequency will be superposed on vibration $\boldsymbol{F} \cdot \overrightarrow{OA}$. At $\boldsymbol{x}_0 = \boldsymbol{F} \cdot \overrightarrow{OA}$, the superposition (vectors added) of the two vibrations will be equivalent to connecting two originally not intersecting curves $H(\omega)$ and $G(\omega)$ at the regenerative chatter frequency, thus the forced regenerative chatter is induced. Its amplitude is $\boldsymbol{F} \cdot \overrightarrow{OA} + \boldsymbol{F} \cdot \overrightarrow{AB} = \boldsymbol{F} \cdot \overrightarrow{OB}$, the amplification ratio of the chatter to the forced vibration $\boldsymbol{x}_0$ is

$$\eta = |\boldsymbol{F} \cdot \overrightarrow{OB}|/|\boldsymbol{x}_0| = |\overrightarrow{OB}|/|\overrightarrow{OA}| \tag{11}$$

For a given machine tool structure and cutting process, $G(\omega)$ and $H(\omega)$ are determined, then points A and B are determined (correspondingly) by $\omega_0 = \omega_n$, thus vector $\overrightarrow{AB}$ is determined uniquely. The result shows that, when $\omega_0 = \omega_n$, whatever the magnitude and the phase of x_0 are, so long as x_0 exists, the amplitude and phase of F is determined by x_0. So that the amplitude ratio and the phase difference of the both may be adjusted automatically to satisfy the condition $|x_0|/F = |\overrightarrow{AB}|$, inducing the forced regenerative chatter. Obviously, the closer the two curves $H(\omega)$ and $G(\omega)$ in Fig. 1 are to each other, the greater will the η value become. In the limit case $|\overrightarrow{AB}| = 0$, we have $\eta = \infty$, a result in agreement with the above-mentioned analysis. That is to say, the lower the stability of the machine tool cutting system itself, manifested as the smaller the value of γ in Eq. (7) and the closer the $G(\omega)$ and $H(\omega)$ are to each other as shown in Fig. 1, then the more seriously to occur will the forced regenerative chatter be and the greater the amplification of vibration in cutting process to the non-cutting state.

2 Judgement for the Forced Regenerative Chatter

2.1 The Spectral Characteristic of the Forced Regenerative Chatter

The type MX-4 lathe is one of the good machines in The Second Automobile Works of China. As the connecting rod necks of two crankshafts by a widened-edged would be machined by the lathe, a violent vibration occurs in the lathe during turning and vibration wave occur on surfaces of the connecting rod necks. According to the principle of regenerative chatter, it has been tentatively determined that this vibration is regenerative chatter. From the judgement made later in this section, the chatter in the MX-4 lathe is mainly forced regenerative chatter with a frequency of 530 Hz. Fig. 2 shows the acceleration spectrum composed of 60 spectral graphs in an entire cutting process of the lathe. Throughout the entire cutting process a spectral peak with a frequency of 530 Hz exists from the beginning to the end and its position remains unchanged whereas another spectral peak with a frequency of 705 Hz gradually shifts toward 530 Hz. After the 40th

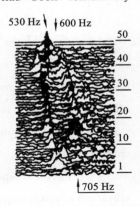

Fig. 2

spectral graph, the shifting of the spectral peak with 705 Hz becomes faster and faster until it is at 600 Hz when the truning ends at the 48th spectral graph. This clearly shows that the frequency of the forced regenerative chatter will not change with the passage of the machining time and it can even absorb the vibration energy of the other frequency components to join it to make the chatter more violent. This is because of the fact that the vibration energy will shift in the process of establishment and development of a pure regenerative chatter and in most cases the main frequency of the chatter shifts from higher to lower frequencies. However, the forced regenerative chatter is characterized by a forced vibration and the frequency of the forced excitation in a machine tool is invariable, therefore, the frequency of the forced regenerative chatter will not change with the passage of the machining time whereas the frequency of a pure regenerative chatter components may shift toward the frequency of the forced regenerative chatter when the shifting takes place. This result is a very important characteristic of the forced regenerative chatter and can be used in the judgement of the forced chatter.

2.2 Judgement of Forced Regenerative Chatter

According to the mechanism of forced regenerative chatter and the above, the judgement of the forced regenerative chatter can be made in the following steps.

(1) Judge whether it is a forced vibration with ordinary methods to avoid mixing up a forced vibration with the forced regenerative chatter.

(2) Judge whether it is a pure regenerative chatter.

(3) Measure the vibration signals respectively in the idle running state and cutting state of the machine tool and make spectrum analysis. Examine whether there exists a frequency ω_0 in the frequency components of forced vibration in the idle running, which is close to ω_n, the main frequency of cutting vibration. If such a ω_0 exists and the amplification ratio η is extremely great, the forced regenerative chatter is probable.

(4) Perform spectrum analysis for the vibration signal (preferably the vibration acceleration signal) of the entire cutting process. If the main spectral peak with frequency $\omega_n \approx \omega_0$ does not shift with the passage of machining and

other spectral peaks shift toward ω_n, then it can be judged to be forced regenerative chatter. The advantage of adopting the vibration acceleration signal consists in that the spectral graph of this kind of signal can give a clear depiction for the higher frequency components of vibration and can represent the shifting of all spectral peaks clearly.

Take the judgement of the MX-4 lathe as an example. The regenerative chatter has already been tentatively demonstrated as the above. Fig. 3 gives a spectrum of vibration acceleration measured in the feeding direction of the tool holder in idle running of the lathe, which shows that there is a forced vibration with a frequency of 532.5 Hz in the lathe. Fig. 4 gives a spectrum of vibration acceleration signal measured at the same point in cutting state, which shows that the frequency of the highest spectral peak is 530 Hz. Moreover, the amplification ratio η is high up to 196 after a calibration is performed to the two spectrums. As the frequency 532.5 Hz of forced vibration changes in idle running by changing the rotative velocity of the spindle, but the spectrum peak of 530 Hz still appears in cutting state, the possibility of forced vibration is ruled out. Finally, based on the spectral characteristic shown in Fig. 2, it is judged that the vibration of the MX-4 lathe is mainly forced regenerative chatter.

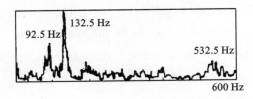
Fig. 3

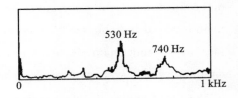
Fig. 4

3 The Control Strategy of Forced Regenerative Chatter

According to the mechanism of forced regenerative chatter, as long as the regenerative chatter frequency of a machine tool cutting system is not close to a forced vibration frequency, the goal of controlling the forced regenerative chatter can be attained. Therefore, there are two way to control the forced regenerative chatter. One is changing the main mode of the regenerative chatter in machine tool cutting system so as to make the frequency of regenerative chatter shift from lower frequency to higher frequency. This is equivalent to reducing the $G(\omega)$ circle in Fig. 1, so that ω_n will be far apart from ω_0. This can be realized by increasing the stiffness or damping of the relevant parts of the machine tool. This is the strategy of shifting main mode of chatter, which in essence agree with ordinary strategies of chatter control. The other one is removing the forced excitation, which is equivalent to remove the vector $\overrightarrow{AB}$ in Fig. 1. For the case what the forced excitation is in main transmission system, the forced regenerative chatter can be controlled by changing the rotative velocity of the spindle to bypass the forced excitation or to change the frequency of the forced excitation. This is the strategy of removing forced excitation, which in essence agrees with ordinary strategies of forced vibration control. The two strategies and their control effects would be investigated as below.

3.1 The Strategy of Shifting Main Mode of Chatter

A machine tool cutting system in cutting state is a multivariable system. For a pure regenerative chatter, the chatter is the superposition of the modal vibrations of the system. Except the multifrequency chatter, the chatter is determined mainly by a certain order modal vibration and this order mode, called main mode or instable mode, is generally a lower order mode of the system. The idea of shifting main mode of chatter consists in increasing the stiffness or damping of the system to enable chatter to be mainly determined by a certain higher order modal. In this way, ω_n, the main frequency of chatter is increased and become far higher than ω_0, the frequency of forced excitation, thus, the forced regenerative chatter is restrained. It should be pointed out that, when the control strategy is adopted, the vibration of the mode close to ω_0 still exerts an amplifying effect on forced vibration although this mode is no longer the main mode of chatter, but forced regenerative chatter is, of course, not the main component.

It has been established from an impact test and a finite element analysis of the workpiece system that the chatter with frequencies 530 Hz and 704 Hz is mainly influenced by the workpiece system of the MX-4 lathe. According to the strategy of shifting main mode of chatter, the clearances between central supports and workpieces are adjusted so that the support stiffness and damping of the workpieces system are increased and the frequency of main mode of chatter is shifted from 530 Hz to 740 Hz. As 740 Hz, the main frequency of vibration, now is much higher than 532.5 Hz, the frequency of forced excitation, the forced regenerative chatter with frequency 530 Hz is restrained.

Fig. 5 shows the spectrum of vibration acceleration after the adoption of this strategy. Compared with Fig. 4, the frequency of the highest spectral peak becomes 740 Hz and $\eta = 102.73$ at frequency 530 Hz, which is 47.6% lower than $\eta = 196$ in Fig. 4. In the past few years, the central supports of the MX-4 lathe were maintained mainly using this strategy and a satisfactory result were achieved. Especially the MX-4 lathe has been working in a normal state and its machining quality has been restored to the level in its initial period of operation since this strategy was adopted in March, 1990. As a matter of fact, the spectrum of forced regenerative chatter shown in Fig. 4 was exactly made during the period when serious faults of crankshaft-broken occurred in the lathe on December 17, 1989 while the spectrum of restraining the forced regenerative chatter shown in Fig. 5 was made on May 2, 1990 right after the central supports were repaired. This shows that the analysis and control strategy described in this paper are correct.

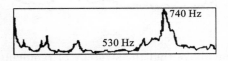

Fig. 5

It can be seen from above that the control strategy of shifting main mode of chatter is fairly effective in the control of forced regenerative chatter but can not completely eliminate it. Moreover, it is in general rather difficult for ordinary universal machine tools to shift the main mode of chatter by increasing the stiffness and damping of a system despite the fact that it is convenient to implement this strategy on the MX-4 lathe.

3.2 The Strategy of Removing Forced Excitation

The idea of removing forced excitation is very simple. It consists in removing the forced excitation with a frequency ω_0 close to the frequency ω_n of the main mode of chatter. Thus, when the condition of regenerative chatter is not reached, no previously mentioned amplifying effect will occur and the system can operate stably. When the conditions for forming regenerative chatter are available, there will only be pure regenerative chatter and the main frequency of the chatter is determined by the characteristics of the system itself rather than by the forced excitation, and no amplifying effect will occur either.

According to this strategy, change the transmission ratio of pulleys in the MX-4 lathe so that the rotative speeds of workpieces change from 110 r/min to 135 r/min and the frequency of the forced excitation becomes 653.5 Hz from 532.5 Hz. As the frequency 653.5 Hz is far apart from 530 Hz and 740 Hz, the two frequencies of chatter, this effect is equivalent to removal of the forced excitation with frequency 532.5 Hz. Fig. 6 shows the spectrum of vibration acceleration after adoption this strategy. Compared with Fig. 4 and Fig. 5, the frequency of the highest spectral peak of chatter in Fig. 6 is 740 Hz and no spectral peak appears at 530 Hz, showing that the forced regenerative chatter is completely eliminated.

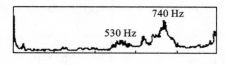

Fig. 6

It can be seen from above that the strategy of removing forced excitation can completely eliminate the forced regenerative chatter and it is convenient for ordinary universal machine tools to bypass a forced excitation or to shift the frequency of a forced excitation by changing the revolution speed of the spindle when the forced excitation is in the main transmissions of a machine. Therefore, as a whole, this strategy is superior to the strategy of shifting main mode of chatter. Nevertheless, careful consideration should be given to the problems cropped up owing to the change of speed. For instance, this strategy is only limited to test purpose in MX-4 lathe because of other problems that are involved.

4 Conclusions

(1) Forced regenerative chatter is a resonance generated when ω_0, the forced excitation frequency, is close to ω_n, the instable modal frequency of the regenerative chatter in a machine tool, manifested as vibration of machine tool in idle running is amplified in cutting. For the MX-4 lathe, when the clearances between the central supports and workpieces are greater, the vibration is mainly the forced regenerative chatter and its amplification ratio η is as high as 196.

(2) One of the characteristics of forced regenerative chatter is that its frequency does not change with the passage of machining time and it can even absorb the vibration energy of other frequency components to join it. This characteristic can be used in the judgment of forced regenerative chatter.

(3) The control strategy of shifting main mode of chatter consists in increasing the ω_n to make it far apart form ω_0, thereby restraining the forced regenerative chatter. This strategy has already been used successfully for the MX-4 lathe, the restraining effect being 47.6%. However, the forced regenerative chatter can not be eliminated by this strategy.

(4) The control strategy of removing forced excitation consists in removal of the forced excitation with frequency ω_0 or setting ω_0 far apart from ω_n, thereby eliminating forced regenerative chatter. This strategy can conveniently realized by changing the speed and is superior to the above-mentioned strategy. But a careful consideration should be given to other problems that may crop up owing to the change of speed.

References

1 星铁太郎. 机械加工ゾゾり现象——解析と对策. 工业调查会, 1977.

2 星铁太郎. 机械加工の振动解析. 工业调查会, 1990.

3 Tobias S A. *Machine Tool Vibration*. Blachie, 1965.

4 Yang Xuguang, Chen Jiwu. *An Analysis on the Dynamic Characteristics of Cutting Process and the Stability of Machine Tool*. Machine Tool Vibration and Noise, 1984, (1): 17-29

5 Leonard Meirovitch. *Elements of Vibration Analysis*. McGraw-Hill Book Company, 1975.

(原载 Chinese Journal of Mechanical Engineering, Volume 5, Number 1, 1992)

智能制造技术与智能制造系统的发展与研究

杨叔子 丁 洪

提 要 评述了智能制造技术与智能制造系统,指出了智能制造确系21世纪的制造技术,分析了智能制造在发展中的问题,提出我国智能制造的近期研究重点应为其关键基础技术。

关键词:智能制造,智能制造技术,智能制造系统,智能机器,集成化,智能化

1 智能制造系统的研究背景与发展现状

近来年,人们对制造过程的自动化程度赋予了极大的研究热情,这是因为从1870年到1980年间,制造过程的效率提高了20倍,而生产管理效率只提高了1.8～2.2倍,产品设计的效率只提高了1.2倍。这表明体力劳动通过采用自动化技术得到了极大的解放,而脑力劳动的自动化程度(其实质是决策自动化程度)则很低,制造过程中人的因素尚未得到充分的认识,人尚未真正地从复杂的生产过程中解放出来,各种问题求解的最终决策在很大程度上仍依赖于人的智慧。因而,人类群体所面临的众多问题(包括社会问题、生理问题等)在制造过程中都有所反映。面对批量小、品种多、质量高、更新快的产品市场竞争要求以及各种社会因素的综合影响,制造过程的自动化程度的提高面临众多问题,譬如:(1)专家人才的短缺和转移致使一些专门技能不能及时或长久地得到提供;(2)现代制造过程中信息量大而繁杂,传统的信息处理方式已不能满足要求,大量的信息资源需要开发与共享;(3)对制造环境柔性要求更高,决策过程更加复杂,决策时间要求更短;(4)制造过程的自动化程度受制于制造系统的自组织能力,即智能水平;(5)现代生产要求专家们在更大范围内进行更及时的合作,小到一个企业内部的各个生产环节,大至一个国家甚至世界范围内的工业界中的众多企业之间。各种

迹象表明,"我们正处在制造历史上的一个危险时期"。幸运的是,计算机与计算机科学以及其他高技术的发展,通过集成制造技术、人工智能等而发展起来的一种新型制造工程——智能制造技术(intelligent manufacturing technology,IMT)与智能制造系统(intelligent manufacturing system,IMS)使我们有可能走出这个危机,"带来真正的第二次工业革命"。这是因为,制造过程所面临的众多问题的核心是"制造智能"(manufacturing itelligence)和制造技术的"智能化"(intellecturalization)。

IMT 是指在制造工业的各个环节以一种高度柔性与高度集成的方式,通过计算机模拟人类专家的智能活动,进行分析、判断、推理、构思和决策,旨在取代或延伸制造环境中人的部分脑力劳动;并对人类专家的制造智能进行收集、存储、完善、共享、继承与发展。未来工业生产的基本特征应该是知识密集型,制造自动化的根本是决策自动化。

目前,IMT & IMS 的研究正迅速受到众多国家的政府、工业界和科学家们的广泛重视,研究方向从最初的"人工智能在制造领域中的应用"发展到今天的 IMS,研究课题涉及的范围由最初仅一个企业内部的市场分析、产品设计、生产计划、制造加工、过程控制、材料处理、信息管理、设备维护等技术型环节的自动化,发展到今天的面向世界范围内的整个制造环境的集成化与自组织能力,包括制造智能处理技术、自组织加工单元、自组织机器人、智能生产管理信息系统、多级竞争式控制网络、全球通信与操作网等,前后历时还不到 10 年时间。这期间,美、英等国都有专著介绍制造智能,如智能制造研究领域的首本专著于 1988 年出版[1],探讨智能制造内涵与前景,定义制造智能的目的是"通过集成知识工程、制造软件系统、机器人视觉和机器人控制来对制造技工们的技能与专家知识进行建模,以使智能机器能够在没有人工干预的情况下进行小批量生产。国际知名学者共同创办的刊物 Int. J. of Intelligent Manufacturing 也已于 1990 年问世。美国国家标准局自 20 世纪 80 年代中期开始的著名的 AMRF 工程的三大任务之一即是"为下一代的以知识库为基础的自动化制造系统(即 IMS)提供研究与实施设施,AMRF 在开展各项课题时,均充分考虑加入有关人工智能的内容,实验设备亦留有扩展的余地"。英国 Birmingham Alcan 铅制件公司所属的 Kitts Green 工厂正在执行一项投资 80 万英镑的研究课题,即"具有思维功能的工厂"。在智能制造的研究与发展计划方面,最引人注目的是日本倡导的智能制造国际合作研究计划,其最终目的是研究开发出能使人和智能设备都不受生产操作和国界限制的彼此合作的系统。

人工智能在制造领域中的应用与 IMT & IMS 的一个重要区别在于：IMT & IMS 首次以部分取代制造中人的脑力劳动为研究目标，而不再仅起"辅助或支持"作用，在一定范围还需要能独立地适应周围环境，开展工作。值得指出的是，CIMS 也是面向制造过程自动化的系统，与 IMS 密切相关但又有区别。CIMS 强调的是企业内部物料流的集成和信息流的集成，而 IMS 强调的则是更大范围内的整个制造过程的自组织能力。从某种意义而言，后者难度更大。CIMS 中的众多研究内容是 IMS 的发展基础，而 IMS 也将对 CIMS 提出更高的要求。集成是智能的基础，而智能也将反过来推动更高水平的集成。目前，CIMS 在工业发达国家已进入工业实践阶段，而 IMT & IMS 尚处于概念研究和实验研究阶段。IMT & IMS 的研究成果将不只是面向 21 世纪制造业，不只是促进 CIMS 达到高程度的集成，而且对于 FMS、MS、CNC 以至一般的工业过程的自动化或精密生产环境而言，均有潜在的应用价值。

国内在智能制造技术与系统方面的绝大多数研究工作，目前还处在探讨人工智能在制造领域中应用的阶段。几年来，开发出了类型众多、水平各异的面向制造过程中特定环节、特定问题的"智能化孤岛"，诸如专家系统、基于知识的系统和智能辅助系统等，而对制造环境的全面"智能化"研究工作还处于刚刚起步阶段。譬如，笔者与合作者从 1986 年即开展了人工智能与制造领域中的应用研究工作，研究课题涉及范围先后有智能监测与诊断系统、制造过程的智能控制、智能管理信息与生产规划系统、基于知识的数据库技术、制造质量信息的智能处理系统、FMS 作业的智能调度与控制、智能预测技术、产品设计与制造中可视（几何）知识处理技术等。并在国家自然科学基金委员会 1988 年组织的"机械制造的未来"的研讨会上首次探讨了"智能制造"的研究问题，并密切注意国际上这一领域的研究动态，对本发展方向的研究难点和存在的问题有一定深度的认识和体会。

总之，智能制造是 21 世纪的制造技术，作为其特征的双 I（integration & intelligence）将是 21 世纪制造业赖以行进的基本轨道。从更深刻的意义上讲，智能制造是从信息时代走向智能时代面临的第一个严重任务。

2　存在的问题

总的说来，目前 IMS 的研究仍处在人工智能在制造领域中应用的阶段，研究课题涉及市场分析、产品设计、制造过程控制、材料处理、信息管理、设备维护等众多方面，研究工作取得了丰硕的成果，开发了种类繁多的面向特定领域的专家系统、基于知识的系统和智能辅助系统,甚至智能加工工作站(IMW),形成了一系列"智能化孤岛"(islands of intelligence)。这中间包括 CIMS 研究中所取得的有关进展。然而,随着研究与应用工作的深入,人们逐渐地认识到自动化程度的进一步提高依赖制造系统的自组织能力,研究工作还面临着一系列理论、技术和社会问题,问题的核心是"智能化"。一般说来,现代工业生产作为一个有机的整体要受技术(包括生产系统)、人(包括间接影响生产过程的社会群体)和经济(包括市场竞争和社会竞争)三方面因素的制约。从技术的角度来看,对一个企业来说,市场预测、生产决策、产品设计、原料订购与处理、制造加工、生产管理、原料产品的储运、产品销售、研究与发展等环节彼此相互影响,构成产品生产的全过程。该过程的自动化程度取决于各环节的集成自动化(integrated automation)水平,而生产系统的自组织能力取决于各环节的集成智能(integrated intelligence)水平。目前,尚缺乏这种"集成"制造智能的技术,这也是目前"并行工程"的研究重点。

由日本提出的国际合作研究计划对 IMS 的解释可看出,IMS 的研究包括三个基本方面:智能活动、智能机器和两者的有机融合技术,其中智能活动是问题的核心。在 IMS 研究的众多基础技术中,制造智能处理技术(manufacturing intelligence processing technology)是最为关键和迫切需要研究的问题之一,因为它负责各环节的制造智能的集成和生成智能机器的智能活动。对一个国家甚至世界范围内的工业界来说,众多企业之间有着密切的联系,譬如,采用相同的生产设备和系统,有着类似的生产控制与管理方式,生产不同产品的企业之间存在业务往来等等。这中间存在的突出问题是产品和技术的规范化、标准化和通用化,信息自动交换形式与接口以及制造智能共享等。

从人的因素方面来看,它包括以下四个方面。其一,企业内部负责各个环节的专家和技术人员有着各自不同的知识背景和解决问题的策略,他们应该"坐"在一起,通过相互之间充分的合作、协商与理解,"并行"地开展制造过程中各环节的工作,把以后可能出现的"隐患"和"反复"降低到最低程度。其二,人们参与制造过程的智能行为和知识存在着多种层次水平、多种类型。因而要采用多种表示方式。其三,参与制造过程的群体,作为社会中的一子集,受社会发展变更的影响,这种影响都将对制造过程产生既有积极又有消极的作用。最后,人与人之间存在生活、语言、社会背景等方面的差别。总之,人的因素对现代生产的自动化程度有着关键作用。事实上,在 20 世纪 70 年代末和 80 年代初,人们已开始认识到人的因素在现代工业生产中的作用,IFS(英国出版公司)于 1984 年就首次发起了第一届"制造中人的因素"研讨会,目的在于提高人们对制造环境中人的因素及其所起作用的认识。事实证明，人的因素是 IMS 中制造智能的重要来源。

从经济因素来看，它包括以下三个方面。第一，IMS 系统的主要目标之一是全面提高制造过程的生产与经济效益，它将把制造过程自动化的概念更新和拓宽到"集成化"和"智能化"的高度，从而具有更强的市场竞争能力。但如何设定和评价 IMS 的各项经济性指标和性能则是一个问题。第二，目前，在工业发达国家普遍存在着劳动力昂贵，所占生产成本的比例越来越高的问题。从当前的经济利益出发，大量的制造企业被转移至发展中国家，致使生产技术和劳动者因素等方面受到牵制，存在丧失他们产品市场竞争力的危险。这也是智能制造国际合作研究计划提出的重要原因之一。此外，在发达国家，工人的知识和专业素质的迅速提高，使得他们纷纷离开工厂，去寻找其他更合适的工作，因而使得发达国家的熟练工人和技术人员缺乏。在我国，企业与技术转移的情况目前尚不严重，但存在一个逆问题，那就是如果发达国家一旦拥有了 IMS，而我们在这方面与他们相差甚远的话，那时，我们将面临失去更多的与他们竞争机会的危险。因为 IMS 是 21 世纪的制造技术，这些发达国家将不再"依赖"发展中国家的"廉价"劳动力。另一方面，专业人员与技术力量缺乏的问题对于我国来说，情况尤其严重。因此，我国也只能开发出具有自身特色的 IMS，提高企业的经济效益，方能在 21 世纪制造业中争得更多的竞争机会。最后，在当前企业的生产中，产品制造的各个环节与产品的销售部门事实上是相脱节的，而企业的经济效益在很大程度上取决于其销售部门的工作成果，销售部门的工作人员的才智被推崇为"推销艺术"。因此，如何使销售部门"智能化"，如何开发企业的"销售智能"，并融合于制造智能中，也是一个有待研究的问题。

3 研究方向与课题

根据国内现有的工作基础和国家的需要，以及 IMT & IMS 研究与开发工作的特点，我们认为近期的研究点应该放在 IMT & IMS 的关键基础技术上，它主要包括以下内容。

3.1 智能制造系统理论基础与设计技术

IMS 的概念正式提出至今仅二三年时间。作为制造工程中的一个全新的概念，IMS 理论基础与体系尚未完全形成，它的精确内涵和设计技术亟待进一步研究，具体研究内容应包括：

3.1.1 体系结构与发展战略

需要建立 IMS 统一的概念体系,研究 IMS 的系统组成和发展方向以及跟踪国际上该领域的研究前沿。

3.1.2 开发环境与设计方法学

IMS 的开发与设计方法将有别于现有任何制造系统的设计方法,因为 IMS 是面向整个制造过程的系统和各个环节的"智能化"的。因此,有必要研究 IMS 的设计策略和开发环境(包括开发语言、操作系统、开发工具等)。必须强调 IMS 设计过程的标准化、模块化和通用化。

3.1.3 评价技术

研究制造过程中的设计评价、生产评价、材料评价、管理评价、市场评价、经济评价、报价评价和功能评价等问题。

3.2 制造智能理论及处理技术

现代工业生产作为一个有机整体不仅是指各制造环节之间存在的技术型联系,而且还表现在人类专家的制造智能的统一体特性方面。制造智能理论及处理技术就是要研究整个制造环境中的各种智能源的开发、描述、集成、共享与处理,最后生成智能机器的智能活动,具体研究内容如下。

3.2.1 制造环境的描述与建模

研究描述制造环境的一致性概念体系、制造过程建模,以及影响制造过程的多因素分析与不确定性处理。

3.2.2 制造智能处理技术

重点研究制造智能源的开发与获取、制造智能的表示、制造智能的集成与共享。

3.2.3 智能活动的生成与融合

研究智能活动的生成策略、智能活动的机器化技术。

3.3 智能制造单元技术的集成

近 10 年来,人工智能在制造领域中的应用研究取得较大进展,建立了一些智能制造单元技术。为了应用于实际制造过程和面向 21 世纪制造工业,这些单元技术除了需要进一步完善与发展外,更重要的是研究如何集成这些单元技术。

3.3.1 并行智能设计

并行工程方法学这一概念是 1986 年由美国国防部定义,并首先应用于美国军事武器系统开发计划 DOS CALS 的。为了在制造过程的设计阶段能有效地模仿由来自各环节制造专家组成的专家组(expert team)的智能行为,集成和共享各环节与各方面的制造智能,并行地开展产品环节的设计工作,必须研究并行智能设计的支撑环境、产品描述的统一模型、设计智能交互和并行智能设计方法学。

3.3.2 生产过程的智能调度、规划、仿真与优化

现代生产过程要面临多信息源、多因素、多对象的及时处理问题,生产过程的调度与规划中的智能决策问题的研究是迫在眉睫的。仿真与优化是实现设计和过程评估的有效途径。目前,更强调对设计、制造、装配、使用、维修等过程的优化与动态仿真。

3.3.3 产品质量信息的智能处理系统

研究整个制造过程的"全质量"(total quality)模型和建立相应的质量数据库,研究质量状态的智能决策和质量过程的智能控制。

3.3.4 制造过程与系统的智能监视、诊断、补偿与控制

研究面向在强干扰、多因素条件下监视与诊断模型,研究制造过程的动态辨识与自适应技术。

3.3.5 生产与经营管理的智能决策系统

研究多因素、多目标智能决策模型,研究生产过程的实时跟踪技术,研究产品市场评估与预测模型。

3.4 知识库系统与网络技术

知识库系统与信息网络技术是制造过程的系统与各环节"集成智能化"的支撑,在 IMT & IMS 研究中占有重要地位。

3.4.1 分布式异构联想知识库系统

研究知识库异构、知识库分布式策略与维修、知识库联想和分布数据库技术。

3.4.2 信息控制与网络通信技术

研究 IMS 中各种信息的交换接口、网络通信技术、系统操作控制策略。

3.5 智能机器的设计

智能机器是 IMS 中模仿人类专家智能活动的工具之一,是新一代的制造工具,因而,研究智能机器的设计方法及其相关技术将有划时代的意义。

3.5.1 机器人智能技术

智能机器人将在 IMS 中占有重要的地位,主要体现在机器的视觉和机器人控

制两个方面,有必要研究智能机器眼(视觉)、信息感知与智能传感器、智能机器手(控制)和智能机器的自适应定位与夹具设计等技术。

3.5.2 机器自学习与自维护技术

研究智能机器的自适应学习模型,系统误差的自动恢复与维护技术。

3.5.3 智能制造单元机的设计与制造

研究智能制造单元机的结构组成与设计方法、新型材料的应用技术。

3.6 制造中人的因素

IMS 的宗旨之一就是减轻人类制造专家的艰苦的脑力劳动负担,因此,与脑力劳动有密切联系的制造中人的因素理应受到充分的重视,研究内容如下。

3.6.1 人-系统柔性交互技术

研究人-系统柔性、联想、容错交互模型以及交互环境。

3.6.2 未来制造环境的设计

研究人在未来制造环境中的地位和作用以及未来舒适、友好的制造环境的设计。

3.6.3 人才培养与教学系统

研究面向 IMT & IMS 的人才培养计划,研制教学示范系统。

参 考 文 献

[1] Wright P K, Bourne D A. *Manufacturing Intelligence*. Addison—Wesley,1988.

[2] Bernold T. *AI in Manufacturing*. McGraw—Hill,1985.

[3] Kusiak A. *Intelligent Manufacturing Systems*. Prentice—Hall,1990.

[4] Milacic V R. *Intelligent Manufacturing Systems* Ⅰ,Ⅱ,Ⅲ,Elsveier,1988,1989,1990.

[5] 杨叔子,丁洪. 机械制造的发展及人工智能的应用. 机械工程,1988(1):32～34.

[6] 杨叔子,余俊,丁洪等. 产品设计、制造、维护的智能辅助技术. 机械工程,1990(3):2～6.

[7] 杨叔子. 机械制造走向智能化. 机械工程,1990(增刊):3～8.

[8] 《机械制造走向 2000 年——回顾、展望与对策》,国家自然科学基金委员会,1989.

(原载《中国机械工程》1992 年第 2 期)

两类小波函数的性质和作用

何岭松　吴　波　吴　雅　杨叔子

提　要　本文简要地介绍了信号的时-频两维分析的目的、意义,分析对比了小波分析和短时傅里叶分析、Wigner 谱分析的异同及主要优点。简述了离散小波变换、离散二进小波变换和离散二进正交小波变换的数字实现方法。文中详细论述了反对称小波和对称小波两类不同小波函数的性质及对小波变换的影响,并分析了它们在异常信号检测中的不同作用。最后介绍了离散二进小波变换滤波器的设计方法,给出了小波互补滤波器的幅频特性公式,并列举了两个滤波器设计实例。

关键词:信号检测,滤波器,时-频分析,小波分析

1　引　言

经典的信号分析方法主要是在时域或频域进行,时域分析简单、直观,而频域分析则更加准确、精练,许多在时间域内看不清楚的问题转换到频域一下就变得很清晰。例如,图 1 是一受白噪声干扰的周期信号,其时域波形杂乱无章,而频谱则很简洁。

但是用傅里叶变换计算信号的频谱时,假定信号是平稳的

$$F(f) = \int_{-\infty}^{+\infty} f(t) e^{-j2\pi ft} dt \tag{1}$$

频谱 $F(f)$ 反映的是从"全局"角度看信号在整个时间轴上的频率构成。对齿轮、滚动轴承表面缺陷引起的调幅振动,机组启、停过程的暂态振动和机床的切削颤振,振动信号的幅值特性和频率特性随时间在不断地改变,单纯的时域分析和频域分析都不能反映信号的特征,必须将两者结合起来才能有效地对信号进行分

析。图 2 是一时变信号的时-频两维表达,在时间-频率相平面上,清楚地反映了信号的时变特征,这一点从时域曲线或频谱上是很难做出判断的。

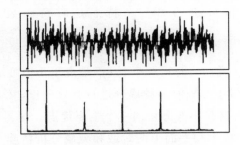

图 1　信号的时域和频域表达

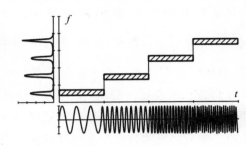

图 2　信号的时-频两维表达

信号的时-频分析方法很多[1-3],其中常用的有短时傅里叶分析、Wigner 谱分析和小波分析三类。短时傅里叶分析定义为

$$\text{STFT}(t,f) = \int_{-\infty}^{+\infty} f(\tau)h(\tau-t)\text{e}^{-\text{j}2\pi f\tau}\text{d}\tau \tag{2}$$

它通过对信号加一滑移时窗 $h(\tau-t)$ 对信号进行分段截取,将其化为若干段局部平稳的信号,对它们分别取傅里叶变换后,得到一组信号的"局部"频谱,从不同时刻的"局部"频谱的差异上,可以看出信号的时变特征。该法简单易行,但分析精度不高,其时域精度要求和频域精度要求相互矛盾,要提高频域分辨率就得加大时窗长度,而时窗增加信号的"局部"性又减弱,造成时域分析精度下降,另外,时窗越长,信号的"局部"平稳性也难以保证。

Wigner 谱直接定义为信号的时-频两维联合分布函数

$$WD(t,f) = \int_{-\infty}^{+\infty} f^*\left(t-\frac{\tau}{2}\right)f\left(t+\frac{\tau}{2}\right)\text{e}^{-\text{j}2\pi f\tau}\text{d}\tau \tag{3}$$

式中,"*"表示共轭。Wigner 谱克服了短时傅里叶分析的上述缺点,能更有效地对时变信号进行分析。但 Wigner 谱分析也有其明显的不足之处,由于时移因子 $\left(\frac{\tau}{2}\right)$ 的作用、离散 Wigner 谱的周期为 π,信号必须以大于乃奎斯特采样频率两倍以上的频率进行采样,或者采用原信号的解析信号[7]进行分析;另外对多频率分量构成的信号,Wigner 谱中存在着严重的干涉现象。图 3 是两个稳态正弦信号的 Wigner 谱,谱图的中间一项就是干涉成分。

虽然干涉可以通过加谱窗进行抑制,但那也是以牺牲时域分析精度为代价的。小波分析是近年来出现的一种新的时-频分析方法,它定义为

$$Wf(t,s) = \frac{1}{\sqrt{s}}\int_{-\infty}^{+\infty} f(\tau)\psi\left(\frac{\tau-t}{s}\right)\text{d}\tau \tag{4}$$

与前两种方法不同的是小波变换用尺度算子代替了频率移动算子,将时间-频率相平面换为时间-尺度平面,尺度轴上的小尺度对应着频率轴上的高频,尺度轴上的大尺度则对应着频率轴上的低频。小波分析的另一个特点是时窗函数 $\psi\left(\frac{\tau-t}{s}\right)$ 是一变特性窗,在时间-尺度相平面上信号的高频段,s 值小,时窗长度短;而在时间-尺度相平面上信号的低频段,s 值大,时窗长度长;正是由于小波变换时窗特性可调这一特点,使其既能对信号中的短时高频成分进行有效分析,又能对信号中的低频缓变成分进行精确估计,这是小波分析的一个主要优点。图 4 给出了短时傅里叶分析和小波分析在时间-频率相平面上的时域和频域分析精度分布情况。

图 3　Wigner 谱的干涉现象

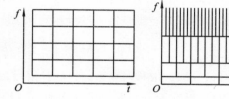

图 4　短时傅里叶分析和小波分析的时-频分析精度

小波分析的性能与时窗函数 $\psi(t)$,也就是小波函数的性能有很大关系,小波函数不同,则小波变换的分析速度和精度也不相同,另外对分析结果的解释也有很大差异。下面对目前工程中广泛应用的对称小波和反对称小波的性质及在小波变换中的作用进行介绍。

2　小波变换和快速算法简介

所谓小波(wavelet)是指其时域波形和频域波形都是有限的一类函数,小波函数 $\psi(t) \in L^2(R)$($L^2(R)$ 是可测平方可积一维函数 $f(t)$ 的向量空间)满足

$$\int_{-\infty}^{+\infty} \psi(t) \mathrm{d}t = 0 \tag{5}$$

定义小波函数 $\psi(t)$ 的展缩小波为

$$\psi_s(t) = \frac{1}{\sqrt{s}} \psi\left(\frac{t}{s}\right) \tag{6}$$

代入式(4),小波变换可以简记为

$$Wf(t,s) = f(t) * \psi_s(t) \tag{7}$$

当小波函数满足

$$C_\psi = \int_0^{+\infty} \frac{|\hat{\psi}(\omega)|^2}{\omega} d\omega < +\infty \tag{8}$$

时,小波变换是可逆的,其逆变换公式为

$$f(t) = \frac{1}{C_\psi} \int_{-\infty}^{+\infty} \int_0^{+\infty} Wf(\tau,s)\psi_s(t-\tau) ds d\tau \tag{9}$$

式中,$\hat{\psi}(\omega)$ 为 $\psi(t)$ 的傅里叶变换。

为了便于计算必须对式(7)进行离散化处理,设信号 $f(x)$ 的离散采样序列为 $f(m)$,展缩小波 $\psi_s(t)$ 的离散采样序列为 $\psi_s(k)$,则有离散小波变换公式

$$Wf(n,s) = \sum_{m=-\infty}^{+\infty} f(m-n)\psi_s(m) \tag{10}$$

由于计算机所能处理的数据长度是有限的,用窗函数 $h(l)$,$h(l) = 0$,当 $|l| \geqslant L$,对数据进行截断,有

$$Wf(n,s) = \sum_{m=-L+1}^{L-1} h(m)f(m-n)\psi_s(m) \tag{11}$$

上式可以按圆卷积方式用快速傅里叶变换(FFT)算法进行计算。

离散小波变换的缺点是计算量太大,为了提高计算效率 Mallat[4] 提出了离散二进小波算法,其特点是将变量 s 按二进取值

$$s = 2^j, 1 \leqslant j \leqslant J \tag{12}$$

相应的小波变换称为离散二进小波变换。为了不漏掉信号中的任一频率成分,小波 $\psi(t)$ 按二进方式展缩后形成的小波基,其频率通带必须能覆盖整个频率轴

$$\sum_{j=-\infty}^{+\infty} |\hat{\psi}(2^j\omega)|^2 = 1 \tag{13}$$

为便于公式推导,引入一个平滑函数 $\varphi(t)$,它满足

$$|\hat{\varphi}(\omega)|^2 = \sum_{j=1}^{+\infty} |\hat{\psi}(2^j\omega)|^2 \tag{14}$$

定义平滑算子

$$Sf(n,2^j) = f(n) * \varphi_{2^j}(n) \tag{15}$$

由式(14)有

$$|\hat{\varphi}(\omega)|^2 = \sum_{j=1}^{J} |\hat{\psi}(2^j\omega)|^2 + |\hat{\varphi}(2^J\omega)|^2 \tag{16}$$

利用巴士瓦定理得

$$\|Sf(n,1)\|^2 = \sum_{j=1}^{J} \|Wf(n,2^j)\|^2 + \|Sf(n,2^J)\|^2 \tag{17}$$

式(17)变形后有
$$\|Wf(n,2^j)\|^2 = \|Sf(n,2^{j-1})\|^2 - \|Sf(n,2^j)\| \tag{18}$$
这说明任一阶小波系数可以由对应的相邻两阶平滑系数得到。

为计算方便,引入一对互补的低通和高通滤波器 $\hat{H}(\omega), \hat{G}(\omega)$
$$\left. \begin{array}{l} \hat{\varphi}(2\omega) = \hat{H}(\omega)\hat{\varphi}(\omega) \\ \hat{\psi}(2\omega) = \hat{G}(\omega)\hat{\varphi}(\omega) \\ |\hat{G}(\omega)|^2 + |\hat{H}(\omega)|^2 = 1 \end{array} \right\} \tag{19}$$

由平滑算子和小波变换的傅里叶变换
$$\left. \begin{array}{l} \hat{S}f(\omega,2^j) = \hat{\varphi}(2^j\omega)\hat{f}(\omega) \\ \hat{W}f(\omega,2^j) = \hat{\psi}(2^j\omega)\hat{f}(\omega) \end{array} \right\} \tag{20}$$

可以很容易地得到小波变换的递推计算公式
$$\left. \begin{array}{l} \hat{S}f(\omega,2^j) = \hat{S}f(\omega,2^{j-1})\hat{H}(2^{j-1}\omega) \\ \hat{W}f(\omega,2^j) = \hat{S}f(\omega,2^{j-1})\hat{G}(2^{j-1}\omega) \end{array} \right\} \tag{21}$$

变换到时域有
$$\left. \begin{array}{l} Sf(n,2^j) = Sf(n,2^{j-1}) * h_{j-1} \\ Wf(n,2^j) = Sf(n,2^{j-1}) * g_{j-1} \end{array} \right\} \tag{22}$$

同理可得到小波逆变换的递推计算公式
$$Sf(n,2^{j-1}) = Sf(n,2^j) * \bar{h}_{j-1} + Wf(n,2^j) * \bar{g}_{j-1} \tag{23}$$

式中,h_0 和 g_0 为滤波器 $\hat{H}(\omega)$ 和 $\hat{G}(\omega)$ 的单位脉冲响应,h_j 和 g_j 则可在 h_0 和 g_0 的每两个系数间补 2^{j-1} 个零系数得到
$$\bar{h}(n) = h(-n), \quad \bar{g}(n) = g(-n)$$

按式(22)和式(23),利用数字滤波技术,可以快速地对信号进行离散二进小波变换和逆变换。

如果变换中采用的小波函数是正交小波,其展缩平移小波 $(\sqrt{2^{-j}}\psi_{2^j}(t-2^{-j}n))_{(j,n)} \in \mathbf{Z}^2$($\mathbf{Z}^2$ 为整数二次方空间)构成一个正交基,可以采用正交分解技术将信号分解为各个正交小波分量的加权和。离散二进正交小波变换快速算法[5]的推导过程和上述的离散二进小波变换快速算法的推导过程很相似,递推公式形式也相同,但滤波算法有些区别,正变换递推公式为
$$Sf(n,2^j) = \sum_{k=0}^{M-1} h(k)Sf(2n+k,2^{j-1}) \tag{24}$$

$$Wf(n,2^j) = \sum_{k=0}^{M-1} g(k) Sf(2n+k, 2^{j-1}) \qquad (25)$$

逆变换递推公式为

$$Sf(n,2^{j-1}) = \sum_{0 \leqslant n-2k < M} h(n-2k) Sf(k,2^j) + \sum_{0 \leqslant n-2k < M} g(n-2k) Wf(k,2^j)$$
$$(26)$$

式中,$\{h(k)\}_{k=0}^{M-1}$,$\{g(k)\}_{k=0}^{M-1}$ 为共轭滤波器 $\hat{H}(\omega)$ 和 $\hat{G}(\omega)$ 的单位脉冲响应。$\hat{H}(\omega)$ 和 $\hat{G}(\omega)$ 满足

$$\left.\begin{aligned}\hat{\varphi}(\omega) &= \prod_{p=1}^{+\infty} \hat{H}(e^{-j2^{-p}\omega}) \\ \hat{\psi}(2\omega) &= \hat{G}(e^{-j\omega}) \hat{\varphi}(\omega) \\ |H(e^{-j\omega})|^2 + |H(-e^{-j\omega})|^2 &= 1 \\ G(e^{-j\omega}) &= e^{-j\omega} H^*(e^{-j\omega})\end{aligned}\right\} \qquad (27)$$

在上述两种快速算法中,离散二进小波变换快速算法将信号在时间-尺度相平面上展开为一组带通信号 $\{Wf(n,2^1)_{n=0}^{N-1}, \cdots, Wf(n,2^J)_{n=0}^{N-1}, Sf(n,2^J)_{n=0}^{N-1}\}$,其中的每一个数据对应着相平面上的一个点,描述了信号的时-频分布特性。离散二进正交小波变换把信号分解为一组长度按二进方式递减的带通信号 $\{Wf(n,2^1)_{n=0}^{\frac{N}{2}-1}, \cdots, Wf(n,2^J)_{n=0}^{\frac{N}{2^J}-1}, Sf(n,2^J)_{n=0}^{\frac{N}{2^J}-1}\}$,其系数也对应于相平面上的各个点,但由于在正交分解过程中,信号不断地按二点抽一的方式重采样,时间轴在不断地压缩,因此,各阶小波系数和平滑系数的时间刻度比例是不同的。在这两种方法中,离散二进小波变换更多地用于信号异常成分检测、时-频两维滤波等信号的时变特性分析,而离散二进正交小波变换则主要应用于图像处理中图像的多精度分解和数据压缩。

3 两类小波函数的性质

小波函数一般不直接定义,而是由平滑函数 $\varphi(t)$ 导出,平滑函数满足

$$\left.\begin{aligned}\int_{-\infty}^{+\infty} \varphi(t) dt &= 1 \\ \lim_{t \to \pm\infty} \varphi(t) &= 0\end{aligned}\right\} \qquad (28)$$

例如高斯函数和三次样条函数就是典型的平滑函数。令该函数是二次可微的,其一阶导数 $\psi^a(t)$ 和二阶导数 $\psi^b(t)$ 为

$$\left.\begin{aligned}\psi^a(t) &= \frac{\mathrm{d}\varphi(t)}{\mathrm{d}t} \\ \psi^b(t) &= \frac{\mathrm{d}^2\varphi(t)}{\mathrm{d}t^2}\end{aligned}\right\} \quad (29)$$

由式(28)得

$$\left.\begin{aligned}\int_{-\infty}^{+\infty}\psi^a(t)\mathrm{d}t &= 0 \\ \int_{-\infty}^{+\infty}\psi^b(t)\mathrm{d}t &= 0\end{aligned}\right\} \quad (30)$$

因此，$\psi^a(t)$ 和 $\psi^b(t)$ 都是小波函数。图5是三次样条函数和其一阶导数、二阶导数的波形。

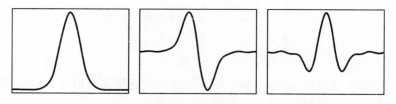

图 5　$\varphi(t)$、$\psi^a(t)$ 和 $\psi^b(t)$ 的波形

由于它们的图形特征，分别称 $\psi^a(t)$ 为反对称小波，$\psi^b(t)$ 为对称小波。不难证明，如果平滑函数是 n 次可微的，则其 n 阶导数是小波函数。

反对称小波的小波变换为

$$W^a f(t,s) = f(t) * \psi_s^a(t) = f(t) * \left[s\frac{\mathrm{d}\varphi_s(t)}{\mathrm{d}t}\right] = s\frac{\mathrm{d}}{\mathrm{d}t}[f(t) * \varphi_s(t)] \quad (31)$$

对称小波的小波变换为

$$W^b f(t,s) = f(t) * \psi_s^b(t) = f(t) * \left[s^2\frac{\mathrm{d}^2\varphi_s(t)}{\mathrm{d}t^2}\right] = s^2\frac{\mathrm{d}^2}{\mathrm{d}t^2}[f(t) * \varphi_s(t)] \quad (32)$$

它们分别正比于平滑算子作用后的平滑信号的一阶和二阶导数。由函数导数的知识可以知道，采用对称小波和反对称小波的小波变换的结果是完全不同的，例如 $W^a f(t,s)$ 中的极值点对应的是 $W^b f(t,s)$ 中的零点。这一点在使用中一定要注意，否则会得出一些错误结论。

4　两类小波函数在异常信号检测中的作用

异常信号是指信号中突变的不规则成分，如设备探伤中裂纹产生的反射信号，零件表面缺陷产出的冲击信号等。异常信号的位置、强弱反映了故障的位置和程度,因此,异常信号的检测具有重要的实用价值。

异常信号分边缘跳变和峰值跳变两类,对边缘跳变检测问题,在理想情况下可以认为是信号在异常处叠加了一个阶跃函数。对阶跃函数 $u(t)$ 取小波变换

$$WU(t,s) = u(t) * \psi_s(t) \tag{33}$$

若选用的是反对称小波函数,有

$$WU(t,s) = s\frac{\mathrm{d}}{\mathrm{d}t}[u(t)*\varphi_s(t)] = s \cdot \varphi_s(t) \tag{34}$$

若选用的是对称小波函数,有

$$WU(t,s) = s^2\frac{\mathrm{d}}{\mathrm{d}t}[u(t)*\varphi_s(t)] = s^2\varphi_s^a(t) \tag{35}$$

由平滑函数和其一阶导数的形状可知,在信号中边缘跳变点处,反对称小波变换产生最大值,而对称小波变换产生过零值,其情形如图 6 所示。

对峰值跳变检测问题,在理想情况下可以认为是信号在异常处叠加了一个脉冲函数。对脉冲函数 $\delta(t)$ 取小波变换

$$W\delta(t,s) = \delta(t) * \psi_s(t) \tag{36}$$

若采用的是反对称小波函数,有

$$W\delta(t,s) = s \cdot \frac{\mathrm{d}}{\mathrm{d}t}[\delta(t)*\varphi_s(t)] = s \cdot \psi^a(t) \tag{37}$$

若采用的是对称小波函数,有

$$W\delta(t,s) = s^2\frac{\mathrm{d}^2}{\mathrm{d}t^2}[\delta(t)*\varphi_s(t)] = s^2\psi^b(t) \tag{38}$$

由小波 $\psi^a(t)$ 和 $\psi^b(t)$ 的形状可知,在信号中峰值跳变处,反对称小波变换产生过零值,而对称小波变换产生最大值,其情形如图 7 所示。

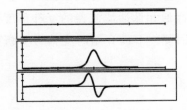

图 6　阶跃函数和其两类小波变换波形

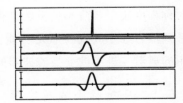

图 7　脉冲函数和其两类小波变换波形

从理论上讲,用两类小波变换后的最大值或过零值特征来判断信号中的异常点,其作用是等价的;但在实际应用中,最大值法更易确认,不易受信号中噪声成分的干扰。因此,从这个角度上看,边缘检测应选用反对称小波函数来进行小波变换,而峰值检测则应选用对称小波函数来进行小波变换。

为了让大家对两类小波函数的差异有一个直观的印象,图8、图9分别给出了一个含有不同程度的边缘跳变和峰值跳变的典型信号的反对称小波变换和对称小波变换结果。

从图中可以看出,对信号中两个不同异常程度的边缘跳变成分,用反对称小波变换的最大点图可以准确判定,而利用对称小波变换的结果进行分析则较困难;反之,对信号中两个不同异常程度的峰值跳变成分,结论则正相反;这和前面的理论分析结果是一致的。

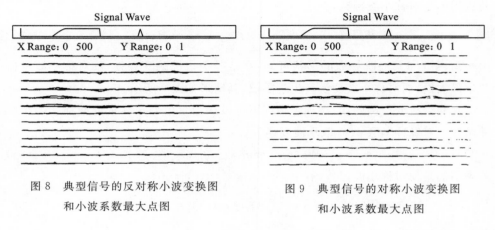

图8 典型信号的反对称小波变换图和小波系数最大点图

图9 典型信号的对称小波变换图和小波系数最大点图

5 小波滤波器设计

滤波器设计是小波分析中的关键一环,滤波器的性能和质量确定了小波变换的性能和质量。滤波器系数的长短决定了小波变换的运算量,滤波系数的对称和反对称程度决定各阶小波系数是否产生相移,而高通和低通滤波器的互补程度则关系到信号经小波变换后是否能经逆变换完全回复。下面介绍时变信号分析中常用的离散二进小波变换滤波器的设计方法。

5.1 反对称小波滤波器

小波滤波器在性能上必须满足式(19)定义的互补条件,对反对称小波,它是平滑函数的一阶导数,有

$$\left.\begin{array}{l}\psi(t) = \dfrac{\mathrm{d}\varphi(t)}{\mathrm{d}t} \\ \hat{\psi}(2\omega) = j2\omega\hat{\varphi}(2\omega)\end{array}\right\} \quad (39)$$

代入式(19)得

$$\left.\begin{array}{r}\hat{G}(\omega) = j2\omega\hat{H}(\omega) \\ |\hat{G}(\omega)|^2 + |\hat{H}(\omega)|^2 = 1\end{array}\right\} \quad (40)$$

解方程组后有

$$\left.\begin{array}{r}|\hat{H}(\omega)|^2 = \dfrac{1}{1+4\omega^2} \\ |\hat{G}(\omega)|^2 = \dfrac{4\omega^2}{1+4\omega^2}\end{array}\right\} \quad (41)$$

这是反对称小波互补滤波器 $\hat{H}(\omega)$ 和 $\hat{G}(\omega)$ 应满足的幅频条件,不难看出它们分别是一阶巴特沃斯低通和高通滤波器的特例。巴特沃斯低通滤波器的一般形式是[6]

$$|\hat{H}_B(\omega)|^2 = \dfrac{1}{1+\in^2\left(\dfrac{\omega}{\omega_c}\right)^{2N}} \quad (42)$$

式中:N 为滤波器的阶数;ω_c 为截止频率;$\in$ 为增益参数。

5.2 对称小波滤波器

对于对称小波,它是平滑函数的二阶导数,有

$$\left.\begin{array}{r}\psi(t) = \dfrac{\mathrm{d}^2\varphi(t)}{\mathrm{d}t^2} \\ \hat{\psi}(2\omega) = -4\omega^2\hat{\varphi}(2\omega)\end{array}\right\} \quad (43)$$

代入互补条件公式(19),有

$$\left.\begin{array}{r}\hat{G}(\omega) = -4\omega^2\hat{H}(\omega) \\ |\hat{G}(\omega)|^2 + |\hat{H}(\omega)|^2 = 1\end{array}\right\} \quad (44)$$

解方程组后得

$$\left.\begin{array}{r}|\hat{H}(\omega)|^2 = \dfrac{1}{1+16\omega^4} \\ |\hat{G}(\omega)|^2 = \dfrac{16\omega^4}{1+16\omega^4}\end{array}\right\} \quad (45)$$

这是对称小波滤波器应满足的幅频条件,可以看出它们分别是二阶巴特沃斯低通和高通滤波器的特例。

上述结论可以推广,若小波函数是平滑函数的 N 阶导数,则其互补滤波器是 N 阶巴特沃斯滤波器的特例。

5.3 滤波器设计实例

有了滤波器的幅频特性以后,就可以采用传统的递归滤波器或非递归滤波器的设计方法来求取小波滤波器的系数。但传统的方法设计重点放在缩短过渡带宽度、保持线性相位和抗频混等方面,所设计的小波滤波器在对称性或反对称性和互补性方面还不够满意,有必要进一步研究专门的小波滤波器设计算法。

表1和表2是笔者采用双线性变换法设计的一个反对称小波滤波器和一个对称小波滤波器的系数。

表 1 反对称小波滤波器系数

$h(n)$	$g(n)$
$h[-4]=0.0$;	$g[-4]=0.0$;
$h[-3]=0.0$;	$g[-3]=0.0$;
$h[-2]=0.0$;	$g[-2]=0.0$;
$h[-1]=0.33333333$;	$g[-1]=0.66666666$;
$h[0]=0.44444444$;	$g[0]=-0.44444444$;
$h[1]=0.1481481$;	$g[1]=-0.148148$;
$h[2]=0.0493827$;	$g[2]=-0.0493827$;
$h[3]=0.0164609$;	$g[3]=-0.0164609$;
$h[4]=0.0054869$;	$g[4]=-0.0054869$;

表 2 对称小波滤波器系数

$h(n)$	$g(n)$
$h[-4]=0.0$;	$g[-4]=0.0$;
$h[-3]=0.0$;	$g[-3]=0.0$;
$h[-2]=0.0$;	$g[-2]=0.0$;
$h[-1]=0.2929$;	$g[-1]=0.2929$;
$h[0]=0.5858$;	$g[0]=0.5858$;
$h[1]=0.2426$;	$g[1]=-0.2426$;
$h[2]=-0.1005$;	$g[2]=-0.1005$;
$h[3]=-0.0416$;	$g[3]=0.0416$;
$h[4]=0.01724$;	$g[4]=-0.01724$;

图 10 分别是用这两对滤波器对典型信号进行小波变换的 $Wf(n,2^4)$ 小波系数的波形和由小波系数经逆变换后回复的信号波形,从工程应用的角度看,两对

滤波器在性能上虽然有一点缺陷,但总的来说还是令人满意的。

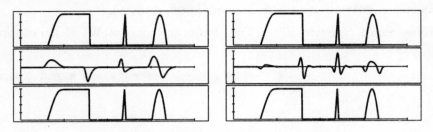

图 10 反对称和对称小波滤波器的性能

参考文献

[1] Cohen L. *Time-frequency distributions—a review*. In: Proc. IEEE, 1989;77(7):941-981.

[2] Rioul O, Vetterli M. *Wavelets and signal processing*. IEEE Sig. Proc. Magazine, Oct. 1991:14-38.

[3] Hlawatsch F, Boudreaux-bartels G F. *Linear and quadratic time-frequency signal representations*. IEEE Sig. Proc. Magazine, April 1992:21-67.

[4] Mallat S. *Zero—crossings of a wavelet transform*. IEEE Trans. Inform. Theory, 1991;37(4):1019-1033.

[5] Mallat S. *A theory for multiresolution signal decomposition: the wavelet representation*. IEEE Trans. Pat. Anal. Machine Intel. 1,1989;11(7):674-693.

[6] 何振亚.数字信号处理的理论与应用.北京:人民邮电出版社,1983.

[7] 杨福生.随机信号分析.北京:清华大学出版社,1990.

The Properties and Usages of Two Kinds of Wavelet

He Lingsong Wu Bo Wu Ya Yang Shuzi

Abstract   This paper introduces the aim of time-frequency signal analysis, compares the advantages of the Wavelet Transform with the Short-Time Fourier Transform and Wigner Distribution, and gives out the algorithms of Discret Wavelet Transform, Discret Dyadic Wavelet Transform and Discret Dyadic Orthonormal Wavelet Transform. More chiefly, the paper discusses the properties of symmetrical wavelets and anti-symmetrical wavelets, analyzes their roles in wavelet transform and compares their differences on the detection of irregular structures of the signal. The design method of the discret dyadic wavelet filters is also introduced. At last, two filter examples are presented.

Keywords: Signal detection, Filter, Time-frequency analysis, Wavelet transform

（原载《振动工程学报》1993 年第 6 卷第 4 期）

Intelligent Prediction and Control of a Leadscrew Grinding Process Using Neural Networks

Ding Hong   Yang Shuzi   Zhu Xinbiao

Abstract A new approach to intelligent prediction and control of manufacturing processes is studied by the use of neural networks in this paper. Based on a back-propagation mechanism a learning algorithm for multilayer neural networks, the "one-by-one algorithm", is presented which is particularly suitable for forecasting and process control. The differences between the one-by-one algorithm and the back-propagation algorithm in current use are clarified. Then an intelligent forecasting and control architecture for a leadscrew grinding process using the one-by-one algorithm is discussed.

Keywords: Manufacturing process modeling, Neural networks, Backpropagation algorithms, Intelligent control, Leadscrew grinding process control

1 Introduction

The identification, control and prediction of manufacturing processes in current use are almost fully based on mathematical systems theory, which in the past five decades has evolved into a powerful scientific discipline of wide applicability. The best developed aspect of the theory treats systems defined by linear operators using well established techniques based on linear algebra, complex variable theory, and the theory of ordinary linear differential equations. Looking back, the evolution in the control area has been fueled by three major needs: the need to deal with increasingly complex manufacturing systems, the need to accomplish increasingly demanding design requirements, and the need to obtain these requirements, with less precise advanced knowledge of the system and its environment; that is, the need to control the system under increased uncertainty.

Since design techniques for manufacturing systems are closely related to their stability properties, and since necessary and sufficient conditions for the stability of linear time-invariant systems has been discovered over the past century, well-known design methods have been established for such systems. In contrast to this, the stability of nonlinear systems (most manufacturing systems) can be established for the most part only on a system-by-system basis. Hence it is not surprising that design procedures that simultaneously meet the requirements of stability, robustness and good dynamical response are not currently available for large classes of such systems.

In the past three decades major advances have been made in adaptive identification and control for identifying and controlling linear time-invariant systems with unknown parameters, but for nonlinear dynamical manufacturing processes this is still a difficult problem to be studied. Today the need for better control of increasingly complex dynamic systems (such as modern manufacturing systems) under significant uncertainty and nonlinearity has led to a re-evaluation of the conventional control methods. The emergence of artificial intelligence (AI) has also led to a more general concept of control, one which includes higher-level decision making, planning, learning, and uncertain representation, which are capabilities necessary when higher degrees of system autonomy are desirable.

Neural networks appear to offer new, promising ways toward better understanding and perhaps even solving some of the most difficult control problems, such as: modeling, forecasting and controlling nonlinear manufacturing processes. Ydstie[1] discussed the application of adaptive connectionist networks with hidden units (i. e. multilayer neural networks) to process forecasting and controlling. A heuristic, parallel algorithm which used error broadcasting was developed to adapt the representation of advance maps to nonlinear autonomous systems. The simulation examples given in this paper indicate that connectionist networks and parallel computation might hold the promise of solving difficult process control and forecasting problems. Narendra et al.[2] have studied the modeling for identification and control of nonlinear dynamical systems. In the models introduced in that paper multilayer and recurrent networks are interconnected in novel configurations. Simulation results were also promising. As a conclusion, layered neural network models trained according to the back-propagation learning algorithm have been widely simulated for solving nonlinear control problems. However, the back-propagation algorithm, being a sort of gradient descent search, has two major disadvantages:

(1) The number of iterations of the algorithm required for good learning is big, which may pose difficulties for a real-time implementation unless the neuron hardware technique is used.

(2) It lacks the convergence property, i.e., even if there exists a valid solution to the learning algorithm, it is still not guaranteed to find it.

In this paper, a new method for controlling the iteration is proposed. Based on this method, we have developed a "one-by-one" learning algorithm. Using this algorithm, a forecasting and control strategy is constructed for a leadscrew grinding process.

2 One-by-one Learning Algorithm

For a multilayer neural network, suppose that a training set of data $\{X_i \to Y_i\}, i=1,2,\cdots,N$, is given. In the back-propagation algorithm, the error between the targets $Y_{p,j}$ and the network outputs $O_{p,j,L}$ in the pth training sample is defined by

$$E_p(W) = \frac{1}{2} \sum_{j=1}^{N_L} (Y_{p,j} - O_{p,j,L})^2 \tag{1}$$

where N_L is the number of neurons in the output layer. The total error can be expressed as

$$E = \sum_p E_p(W) \tag{2}$$

Two kinds of schemes for error back-propagation are currently used, one uses $E_p(W)$ after the input of each sample to the network, the other uses E after the operation of all samples. The operation criterion is that if the maximum in all $E_p(W)$ s (or the total error E) is less than a given value ε (the iteration precision), the iteration will stop. In our work, an alternative operation criterion is used. If $E_p(W) < \varepsilon$, we will consider the learning successful for the pth sample and $E_p(W)$ will not be back-propagated through the network.

Tab. 1   One-by-one learning algorithm

1. Set the training sample space $U = \{X_p \rightarrow Y_p\}$, $p = 1, 2, \cdots, N$, and choose initial conditions for the weights W
2. for ($p=1; p \leqslant N; p++$) {
3.   do {
4.   for {$j=1; j \leqslant p; j++$} {
5.   forward-computing ( ); {forward computing until the output layer}
6. $E_j(W) = \| Y_j - O_j(W) \|$;
7.   if $E_j\{W\} < \varepsilon$, then goto 10;
8.   justifying ( ); {modifying weights based on the back-propagation mechanism}
9.   errormax = max$\{E_j(W)\}$
10.   }
11.   } while (errormax $> \varepsilon$)
12. }

If there exists a sample $\{X_i \rightarrow Y_i\}$, $E_i(W) > \varepsilon$ then $E_i(W)$ will be back-propagated, and the weights W will be modified. Based on such a method, we have developed a one-by-one learning algorithm for the given samples. Tab. 1 shows a schematic of the algorithm scheme. It can be seen that the working space of training samples $U^* = \{X_j \rightarrow Y_j\}$, $j = 1, 2, \cdots, p$, increases one training sample in each iteration from the initial $\{X_1 \rightarrow Y_j\}$ to U. Compared to the back-propagation (B-P) algorithm, the one-by-one algorithm has the following features:

(1) For a given sample $\{X_j \rightarrow Y_j\}$, if $E_i(W) < \varepsilon$, then the error $E_j(W)$ will not be back-propagated through the network, unlike that in the B-P algorithm, and the network learning will go to the next sample in U^*. Otherwise, the weights will be modified based on the back-propagation mechanism. This will doubtlessly speed up the network learning.

(2) When a new sample is added to the working space U^*, the initial weights of the network in this learning will be given by the results of the former learnt network. This it may improve the convergence of the back-propagation mechanism.

(3) The working space U^* in the one-by-one algorithm is generally much smaller than that in the B-P algorithm. After the learning of a given sample space U^*, a new subspace of training data U will be the focus of learning in the one-by-one algorithm, while the learning space of the B-P algorithm will always include all samples, $U^* + \Delta U$, as shown in Fig. 1.

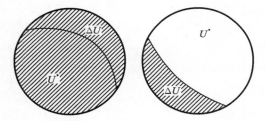

Fig. 1   The focus of learning in the one-by-one and the backpropagation algorithms

(4) The backpropagation algorithm may give the same results under any input sequence of training samples. Theoretically it cannot represent dynamic changes implicitly included among given continuous training data unless a time variable is introduced in the network. Since the learning of new samples in the one-by-one algorithm is based on the results learnt from the last training data, intuitively it can implicitly elicit the time variable in the training data. Therefore, it is more suitable for the forecasting and control of manufacturing processes. Tab. 2 gives the results of learning an XOR problem for four different sequences of input samples with the 2-3-1 mode of network structure by the two algorithms.

Tab. 2   Comparison of the backpropagation and one-by-one algorithms for learning an XOR problem

Algorithm	Samples	Computing time/s	Weights
Back-propagation	{{0,0;1},{0,1;1} {1,0;1},{1,1;0}}	37.78	{−6.07,−3.53,−6.07,−3.53, −7.33,6.93,2.12,5.12,−3.08}

续表

Algorithm	Samples	Computing time/s	Weights
Back-propagation	{{0,1;1},{0,0;0} {1,1;0},{1,0;1}}	31.80	{−5.77,−3.58,−5.70,−3.57, −7.32,6.91,2.05,5.21,−3.08}
	{{1,1;0},{1,0;1} {0,1;1},{0,0;0}}	32.13	{−5.70,−3.58,−5.67,−3.58, −7.31,6.90,2.04,5.23,−3.08}
	{{1,0;0},{0,1;1} {1,0;1},{1,1;0}}	31.80	{−5.70,−3.58,−5.66,−3.58, −7.30,6.89,2.03,5.23,−3.08}
One-by-one	{{0,0;1},{0,1;1} {1,0;1},{1,1;0}}	26.75	{6.18,3.12,6.05,3.11,6.49,−6.58, −1.73,−4.51,−3.02}
	{{0,1;1}{0,0;0} {1,1;0},{1,0;1}}	24.33	{−5.92,3.98,5.15,−3.98,6.60, 5.99,−3.84,−2.08,−3.01}
	{{1,1;0},{1,0;1} {0,1;1},{0,0;0}}	20.87	{−5.16,−3.93,5.62,3.91,−6.43, 6.11,3.51,−1.90,3.23}
	{{1,0;1},{1,1;0} {0,1;1},{0,0;0}}	31.03	{−4.05,−4.33,−4.05,−4.30, −6.70,6.18,1.49,6.60,−2.91}

3 Intelligent Forecasting and Control Architecture

Current studies on neural networks applied in control problems are mainly motivated and constrained by ideas that are familiar in linear system theory and adaptive control, and by simple physical models for simulation. Little progress has been made so far with the methodology and the experience in applying neural networks to manufacturing process control. In this section, we will introduce our work on forecasting and control of a leadscrew grinding process using neural networks based on the one-by-one learning algorithm proposed earlier.

3.1 Measurement and Compensation of Leadscrew Grinding Errors

In general, the machining errors of a precision leadscrew may mainly come from three sources, i.e., kinematic errors in the leadscrew grinding machine transmission train, grinding temperature variation and grinding wheel wear. The comprehensive effects of the three error sources will produce the leadscrew machining errors that can be measured by some precision equipment. Several studies have been performed to reduce the precision leadscrew machining errors by error compensation techniques. Both closed-loop feedback and semiclosed-loop compensation schemes have been used to improve the leadscrew grinding accuracy. These schemes, based on linear control theory, are mostly designed to aim at reducing only one of the error sources.

Fig. 2 presents a block diagram of measurement and compensation for thread grinding machine transmission errors. The phase difference between signals from circular grating and linear grating represent kinematic errors in the transmission system by a certain relationship and will be used to compensate the kinematic errors by a stepping motor[3]. The application of this scheme, although it has successfully obtained a high cost-effectiveness in practice, still faces some problems.

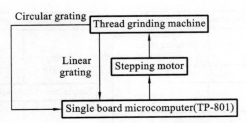

Fig. 2 Measurement and compensation of grinding machine transmission errors

(1) Presently, the scheme can only be used to reduce the transmission errors. Other kinds of leadscrew errors cannot be compensated because of the lack of a nonlinear adaptive model for such a process. It has been proved that the longer the length of leadscrew (e. g. more than two meters), the higher the ratio of errors caused by the grinding temperature variation and the grinding wheel wear as well as other factors in the total errors of the workpiece leadscrew. In order to improve further the leadscrew grinding accuracy, these influencing factors should be taken into account. In our work, we measure the grinding force to monitor the state of grinding wheel wear. The workpiece is cooled by a shower for controlling the temperature variation.

(2) The algorithms in current use are almost all based on linear control

theory or on a system-by-system basis. The non-linearity and uncertainty in the leadscrew grinding procedure cannot be represented and modeled. In this paper, our interest is in the identification of nonlinear dynamic processes of leadscrew grinding which is discussed in Section 3.2.

(3) Generally, the workpiece leadscrew errors should be measured by a precision equipment such as a laser interferometer after each grinding operation. We can obtain the updated state of workpiece precision compared with the requirements. In order to increase the productivity of leadscrew machining, the measured results should quickly be fed back to the machining to guide automatically the next grinding operation because this updated information can theoretically be considered as an output of the grinding process in the last operation and as an input of that in this operation. In our work, we implement this automatic information feedback through a data communication network as shown in Fig. 3.

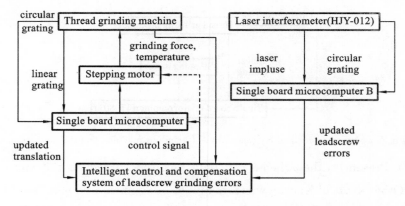

Fig. 3 Intelligent control and compensation of leadscrew grinding errors

3.2 Modelling and Forecasting of Leadscrew Grinding Process

One way of obtaining knowledge about the unknown manufacturing process dynamics is by modeling a flexible-structured system that will imitate the process by adaptively changing its parameters to match the observable process' output signal when driven by the same input. In adaptive control systems one mode of adaptation is to identify the process and feed the reference signal into an inverted process dynamics model in series with the process[4]. In our work we train a neural network model to identify the inverse dynamics of the leadscrew grinding process and then apply it as a controller as shown in Fig. 4. The neural network model here performs a specific form of adaptive control, with the controller taking the form of a nonlinear discrete network and the adaptable parameters being the strength of the synaptic connections to the neurons.

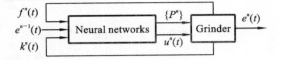

Fig. 4 The operation mode block diagram

Fig. 5 gives a simulation result of the error compensation based on the operation mode in the case of only using $k^n(t)$ and $e^{n-1}(t)$ as inputs to the network. The network modelling is in nature a learning process. Before the network is trained, a training set for the network is built by introducing selected training signals which include the last leadscrew error $e^{n-1}(t)$ with grinding force $f^n(t)$ as well as transmission error $k^n(t)$ as inputs to the network, and control impulses of the stepping motor $u^n(t)$ with grinding parameters $\{P^n\}$ as outputs of the network, and constructing the appropriate input-output transition with $n= 1,2,\cdots,L$. The learning problem is defined now as one of finding the appropriate synaptic weights and thresholds to realize this mapping based on the one-by-one algorithm proposed earlier. The architecture of neural networks in the learning mode is shown in Fig. 6.

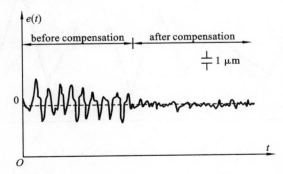

Fig. 5 Simulation results of the error compensation using $k^n(t)$ and $e^{n-1}(t)$

In general, the variation of the transmission error in each grinding operation

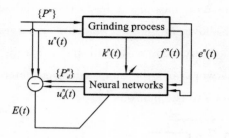
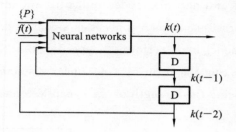

Fig. 6　The learning mode block diagram

Fig. 7　One-step ahead forecasting architecture of transmission errors

for the same machine state is relatively small. It is possible to predict the error one step ahead using neural networks. Fig. 7 indicates the learning forecasting architeture using neural networks for $k(t)$. In practice a three-layer neural network is modeled with the 10-5-1 mode of structure. Each training sample consists of eleven points of the error, with the first ten points as inputs to the network and the last point as its output. After the network is trained in the learning space including ten continuous samples, it will predict the next point of error. Then the next new sample is added to the space from which the first sample is excluded so that a new sample space is constructed. The network will be updated based on the last weights using the one-by-one algorithm. Fig. 8 shows a part of one-step-ahead prediction of the transmission error.

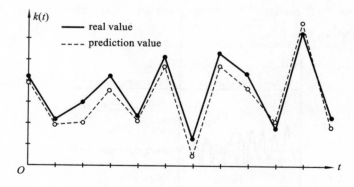

Fig. 8　One-step off-line prediction of $k(t)$

4 Further Work and Conclusions

At present, we have developed a leadscrew error processing system for the data measured by a laser interferometer HJY-012, and have built a whole data communication network for the intelligent compensation and control of leadscrew grinding. Several significant results have been obtained about the learning of neural networks and the one-step-ahead prediction of the grinding process using the network. All this work was performed at and cooperated by the Hanjiang Machine Tool Works. Our further work will focus on training the neural networks in workshop conditions and adapting them in real-time applications.

In this paper a new learning algorithm of neural networks, the one-by-one algorithm, is presented which is particularly suitable for forecasting and process control. An intelligent forecasting and control architecture for a leadscrew grinding process has been introduced. Our experience shows that the application of neural networks to manufacturing process forecasting and control is promising and the greatest difficulties in the application are the convergence property and the speed of learning algorithms. It is hoped that the introduction of neural hardware will improve the situation.

References

1 B. E. Ydstie. "Forecasting and control using adaptive connectionist networks". Chem. Eng, Vol. 14, Nos. 4/5, 1990.

2 K. S. Narendra, et al.. "Identification and control of dynamic systems using neural networks". IEEE Trans. Neural Networks Vol. 1, No. 1, 1990.

3 H. Z. Bin, et al.. "On-line measurement and compensation of transmission errors of leadscrew grinder by a digital filter". Proc. 2nd Int. Conf. on Metal Cutting, 1989.

4 E. Levin, et al.. "Neural network architecture for adaptive system modelling and control". Proc. IJCNN'89, 1989.

(原载 Computers in Industry, Volume 23, Issue 3, 1993)

大直径钢丝绳轴向励磁磁路的研究

谈 兵　杜润生　康宜华　杨叔子

提　要　在分析钢丝绳状态检测已有的励磁方式的性能的基础上,针对大直径钢丝绳的特点,提出了二级励磁磁路结构。分析了二级励磁磁路结构的磁场特性,表明具有二级励磁结构的系统所获取的断丝信号信噪比高,无漏检,为断丝的定量识别提供了保证。

关键词:钢丝绳,励磁方法,断丝,磁场分析

钢丝绳损坏的重要原因之一,是钢丝绳在使用过程中因疲劳和磨损而出现断丝,对断丝定量检测技术的研究已取得很大进展[1-3],并已开发出多种型号的钢丝绳断丝定量检测仪。其励磁结构为单回路励磁结构。但是,对大直径钢丝绳(主要指直径在 $\phi 40$ mm 以上)断丝定量检测技术的研究还刚刚开始。在断丝的漏磁场检测方法中,作为关键环节之一的是励磁磁路结构。如果励磁磁路结构不合理,不能在钢丝绳中形成较为均匀的磁场,则无法可靠、准确地获取钢丝绳的断丝信息。故对大直径钢丝绳状态检测中的励磁磁路结构的研究是非常必要的。

1　钢丝绳励磁结构的磁路分析

钢丝绳轴向局部励磁结构以往多采用单回路和双回路励磁结构,但这两种结构均存在着难以克服的缺陷,即不能在钢丝绳中形成均匀的轴向励磁场。此缺陷当钢丝绳直径小于 30 mm 时不明显,但当钢丝绳直径大于 40 mm 时却很明显。

1.1　单回路励磁结构的磁路分析

当采用稀土永久磁铁对钢丝绳励磁时,励磁磁路采用图 1 所示的结构。磁路

的结构尺寸和磁源确定后,可采用多种方法对磁路各部分的磁特性进行分析计算,在此采用磁导法。一方面是因为分析计算相对简单,另一方面是因为可以直接获得宏观的磁场特征参数。

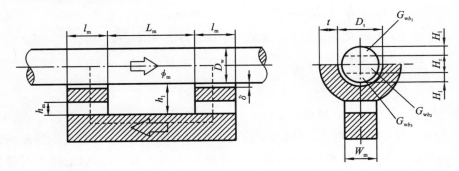

图1 单回路励磁结构

D_w 为钢丝绳公称直径;D_t 为极靴内径;δ 为极靴内侧面与钢丝绳表面间的径向距离 $\delta = (D_t - D_w)/2$;t 为极靴沿径向的厚度;W_m 为稀土磁体的宽度;h_m 为稀土磁体的高度(磁化方);l_m 为稀土磁体沿钢丝绳轴向的长度,L_m 为两极靴内侧面间沿钢丝绳轴向的距离。

图1所示结构的等效磁导模型可如图4左半部所示。忽略两极靴间气隙磁导。在此,把处于两极靴间的一段钢丝绳分成体积相等的三部分,每部分的磁导分别为 $G_{wb_1}, G_{wb_2}, G_{wb_3}$。因钢丝绳是由丝捻制成股,再由股捻成绳的结构,内部仍存在空气隙,丝与股、股与绳之间的空气隙磁导分别为 $G_{w\delta_1}, G_{w\delta_2}, G_{w\delta_1} \approx G_{w\delta_2}$。极靴内侧,侧面与钢丝绳表面间气隙磁导

$$G_\delta = 4\mu_0 \frac{D_w}{2} + \sqrt{\delta\left(\frac{W_m}{2} + \delta\right)} \cdot \ln\frac{W_m/2 + \delta}{\delta}$$

μ_0 为空气磁导率;极靴磁导

$$G_p = \frac{\mu_{rp}(B_p)\left[\pi\left(D_w + \delta + \dfrac{h_m}{2}\right)l_m\right]}{h_m}$$

μ_{rp} 为极靴的相对磁导率,B_p 为极靴内磁感应强度;衔铁磁导

$$G_L = \frac{\mu_{rL}(B_L)W_m l_m}{L_m}$$

μ_{rL} 为衔铁的相对磁导率,B_L 为衔铁内磁感应强度;各部分的磁导

$$G_{wb_1} = G_{wb_2} = G_{wb_3} = \mu_{rw} \cdot \frac{(B_w)\pi D_w^2}{12 L_m}$$

μ_{rw} 为钢丝绳的相对磁导率，B_w 为钢丝绳内磁感应强度，钢丝绳内各部分之间空气隙磁导 $G_{w\delta_1} = G_{w\delta_2} = 0.077\mu_0 H_1$。

根据磁路计算定律，对单回路励磁的等效磁路列出方程组如下：

$$\begin{cases} (R_p + R_L + R_\delta)\phi'_m + R_{wb_1}\phi'_{wb_1} = H_m h_m \\ (R_p + R_L + R_\delta)\phi'_m + (R_{w\delta_1} + R_{wb_2})\phi'_{wb_2} = H_m h_m \\ (R_p + R_L + R_\delta)\phi'_m + (R_{wb_1} + R_{wb_2} + R_{wb_3})\phi'_{wb_3} = H_m h_m \\ \phi'_m - \phi'_{wb_1} - \phi'_{wb_2} - \phi'_{wb_3} = 0 \end{cases}$$

也即 $[R][\phi'] = [F]$。解得：$[\phi'] = [\phi'_m, \phi'_{wb_1}, \phi'_{wb_2}, \phi'_{wb_3}] = [R]^{-1}[F]$，式中 $[R]$ 为磁路中的磁阻，它们分别为对应通路上磁导之倒数；H_m 为磁铁的磁场强度。由于磁路中存在铁磁性材料，其磁导率随 B 非线性变化，所以上述方程组为非线性方程组，需采用迭代的方法，逐步确定磁铁的工作点和磁性材料的磁导率后，采用逼近的方法计算。可得出结论 $\phi'_{wb_1} > \phi'_{wb_2} > \phi'_{wb_3}$。

由于在钢丝绳的各部分中通过的磁通量 ϕ 的不同，导致钢丝绳中形成轴向励磁场不均匀。在钢丝绳截面上远离磁极的部分，所达到的磁感应强度 B_r 小，如果缺陷发生在径向磁场的 B_r 较小的部位，就会发生由于 B_r 太小而引起的漏磁场很微弱，磁敏元件检测不出来，其结果导致漏检。

1.2 双回路励磁结构的磁路分析

由于双回路磁路沿钢丝绳横截面的布置具有对称性，因而可以在一定程度上克服单回路形式的缺点，但由于安装钢丝绳的要求，使得形成磁路场源的部分不能采用整体环绕钢丝绳截面的结构，而需采用分离式结构，这造成两部分磁源的接合部分的磁场相互抵消，形成较弱的励磁场，如同单回路形式一样，也会发生漏检现象。

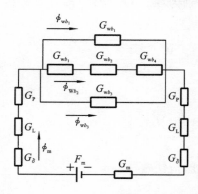

图 2 双回路励磁磁路的等效磁路

双回路励磁磁路的等效磁导模型如图 2 所示。各部分磁导定义与单回路形式相同。根据磁路计算定律，对图 2 所示的等效磁路列出方程组求解，可得出 $\phi_{wb_1} = \phi_{wb_3} > \phi_{wb_2}$ 的结论，即在钢丝绳内部也不能形成均匀的励磁场。

1.3 二级励磁结构的磁路分析

考虑上述两种励磁结构的特点,提出二级励磁磁路结构,如图 3 所示。二级励磁磁路的优点:a.二级励磁磁路对钢丝绳进行励磁,在钢丝绳横截面周向形成较均匀的励磁场 B_r;b.二级励磁磁路结构,无论断丝发生在钢丝绳横截面的任何位置,均能可靠、准确地捕获表面断丝的漏磁场信号,无漏检现象。

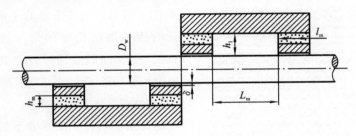

图 3　二级励磁磁路

二级励磁磁路结构的等效磁导模型如图 4 所示。根据磁路计算定律,对图 4 所示的等效磁路左右两回路列方程组:

$$\left.\begin{array}{l}[R'][\phi']=[F']\\ [R''][\phi'']=[F'']\end{array}\right\} \quad (1)$$

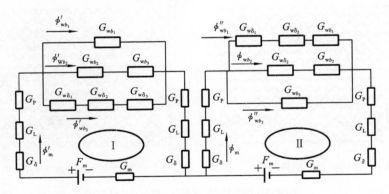

图 4　二级励磁结构的等效磁路

求解方程组(1),得

$$[\phi']=[\phi'_m, \phi'_{wb_1}, \phi'_{wb_2}, \phi'_{wb_3}], \quad [\phi'']=[\phi''_m, \phi''_{wb_1}, \phi''_{wb_2}, \phi''_{wb_3}]$$

其中 $\phi_{wb_1}=\phi'_{wb_1}+\phi''_{wb_1}$,　$\phi_{wb_2}=\phi'_{wb_2}+\phi''_{wb_2}$,　$\phi_{wb_3}=\phi'_{wb_3}+\phi''_{wb_3}$,

可得出 $\phi_{wb_1}=\phi_{wb_3}\approx\phi_{wb_2}$ 的结论,故可在钢丝绳内部形成较为均匀的励磁场。

2 实验分析

为了验证提出的二级励磁磁路结构的优点,对 $\phi 40$ mm 的钢丝绳,用三种形式做了实验。本实验采用 16 片霍尔元件沿钢丝绳周向均匀布置,对 $\phi 40$ mm 的

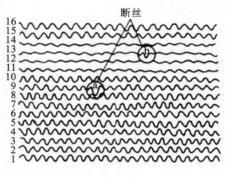

图 5　单回路励磁结构信号

钢丝绳分别进行了测试,信号波形如图 5、图 6、图 7 所示。图 5 表示的是单回路励磁方式的传感器所获得的信号波形。其中 11、12、13、14 路股波信号较之其他各路信号弱,幅值下降了 50% 左右,励磁场不均匀,发生在 13 路上的断丝信号,与其股波信号相比很不明显。图 6 表示的是双回路励磁方式的传感器所获得的信号波形,从中可以看到 1、8、9、16 路股波信号较之其他各路信号弱,幅值下降了 50% 左右,励磁

场不均匀,发生在 9 路上的断丝信号,与其股波信号相比很不明显。

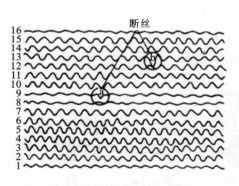

图 6　双回路励磁结构信号

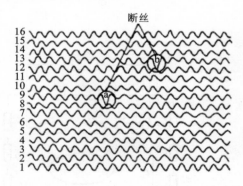

图 7　二级励磁结构信号

图 7 表示的是二级励磁方式的传感器所获得的信号波形。其中 1～16 路股波信号幅值均匀,说明励磁场均匀,发生在 9、13 路上的断丝信号与其各自的股波信号相比很明显。

从以上实验结果分析可知,单回路励磁方式和双回路励磁方式的都会发生励磁不均匀现象,有几路信号较弱。若断丝正好位于磁场较弱的部分,就会因磁场太弱,漏磁场太小,霍尔元件不能准确地捕获到断丝信号,发生漏检现象,或检测到了断丝信号,但无法从股波信号中分离出来,也无法进行定量判别。而二级励

磁方式的传感器,在钢丝绳横截面周向励磁场均匀,无论断丝发生在任何部位,均布于钢丝绳周向的 16 个霍尔元件完全能够把断丝信号检测出来,不会发生漏检现象,而且断丝信号的信噪比高,为钢丝绳断丝的定量判别提供了保证。

参 考 文 献

[1] Li Jinsong, Yang Shuzhi, Lu Wenxiang, et al. *Space-Domain Feature-Based Automated Quantitative Determination of Localized Faults in Wire Ropes*. Material Evaluation. 1990,48(3):336~341.

[2] 王阳生,师汉民,杨叔子等. 检测局部异常信号的一个新特征量. 华中理工大学学报,1988,16(3):62~64.

[3] Kalwa E, Piekarski K. *Qualitative and Quantitative Determination of Densely Occuring Defects in Steel Ropes by Magnetic Testing*. Material Evalution,1988,46(6):767~770.

On the Magnetic Circuit for the Condition Monitoring of Steel Wire Rope of Large Diameter

Tan Bing　Du Runsheng　Kang Yihua　Yang Shuzi

Abstract　Based on the performance of existing magnetic methods for the condition monitoring of the steel wire rope, a new method using a two-step exciting field circuit structure is proposed in the light of the specific features of the steel wire rope with a large diameter. The field characteristics of the magnetic circuit are discussed. Theoretical and experimental investigations show that with a new structure the SNR of the broken wire signal is high and no missing has been found during detection.

Keywords:Steel wire rope, Exciting method, Broken wire, Magnetic field analysis

(原载《华中理工大学学报(自然科学版)》1994 年第 22 卷第 7 期)

金属切削机床切削噪声的动力学研究

吴 雅　柯石求　杨叔子

提　要　根据切削噪声试验结果,提出在强烈切削振动情况下,切削噪声与刀具主切削力方向的振动在时域中具有极强的相关性,在频域中具有极大的频率结构相似性,并进而提出了稳态切削噪声与动态切削噪声的概念,解释了切削噪声发声的物理机理。本文还采用倒谱剔除观测切削噪声信号中的回声信号,以分析原始切削噪声的频率结构;并将切削振动作为主要输入,将切削噪声作为输出,采用二维 ARMAV 模型辨识切削噪声发声系统,辨识结果描述了发声系统对切削振动的传输关系,并能较好地解释原始切削噪声的主要频率成分的来源与传输。

关键词:切削,噪声,振动,动力学

0　前　言

机床的噪声分为结构噪声与切削噪声,现有的大多数研究属于前者,对于后者,这一金属切削过程中所产生的特有噪声,则研究不多。本文以切削噪声试验为基础,根据实验信号进行分析研究。

1　机床动态特性对切削噪声的影响

1.1　切削噪声试验系统

分别在 CW6163,CW6180,CW61100,C630 这 4 台卧式机床上进行切削噪声试验,采用宽刀刃、切入式车削,因为这种车削方式所产生的切削振动与切削噪声较大,便于研究。试验系统如图 1 所示,声级计距刀尖 0.5 m。两个试验系统的切

削条件分别列于表 1。试验前先测量背景噪声与空运转噪声,4 台机床在试验所用主轴转速下的空运转噪声均低于 85 dB(A),而切削状态下的切削合成声至少也高于 95 dB(A),有的还高达 110 dB(A),因此认为切削合成声就是切削噪声。在一次切削过程中,将声级计的最大读数取为切削噪声的声压级。

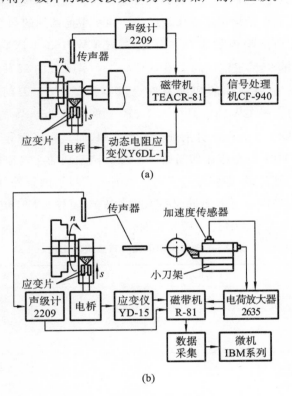

图 1 切削噪声试验系统

(a) CW6163,CW6180,CW61100 车床;(b) C630 车床

表 1 切削噪声试验的切削条件

车床	CW6163,CW6180,CW61100	C630
刀具材料	硬质合金 YG6X	
工件材料	球墨铸铁 Q600-3	45 钢
切削宽度 b/mm	35	20
单边切削余量 Δ/mm	3~4	

1.2 3台车床切削噪声声压级 L 的比较

CW6163、CW6180、CW61100 这 3 台车床的切削方式与切削条件完全一致,仅切削用量与刀具角度有差别。然而,在排除了仪器误差与读数误差后,3 台车床 L 值有很大差别,表 2 示出了这一比较结果。表 2 中前两列的切削条件基本相似,但 CW6180 车床切削噪声的平均声压级 $\bar{L}$ 较 CW6163 车床高出 12.5 dB,且 CW6163 车床 L 的 95% 置信区间上限与 CW6180 车床 L 的 95% 置信区间下限不重合。同样,表 2 中后两列的切削条件也基本相似,仅车刀前角差别较大,但 CW61100 车床的 $\bar{L}$ 较 CW6180 高出 5.7 dB,且两者的 95% 置信区间也不重合。表 2 结果表明。除了 3 台车床分别处于 3 个车间,而车间厂房等环境条件的不同对 L 的测量有影响外,3 台车床切削噪声出现明显差别的主要原因还在于其各自动态特性的影响,即是说,即使是对于同一切削方式与切削条件,所使用的机床不同,所产生的切削噪声也可能不同,例如,在 CW6163 与 CW6180 车床上,就出现了平均声压级相差 12.4 dB 这一很大差别。

表 2 3 台车床切削噪声声压级的比较

机 床	CW6163	CW6180	CW6180	CW61100
切削速度 $v/(\text{m/min})$	31.2	30.8	29.7	29.1
进给量 $s/(\text{mm/r})$	0.200,0.225,0.400	0.225,0.260,0.300	0.260	0.275
车刀前角 $r/(°)$	8,10	6,8,10	7,9	−2,0,2
试验次数	14	19	13	10
平均声压级 $\bar{L}/\text{dB}$	95.1	107.5	107.5	113.2
均方值 σ/dB	2.29	3.35	2.31	0.24

1.3 切削噪声声谱密度的比较

图 2、图 3 分别示出了 CW6163、CW6180、C630 车床的切削噪声谱图。比较这两个图,CW6163 车床切削噪声的第一谱峰频率为 210 Hz,在所分析的 2000 Hz 频率范围内还有数个小峰,噪声能量较为分散。CW6180 车床切削噪声的谱值频率为 830~855 Hz 及其倍频 1650~1690 Hz,噪声能量集中在这两频率范围附近,与图 2 的"分散"形成明显对比。C630 车床切削噪声的谱图形状介于前两者之间,其第 1 谱峰频率为 93.8 Hz,远低于 CW6163 车床的 210 Hz,在所分析的 600 Hz 频率范围内,还有数个小峰,噪声能量不如 CW6163 车床分散,但也不如 CW6180

车床集中。不难理解,出现这种情况的原因在于机床动态特性的不同,使得这 3 台机床对于同一切削方式与切削条件所产生的切削振动不同,从而导致切削噪声声压级 L 的不同与谱图频率结构的不同。

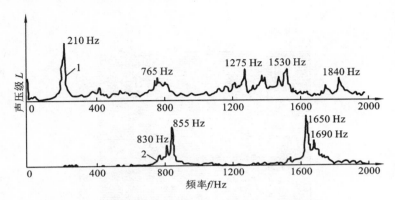

图 2　CW6163 和 CW6180 车床切削噪声平均谱

上图:$r=0°,v=31.2$ m/min,$s=0.4$ mm　　下图:$r=10°,v=31.2$ m/min,$s=0.3$ mm

1—CW 6163 车床($s=0.4$ mm/r)　2—CW6180 车床($s=0.3$ mm/r)

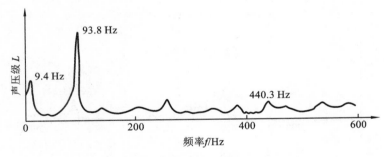

图 3　C630 车床切削噪声谱

($r=15°,v=32.5$ m/min,$s=0.1$ mm/r)

2　切削噪声与切削振动的相对关系

2.1　CW6180 车床切削噪声与切削振动的相关关系

图 4 分别示出了 CW6180 车床同一切削过程的切削噪声与切削振动位移谱阵。切削噪声声压级为 112 dB。仅就谱阵形状来看,两图极其相似,随着切削时间的推移,两图的两个谱峰都由高频向低频移动。就切削过程的任一瞬时来看,

图 5 分别示出了该切削过程的平均谱(16 次平均),可见切削噪声与振动位移两谱图不仅形状一致,而且谱峰位置、谱峰高低都完全对应。该次切削过程的切削噪声与振动位移的瞬时互相关系数均在 0.6 以上,最高达 0.847;平均互相关系数也高达 0.653。

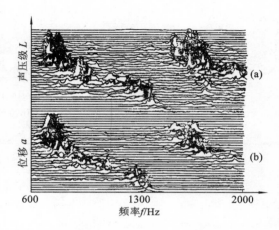

图 4 谱阵(CW6180 车床,$r=15°$,$v=26.7$ m/min,$s=0.2$ mm/r)
(a) 切削噪声;(b) 刀具主切削力方向的振动位移

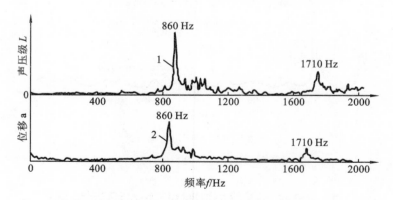

图 5 平均谱
1—切削噪声 2—刀具主切削力方向的振动位移

2.2 C630 车床切削噪声与切削振动的相关关系

图 6 分别示出了 C 630 车床同一切削过程中同时采样计算的切削噪声与振动位移谱阵,该切削过程的切削噪声声压级为 93 dB(A),从谱阵形状来看,图 6 中的两图都很相似,谱峰位置对应。就切削过程的任一瞬时来看,图 7 分别示出了该

切削过程的瞬时谱,可见切削噪声与振动位移的谱图也具有很好的对应性。切削噪声与振动位移的瞬时互相关系数高达 1。然而,对刀架进给抗力方向的振动加速度信号与切削噪声信号进行类似的计算,没有发现这种关系。

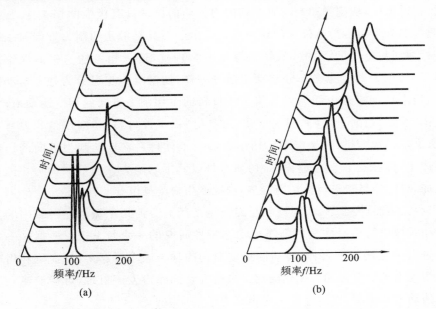

图 6　谱阵(C630 车床,$r=15°$,$v=32.46$ m/min,$s=0.1$ mm/r)
(a) 切削噪声;(b) 刀具主切削力方向的振动位移

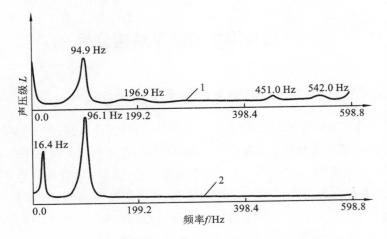

图 7　瞬时谱

1—切削噪声　2—刀具主切削力方向的振动位移

2.3 分析与结论

(1) 事实上,在所试验的 4 台车床的强振切削中,上述事实是普遍存在的。这表明,在时域中,切削噪声与刀具主切削力方向的振动具有极强的相关性;在频域中,两者的频率结构具有极大的相似性。又由于刀具是沿主切削力方向对工件材料进行剪切与分离的,因此,从物理意义上来理解,这一相关性与相似性正好说明,如同切削力具有稳态部分与动态部分一样,切削噪声也可分为稳态与动态两部分,稳态切削噪声由切削刀具与工件材料的主切削力方向的切削运动所产生,主要由"静态"切削条件(如切削方式、刀具材料、工件材料、切削用量和刀具角度等)所决定;而动态切削噪声则由切削刀具与工件材料在主切削力方向的切削振动而产生,即在这一动态过程中,切削刀具以其主切削力方向的振动与切削运动相叠加,对工件材料进行剪切与分离,由此产生切削噪声。

(2) 在强振状态下,切削噪声与刀具主切削力方向切削振动在时域中的相关性和在频域中的相似性是切削噪声动态特性研究的一个新发现。根据这一新发现而进行的上述物理解释,将切削振动与切削噪声联系了起来,从而可利用现有的对切削振动研究得相当成熟的成果来研究切削噪声发声机理,可望能建立切削噪声的动力学模型,以研究切削噪声动力学。

(3) 根据上述相关性与相似性,要减小切削噪声,主要应减小切削刀具与工件在主切削力方向的振动。

3 切削噪声的频率结构分析

3.1 CW6180 车床切削噪声的频率结构分析

对夹持在 CW6180 车床上的工件(车削完毕后的工件)进行冲击试验,计算冲击力与工件振动加速度的奈奎斯特(Nyquist)图,如图 8 所示。可见工件系统的第 1 阶模态固有频率为 872.5 Hz,这一频率与图 5 中的第一阶谱峰频率 860 Hz 基本接近(因为在每次试验中,每个工件的安装、装夹情况以及车削完毕后的工件直径不可能完全相同,所以,试验所得的第 1 阶模态固有频率 872.5 Hz 并不可能精确代表每个工件的情况)。这一事实可理解为:在 CW6180 车床试验中,切削刀具主要以工件系统的第 1 阶模态固有频率在主切削方向上进行振动,并以这一振动与切削运动相叠加,对工件材料进行剪切与分离,由此产生切削噪声。

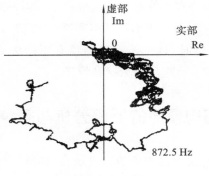

图 8　奈奎斯特图

3.2　C630 车床切削噪声的频率结构分析

对 C630 车床刀架进行冲击激振,图 9 为刀架主切削力方向的冲击响应位移谱。图中示出,刀架主切削力方向的第 1 阶模态固有频率为 96.1 Hz,这一频率与图 6、图 7 所示的谱峰频率 94.9 Hz、96.1 Hz 基本吻合。同上理,这一事实可理解为,在 C630 车床试验中,切削刀具主要以刀架的第 1 阶模态固有频率在主切削力方向进行振动,并以这一振动与切削运动叠加,对工件材料进行剪切与分离,由此产生切削噪声。

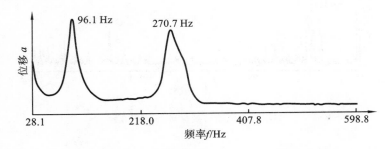

图 9　C630 车床刀架主切削力方向的频率响应

3.3　分析结论

影响动态切削噪声的"动态切削条件"是机床切削系统的动态特性,即对于产生动态切削噪声的刀具主切削力方向的振动,其振动频率主要由机床切削系统中某一薄弱环节的模态固有频率所决定。例如,对于 CW6180 车床试验系统,薄弱环节是工件系统,则由工件系统的模态固有频率所决定;而对于 C630 车床,薄弱环节是刀架,则由刀架的模态固有频率所决定。

同时还需要强调,刀具对工件材料进行剪切与分离的这一振动频率取决于机床切削系统的"系统特性",而并不取决于某单个部件的"部件特性",只是当系统的各阶模态为非密集模态时,这一振动频率才有可能主要取决于某单个部件的"部件特性",本文试验的 4 台车床均为这种情况。

4 切削噪声的倒谱分析与回声剔除

对 C630 车床在图 6 所示过程中的一段观测切削噪声信号计算 FFT 谱,如图 10 所示,求出该谱的倒谱如图 11 所示。倒谱中含有间距大约为 5.0 ms,10.0 ms,24.2 ms,30.0 ms 的等距脉冲,经测量刀尖、声级计与车间厂房墙壁、天花板、地面之间的距离,确定此 4 回声分别主要由车间厂房的天花板、右墙、前墙与地面所反射,5.0 ms,10.0 ms,24.2 ms,30.0 ms 分别是回声与测点处直接声之间的传播时间差。剔除此 4 回声,得到原始切削噪声的谱图如图 12 所示。比较图 10 与图 12,原始切削噪声谱图中仍然保留了观测切削噪声谱图的主要谱峰,去掉了谱图中的毛刺。表明剔除回声后的结果是正确的。

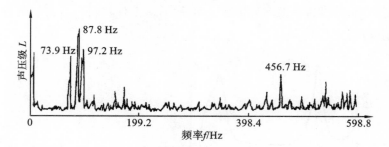

图 10 观测切削噪声的 FFT 谱

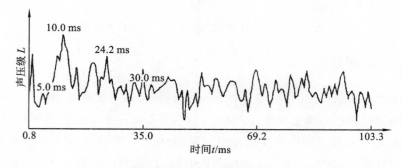

图 11 观测切削噪声的倒谱

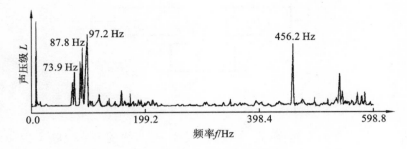

图 12　原始切削噪声的 FFT 谱

对上述结构可作如下分析：

（1）图 12 表明，原始切削噪声具有 97.2 Hz 与 456.2 Hz 这两个主要频率成分，而图 10 在 456.7 Hz 处的谱峰较低，若不剔除回声，则可能忽视原始切削噪声中的这一频率成分。

（2）切削噪声的两个主要频率成分 97.2 Hz、456.2 Hz 分别与刀架 z 方向、工件系统第 1 阶模态固有频率 96.1 Hz、456.2 Hz 相吻合，表明在所计算的时间间隔内，刀架与工件系统都同时影响切削噪声，当然 97.2 Hz 附近的谱峰密集，且谱峰较高，表明刀架 z 向振动的影响较大。

5　切削噪声发声系统的二维 ARMAV 模型辨识

（1）根据前述切削噪声的发声机理，虽然目前尚未根据物理定理建立发声系统的动力学模型，但仍然可以对发声系统进行系统辨识，即将切削振动$\{z_t\}$作为输入，切削噪声$\{n_t\}$作为输出，采用二维 ARMAV 模型进行辨识。二维 ARMAV 模型的意义如图 13 所示（详见文献[3]），$\{z_t\}$与$\{n_t\}$之间的频率响应函数为

$$s(f) = \frac{\phi_{12}(f)}{\phi_{11}(f)} = \frac{\sum_{i=1}^{n} \phi_{12} i e^{-j2\pi fi\Delta}}{1 - \sum_{i=1}^{n} \phi_{11} i e^{-j2\pi fi\Delta}}$$

$\{a_{1t}\}$是除$\{z_t\}$以外的所有影响$\{n_t\}$的总和，视为白噪声。

对 C630 车床同一时间起点采样的刀具主切削力方向的振动位移信号$\{z_t\}$和振动加速度信号$\{a_t\}$分别作为输入，切削噪声信号$\{n_t\}$作为输出，辨识出的 $s(f)$ 分别如图 14 所示。

图 14 中的第 1 条曲线第 1 谱峰频率为 443.0 Hz，而第 2 条曲线中为 89.1 Hz，这一差别是由于图 14 所示的 $s(f)$ 的意义不同：曲线 1 是以振动位移$\{z_t\}$为输

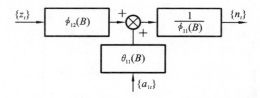

图 13　二维 ARMAY 模型的系统结构

入,而曲线 2 是以振动加速度 $\{a_t\}$ 为输入,因此,两曲线的差别反映了振动位移和振动加速度对发声系统作用方式的差别。这后一点还可由以下分析进一步证实。

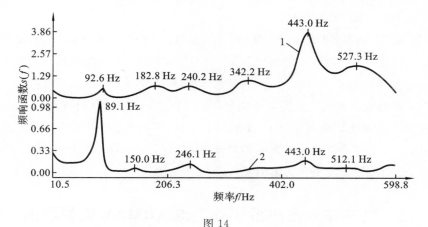

图 14

1—$\{z_t\}$ 输入、$\{n_t\}$ 输出时的 $s(f)$　2—$\{a_t\}$ 输入、$\{n_t\}$ 输出时的 $s(f)$

（2）图 14 中,曲线 1 第 1 谱峰频率 443.0 Hz 接近于工作系统的第 1 阶模态固有频率 485.2 Hz,而曲线 2 中,第 1 谱峰频率 89.1 Hz 接近于刀架 z 向的第 1 阶模态固有频率 96.1 Hz。这一结果可理解为,发声系统对于 443.0 Hz 左右的振动位移与 89.1 Hz 左右的振动加速度的传输能力较强,从而切削噪声中由工件系统动态特性所影响的高频成分主要由振动位移激励发声系统而产生,而切削噪声中由刀架 z 向动态特性所影响的低频成分则主要由振动加速度激励发声系统而产生。

（3）进一步分析该两曲线的谱峰值,图 14 中,曲线 1 的 92.6 Hz 处的谱峰值为 0.64,443.0 Hz 处的谱峰值为 3.86,后者是前者的 6 倍;曲线 2 中,89.1 Hz 处的谱峰值为 0.98,443.0 Hz 处的谱峰值为 0.19,前者是后者的 5 倍。因此,也表明上述结论是正确的。

（4）上述结论还能较好地解释图 12。正是由于发声系统对于高频振动位移与低频振动加速度的传输,使得原始切削噪声谱图 12 的高、低频谱峰都较高。至此可以看出,利用二维 ARMAV 模型对发声系统进行辨识,能较深入地研究发声系统对于切削振动的传输关系,从而表明,采用倒谱分析与二维 ARMAV 模型的系统辨识相结合的方法研究切削噪声的发声机理与发声系统是可行的,当然,这一研究仅仅是切削噪声动力学研究的一个开端,这一方法的可行性尚需进一步实践所证实。

参 考 文 献

[1] 一宫亮一,相地诚,切削加工における骚音的解析(第 1 报)——单一刃工具における金属切削时的音压しべル.精密机械,1974(3):95~100.

[2] Favareto M, Borzati L. Typical whittle liminated in turning operation. RTM. Vieo. Canawese, 1974.

[3] 杨叔子,吴雅等.时间序列分析的工程应用(下册).武汉:华中理工大学出版社,1992.

Study on Cutting Noise Dynamics of Machine Tools

Wu Ya Ke Shiqiu Yang Shuzi

Abstract According to the results of the cutting noise experiments, a new discovery is put forward. The discovery can be briefly stated as this: in the time domain, the cutting noise and the cutting vibration in the main cutting direction have very close relativity; in the frequency domain, they have very great similarity in the frequency structure. Based on the new discovery, a concept about static cutting noise and dynamic cutting noise is presented. With the concept, the physical mechanism of the cutting noise can be explained and the relationship between the influence of the cutting vibration on the cutting noise and the influence of this dynamic behavior of a machine tool on the cutting noise can be studied. In this paper, the echoes are deleted from the cutting noise signals by means of cepstrum analysis. The purpose of deleting the echoes is to analyze the frequency structure of the pure cutting noise. With the help of 2-D ARMAV model the generating noise system in which the cutting vibration is taken as its main input and the cutting noise as its output can be identified. The identification states the transmission relationship of the generating noise system and the cutting vibration. It also gives a satisfactory explanation to the source and the transmission of the main frequencies of the pure cutting noise.

Keywords:Cut,Noise,Vibration,Dynamics

(原载《机械工程学报》1995 年第 31 卷第 5 期)

BP 网络的全局最优学习算法

徐宜桂　史铁林　杨叔子

Abstract　A new global optimization training algorithm for BP network is presented in this paper. This algorithm can solve some knotty problems such as local minimization and network oscillation, which result from the traditional gradient search algorithm, and have more high training efficiency, more simple and feasible application procedure than Simulated Annealing and Genetic Algorithm. As an example, a 2-2-1 neural network on XOR problem is trained with this algorithm, and the results are satisfactory.

Keywords：BP network, Global optimization, Training algorithm

一、概　　述

　　自从 1985 年，Rumechere 和 McLelland 领导的 PDP 研究小组提出多层前馈网络的误差反传训练算法（即 BP 网）以来，BP 网络已成功地解决了语言识别、过程监控等工程领域中的大量问题，成为目前众多神经网络中应用最为广泛的代表性网络之一。然而，由于 BP 网络大多采用的是沿梯度下降的搜索求解算法，这就不可避免地出现了网络学习收敛速度慢以及容易陷于局部极小等问题。此外，在具体实施过程中，有关参数如训练速率 η 和冲量系数 α 的选取，只能凭实验和经验确定，而且一旦取值不当，又会引起网络振荡，甚至导致网络麻痹以致不能收敛。

　　模拟退火和遗传算法可在一定程度上克服上述缺陷。但模拟退火方法在使用过程中要解决一些目前尚无理论指导依据的问题，例如，如何产生新的搜索状态、如何确立新状态的接受标准、如何确定退火温度的下降过程等等，且退火结束后以多大概率稳定在全局最优解也缺乏有力的理论依据。而遗传算法在使用时则要解决诸如如何找到通用且有效的编码方法、如何定义适应度函数、如何确定群体规模值,以及交叉、变异发生概率值等参数取值问题。所有这些问题直接影响到模拟退火和遗传算法应用的成功与否,但又只能凭使用者的经验或经过实验确定,这就难以为一般工程人员所接受和使用。而且模拟退火和遗传算法所花费的学习时间也是相当长的。

因此,寻求一种具有明确的理论指导依据,比较规范的应用步骤,较高的学习效率,以及能够保证得到全局最优解的 BP 网络学习算法,无疑有着十分重要的意义。本文即是对此进行了一个初步探索。其基本思想是:首先建立一个比较合理的 BP 网络误差度量函数 $E(W)$,W 为网络权值组成的向量,进而采用目前已经比较成熟的求非线性函数总体极值的方法求解 $E(W)$,得到 W 值,从而完成 BP 网络的学习过程。

二、BP 网络的学习过程

BP 网络的学习过程由正向传播和反向传播组成。正向传播时,将训练样本集中的任一样本置于网络的输入层,经隐含层处理,在输出层得到该样本对应的输出值。反向传播时,则是根据输出值与期望值的误差调整网络的各个权值,以使其输出值与期望值误差在允许范围,整个学习过程按正向传播和反向传播循环反复进行,直至网络收敛。

按照正向传播过程,在网络输出层第 j 个节点的输出值可表示为(以图 1 所示的二层网络为例):

$$y_j^k = F\left(\sum_h W_{hj} F\left(\sum_i W_{ih} X_i^k\right)\right) \tag{1}$$

式中:X_i^k 表示训练样本中第 k 个样本的第 i 个分量;W_{ih}、W_{hj} 分别表示输入层与隐含层、隐含层与输出层之间的连接权值;$F(\cdot)$ 为节点转移函数,一般为 sigmoid 函数。

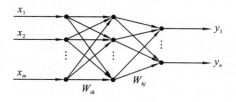

图 1　BP 网络

一般来说，y_j^k 与其相应的期望输出值 T_j^k 之间存在误差。故定义

$$E_k(\boldsymbol{W}) = \frac{1}{2}\sum_j (T_j^k - y_j^k)^2 \qquad (2)$$

为第 k 个样本的误差函数，当样本集中有 p 个样本时，则总的系统误差为

$$E(\boldsymbol{W}) = \sum_k E_k(\boldsymbol{W})/p = \frac{1}{2p}\sum_k\sum_j (T_j^k - y_j^k)^2 \qquad (3)$$

为了表示输出层各节点输出的相对误差大小，一般将上式改写为

$$E(\boldsymbol{W}) = \frac{1}{2p}\sum_k\sum_j (1 - y_j^k/T_j^k)^2 \qquad (4)$$

显然这是一个非线性函数，BP 网络的学习过程（即求解权值 W_{ih}、W_{hj} 的过程）实质上就是非线性函数的极值求解过程。因而可以用求非线性函数总体极值的方法求解式(4)，从而得到网络的全局最优解 $\boldsymbol{W}^*$。

三、非线性函数总体极值求解算法

目前，求解非线性函数总体极值的代表算法有隧道函数法[1]、填充函数法[2]、压缩变换法[3]，以及基于 Monte-Carlo 模拟计算的逐步逼近法[4,5]等。考虑到逐步逼近法算法相对简单，便于工程应用，故可作为求解式(4)的总体极值即 BP 网络全局最优解的算法。

根据文献[4]、[5]，逐步逼近法的基本思想如下：

首先定义 $\Omega_k = \{\boldsymbol{W} \mid E(\boldsymbol{W}) \leqslant c_k\}$ 为函数 $E(\boldsymbol{W})$ 在有界区域 G 上对应于某一实数 c_k 的有效作用区域。由于 $E(\boldsymbol{W})$ 为非线性函数，有

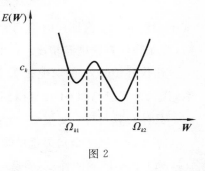

图 2

多个极小点，故可以是由多个不相连的区域组成。如图 2 所示，Ω_k 即由 Ω_{k1} 和 Ω_{k2} 组成。显然若有 $c_i \geqslant c_j$，则有 $\Omega_i \geqslant \Omega_j$。

若 Ω_k 非空，则 $E(\boldsymbol{W})$ 在 Ω_k 上的均值可表示为

$$u_k(E(\boldsymbol{W}), c_k) \approx \int_{\Omega_k}\cdots\int E(\boldsymbol{W})\mathrm{d}\boldsymbol{W} \Big/ \int_{\Omega_k}\cdots\int \mathrm{d}\boldsymbol{W} \leqslant c_k \qquad (5)$$

进一步若令 $c_{k+1} = u_k(E(\boldsymbol{W}), c_k)$。当取 $k=1,2,\cdots$ 时，则可得到单调下降的实数序列 $\{c_k\}$，有效作用区域序列 $\{\Omega_k\}$ 和均值序列 $\{u_k\}$。

$$\left.\begin{aligned} c_k &= u_{k-1}(E(\boldsymbol{W}), c_{k-1}) \\ \Omega_k &= \{\boldsymbol{W} \mid E(\boldsymbol{W}) \leqslant c_k\} \\ u_k &= \int\cdots\int_{\Omega_k} E(\boldsymbol{W}) \mathrm{d}\boldsymbol{W} \Big/ \int\cdots\int_{\Omega_k} \mathrm{d}\boldsymbol{W} \end{aligned}\right\} \quad (6)$$

$k=1,2,\cdots$ 令 $k \to \infty$，则有

$$\left.\begin{aligned} \lim_{k\to\infty} c_k &= c^* \\ \lim_{k\to\infty} \Omega_k &= \Omega^* \\ \lim_{k\to\infty} u_k &= u^*(E(\boldsymbol{W}^*)), c^*) \end{aligned}\right\} \quad (7)$$

可以证明

$$u^*(E(\boldsymbol{W}^*), c^*) = c^* \quad (8)$$

在上两式中，u^* 就是 $E(\boldsymbol{W})$ 在 G 上的总极小值，Ω^* 是其总极小点集，$\boldsymbol{W}^*$ 即为 BP 网络全局最优解。

具体计算时，$u_k(E(\boldsymbol{W}), c_k)$ 一般采用 Monte-Carlo 方法计算，若采用 Monte-Carlo 求积的平均值法，则表达式(5)可改写成：

$$u_k(E(\boldsymbol{W}), c_k) = \frac{1}{N} \sum_{i=1}^{N} E(\boldsymbol{W}_i) \quad (9)$$

式中，$\boldsymbol{W}_i$ 是第 i 次抽样时得到的在区域 Ω_k 上均匀分布的随机向量；N 为抽样次数，为兼顾计算速度和精度，N 可随着向总极值点的逼近而逐渐增大。

迭代开始时 c_0 的取值按如下方法确定：将权向量中各分量的初始化随机小数，代入公式(1)、(4)中计算得到的 $E(\boldsymbol{W})$ 值作为 c_0。

具体计算过程参见文献[4,5]等，不再赘述。

至此，BP 网络全局最优的学习算法可归纳如下：

（1）根据问题的性质，构造一个合适的 BP 网络结构；

（2）根据网络结构，按式(5)形式构造系统误差度量函数 $E(\boldsymbol{W})$；

（3）用上述非线性函数总体极值求解方法求解 $E(\boldsymbol{W})$，得到网络权值的全局最优解 $\boldsymbol{W}^*$。

显然，与传统的沿梯度下降算法相比，不存在局部极小问题，也不存在由于 η、α 等参数选择不当而导致的网络振荡甚至不能收敛等问题；与模拟退火和遗传算法相比，不存在大量需要经验式实验才能确定的问题，且应用方便，一旦形成误差函数 $E(\boldsymbol{W})$，即可由非线性函数总体极值求解程序实现计算机自动求解。

为了检验上述全局最优求解算法的有效性，以在神经网络研究中具有重要意

义的异或(XOR)问题为例进行对比计算。

结果表明,当需要处理的输入样本与训练集中的标准输入样本相差不大(10%以内)时,全局最优解的计算结果精度高于梯度法计算结果的精度。但随着处理样本相对标准输入样本变形程度的增加,全局最优解计算结果相对于相应标准输入样本期望输出值的误差急剧增大。这就说明,在处理输入输出非线性映射关系等问题时,全局最优解可望获得更精确的结果(因为全局最优解的计算结果对输入样本相对标准样本的变形程度较敏感)。但在处理分类问题时,特别是在处理样本偏离标准样本较大的情况下,采用全局最优解就不一定合适,因为此时有可能得出完全相反的结论。

参 考 文 献

[1] A. Torn and A. Zilinskas. *Global Optimization*. Lecture Notes in Computer Science,1989.

[2] 葛人溥,实用非线性最优化方法,高等教育出版社,1984.

[3] 阳明盛,求解总极值问题的压缩变换法,ORSC'92文集,成都科技大学出版社,1992.

[4] 郑权等,一个求总极值方法的构造和实现,自然杂志,1978,NO.1-2.

[5] 郑权等,关于总极值问题的最优性条件,高等学校计算数学,NO.3,1981.

[6] 殷勤业等,模式识别与神经网络,机械工业出版社,1992.

[7] 周继成,人工神经网络,科学普及出版社,1993.

[8] L. C. W. Dixon & G. P. Szergo. *Towards global optimization*. 2nd, Amsterdam: North-Holland,1978.

[9] 史忠植,神经计算,电子工业出版社,1993.

(原载《计算机科学》1996年第23卷01期)

基于神经网络的结构动力模型修改和破损诊断研究

徐宜桂　史铁林　杨叔子

提　要　提出了一种基于神经网络的结构动力模型修改和破损诊断方法,讨论并解决了该法实施中的若干技术问题,如神经网络的拓扑结构和快速算法、结构模态特性的量化比较、结构破损特征信息的提取和处理、计算结果精度的提高等等。数值算例表明:该法对于结构动力模型的修改以及结构破损部位和程度的诊断非常方便和有效。

关键词:神经网络,结构动力分析,模型修改,结构诊断

引　言

目前用于动力模型修改的典型方法,有矩阵摄动法、误差矩阵范数极小化法以及元素修正法等等,这些方法在使用中均存在一些不足之处[1]。为此,本文提出了一种基于神经网络的结构动力模型修改方法,其基本思想是通过神经网络建立结构物理、几何、边界条件等建模参数与结构模态参数之间的非线性映射关系,然后通过实测的结构模态参数,依据这种非线性映射关系,得到结构与之相应的物理、几何及边界条件,完成结构动力模型修改。由于该法直接修改建模参数,因而不仅可以避免传统方法中出现的一些问题,而且计算过程简单。

在动力模型修改的基础上,进行结构破损诊断是国内外研究的又一热点和难题。不少学者如 Yao(1985),Stepheas(1985),Woff.T(1989),Stubbs.N(1990)等都在此方面做了大量工作,理论和实验研究成果在历届 MAC 会议上均有报道。但由于现有方法一般采用频率或振型作为表征结构损坏部位和大小的敏感参数,因而应用起来不太有效。为此,本文提出了一种基于神经网络的结构破损诊断方

法,并应用结构第一阶振型曲率作为表征损伤部位和大小的敏感参数。具体做法是:先应用神经网络建立破损结构第一阶振型曲率差值(相对无破损结构)与结构破损部位和大小的非线性映射关系,再将测试到的结构第一阶振型曲率送入诊断神经网络,经网络映射即可得到该结构的破损部位和大小。

1 非线性映射用神经网络结构及其快速算法

BP 神经网络(图 1)由于具有较强的非线性映射能力而被广泛用于结构分析中,根据经验和算例[4,5],网络一般取 2 个隐含层,而隐含节点数则凭经验并结合现场试验确定。

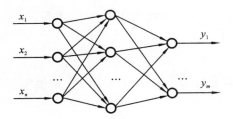

图 1　BP 网络

为了避免 BP 网络传统学习算法引起的网络麻痹或振荡问题,当系统误差 E 采用平方和形式时,将网络输出层误差[6]

$$\delta_j^k = F'(s)(T_j^k - y_j^k) \tag{1}$$

重新定义为

$$\delta_j^k = (F'(s) + r)(T_j^k - y_j^k) \tag{2}$$

其中 $F'(s)$ 为节点转移函数(sigmoid 函数)的导数,T_j^k, y_j^k 分别为第 k 个训练样本中第 j 个节点的期望和实际输出值,r 为可调的修正系数,取值 $0 \sim 0.15$。由于 r 能在训练中自动调整大小,避免 δ_j^k 有时过大(振荡)和过小(麻痹)等情况,提高了学习效率,如在算例 1 中,采用传统算法和式(2)分别迭代 5772 次和 1964 次收敛,效率提高近 3 倍。

2 基于神经网络的结构动力模型修改

根据上述思想,进行动力模型修改时,对物理、几何参数 X 还需按式(3)进行规范处理

$$\overline{X} = \frac{X - X_{\min}}{X_{\max} - X_{\min}} \tag{3}$$

对频率 ω、振型 ϕ 同样需按式(4)、(5)进行规范处理

$$\overline{\omega} = \frac{\omega - \omega_0}{\omega_{\max} - \omega_0} \tag{4}$$

$$\overline{\phi} = \frac{\phi^T \phi_0}{\sqrt{\phi^T \phi} \sqrt{\phi_0^T \phi_0}} \tag{5}$$

其中 ω_0、ϕ_0 分别为参考频率和振型值。

为了验证和进一步提高模型修改精度,可将网络映射得到的模型参数修改值,重新输入有限元程序,计算相应的 ω、ϕ 值,将该值与实测值进行比较,若两者误差在允许范围内,表明模型修改是成功的。否则,以上述有限元输入输出值组成一新的网络训练样本,加入到原训练样本集中,重新训练网络,直至误差满足要求为止。

算例 1

图 2 给出了本实验室内一个钢筋混凝土简支梁模型。几何尺寸及截面特性如表 1 所示。

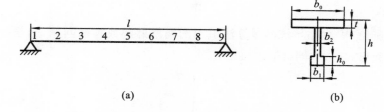

图 2 梁计算模型

(a) 单元划分;(b) 截面形状

表 1 梁几何尺寸及截面特性表

L/cm	b_0/cm	b_1/cm	b_2/cm	h_0/cm	h_1/cm	t/cm	A/cm²	I/cm⁴
245	222/222	67.18/57.76	30/30	60/60	260/260	25.31/25.31	148.9/143.2	12600/11690

注:"/"号下为梁单元①、⑧的值,"/"号上为其余梁单元值,以后同。

取弹性模量 $E = 2.6 \times 10^5$ kg/cm² 不变,再分别取:

I:12600/11590　　　14490/13443　　　16664/15460

A:149/143　　　　　172/165　　　　　195/187

ρ:0.0025　　　　　0.0035　　　　　0.0040

组合得 27 组建模参数,应用 Super SA P91 有限元程序,计算得到相应的 27

组梁弯曲振动 $\omega, \phi_i (i=1,2,\cdots,7)$ 值。以 $\overline{\omega_1}$、$\overline{\phi_1}$ 为输入，$\overline{EI}$、$\overline{EA}$ 和 $\overline{P}$ 为输出，组成 27 组网络训练样本。

取网络结构为 2-9-6-3，$\eta=0.5, \alpha=0.4, r=0.1, E=0.01$，使用上述训练样本迭代 1964 次网络收敛。

为测试网络精度，取 3 组测试数据：
1. $E=2.6\times10^5$　$I=13860/12860$　$A=152/146$　$\rho=0.0032$
2. $E=2.6\times10^5$　$I=15120/14028$　$A=169/162$　$\rho=0.0027$
3. $E=2.6\times10^5$　$I=15851/14706$　$A=161/154$　$\rho=0.0030$

按上述方法得到 3 组 $\overline{\omega_1}$、$\overline{\phi_1}$ 值，送入神经网络，结果如表 2 所示。可见完全满足工程要求。

表 2　测试样本输入数据及其计算结果和误差

序号	输入值		输出值			计算值			误差/%		
	$\overline{\omega_1}$	$\overline{\phi_1}$	$\overline{EI}$	$\overline{EA}$	$\overline{\rho}$	I	A	ρ	$\left\|\frac{\Delta I}{I}\right\|$	$\left\|\frac{\Delta A}{A}\right\|$	$\frac{\Delta\rho}{\rho}$
1	0.63896	0.87511	0.16193	0.08007	0.39129	13258/12300	152.7/146.5	0.00308	4.35	0.44	3.53
2	0.44101	0.90357	0.62782	0.45386	0.07917	15151/14056	169.9/162.9	0.00262	0.20	0.52	3.01
3	0.45401	0.87824	0.89219	0.30793	0.43144	16225/15053	163.2/156.6	0.00315	2.37	1.34	4.91

3　基于神经网络的结构破损诊断

基本思想与实施步骤和动力模型修改方法类似。此时网络的输入值为有破损和无破损结构振型曲率 c'_i, c_i 的绝对差值 d_i

$$d_i = |c'_i - c_i|$$
$$c_i = \frac{\phi_{i+1} - 2\phi_i + \phi_{i-1}}{l_i^2}, i=1,2,\cdots,m-2 \tag{6}$$

式中：ϕ_i 是振型 ϕ 第 i 个元素值；l_i 是第 i 个单元的长度；m 为计算模型节点的总数。网络的输出值则为结构破损部位和程度的量化值，如某一组 d_i 值对应的结构破损部位为单元 e，破损程度为刚度下降 $\alpha\%$，则对网络输出的第 e 个节点取值 $\alpha/100$，其余节点取值为 0（0 表示无破损）。同样对网络诊断的结果需要进行检验，方法同前。

这里需要说明的是，以 d_i 作为表征结构破损部位和程度的特征参数，首先是由 Pandey A. K. (1991) 提出的[2]。但本文在研究过程中发现：只有第一阶振型曲率差值可以准确反映破损部位和大小。所得结论与 Pandey A K 的结论有些差异（参见图 3 与文献[2]中的计算结果）。

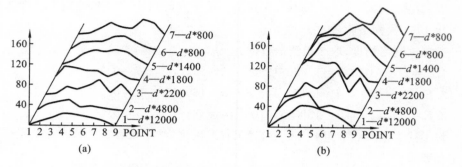

图 3　不同破损状态下各阶振型曲率差值 d 的变化（d 值放大倍数见图示）
(a) EL E.4,15%；(b) EL E.4,30%

算例 2

计算模型如图 2 所示。结构无破损时的参数为：$E=2.6\times10^5$ kg/cm^2，$I=12600/11690$ cm^4，$A=149/143$ cm^2，$\rho=0.0025$ kg/cm^3。假设结构上 1~8 个单元刚度分别下降 25%、40%，得到 16 种破损情况。用有限元法计算得到相应的 16 组频率 ω 和振型 ϕ 值，组成 16 组训练样本。

取诊断神经网络结构为 9-19-16-8，$\eta=0.5$，$\alpha=0.4$，$r=0.1$，$E=0.0004$，迭代 2384 次网络收敛。

为测试网络的诊断能力，假设 4 种结构破损状况：①EL E.3,20%，②EL E.4,35%，③EL E.6,5%，④EL E.7,10%。按上述方法得到相应的 4 组 d_i 值，送入网络，第一次诊断结果的最大误差为 67.25%（情况④），且情况③的破损部位诊断错误。为此，补充 2 个训练样本（由情况①、情况④的第一次诊断值产生），重新训练网络，网络第二次诊断的结果如表 3 所示，可见除第 2 种情况（因刚度下降太小）外，其余诊断结果完全满足工程要求。

表 3　四种破损状况的第二次网络诊断结果

破损状况	网络输出值		破损部位		破损程度		
	最大值单元	最大值	诊断单元	实际单元	诊断破损值	实际破损值	误差/%
EL E.3,20%	3	0.39461	3	3	19.73%	20%	1.35
EL E.4,35%	4	0.74006	4	4	37.00%	35%	5.71
EL E.6,5%	6	0.08604	6	6	4.30%	5%	14.0
EL E.7,10%	7	0.18461	7	7	9.23%	10%	7.70

注：EL E.3,20% 表示破损单元为 3，刚度下降 20%。

4 结 论

(1) 理论分析和数据计算表明：本文提出的基于神经网络的结构动力模型修改和结构破损诊断方法，以及该法实施过程中的若干技术处理是实用可行的。

(2) 本文提出的基于神经网络的动力模型修改方法，不仅避免了现有各种修改方法中的一些棘手问题，而且计算过程简单，实施方便(仅需测取结构的第一阶频率和振型)，物理意义明确。特别是对结构系统性质(线性或非线性)、模态性质(实模态或复模态)，该法均未提出任何限制条件，因而可以方便地推广到复模态甚至非线性系统的动力模型修改中。

(3) 本文提出的基于神经网络的结构破损诊断方法，不仅能诊断出结构的破损部位，而且还能诊断出结构破损程度的大小。提出的以第一阶振型曲率差值作为表征结构破损部位和程度的特征参数，物理意义明确，实测方便。

参 考 文 献

[1] 顾松年. 结构动力修改的发展和现状. 机械强度, 1991; 13(1): 1-9.

[2] Pandey A K. *Damage detection from changes in curvature mode shapes*. J. sound & vibration, 1991; 145(2): 321-332.

[3] Agbabian M S. *Detection of structural changes in a bridge model*. Bridge Evaluation. Repair and Rebabilization, Nowak, A. S. (ed), 1990.

[4] Pandey P C. *Nonlinear analysis of plates using artificial neural networks*. J. Structural Engineering, 1994; 21(1): 65-78.

[5] Wu X. *Use of neural networks in detection of structural damage*. Computer & Structures, 1992; 42(2): 649-659.

[6] 冯昭志. 神经网络快速学习算法及其硬件实现研究. 武汉, 华中理工大学博士论文, 1994.

[7] Richardson M H. *Determination of modal sensitivity functions for location of structural faults*. In: Proceedings of 9th MAC, Florence, Italy, 1991.

Structural Dynamic Model Modification and Damage Diagnosis Method Based on Neural Networks

Xu Yigui Shi Tielin Yang Shuzi

Abstract A new structural dynamic modal modification and structural damage diagnostic method based on neural networks is presented in this paper, and some keys problems related to the method such as the neural networks topologic architecture and fast learning algorithms, the quantitative comparison of structural modal properties, the extraction and processing of characteristic information on faults location and extent, the improving of computing precision etc, are discussed and solved, computing results show that the dynamic modal modification method based on neural networks have not some difficult problems which must be dealt with when using other modification method, and have high modification precision. The diagnostic method based on neural networks can determinate precisely the faults extents as well as faults location.

Keywords: Neural networks, Dynamic structural analysis, Modal modification, Structural damage diagnosis

(原载《振动工程学报》1997 年第 10 卷第 1 期)

A CORBA-based Agent-driven Design for Distributed Intelligent Manufacturing Systems

Lei Ming   Yang Xiaohong   Mitchell M. Tseng
Yang Shuzi

Abstract An agent-oriented methodology is presented for representation, acquisition, and processing of manufacturing knowledge along with analysis and modeling of an intelligent manufacturing system (IMS). An intelligent manufacturing system adopts heterarchical and collaborative control as its information system architecture. The behavior of the entire manufacturing system is collaboratively determined by many interacting subsystems that may have their own independent interests, values, and modes of operation. The subsystems are represented as agents. An agent's architecture and task decomposition method are presented. The agent-oriented methodology is used to analyze and model an intelligent machine cell. An intelligent machine center is considered as an autonomous, modular, reconfigurable and fault-tolerant machine tool with self-perception, decision making, and self-process planning, able to cooperate with other machines through communication. The common object request broker architecture (CORBA) distributed software control system was developed as a simple prototype. A case study illustrates an intelligent machine center.

Keywords: Agent-based analysis, Autonomous, Cooperation, CORBA, Intelligent manufacturing system

1 Introduction

The new generation of advanced manufacturing systems is forcing a shift from mass production to mass customization (Tseng et al.,1997a,b),the ability to manufacture in small batches. To achieve this, it is becoming increasingly important to develop manufacturing systems and equipment control architectures that are modifiable, extensible, reconfigurable, adaptable, and fault tolerant. Centralized control architectures have given way to hierarchical schemes, where higher levels of the hierarchy control the lower level via a master-slave relationship. However, to achieve even greater levels of reconfigurability and adapt-ability, the newer manufacturing systems are adopting heterarchical control structures that are made up of multiple, distributed, locally autonomous entities, thus allowing a cooperative approach to global decision making (Ditts and Whorms,1993).

Conventional flexible manufacturing systems imply centralized or fixed hierarchical control. Work cell functionality and product description are fixed for optimal or near-optimal performance based on a narrow product mix. Given current trends, where customer demands cause large variations on product mixes, and new technology is continually introduced into the manufacturing process, a predetermined fixed hierarchy is not likely to produce good performance over extended periods of time. This motivates research on dynamically reconfigurable hierarchies in order to meet customer demands while striving to maintain high productivity in the manufacturing system.

The next generation of advanced manufacturing systems need to incorporate three characteristics:

(1) More distributed (heterarchical) control to achieve fault tolerance and robustness;

(2) Work cell and machine flexibility, the ability to change and reconfigure work cell functionality to handle a greater variety of product mixes effectively;

(3) Product flexibility adapts part selection and assembly sequencing to the current system configuration.

Other factors that need to be addressed are dynamic reconfiguration to accommodate unusual events, such as machine breakdowns and the introduction of new technology and processes as they become available. The goal is to maintain effective and efficient manufacturing operations with minimum downtime in reconfiguring, replanning, and rescheduling manufacturing operations.

IMS TC-5 proposes the concept of a holonic manufacturing system (HMS) as a framework for allowing adaptive reconfiguration (http://www.ims.org/holonic.htm). An intelligent manufacturing system (IMS) can be regarded as the complex whole of various subsystems (http://www.ims.org/). Because of the rapidly changing market, the subsystems must cooperate very well with each other to achieve high efficiency under small-lot or job-lot production. The manufacturing is thus the process of distributed cooperative problem solving by exchanging and sharing of materials, energy and information between the subsystems. Before the realization of such a large intelligent system, proper methods should be used for system analysis and modeling. A few methodologies for manufacturing system analysis, such as GRAI (Pun, 1986) and IDEF (Mackulak, 1984), had been used during research on computer integrated manufacturing systems (CIMS). These methodologies were useful in developing CIMS, but have a low efficiency for IMS because of the system integration on the level of intelligence. Although an organism-oriented modeling approach (Lefrancois and Montreuil, 1994) had been proposed to design the manufacturing system tools are still lacking for analysis and processing of the manufacturing knowledge.

The concept of an (intelligent) agent is proposed as the essential technique for IMS analysis and modeling. An agent-oriented methodology is proposed for representation, acquisition, and processing of manufacturing knowledge along with analysis and modeling of the IMS. The manufacturing system is viewed as an organization composed of intelligent entities called agents. Section 2 describes the conception and architecture of the agent; Section 3 describes the agent-oriented analysis and modeling methodology. Section 4 introduces a CORBA-based manufacturing cell control environment. Section 5 describes a case study for implementing an intelligent machine center.

2 Conception and Architecture of an Agent

The term *agent* is mostly used in the artificial intelligence (AI) domain as a high-level concept for analyzing AI problems. Usually an agent means an entity that can perform a task continuously and autonomously in the non-determinacy environment where there exist other processes and entities. Although the term *agent* is used frequently, it has no universal meaning, definition and structure. Shoham (1993) defined an agent to be any hardware or software component that has such mental states as beliefs, capabilities, choices, and commitments. Genesereth et al. (1987) suggested that an agent can be analyzed by its state activities and perception. For representing and analyzing the IMS and considering its realization, the following definition is appropriate: an agent is an entity that can perform some tasks and achieve the predetermined goal autonomously. The autonomy in the definition does not mean the agent must be fully independent, but means that it can automatically ask for assistance and services from other agents whenever necessary. According to this definition, human experts, intelligent CAD systems, and intelligent machining cells are all agents of the IMS.

An agent in the IMS consists of two parts: the set of tasks and the set of activities. The task is the work to be done or the activities to be performed. For each agent there exists a set of tasks it is capable of performing. Every task is given a name as its only label when it is to be processed. The names of all the tasks form a task list. From the task list, other agents get information about what tasks an agent can fulfil. The performance of a task can usually be described as a series of activities that begins with a start-event and is directed toward a goal-event.

The agent in the IMS should satisfy the following requirements:

(1) Each agent has the knowledge necessary for performing its tasks. Most knowledge exists when the agent is constructed. The agent can also get knowledge by self-learning or by learning from other agents. The agent must process its knowledge to select proper actions. Knowledge may be expressed

implicitly, perhaps in the architecture of the networks, the weights of the links, the algorithms, the software, or the computing hardware. Knowledge may also be explicitly represented by the data structures, e. g. , as facts or frames.

(2) Each agent has an interface to interact with the environment and other agents. The agent may ask for assistance from or give assistance to other agents including human experts. Furthermore, as the subsystem of an intelligent system, the agent must cooperate with other agents so as to integrate all the subgoals to the system goal. Agents must have the ability to cooperate. Cooperation is one of the tasks of all agents.

(3) Each agent is hierarchically constructed; this means an agent may consist of finite subagents or it may be the subagent of finitely many other agents.

An agent has some aspects in common with an object of the object-oriented method. From the engineering viewpoint, an agent is a specialization of an object (Baker, 1996). The object-oriented method can be used as the operational model for an agent. Frames are used here to represent the agent class and instance agents. In particular, the dynamic slots are the activities. The architecture of the agent is presented as follows:

(1) An agent consists of a set of tasks and a set of activities.

(2) A task is represented by the task frame, which includes task name, activity names, start-event, goal-event, task state, task priority, and time requirement.

(3) Activities are divided into three types: domain activities, interfacing activities, and interchanging activities. The domain activities are limited within the agent. The interfacing activities are for asking assistance from other agents; the interchanging activities are for giving assistance to other agents.

(4) Activities are composed of knowledge and knowledge processing units.

(5) The task frames are accessible to the environment, and the activities are encapsulated within the environment.

(6) Agents interact with the environment by message passing.

The difference between an agent and an object is that the agent is autonomous and cooperative. The commonly used object-oriented method cannot support an agent. An independent task scheduling mechanism is provided for the object acting as an agent. It is used to interpret and process the messages. Agents are divided into several groups to support the cooperation within the groups.

3 Agent-oriented Analysis and Modeling of IMS

The IMS can be viewed as the organization of a finite set of agents according to the previous definition of an agent. The system analysis may begin with task decomposition, because every agent can perform a set of tasks. As a result, a set of primary tasks of the IMS is obtained, from which a set of agents that forms the IMS can be developed. Based on the architecture, the system analysis can be carried out hierarchically. For instance, the manufacturing system can be decomposed into business and management, research and development, design, manufacturing, and marketing; then manufacturing can be decomposed into scheduling, part manufacturing, production control, and assembly.

The agent-oriented methodology for analysis and modeling is a four-step procedure:

(1) Decomposing the system tasks;
(2) Forming a set of primary agents which can cover the whole system;
(3) Realization of each agent;
(4) Organization of the primary agents to construct high-level agents until the original system becomes an agent.

When these procedures have been completed, we get the agent models of the system for all levels of the architecture. The following principles must be satisfied when the primary agents are formed from the decomposed tasks and when the high-level agents are organized:

(1) An agent must be logically and physically practical.
(2) Every primary task of the system must be covered by at least one agent.
(3) Interaction exists within the agents so that the local function of each agent can be synthesized properly to support the performance of the system.

As an example, consider the methodology for designing an intelligent machining center (IMC). The IMC (Fig. 1) is an intelligent machining agent in the IMS. It receives the machining tasks from other agents and determines whether the tasks are suitable according to its abilities and current state. Then it negotiates with the agents in the same machining group if it is necessary to adjust the job load (Fig. 2). When the job is settled, the IMC inspects whether the workpiece is correctly prepared and set up before scheduling the operation plans. During the machining process, multisensors are used to monitor the state of the IMC and the environment. The activities of the IMC are controlled in-process on the basis of state variables.

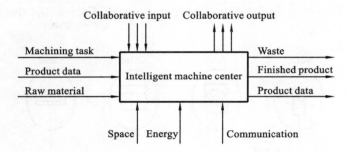

Fig. 1 Intelligent machine center

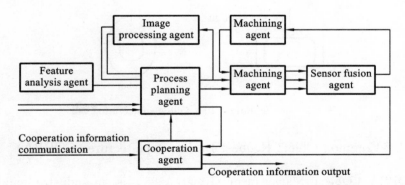

Fig. 2 Relationships between agents in the intelligent machine center

4 A CORBA-based Manufacturing Cell Control Environment

4.1 Distributed Objects

Distributed object technology allows computing systems to be integrated such that objects or components work together across machine and network boundaries. Examples of current distributed object or component technologies include CORBA (Object Management Group, 1995), OLE (Orifali and Harkey, 1996). A distributed object is not necessarily a complete application but rather a reusable, self-contained piece of software that can be combined with other objects in a plug-and-play fashion to build distributed object systems (Fig. 3). A distributed object can execute either on the same computer or on another networked computer with other objects. Thus a client object may make a request of a server object and the operation proceeds unaffected by their respective locations.

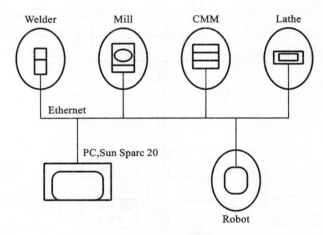

Fig. 3　Configuration of an agile manufacturing cell

4.2　Common Object Request Broker Architecture (CORBA)

CORBA (Object Management Group, 1995) is an industry middleware standard for building distributed, heterogeneous, object-oriented applications. CORBA is specified by the Object Management Group (OMG), a non-profit consortium of computer hardware and software vendors. CORBA currently provides the best technical solution for integrating our manufacturing test bed; it is robust, heterogeneous, interoperable, multiplatform, and multivendor supported. Fig. 4 shows the CORBA reference model. The object request broker (ORB) is the communication hub for all objects in the system; it provides the basic object interaction capabilities necessary for components to communicate.

When a client invokes a method on a remote object, the ORB is responsible for marshaling arguments for the call, locating an object server, physically transmitting the request, unmarshaling the arguments in a format required by the server, marshaling return values or exceptions for the response, transmitting the response, and unmarshaling the response at the client end.

CORBA object services in Fig. 4 are common services needed by a distributed object system; these services add functionality to the ORB. CORBA object services include standards for object life cycle, naming, persistence, event notification, transactions, and concurrency. Common facilities provide a set of general-purpose application capabilities for use by object systems, including accessing databases, printing files, document management, and electronic mail in a distributed system. Finally, application objects in Fig. 4 are the developed software which make use of the other three categories.

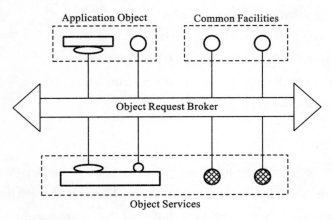

Fig. 4　CORBA object model

4.3　OMG Interface Definition Language (IDL)

The OMG interface definition language (IDL) is used to define interfaces in CORBA. An IDL interface file describes the data types and methods or operations that a server provides for an implementation of a given object. IDL is not a programming language; it describes only interfaces and it has no implementation-related constructs. The OMG does specify mappings from IDL to various programming languages, including C, C++, and Smalltalk. IDL is used here to show interfaces to manufacturing devices, the task sequencer, and other CORBA objects in the system. All the manufacturing implementations were written in C++, though other languages (e. g. , Tcl and Visual Basic) were chosen to write client-only applications using CORBA objects. Even without a detailed description, the OMG IDL language is fairly readable and we present an informal introduction by describing a simple interface within an environment called **INotify**. As its name suggests, this interface is used to communicate notifications of events. For example, when some asynchronous process is requested, success or failure may be later reported back through an object **INotify** interface. The OMG IDL code for **INotify** interface is as follows:

    interface INotify {

    readonly attribute string str;

    readonly attribute boolean is Empty;

    readonly attribute long ExceptCode;

    readonly attribute string ExceptString;

    oneway void SetString (in string str);

    oneway void SetExcept(in long code, in string str);

    oneway void Clear();

    INotify New();

    };

INotify is declared as an interface in the first line of the file. A description of the interface is provided in subsequent lines, between the curly braces. This interface consists of four attributes followed by four operations. Attributes correspond conceptually to state variables for the object. For example, the first attribute listed, **str**, is declared to be of type string. Thus each **INotify** object has a string of text called **str** that can be queried. The leading **readonly** keyword in the declaration of **str** indicates that its value cannot be directly modified. Without this keyword, the value of the attribute could be set as well as queried. Subsequent attribute declarations use some of the other basic types available in IDL. These include long, short, boolean, and double. New types can also be defined as structures, arrays, and sequences. And interfaces themselves represent types. Thus, an attribute might be of type **INotify**, for example.

4.4 Interfaces for Manufacturing Devices

We have used OMG IDL to specify standard software interfaces to the various manufacturing devices in the agile manufacturing cell. A goal of this effort is illustrated in Fig. 5. At the bottom of the figure are the various machine tools. On the PC associated with each machine tool, we implement software which provides the OMG IDL interface. By manipulating the software interface, a client program can control the corresponding machine. With this, then, we can write client software illustrated in Fig. 5 as the cell management software components, which controls the manufacturing activities.

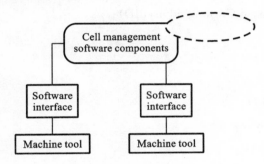

Fig. 5 Cell management software

Thus, each of the physical manufacturing objects in the agile manufacturing cell is controlled by a corresponding CORBA software object. In spite of the apparent differences among the various devices (lathe, robot, storage table, etc.), they all support the same software interface: an OMG IDL interface called IDevice. This section presents a description of our IDevice interface. For brevity and for clarity, an abbreviated IDL will be shown.

The first declaration line for interface IDevice illustrates the inheritance feature of OMG IDL. The declaration of IDevice as an interface is followed by a colon, then a list of other interface names, starting with IBaseDev. IDevice therefore inherits operations and attributes from the following other interfaces:

(1) IBaseDev—naming and operational status for the machine;

(2) IAllocaDev —controlling access to the machine;

(3) IRunDev —running processing activities on the machine (machining a

part);

(4) IMovePart — transfering material into and out of the machine.

Here is some of the relevant IDL code:

```
interface IBaseDev
{
readonly attribute string DeviceClass;
readonly attribute string DeviceID;
readonly attribute string VersionsString;
readonly attribute string Status;
long Advise (in Inotify hnotify);
boolean UnAdvise (in long advisedID);
};
interface IRunDev
{
boolean Pause();
boolean Resume();
boolean Abort();
};
interface IMovePart
{
boolean SetPartner();
boolean TakeFromPartner();
boolean GiveToPartner ();
};
interface ICellSeq
{
long AddJob (in string JobName, in string JobText, in string PartName, in Inotify WhenDone);
    void Pause(in long JobID);
    void Resume(in long JobID);
    void Abort(in long JobID);
    boolean DeletJob (in long JobID);
```

boolean AddDevice

};

4.5 Integration of Devices in a Manufacturing Cell

As a research facility, IDevice objects were among the first software modules created for this project. Our experience has been that the IDevice design and implementation is sufficiently flexible and adaptable to accommodate both the addition of new machine tools and the evolution of more sophisticated cell management software.

As a CORBA object itself, the cell sequencer is also a client to the IDevice objects. This configuration can be seen in Fig. 6. The cell sequencer dynamically attaches to devices, hence there is no need to recompile when a new machine tool comes on-line or a machine tool disappears. The sequencer accepts jobs, dispatches tasks in the cell, prevents deadlock situations and guards against starvation of any single job. An abbreviated IDL code for the sequencer **ICellSeq** is as follows:

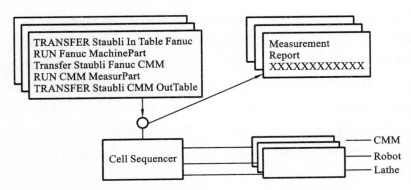

Fig. 6　Cell sequencer manipulating several IDevice objects

interface ICellSeq

{

long AddJob (in string JobName,

in string JobText, in string PartName, in INotify WhenDone);

void Pause(in long JobID);

void Resume(in long JobID);

```
void Abort(in long JobID);
boolean DeleteJob(in long JobID);
boolean AddDevice(in string DevName,in IDevice Dev);
boolean AddRobotDevice(in string DevName,in Device Dev);
boolean RemoveDevice(in string DevName);
IDevice QueryDevice(in string DevName);
};
```

A new job can be added to the sequencer with the **AddJob**() operation. This operation takes three input strings as arguments: **JobName**, a unique name within the cell; **JobText**, a script of high-level instructions to be accomplished in the cell; and **PartName**, which identifies the part to be manufactured. The **AddJob**() operation also takes an **INotify** object reference so that the sequencer's client can be notified of job completion or any error conditions encountered. The return value of the **AddJob**() operation is of type long, indicating the assigned **JobID** given by the sequencer; this **JobID** can then be used to **Pause**(), **Abort**(), **Resume**(), or **Delete**() a job in the sequencer, even while the task dispatcher is operating. The task sequencer parses **JobText**, written in a simple scripting language developed for our manufacturing cell. The language supports manufacturing setup operations, material transfer between two **IDevice** objects, access to **IDevice** objects, and program execution at **IDevice** objects. Though the cell sequencer coordinates cell activities, many operations can be accomplished intelligently by the devices. For example, we have mentioned that all material transfer is performed as peer-to-peer object interaction, in dependent of the supervisory control of the task sequencer.

There are four operations in **ICellSeq** which allow a client to manipulate devices known to the sequencer: **AddDevice**(), **AddRobotDevice**(), **RemoveDevice**(), and **QueryDevice**(). Notice that there are two different operations to add a device to the sequencer: the **AddRobotDevice**() is necessary to distinguish robot and transport vehicles from all other manufacturing and storage devices known by the sequencer. The sequencer must know if a device is a transport device in order to prevent certain deadlock conditions in the cell. By including operations for dynamically adding and removing devices in a cell sequencer, the sequencer will never have to be recompiled or restarted when a new **IDevice** object is available on the network. In theory, this architecture supports a cell sequencer remotely dispatching jobs to any **IDevice** objects.

The current task dispatcher is quite simple, with no scheduling optimization criteria. This task dispatcher will be used by a smart scheduling object, **ICellSched**. This elaborate scheduler will call **ICellSeq** to dispatch jobs once it has optimized on time, cost, and priority values on jobs. We expect the current interface and implementation of **ICellSeq** to remain as described above. Allowing for this type of growth is a strength of using this distributed object architecture.

5 Intelligent Control for an Open Architecture System

Open architecture controllers are designed to eliminate the problem of implementation by creating a flexible control system which can be attached to a wide variety of machine tool systems. The open architecture control system for the machining process is designed to replace or supplement an existing CNC controller while leaving the original axis, spindle drive-motors and supporting electronic interfaces intact. Open architecture controllers are usually constructed using standard minicomputers, usually running in a DOS or UNIX environment such as an IBM-type personal computer or workstation.

Several open architecture controller schemes have been introduced. Altintas proposed a hierarchical open architecture CNC (HOAM-CNC) which provided an open platform for implementing machining algorithms (Altintas, 1994; Altintas, 1996; Altintas and Newell, 1996). This system used a 486 PC with an off-the-shelf digital signal processing (DSP) board. Its multiprocessor setup consisted of a system master, which maintained overall control of the system; a CNC master, which maintained the CNC functions of workpiece position and velocity; and individual machine axis controllers, which were designed in-house. Mosaic, a system of Sun workstations based on a VME bus, controls a tree-axis machining center using off-the-shelf components for the axis and spindle controllers (Greenfield, 1989). The system also integrates touch probes, basic adaptive control for cutting, and interfaces with programs such as Machinist, an expert system for setup and clamping procedures.

The implementation of control strategies to machining processes often requires the use of application-specific knowledge and procedures, known as skilled technology. An intelligent machining will normally succeed when the following procedure is observed:

(1) Feed is maximized such that tool breakage is avoided.

(2) Depth of cut is maximized within the chatter limit.

(3) Velocity is adjusted to achieve acceptable tool life.

The objective of the controller is to increase the utilization of the machine tool by coordinating event-design chatter control and adaptive force control subsystems. This is accomplished by using an intelligent agent (Lei, 1995; Lei et al., 1996).

The overall control system consists of several control tasks operating under a small multitasking kernel. A realtime inference engine, operating as a high priority task, regularly evaluates events and performs coordination and recovery operations. A frame-based knowledge base is used to store and process information. The use of an intelligent agent is shown to achieve a coordinated and unified control objective.

The system is implemented on a vertical milling machine center having a 7.5 kW spindle motor. A two-axis numerical controller coordinates the tool movement. Tool force measurements are taken from load cells at the tool post, and vibration signals are taken from accelerometers at the tool post. These signals are digitally sampled by the control microcomputer.

The numerical controller has an optional programmable logic controller (PLC) that allows for NC programming using variables in place of geometric coordinates. A dedicated interface allows for tool paths to be commanded externally.

Variation in the feed rate is achieved using a feed-rate override circuit. A 4-bit binary signal applied to the override circuit will vary the actual tool feed rate from 0 to 150% of the programmed value. This variation occurs in discrete steps of 10%. The knowledge-based controller resides in a 32-bit microprocessor-based computer. These symbolic algorithms encode a *priori* defined reasoning about indicated process system conditions and the actions required to maintain a desired performance. At the lowest level of intelligence are the process level tasks. These include individual loop controllers and more computationally demanding signal analysis algorithms. Such operations involve numeric data processing, as opposed to the symbolic processing performed at the highest level.

The controller software is executed under control of a real-time, multitasking operation system. Utilities are provided for task synchronization and intertask communication. The knowledge base exists as a global blackboard entity. The remaining elements of the system are realized as four separate concurrent tasks:

(1) The intelligent monitor's real-time inference engine;

(2) A signal processing task for force and vibration monitoring;

(3) A force controller task;

(4) A task to send tool path commands to the CNC.

An approach to achieving these goals would involve the separation of knowledge representation issues, and event evaluation and recovery action issues. The frame-based approach to knowledge representation offers many advantages for control applications. However, the inherent backtracking inference mechanism and its associated difficulties in real time imply that knowledge evaluation should be separated from the frame-based knowledge base. In the system presented here, the intelligent system supervisor is therefore divided into three distinct parts:

(1) A frame-based knowledge base for maintenance of system knowledge and evaluation of single events related to signal or alarm conditions. The evaluations convert the numeric process data into a symbolic form required by the inference engine. Such actions require low system overhead and represent the minimal extent of real-time activities in the knowledge base.

(2) A logic structure relating each of the signal or alarm events in the knowledge base to the recovery actions required to bring about stable and coordinated system operation.

(3) An inference engine to direct searching of the logic structure. Rules are evaluated symbolically and the associated recovery actions are systematically executed. As a task in the multitask system, the inference engine executes concurrently with the conventional control systems.

These are implemented in the C++ programming languages for fast and efficient execution. The uses of objected-oriented C allowed for the development of a frame-based knowledge base while retaining the execution speed of C.

6 Conclusion

A distributed object CORBA framework has been presented for the management of a manufacturing cell. This architecture is robust, allowing for easy addition, deletion, and updating of manufacturing devices in a plug-and-play manner. Several different implementations were shown for a CORBA **IDevice** interface and the corresponding C++ class, **CDevice**. These well-defined common interfaces hide from the cell management software their implementation details, including the machine controller and the communications to the machine tool. The **IDevice** interface also supports automated and manual machine operations and/or material transport. The **CDevice** class allows a wide range of machining, storage, or transport devices to be encapsulated. A CORBA interface to a cell sequencer has been defined and implemented. The cell sequencer manipulates **IDevice** objects and coordinates manufacturing jobs on a shop floor. CORBA enhances the system integration because it is an industry standard for interoperable, distributed objects across heterogeneous hardware and software platforms. In time, as commercial software vendors provide CORBA interfaces to various software components, it will be easy to integrate them with the software developed here. The resulting architecture is scalable across a wide-area enterprise (Mratin and Spooner, 1997; Tseng et al., 1997a). Our future work includes adding support for a more information-intensive environment available at the shop floor. Intelligent cell management components must have access to design and planning information in order to make well-informed decisions about scheduling jobs and managing manufacturing devices. The bidding scheme based collaborative control system (Tseng et al., 1997a, b) can be implemented with this kind of information infrastructure.

References

1 Altintas, Y. (1994) *A hierarchical open architecture CNC system for machine tools*. Annals of CIRP, 43(1), 349-354.

2 Altintas, Y. (1996) *Modular CNC design for intelligent machining, part 2: modular integration of sensor-based milling process monitoring and control tasks*. Journal of Manufacturing Science and Engineering, 118(4), 514-521.

3 Altintas, Y. and Newell, N. (1996) *Modular CNC design for intelligent machining, part 1: design of hierarchical motion control module for CNC machine tools*. Journal of Manufacturing Science and Engineering, 118(4), 506-513.

4 Baker, A. D. (1996) Metaphor or reality: a case study where agents bid with actual costs to schedule a factory, in Market-Based Control, Clearwater, S. H. (ed.), World Scientific, London, pp. 184-223.

5 Ditts, D. and Whorms, H. (1993) *The evolution of control architectures for automated manufacturing systems*. Journal of Manufacturing Systems, 12(1), 79-93.

6 Duffe, N. A. (1990) *Synthesis of heterarchical manufacturing systems*. Computers in Industry, 14, 167-174.

7 Genesereth, M. R. (1987) *Logical Foundations of Artificial Intelligence*. Morgan Kaufmann, San Francisco, CA.

8 Greenfeld, I. (1989) *Open system machine controllers—the MOSAIC concept and implementation*. Transactions of ACM, 18, 91-97.

9 Lefrancois, P. and Montreuil, B. (1994) *An organism-oriented modeling approach to support the analysis and design of manufacturing systems*. Journal of Intelligent Manufacturing, 5(4), 121-142.

10 Lei, M. (1995) Research on intelligent cutting monitoring and manufacturing intelligence cooperating, PhD Dissertation, Huazhong University of Science and Technology.

11 Lei, M., Tseng, M. M., Yang, X. H. and Yang, S. Z. (1996) Agent-oriented analysis methodology in intelligent manufacturing system in Proceedings of the Fourth International Conference on Control, Automation, Robotics and Vision, Singapore, World Scientific, pp. 1299—1303.

12 Lin, G. Y. and Solberg, J. J. (1992) *Integrated shop floor control using autonomous agents*. HE Transactions, 24(3), 57-71.

13 Luscombe, A. M., Toncich, D. J., Thompson, W. and Dluzniak, R. (1984) *A new type of machine control system to replace traditional CNC*. International Journal of Advanced Manufacturing Technology, 9(6), 369-374.

14 Mackulak, G. T. (1984) *High-level planning and control: an IDEF0 analysis for airframe manufacturing*. Journal of Manufacturing Systems, 3(2), 121-132.

15 Mratin, H. and Spooner, D. L. (1997) *Data protocols for the industrial virtual enterprise*. IEEE Internet Computing, 1(1) 20-30.

16 Object Management Group (1995) *The Common Object Request Broker: Architecture and Specification*. OMG Technical Document PTC/96-03-04, Framingham, Massachusetts.

17 Orifali, R. and Harkey, D. (1996) *The Essential Distributed Objects Survival Guide*. John Wiley, New York.

18 Pun, L. (1986) *Artificial Intelligence Systems and Applications to Industrial Production Systems*. Plenum, New York.

19 Rober, S. J. and Shin, Y. C. (1995) *Modeling and control of CNC machines using a PC-based open architecture controller*. Mechatronics, 5(4), 401-420.

20 Shoham, Y. (1993) *Agent oriented programming*. Artificial Intelligence, 60(1), 51-92.

21 Staff Writer (1990) *Concept of IMS and its stagnant progress*. Techno Japan, 23(11), 48-63.

22 Tseng, M. M., Lei, M., and Su, C. J. (1997a) *A collaborative control system for mass customization manufacturing*. Annals of CIRP, 46(1), 373-376.

23 Tseng, M. M., Lei, M., Su, C. J. and Wang Wei (1997b) *A marketlike coordination for mass customization manufacturing in Proceedings of the 6th Industrial Engineering Research Conference*, Miami Beach, FL, pp. 972-977.

24 Wright, P. K. (1994) *Principles of OpenArchitecture Manufacturing*, ESRC 94-26, Engineering Systems Research Center, University of California at Berkeley.

(原载 Journal of Intelligent Manufacturing, (1998)9)

杨叔子科技论文选（下）

涂又光 题

杨叔子

图书在版编目(CIP)数据

杨叔子科技论文选(上、下)/杨叔子. —武汉：华中科技大学出版社，2012.9
ISBN 978-7-5609-8363-9

Ⅰ.杨…　Ⅱ.杨…　Ⅲ.机械学-文集　Ⅳ.TH11-53

中国版本图书馆 CIP 数据核字(2012)第 209443 号

杨叔子科技论文选(上、下)　　　　　　　　　　　　　　　　　　　　　杨叔子

策划编辑：熊新华　杨　玲
责任编辑：包以健
责任校对：张　琳
责任监印：周治超
出版发行：华中科技大学出版社(中国・武汉)
　　　　　武昌喻家山　邮编：430074　电话：(027)81321915
录　　排：华中科技大学惠友文印中心
印　　刷：湖北新华印务有限公司
开　　本：710mm×1000mm　1/16
印　　张：38　插页:4
字　　数：675 千字
版　　次：2012 年 9 月第 1 版第 1 次印刷
定　　价：100.00 元(含上、下册)

本书若有印装质量问题，请向出版社营销中心调换
全国免费服务热线：400-6679-118　竭诚为您服务
版权所有　侵权必究

目录

三支承主轴部件静刚度的分析与讨论 /1
δ函数在机械制造中的应用 /19
A Study of the Static Stiffness of Machine Tool Spindles /30
平稳时间序列的数学模型及其阶的确定的讨论 /57
时序建模与系统辨识 /64
金属切削过程颤振预兆的特性分析 /74
机械设备诊断学的探讨 /86
灰色预测和时序预测的探讨 /97
Quantitative Wire Rope Inspection /106
钢丝绳断丝定量检测的原理与实现 /116
复杂系统诊断问题的研究 /126
Plant Condition Recognition—A Time Series Model Approach /136
Space-domain Feature-based Automated Quantitative Determination of Localized Faults in Wire Ropes /150
基于深知识的多故障两步诊断推理 /168
机床切削系统的强迫再生颤振与极限环 /176
机械设备诊断学的再探讨 /185
机械设备诊断策略的若干问题探讨 /194
Forced Regenerative Chatter and its Control Strategies in Machine Tools /201
智能制造技术与智能制造系统的发展与研究 /212
两类小波函数的性质和作用 /220
Intelligent Prediction and Control of a Leadscrew Grinding Process Using Neural Networks /233
大直径钢丝绳轴向励磁磁路的研究 /244
金属切削机床切削噪声的动力学研究 /250
BP网络的全局最优学习算法 /263
基于神经网络的结构动力模型修改和破损诊断研究 /268

A CORBA-based Agent-driven Design for Distributed Intelligent Manufacturing Systems /275

内燃机气缸压力的振动信号倒谱识别方法 /295

A Novel Co-based Amorphous Magnetic Field Sensor /302

基于高阶统计量的机械故障特征提取方法研究 /312

基于因特网的设备故障远程协作诊断技术 /318

磁性无损检测技术中磁信号测量技术 /323

分布式网络化制造系统构想 /332

网络化制造与企业集成 /342

磁性无损检测技术中的信号处理技术 /351

Intelligent Machine Tools in a Distributed Network Manufacturing Mode Environment /362

AR 模型参数的 Bootstrap 方差估计 /387

虚拟制造系统分布式应用研究 /392

Wigner-Ville 时频分布研究及其在齿轮故障诊断中的应用 /400

先进制造技术及其发展趋势 /409

Feature Extraction and Classification of Gear Faults Using Principal Component Analysis /420

制造系统分布式柔性可重组状态监测与诊断技术研究 /434

基于 Markov 模型的分布式监测系统可靠性研究 /444

再论先进制造技术及其发展趋势 /456

基于神经网络信息融合的铣刀磨损状态监测 /464

以人为本——树立制造业发展的新观念 /472

走向"制造-服务"一体化的和谐制造 /480

Kinematic-parameter Identification for Serial-robot Calibration Based on POE Formula /491

高端制造装备关键技术的科学问题 /524

后记 /536

论文附录 /541

内燃机气缸压力的振动信号倒谱识别方法

刘世元　李锡文　杜润生　杨叔子

提　要　分析了内燃机缸盖系统的传递特性,提出了利用缸盖表面振动信号识别气缸压力的倒谱分析方法。通过倒谱开窗并进行平滑处理,使测量的振动信号和计算的传递函数更具鲁棒性,可以消除传感器测点位置和内燃机运行工况等敏感因素的影响。实验及分析结果表明了这种方法的可行性和有效性。

关键词:内燃机,气缸压力,振动,倒谱分析

目前,测量气缸压力的方法是在内燃机气缸盖上打孔,用压力传感器直接进行测量,但这种测量方法难以投入实际应用。内燃机表面振动信号的获取则方便可靠,通过适当处理可以用于识别和反演气缸压力。一种简单的方法是设计一个逆滤波器[1],其值是缸盖传递函数的倒数。利用这种逆滤波器法获得的传递函数,只要幅值低点存在微小的变化,都将导致反演信号的严重失真。由于振动弹性波在传递通道间的消散和混响[2],测量的振动信号及由此获得的传递函数易于受传感器安装位置和内燃机运行工况等敏感因素的影响[3]。

本文提出倒谱变换及开窗处理的方法,获得平滑且具鲁棒性的传递函数,用于气缸压力信号的识别。这种方法消除了上述敏感因素的影响,为内燃机工作过程的振动诊断奠定了基础。

1　内燃机缸盖振动特性

工作状态下内燃机缸盖系统主要承受的激励源有 5 个:进、排气门开启和关闭的 4 个冲击力以及缸内气体压力。可以认为这些激励源彼此之间是线性无关的,缸盖系统简化为一个多输入、单输出的线性系统[4],其输入和输出关系满足:

$$a(t) = p(t) * h_g(t) + x_{v1}(t) * h_{v1}(t) + x_{v2}(t) * h_{v2}(t)$$
$$+ x_{v3}(t) * h_{v3}(t) + x_{v4}(t) * h_{v4}(t) + n(t)$$

式中，$a(t)$ 为缸盖表面振动信号；$p(t)$ 为气缸压力信号；$x_{v1}(t) \sim x_{v4}(t)$ 分别为进气门开启、排气门关闭、进气门关闭和排气门开启的激励信号；$h_g(t)$，$h_{v1}(t) \sim h_{v4}(t)$ 分别为缸盖表面对上述各激励信号的响应函数；$n(t)$ 为噪声信号。对多缸内燃机而言，各缸按一定的发火顺序依次工作，缸盖振动响应是各缸响应的总和。

上述各激励信号之间具有严格的相位关系，而且进、排气门开启的的激励信号很小，可以在时域内进行分离。对于气缸压力的识别和反演问题，只取燃爆上死点附近的一段信号进行分析是可行的，此期间也不存在进、排气门关闭激励。

2 气缸压力的倒谱识别方法

在燃爆上死点附近，只考虑 $a(t)$ 和 $p(t)$ 的关系有
$$a(t) = p(t) * h_g(t) \tag{1}$$

从 $a(t)$ 中识别和分离出 $p(t)$ 的过程，实质上是一个解卷积的过程。时域解卷积是相当困难的，一般转换为频域内进行，即
$$A(f) = P(f) H_g(f) \tag{2}$$

式中，$A(f)$ 和 $P(f)$ 分别为 $a(t)$ 和 $p(t)$ 的傅里叶变换；$H_g(f)$ 为缸盖传递函数，是 $h_g(t)$ 的傅里叶变换。在此基础上，设计一个逆滤波器 $H_f(f)$，其幅值是 $H_g(f)$ 幅值的倒数，相位是 $H_g(f)$ 相位的负值，即满足 $H_f(f) = 1/H_g(f)$。那么，根据系统的输出信号就能准确地反演输入信号，即 $A(f) H_f(f) = P(f) H_g(f) H_f(f) = P(f)$。获得 $P(f)$ 后，再经傅里叶逆变换即得 $p(t)$。

在实际应用中，如果逆滤波器通过传递函数的简单反转得到，则传递函数幅值中若干尖锐的低点，将成为逆滤波器幅值中同样尖锐的峰值。传递函数幅值的这些低点只要存在微小的变化，都将在逆滤波器中放大，并造成反演信号的严重失真。事实上，振动传感器的安装位置、内燃机循环间波动和任何微小的测量误差，都将对振动信号的测量产生影响，进而严重影响传递函数的这些低点。因此，为了消除这些敏感因素，最好采用非线性逆滤波，倒谱开窗是一种有效的实现方法[3]。

时域信号 $x(t)$ 的复频谱为 $X(f)$，其复倒谱 $C_x(\tau)$ 定义为复频谱的复对数的傅里叶逆变换：

$$C_x(\tau) = F^{-1}[\ln X(f)]$$

式中，$F^{-1}[\cdot]$ 表示进行傅里叶逆变换。频谱 $X(f)$ 包括幅值谱 $|X(f)|$ 和相位谱 $\phi(f)$：

$$X(f) = F[x(t)] = |X(f)|\exp(-\mathrm{j}\phi(f))$$

式中，$F[\cdot]$ 表示进行傅里叶变换。倒谱也包括幅值倒谱 $C_x(\tau)$ 和相位倒谱 $C_\phi(\tau)$：

$$C_x(\tau) = C_{|X|}(\tau) + C_\phi(\tau)$$

式中，$C_{|X|}(\tau) = F^{-1}[\ln|X(f)|]$；$C_\phi(\tau) = F^{-1}[-\mathrm{j}\phi(f)]$。式(1)的时域卷积运算和式(2)的频域乘积运算，在倒频域内变成了加法运算：

$$C_a(\tau) = C_h(\tau) + C_p(\tau)$$

对于任一具体的内燃机，都可以先进行实验，根据测得的 $p(t)$ 和 $a(t)$，通过倒谱变换得出其传递函数的倒谱 $C_h(\tau)$。此后，就可以在内燃机状态监测中利用已知的 $C_h(\tau)$，从测量的 $a(t)$ 中识别出 $p(t)$。具体方法是：先对 $a(t)$ 进行倒谱变换并开窗处理，减去 $C_h(\tau)$ 后得到气缸压力信号的倒谱 $C_p(\tau)$；接着进行傅里叶变换并取指数，得到气缸压力信号的频谱 $P(f)$；最后进行傅里叶逆变换，就获得了气缸压力信号的时域波形 $p(t)$。计算过程如图1所示，虚框内给出的是通过先期实验分析获取缸盖传递函数倒谱的过程。

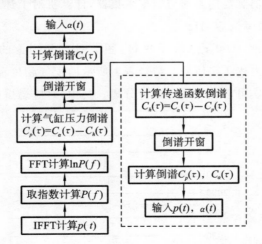

图1 气缸压力波形的缸盖振动信号倒谱识别过程

3 实验与讨论

实验在一台四缸四冲程水冷柴油机(4135D-5型)上进行,发火顺序为1—3—4—2,缸径135 mm,行程150 mm,额定功率和额定转速分别为73.6 kW和1500 r/min。配气相位和喷油提前角(本文中的角度均指曲轴转角)如表1所示,它们为缸盖振动信号的识别提供重要依据。

表1 配气门相位和喷油提前角

进气门开启角	上止点前 20°±6°
进气门关闭角	下止点后 48°±6°
排气门开启角	下止点前 48°±6°
排气门关闭角	上止点后 20°±6°
喷油提前角	上止点前 26°~29°

内燃机曲轴飞轮上装有磁电传感器,可以获取上止点参考信号;对第1缸缸盖打孔并安装石英压电传感器,可以测量气缸压力信号;第1缸缸盖上装有两个加速度传感器,分别靠近进气门和排气门,可以测量两个测点A和B处的振动信号。通过电荷放大器和高速多通道数据采集器,计算机对上述4路信号进行同步采集,每路的采样频率为50 Hz。

实验在额定转速1500 r/min下进行,图2是实际测量的时域波形。其中,(a)~(d)分别对应上述4路信号。结合表1所示的配气相位和喷油提前角,可以发现各激励信号确实可以在时域内进行分离。其中,对应燃爆压力段的缸盖振动

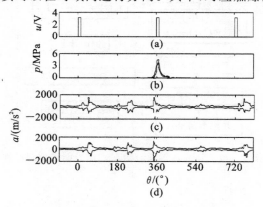

图2 气缸压力与缸盖振动时域波形

信号从开始喷油时明显增大,约持续 90°曲轴转角。对于气缸压力的识别和反演问题,可以从燃爆上死点前 30°处开始抽区间采样,为便于使用快速傅里叶变换,每次采样长度取为 512 个点,相当于 10.24 ms 时间和 92.2°曲轴转角。

根据测点 A 处振动信号计算的缸盖传递函数如图 3 所示。其中:(a)为采用频域方法直接计算的结果;(b)和(c)为倒谱开窗处理后的结果,开窗宽度分别为 64 和 32 个点,采用简单的矩形窗,保留低倒频率部分,去掉高倒频率部分,分别相当于去掉原时域信号中 1.28 ms 和 0.64 ms 的时延部分。结果表明,频域计算得到的频谱中含有很多尖锐的低点和峰值;进行倒谱开窗后,这些低点和峰值变得平顺而光滑,而且开窗宽度越窄,频谱越光滑,但也失去了频谱中的一些细节,因此开窗宽度应权衡选择。

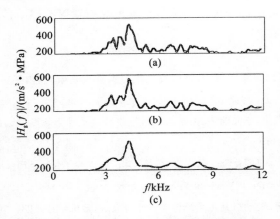

图 3 从测点 A 处获得的缸盖传递函数

获得缸盖传递函数的倒谱后,就可以从其他表面振动的测量信号中识别出气缸压力。图 4 给出了用测点 A 处振动信号识别的气缸压力波形,采用的传递函数倒谱是先期从测点 A 处获得的。图中,(a)为实际测量结果,(b)为采用逆滤波器法获得的结果,(c)为倒谱开窗处理后的结果。

从图 4 中可以看出,采用逆滤波器法时,传递函数未经光滑处理,识别的气缸压力波形与测量波形相差很大。相反,经过倒谱开窗处理后获得的波形与测量波形则非常相近,两者之间的误差是由于倒谱开窗处理造成的,虽然因此失去了原测量压力波形中的某些细节成分,但识别波形随时间变化的趋势依然明显,可以进一步从中获取确定内燃机状态的信息。

此外,利用测点 A 处先期获得的传递函数倒谱,对测点 B 处振动信号也进行了气缸压力识别(图形略),获得的识别波形与从测点 A 处获得的识别波形几乎相

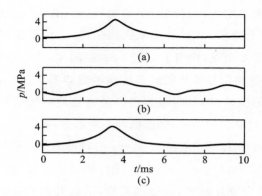

图 4 用测点 A 处振动识别的气缸压力

同,说明倒谱识别方法确实可以消除传感器测点位置等敏感因素的影响。

参 考 文 献

[1] 郝志勇,舒歌群,薛远等.内燃机气缸压力振动识别研究.内燃机学报,1994,12(1):43～48.

[2] 耿遵敏,宋孔杰,李兆前等.关于柴油机振声特点及动态诊断方法的研究与探讨.内燃机学报,1995,13(2):140～147.

[3] Kim J T, Lyon R H. *Cepstral Analysis as a Tool for Robust Processing, Reverberation and Detection of Transients*. Mechanical System and Signal Processing,1992,6(1):1～15.

[4] 周轶尘,彭勇.发动机缸盖系统振动特性研究.内燃机学报,1988,6(1):49～56.

Identification of Engine Cylinder Pressure from Vibration Signals by Cepstral Analysis

Liu Shiyuan Li Xiwen Du Runsheng Yang Shuzi

Abstract A unique signal processing technique has been developed for identification of engine cylinder pressure from cylinder head vibration signals. By cepstral windowing, the smoothed and robust transfer function of cylinder head is obtained and will help to construct engine cylinder pressure. Experimental results in a real-world engine are presented and used to verity the theoretical developments. It is shown that this technique is practical to implement and provides the basis for engine diagnosis by vibration signals.

Keywords: Internal combustion engine, Cylinder pressure, Vibration, Cepstral analysis

(原载《华中理工大学学报(自然科学版)》1998年第26卷第6期)

A Novel Co-based Amorphous Magnetic Field Sensor

Haixia Zhang   Yingjun Zhao   Shuzi Yang   Hejun Li

Abstract A novel Co-based amorphous magnetic field sensor was developed in this paper. It is a combination of two identical coils which are wrapped in one amorphous core, connected in anti-series, and they are induced by two identical pulse current sources. This structure is simple to construct and easy to optimize. Two optimizing methods: negative feedback coil and bias magnetic coil, greatly improved the sensors properties (linearity, sensitivity and dynamic range). At present, it is used in testing micro flaws in pipelines.

Keywords: Magnetic field sensor, Amorphous alloy, Negative feedback, Bias magnetic field

1 Introduction

Due to their superior soft magnetic properties, amorphous magnetic alloys have been considered as ideal active materials for high-performance magnetic sensors, such as MES (magnetoelastic sensors)[1], MI (magneto-impedance) element[2], and other kinds of sensor; these sensors play important roles in Non-destructive Testing (NDT). In our previous research, we paid more attention to pulse-induction amorphous magnetic field sensor[3], which is simple and easy to operate. According to pick-up techniques, this kind of sensor can be divided into two types: the first one has single coil, which is supplied by two opposite pulse current alternately; the second one is constructed by two separate coils, each of them having its own magnetizing circuit and pick-up circuit. Both of these schemes have disadvantages with respect to this principle [4]: the error of the single one comes from different operating time, the double coil's error source is the coil structure's difference. On the other hand, these sensors are difficult to be optimized. In order to improve the sensors properties, we put forward a novel structure: two coils wrapped on the same core, and connected in anti-series. The structure is novel in two ways. First, the sensor avoids the structural error based on the above principle. Second, this structure is simple and suitable for optimization. In this paper, we first introduce this sensor's structure. Then, we analyze the transient processing of the emf in coil, and find out the ideal testing point. After that, we put forward the optimizing techniques. At the end, the experimental results and its application in engineering testing show good agreement with the above analysis.

2 Principle

The principle of this kind of magnetic field sensor is illustrated in Fig. 1.

Coils 1 and 2 are two identical cols, which are wrapped over amorphous core, connected in anti-series; their joint terminal is connected at ground. Two identical pulse current sources $S_1(I_1)$, $S_2(I_2)$ supply coils 1 and 2, respectively. Therefore, the two coils are identical in structure and magnetize the amorphous core in opposite directions. According to the Faraday's law of induction, the emf in coil 1 is:

$$V_1 = -N\frac{d\phi_1}{dt} = -NA\frac{dB_1}{dt} = -NA\frac{d[\mu_0(H_1+M_1)]}{dt} = -NA\mu_0\left[1+\frac{dM_1}{dH_1}\right]\frac{dH_1}{dt} \quad (1)$$

where ϕ is the magnetic flux of the amorphous core (Wb); $\mu_0 = 4\pi \times 10^{-7}$ (H/m) is the permeability of air; N is the turns of coils; A is the area of the amorphous core (m^2); B is the magnetic induction of the amorphous core (A/m); H is the magnetic field (A/m); M is the magnetization of the amorphous core (A/m).

Because

$$H_1 = H_{el} + H_d, \quad H_{el} = \alpha\frac{NI_1}{l_m}, \quad \frac{dH_{el}}{dt} \gg \frac{dH_d}{dt} \quad (2)$$

Here, H_d is the external (detected) magnetic-field (A/m); H_{el} is the magnetic-field induced by pulse-current I_1 (A/m); α is the revised factor of the leakage magnetic flux in air; l_m is the length of amorphous core (m).

Therefore

$$V_1 = -NA\mu_0 \left[1 + \frac{dM_1}{dH_1}\right]\frac{dH_{el}}{dt} = V_0 + \Delta V_i = V_0 + V_0 \frac{dM_1}{dH_1} \quad (3)$$

Here, $V_0 = -NA\mu_0 (dH_{el}/dt)$, which is a constant parameter independent of the external magnetic field H_d; we call it the zero magnetic-field voltage.

According to the same principle, the emf in coil 2 is:

$$H_2 = H_{e2} - H_d, \quad H_{e2} = \alpha \frac{NI_2}{l_m} = H_{el}$$

$$V_2 = -NA\mu_0 \left[1 + \frac{dM_2}{dH_2}\right]\frac{dH_{el}}{dt} = V_0 + \Delta V_2 = V_0 + V_0 \frac{dM_2}{dH_2} \quad (4)$$

As seen in Eq. (3) and Eq. (4), $V_0 (dM_1/dH_1)$ and $V_0 (dM_2/dH_2)$ reflect the change of H_d. Fig. 2 is the magnetic curve (M-H) of amorphous alloy. If the sensor's operating point is within the nonlinear range in the M-H curve, especially when

$$M = aH + bH^2 \Rightarrow \frac{dM}{dH} = a + 2bH \quad (a, b \text{ are constants}) \quad (5)$$

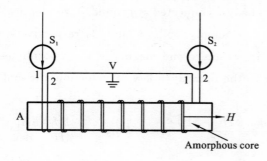

Fig. 1 The scheme of the double coil sensor

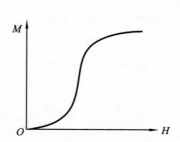

Fig. 2 The magnetic curve (M-H) of amorphous alloy

from Eq. (3) and Eq. (4):

$$V_1 - V_2 = \Delta V_1 - \Delta V_2 = 2bV_0 (H_1 - H_2) = -4bNA\mu_0 H_d \frac{dH_{el}}{dt} \quad (6)$$

From Eq. (6), we can get the signal of the external magetic field H_d.

3 Signal's Pick-up

However, the emf in coil is changing complexly in one pulse cycle[5], as shown in Fig. 3. Which is the best testing point and how to pick it up are two key clues in this sensor. We studied its transient processing deeply. The pulse current can be described as follows:

$$S(t) = \begin{cases} I_0(t), & nt_1 \leqslant t \leqslant nt_1 + t_0 \\ I_0(t) - I_0(t - t_0), & nt_1 + t_0 \leqslant t \leqslant (n+1)t_1 \end{cases} \quad (7)$$
$$n = 0, 1, 2, \cdots$$

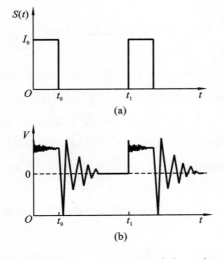

Fig. 3 The transient processing of the emf in coil
(a) The pulse current source $S(t)$; (b) The emf in pulse-reduction coil

Due to Eq. (6), the emf's transient processing can be divided into two periods.

(1) $nt_1 \to nt_1 + t_0$: the coil is excited by the pulse current, its emf arrives at its positive maximum rapidly, and the amorphous core reaches its operating point $H = H_e + H_d$ quickly; then the circuit goes into stability, and coil stores energy. During this processing, the amorphous core is induced by the induced magnetic field H_e and the external magnetic field H_d simultaneously. Both of these changes greatly affect the emf, therefore the change of H_d cannot be ignored; the emf is

not valuable for detecting H_d.

(2) $nt_1+t_0 \to (n+1)t_1$: in this period the pulse current is switched off, the inductance in coil is greatly changed, and the energy that was stored at period (1) is released rapidly, the emf jumps to its negative maximum immediately. At the very beginning of this period, the change of H_e is much higher than that of H_d, according to Eq. (2); the negative maximum emf is suited for testing H_d.

Thus, the pick-up circuit of this sensor is a negative peak detector; the subtraction of two maximum emf's cancels the induced magnetic field and describes the external magnetic field H_d; H_d/max (emf) is the sensitivity of the sensor.

4 Optimization

Using the subtraction of the two coils' max (emf) as the sensors' output, the whole system is an open loop system. It is greatly affected by external/inner disturbance, unable to adjust output error. Furthermore, since the sensor's initial, its dynamic response is slower than the change of inputs. So, the sensor has high sensitivity around zero magnetic field, but its dynamic range is too narrow to be used in real testing and its linearity is not very good. In order to improve the sensor's properties, we add negative feedback coil and magnetic bias coil, as shown in Fig. 4.

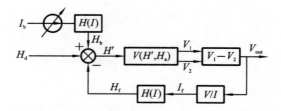

Fig. 4 The optimization scheme

H_f—Negative feedback magnetic field H_b—Bias magnetic field H_d— External magnetic field
H_e—Pulse current excited magnetic field

4.1 Negative Feedback Coil

The feedback coil, which is the same as the two pick-up coils, is wrapped over the amorphous core. The output voltage signal V_{out} passes through an inverting voltage/current (V/I) convener to be transferred to I_f, which is proportional to V_{out} and has converse direction. I_f excites the feedback coil, produces feedback magnetic field H_f, which is changing with the external magnetic field H_d, balances the magnetic field of the amorphous core, makes the whole system as a close-loop system, controls the affects of distributions in sensor, improves the sensor's linearity and widens its linearity range. The negative feedback is deeper, the linearity is better, the sensitivity is lower. So, the ideal depth of the negative feedback is decided not only by linearity but also by sensitivity.

4.2 Magnetic Bias Coil

After adding the negative feedback coil, we add bias coil on the core to widen the sensor's dynamic testing range. The bias coil is the same as the above three coils. It is excited by an external voltage which can be adjusted by smoothly variable resistor. The bias magnetic field moves the operating point of the amorphous core, thus the dynamic range of the sensor is widened. In real application, this magnetic field may be adjusted to cancel the background magnetic field.

5 Experiment and Application

We have done experiments with this sensor in a system. The experimental sensor consists of four coils of 250 turns of enameled Cu wire wrapped on a 15 mm×2 mm ϕ former, containing a 12 mm×1.0 mm×0.05 mm core; the material of the core is as-cast amorphous ribbon, 2705 M (Allied-Signal). The frequency of the pulse-current source is 20 kHz, the magnitude is 200 mA. The external magnetic field is supplied by a couple of Helmholtz coils. The experimental circuit is shown in Fig. 5.

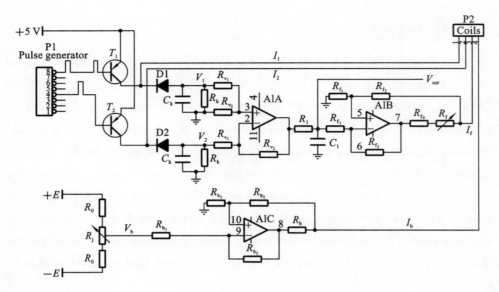

Fig. 5 The experiment circuit

The high-frequency pulse signal, which comes from a pulse generator, induces the transistors T_1, T_2 respectively; the transistors are alternatively conducting and nonconducting, so the pulse currents I_1, I_2 excite the two pick-up coils. D_1, C_k, R_k and D_2, C_k, R_k are two identical negative peak detectors to get the negative maximum emf of each coil V_1, V_2; then they pass through A1A, where $V_0 = V_2 - V_1$. Following the low-frequency filter R_1, C_1 is the output voltage V_{out}.

The negative feedback circuit is such that V_{out} passes through a voltage/current converter A1B, $I_f = -\{R_{f2}/[R_{f1}(R_f + R_{f0})]\}V_{out}$; the depth of the negative feedback is $D = R_{f2}/[R_{f1}(R_f + R_{f0})]$, adjusted by R_f. I_f excites the feedback coil, induces the feedback magnetic field which balances the disturbance of the whole magnetic field.

Bias magnetic field is adjusted by R_j. V_b is changing with R_j. A1C is another V/I converter; V_b is transferred to I_b which excites the bias coil. This magnetic field cancels the constant background magnetic field.

Comparing the outputs of the novel structure with the old one[3], we found that the novel one has wider testing range and better linearity than the old one, but the old one has high sensitivity around zero magnetic field. The reason for this problem is that two excited magnetic fields of the novel sensor will be overlapped during testing; this overlap has even distribution, reduces the output signal, lessens the sensor's sensitivity.

The main results of the experiment are shown in the following plots.

Fig. 6 illustrates the effect of the feedback coil. The linearity of the sensor is getting better and its sensitivity is getting lower while the depth of feedback (D) is getting deeper, the feedback current is getting stronger. The ideal D is around 0.01.

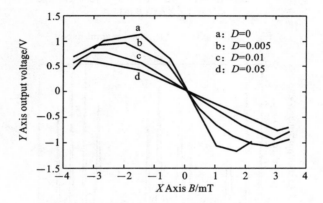

Fig. 6 The effect of the feedback coil

Fig. 7 shows the effect of bias coil. Due to the change of the bias magnetic field, the operating point of the core is moving, output plots moving parallel, so, the dynamic range of the sensor is wider. It keeps good sensitivity and linearity during ±20 mT.

After these theoretical experiments, we used it in testing micro defects on pipeline by the magnetic flux leakage (MFL) nondestructive testing technique. Our experimental pipeline is 70～90 mm diameter, 6 mm in wall thickness, 250 mm in length, with 10 transverse surface flaws; 0.2～0.5 mm in depth, 5～10 mm in length. Using permanent magnet yoke magnetize the wall of pipe, and flaws on pipeline cause magnetic flux leakage which can be sensed by this magnetic sensor. Fig. 8 shows the testing data. According to this result, the micro flaws in pipeline can be detected easily.

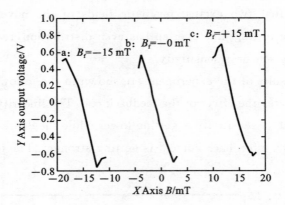

Fig. 7　The effect of the bias magnetic coil

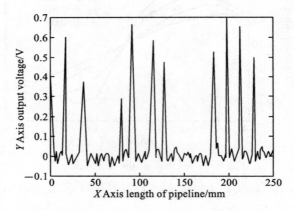

Fig. 8　The experimental data of testing micro flaws in pipeline

6　Conclusions

In this paper, we put forward a novel structure of pulse induction magnetic field sensor, which avoids the error source of traditional schemes, has high linearity and wide testing range. Two optimization techniques are presented to improve its properties: adding negative feedback coil to make its linearity better, and adding magnetic bias coil to move the working point of the amorphous core and widen its dynamic range. The experiments of this sensor illustrate that the novel structure sensor has not only high sensitivity, linearity and stability, but also wide testing range (± 20 mT). Its successful application in testing micro flaws on pipeline made us believe that it can be used in engineering testing field in the future.

Acknowledgements

This project is sponsored by National Scientific Foundation of China (No.: 59475089).

References

1 V. E. Makhotkin, B. P. Shurukhin, V. A. Lopatin, P. Yu. Maichukov, Yu. K. Levin. *Magnetic field sensors based on amorphous ribbon*. Sensors and Actuators A 25-27 (1991) 759-762.

2 K. Mohri, L. V. Panina, T. Uchiyama, K. Bushida, M. Noda. *Sensitive and quick response micromagnetic sensor utilizing magneto-impedance in Co-rich amorphous wires*. IEEE Trans. Magn. 31 (2) (1995) 1266-1275.

3 Y. Zhao, K. Yang, S. Yang, *The pulse-induced magnetic field sensor*. Instrum. Tech. Sensor (Chin.) (5) (1993) 4-6.

4 H. Zhang, Y. Zhao, S. Yang, *The calculation and testing scheme of pulse-induced magnetic field sensor*. Instrum. Tech. Sensor (Chin.) (6) (1997) 5-9.

5 H. Zhang, Y. Zhao, S. Yang, *The electrical model of pulse-induced magnetic field sensor*. Proceedings of FENDT '97, Korea.

（原载 Sensors and Actuators A:Physical, Volume 69, Issue 2, August 1998）

基于高阶统计量的机械故障特征提取方法研究[*]

张桂才　史铁林　杨叔子

提　要　对高阶统计量用于机械故障特征提取进行了研究。首先利用 Hilbert 变换构造原始信号的解析信号，求取信号的包络，然后计算包络信号的高阶统计量。研究表明，用高阶统计量提取信号特征，可以容易地将正常齿轮信号和齿轮裂纹、断齿的信号分离。

关键词：高阶统计量，特征提取，机械故障诊断

在机械故障诊断中，故障信号的提取是一个关键问题。在实际应用中，测取的信号中常含有随机噪声，信号的信噪比很低。当机器发生早期故障时，其微弱的故障信息完全淹没在噪声中，给信号特征的提取带来了很大困难。

在齿轮、轴承等零件的故障诊断中，常常应用包络分析提取冲击信号，但当信噪比很低以及故障信息十分微弱时，包络分析往往也难以获得满意的效果。本研究以齿轮为对象，在提取包络信号的基础上，进一步计算包络信号的高阶统计量估计值，成功地将齿轮正常、裂纹和断齿的信号进行了分离，效果十分显著。

1　高阶统计量理论简介

对于高斯信号，其统计特性可由其均值（一阶矩）和方差（二阶矩）来描述，但对于非高斯信号，就需要用更高阶的统计量才能完整描述其统计特性[1]。设 $x(n)$ 为离散时间实值平稳随机过程，其二、三、四阶矩分别定义为：

---

[*] 国家"九·五"攀登计划预选资助项目（编号：PD9521908）。

$$m_{2x}(i) = E[x(n)x(n+i)] \tag{1}$$
$$m_{3x}(i,j) = E[x(n)x(n+i)x(n+j)] \tag{2}$$
$$m_{4x}(i,j,k) = E[x(n)x(n+i)x(n+j)x(n+k)] \tag{3}$$

若 $x(n)$ 为零均值随机过程,则其二、三、四阶累积量分别定义为:

$$c_{2x}(i) = m_{2x}(i) = E[x(n)x(n+i)] \tag{4}$$
$$c_{3x}(i,j) = m_{3x}(i,j) = E[x(n)x(n+i)x(n+j)] \tag{5}$$
$$\begin{aligned}c_{4x}(i,j,k) = &m_{4x}(i,j,k) - m_{2x}(i)m_{2x}(j-k) \\ &- m_{2x}(j)m_{2x}(k-i) - m_{2x}(k)m_{2x}(i-j)\end{aligned} \tag{6}$$

由式(1)~式(6)可知,零均值随机过程的二、三阶累积量分别与它的二、三阶矩相等,但更高阶的累积量与相应阶次的矩是不相等的。高阶累积量可由相应阶次及低阶次矩表达;反之亦然。

对于零均值高斯随机过程 $x(n)$,其累积量和矩有以下结论[2]:$c_{1x} = 0, c_{2x} = \sigma^2$,及 $c_{kx} \equiv 0 (k \geqslant 3)$ 及

$$m_{kx}(i_1, i_2, \cdots, i_{k-1}) = \begin{cases} 0, & (k \text{ 为奇数}) \\ 1 \times 3 \times \cdots \times (k-1)\sigma^k, & (k \text{ 为偶数}) \end{cases}$$

式中,σ^2 为方差。即零均值高斯过程三阶($\geqslant 3$)以上的累积量恒等于零,而只有奇数阶次的高阶矩才等于零,偶数阶次的高阶矩不恒等于零,并且偶数阶次的高阶矩归根到底是由其二阶矩(即方差)决定的。因此,应用中常用高阶累积量研究非高斯信号的统计特性。

以上假定高斯过程具有零均值并不失一般性,因为在实际应用中,非零均值随机时间序列可以通过减去其均值估计化为零均值序列。

高阶累积量有一个重要的性质:两个统计独立随机过程之和的累积量等于这两个过程累积量之和。由以上零均值高斯过程三阶以上累积量恒等于零的结论可知,当信号中含有加性高斯有色噪声时,在理论上高阶累积量可以完全抑制噪声的影响,从而提高信噪比。

2 故障特征提取方法及实例

在齿轮和轴承的振动及噪声信号中,调制是普遍存在的现象,当这些零件发生故障时,往往在其振声信号中以调幅、调频、调相的形式表现出来,并且这几种调制形式常常是并存的,包络分析是幅值解调的主要方法之一,但对于早期故障,调制信息是很微弱的,往往被淹没在噪声中,当信噪比很低时,即便是简单的幅值

调制,仅用包络分析也难以奏效。众所周知,当齿轮的轮齿发生早期裂纹时,表现出的调制是很微弱的,裂纹故障特征的提取一直是一个困难的问题。许多学者都对此问题进行过研究,但真正行之有效的方法并不多。

本文将高阶统计量分析和包络分析相结合,来提取齿轮故障信息。具体方法如下。

首先对从齿轮箱上测得的原始振动信号 $x(t)$ 进行零均值化处理,然后实施 Hilbert 变换,得到信号的解析信号:

$$z(t) = x(t) + jH[x(t)]$$

式中,$H[x(t)]$ 即为 $x(t)$ 的 Hilbert 变换。$x(t)$ 的包络信号为

$$Z(t) = |z(t)| = [x^2(t) + (H[x(t)])^2]^{1/2}$$

然后计算 $Z(t)$ 的二、三、四阶矩及累积量 $m_{2z}, m_{3z}, m_{4z}, c_{2z}, c_{3z}$ 和 c_{4z},为简单起见,仅计算零滞后统计量,由于 $m_{2z} = c_{2z}, m_{3z} = c_{3z}$,故实际上只计算四个统计量,即,$m_{4z}(0,0,0), c_{2z}(0), c_{3z}(0,0)$ 和 $c_{4z}(0,0,0)$。

图 1 至图 4 分别为齿轮正常、裂纹、断齿各 20 组振动信号原始信号和包络信号的二、三、四阶累积量及四阶矩的计算结果。

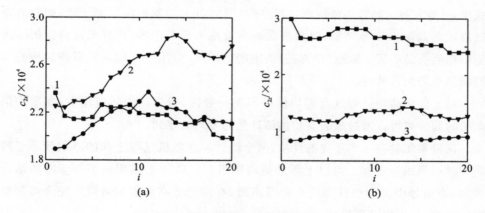

图 1 振动信号的二阶累积量(矩)

(a) 原始信号的二阶累积量(矩);(b) 包络信号的二阶累积量(矩)

1—断齿 2—裂纹 3—正常(图 2~图 4 说明均相同)

由图 1(a) 和图 2(a) 可明显看出,原始信号的二阶累积量(矩)和三阶累积量(矩)无法将正常、裂纹、断齿的信号区分开,而原始信号的四阶矩、四阶累积量和包络信号的二、三、四阶累积量及四阶矩都可以将三种状态的信号区分开来。在图 1(b)、图 2(b)、图 3 和图 4 中,显然断齿时由于很强的冲击脉冲的存在,振动信号严重偏离正态分布,其各阶统计量都比裂纹与正常信号的相应统计量大得多。

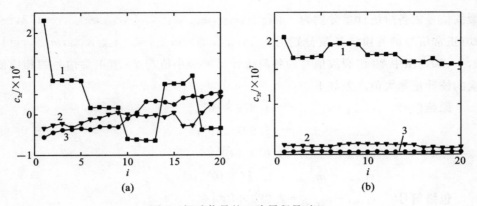

图 2 振动信号的三阶累积量(矩)

(a) 原始信号的三阶累积量(矩);(b) 包络信号的三阶累积量(矩)

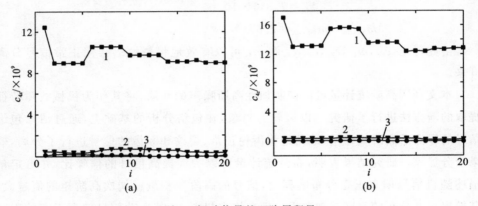

图 3 振动信号的四阶累积量

(a) 原始信号的四阶累积量;(b) 包络信号的四阶累积量

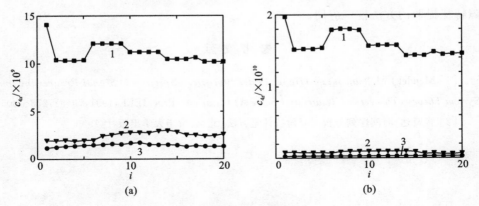

图 4 振动信号的四阶矩

(a) 原始信号的四阶矩;(b) 包络信号的四阶矩

裂纹信号的各阶统计量分别为 $c_{2xi}, c_{3xi}, c_{4xi}, m_{4xi}, c_{2zi}, c_{3zi}, c_{4zi}, m_{4zi}, i=1,2,\cdots,20$；正常信号的各阶统计量分别为 $c_{2ni}, c_{3ni}, c_{4ni}, m_{4ni}, c_{2ni}, c_{3ni}, c_{4ni}, m_{4ni}, i=1,2,\cdots,20$。图中20组裂纹信号的各阶统计量的最小值与20组正常信号的相应阶次的统计量最大值之差如下。

原始信号： $c_{2x\min} - c_{2xn\max} = -1.272 \times 10^3$;

$c_{3x\min} - c_{3xn\max} = -8.997 \times 10^5$;

$c_{4x\min} - c_{4xn\max} = 1.079 \times 10^8$;

$m_{4x\min} - m_{4xn\max} = 1.277 \times 10^8$。

包络信号： $c_{2x\min} - c_{2xn\max} = 2.286 \times 10^3$;

$c_{3x\min} - c_{3xn\max} = 5.203 \times 10^5$;

$c_{4x\min} - c_{4xn\max} = 7.844 \times 10^7$;

$m_{4x\min} - m_{4xn\max} = 2.874 \times 10^8$。

由此可见，$c_{4x}, m_{4x}, c_{2z}, c_{3z}, c_{4z}$ 和 m_{4z} 可以有效地将裂纹信号同正常信号分离开来。

本文利用高阶统计量可以抑制加性高斯噪声的性质，将其作为机械故障特征提取的新方法进行了研究。以齿轮为对象，在包络分析的基础上，通过估计包络信号的零滞后高阶统计量，成功地将齿轮正常、裂纹和断齿的信号进行了分离，效果十分显著。研究结果表明，高阶统计量可以大大提高信号的信噪比，并可定量描述随机信号偏离正态分布的程度，信号的高阶累积量或偶次高阶矩的值越大，其偏离正态分布的程度就越严重。而在许多机器中，其振动信号偏离正态的程度恰恰反映了机器故障的严重程度，因而高阶统计量可以作为信号预处理和机械故障特征提取的方法推广应用。

参 考 文 献

[1] Mendel J M. *Tutorial on Higher-Order Statistics (Spectra) in Signal Progressing and System Theory: Theoretical Results and Some Applications*. Proc. IEEE, 1991, 79(3): 278～305.

[2] 张贤达. 时间序列分析—高阶统计量方法. 北京：清华大学出版社，1996.

A Method for Extracting Mechanical Faults Features Based on Higher-order Statistics

Zhang Guicai   Shi Tielin   Yang Shuzi

Abstract The application of higher-order statistics extracting of mechanical fault features is studied. The analytical signals are acquired by means of Hilbert transformation of the original signals. The enveloped signals are calculated from the analytical signals. The higher-order cumulants and moments of the enveloped signals are estimated. The results of the research show that normal gear signals, cracked gear signals and broken gear signals can be easily separated by using higher-order statistics as the signal features.

Keywords:Higher-order statistics,Feature extraction,Mechanical fault diagnosis

(原载《华中科技大学学报(自然科学版)》1999 年第 27 卷第 3 期)

基于因特网的设备故障远程协作诊断技术[*]

何岭松　王峻峰　杨叔子

提　要　简述了基于因特网的设备故障远程协作诊断技术的国内外研究现状；介绍了远程诊断在开展学术界和企业界新型技术合作中的作用和意义；指出了现阶段实施该技术需要解决的关键问题；对远程诊断的基本概念和系统运作原理也进行了简介。

关键词：故障诊断，远程诊断，因特网，计算机网络

1　远程协作诊断简介

基于因特网的设备故障远程协作诊断是将设备诊断技术与计算机网络技术相结合，用若干台中心计算机作为服务器，在企业的重要关键设备上建立状态监测点，采集设备状态数据；而在技术力量较强的科研院所建立分析诊断中心，为企业提供远程技术支持和保障。生产企业设备运行出现异常，其状态监测服务器立即以工作传票方式向诊断分析服务器申请在线技术援助，同时以电子邮件方式向有关专家发出离线会诊请求；在短时间内调动入网的所有诊断资源，实现对设备故障的早期诊断和及时维修。基于因特网的设备故障远程协作诊断系统如图1所示。

远程协作诊断以因特网为桥梁，跨越企业和研究机构在时间和空间上的距离，学术界可利用网上诊断分析服务器设立技术讲座，对企业技术人员进行理论辅导和技术培训，同时发表自己的最新研究成果供企业使用；而企业界则可利用监测服务器为研究机构提供宝贵的现场经验和数据。双方取长补短，广泛合作，

---

[*] 国家"九五"攀登预选资助项目(PD9521908)。

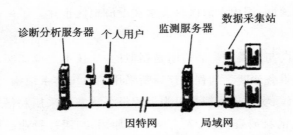

图 1　基于因特网的设备故障远程协作诊断系统

共同发展,这样既解决了生产企业技术力量不足和理论提高的问题,又有利于研究机构更准确、有效地得到企业内设备运行的第一手资料,充实理论和技术研究[1]。因此,远程协作诊断具有重要的理论和实际意义:①实现远程诊断后,可利用诊断协作网对企业技术人员进行培训,提高其理论水平;②企业可申请协作网专家在异地对设备故障进行会诊,提高诊断的准确性和可靠性;③远程诊断可实现全国范围内的诊断知识与诊断数据共享;④远程诊断能切实加强科研院所和生产企业的技术合作;⑤远程诊断能实现对远洋船舶、海洋钻井平台等特殊设备的远程监控和管理,为设备的安全运行提供可靠的技术保障;⑥Internet/Intranet 的互通性,远程诊断可在大到因特网,小到几台 PC 机的企业局域网上实现,便于技术的普及和推广。

在当今信息时代,因特网的迅速发展及其良好的应用前景,使之成为各种信息的载体。基于因特网的远程协作作为 21 世纪的新型合作方式备受学术界和工业界重视[2];因特网模式的开放式软硬件体系结构也得到人们的认同,成为各种系统开发的必然趋势。因此未来的设备故障诊断技术必须和因特网相结合,必须采用开放式体系结构,才会有强大的生命力和广阔的应用前景。

2　远程诊断的国内外发展现状

基于因特网的远程协作诊断研究工作最先是从医学领域开始的,1988 年开放式远程医疗系统的概念在美国提出,人们普遍认为一个开放式远程医疗系统应包括远程诊断、专家会诊、信息服务、在线检测和远程学习几部分。1994 年 9 月 SysOptics 公司在美国国会山庄向克林顿总统演示了一个基于因特网的全国保健试验示范系统;1995 年 1 月美国俄克拉何马州的远程医疗系统投入使用,它把 54 家乡村医院与州中心医院联系在一起,并通过计算机网络将 CT、X 光片等病人临床检验结果送到州中心医院诊断,这样病人在入网的任何一家乡村小医院就诊都能得到专家级的诊断;国内上海医科大学在上海地区也建立了一个类似的远程诊疗系统。

设备故障诊断与人类的疾病诊断是相似的,从技术上说能实现远程医疗诊断也就能实现远程设备诊断。远程医疗诊断采用的系统体系结构、信息传输方法和异地专家会诊组织、实现形式等都可为远程设备诊断所采用和借鉴。但由于重视程度不够和投入的科研资金少、人力不足等原因,与医疗行业已取得的显著成果相比,工业领域的远程诊断工作进展相对较慢。1997年1月,首届基于因特网的工业远程诊断研讨会由斯坦福大学和麻省理工学院联合主办,有来自30个公司和研究机构的50多位代表到会。会议主要讨论了远程诊断系统连接开放式体系、诊断信息规程、传输协议,以及对用户的合法限制,并对未来技术发展作了展望。会上确定由斯坦福大学和麻省理工学院合作开发基于因特网的下一代远程诊断示范系统,该项工作得到了制造业、计算机业和仪表业的 Boeing、Ford、Segate、Intel、SUN、HP 等12家大公司的支持和通力合作,并很快建立了一个限于合作者间的远程诊断示范体系 Testbed。Testbed 采用嵌入式 Web 组网,用实时 JAVA 和 Bayesian Net 实现远程信息交换和诊断推理;从该项目对外开放内容和项目组1997年底的研究总结报告来看,系统离实用还有很大距离,许多研究内容也还只是一个提法。此外,密执安大学也在积极开展针对机械加工的远程诊断和制造系统的研究工作,并在因特网上设立了一个宣传站点。

另外,许多国际组织,如 MIMOSA(Machinery Information Management Open Systems Alliance)、MFTP(Society for Machinery Failure Prevention Technology)、COMADEM(Condition Monitoring and Engineering Management)、Vibration Institute 等,也纷纷通过网络进行设备故障诊断咨询和技术推广工作,并制定了一些信息交换格式和标准。许多大公司在它们的产品中也加入了因特网功能,如 NENTLY 公司的计算机在线设备运行状态监测系统 Data Manager 2000 可以通过网络动态数据交换(Net DDE)的方式向远程终端发送设备运行状态;National Instruments 公司在其虚拟仪器产品 Lab View 中新增了因特网模块,可以通过 WWW、Email、FTP 方式发送测试数据。

国内就目前掌握的资料来看,西安交通大学、上海交通大学和哈尔滨工业大学都在向国外先进水平看齐,已开始或准备开始从事工业领域的远程诊断研究工作;华中理工大学也于1979年初开始了前期研究工作,并于同年11月在因特网上设立了一个远程诊断宣传站点,向国内介绍远程诊断技术,并以技术示范的方式向用户提供十分有限的远程诊断服务。同时 BENTLY、ENTEK、SOLTRAN 等大公司也纷纷将他们最新的网络化设备状态监测产品推向中国市场,这对增进我国学术界和企业界的网络化设备故障诊断意识和提高我国的设备故障诊断水平也起到了积极的促进作用。

3 待解决的关键问题

跨地域远程协作诊断的特点是测试数据、分析方法和诊断知识的网络共享,因此必须使传统诊断技术的核心部分:信号采集、信号分析和诊断专家系统,能够在网络上远程运行。要实现这一步应重点研究和解决如下几个问题。

(1) 必须解决网络环境下运行的远程信号采集、信号分析软件设计,使传统意义上的设备状态监测终端通过 Intranet 和 Internet 从设备运行现场延伸到企业总控室,甚至延伸到数千公里外的有关部门和专家的微机上。

(2) 远程诊断通过网络传输信息,对大量的实时监测数据必须进行处理和取舍,如何保证在远程诊断中传输必要和充分的振动、温度、压力数据和铁谱等信息是一个核心技术问题。

(3) 必须解决基于 Web 数据库的开放式诊断专家系统设计。传统诊断专家系统知识库一般是封闭或半封闭的,必须由设计者构造和修改,其适应性很差。对设计者来说,每一种设备都得设计一个专家系统;对企业来说,生产方式、设备工作环境的变化都可能使原有的诊断专家系统报废。因此设计者的主要任务应该是为用户设计一个简单实用的专家系统框架,而不是填充专家系统的诊断知识;知识库的填充、维护应以系统的使用者为主。例如,我们常用的网上搜索工具 Infoseek 的文献数据库并不是由 Infoseek 公司维护的,它只是提供一个工具和场所,其纪录是由包括我们在内的广大用户填充(submit)和维护的,正是如此它才能保持其文献覆盖各行各业,并不断更新。笔者认为新型的远程诊断专家系统也应该是这样,其诊断知识库应该由广大用户共同建造和维护,而不仅仅由设计者建造和维护。如何设计这样一种新型的诊断专家系统是远程诊断技术能否从设想走向实用的关键。

(4) 要实现诊断知识、理论和技术的共享,必须有一套通用的标准,包括测试数据标准、诊断分析方法标准和共享软件设计标准,因此,标准设计是远程协作诊断的重要内容。

4 结 束 语

远程协作诊断是一个需要社会广泛参与、配合的项目,协作诊断中的测试数据共享、诊断知识共享和分析软件共享都是以企业和科研院所的参与为前提的,没有广泛的参与就谈不上远程协作诊断,因此这项工作需要广大读者的共同关心和支持。笔者于1997年初开始从事设备故障远程协作诊断研究工作,并在因特网上设立了一个功能有限的远程诊断站点:http://202.114.6.48/server/server.htm。欢迎读者访问本站,也欢迎有兴趣的读者和我们一起共同开展此项研究工作。

参 考 文 献

[1] 蔡肖兵,黄昭毅,余琦. 中国设备管理,1995(9):31～33.
[2] 夏峰. 远程医疗系统的建立和发展. 计算机世界,1997.12.29(25).

Internet Based Remote Cooperative Diagnostics for Machine Faults

Abstract　The state and development in studies of internet based remote cooperative diagnostics for machine faults are briefly described in this paper. The new partnerships between academy and industry that are based on this new technique are introduced. Some main problems and barriers that need to further study and solve are discussed. The basic concepts and principles of remote cooperative diagnostics are also presented in this paper.

Keywords:Fault diagnosis,Remote diagnosis internet,Computer network

(原载《中国机械工程》1999年第10卷第3期)

磁性无损检测技术中磁信号测量技术[*]

康宜华　武新军　杨叔子

提　要　论述磁性无损检测技术中磁场信号测量的基本要求、磁场测量原理和元件、磁场测量的方法以及测量探头的设计方法。

关键词： 磁性检测，磁场，探头，磁敏元件

磁场测量探头实现磁场信号的转换，它是磁性无损检测技术的核心，决定着检测电信号的信噪比、分辨力及稳定性等多项性能，进而决定磁性检测装置或系统的性能，不同的磁电转换元件和磁场测量方法将带来不同的探头结构和检测性能指标。本文系统论述了磁性无损检测技术中磁信号测量的基本要求、磁电转换原理和元件、磁场信号的测量方法以及探头结构等。

1　磁场信号测量基本要求

在磁性无损检测中，被测磁场通常是空间三维矢量，单个磁敏元件或检测探头往往测量的是某一点、线或面上的磁场分量或均值。从实际应用来看，磁敏感元器件和磁场测量原理的选择，应综合考虑下述几方面的要求。

（1）灵敏度　根据不同检测目的和检测方法选择最佳的敏感元件。一般而言，随着磁场测量灵敏度的提高，元件和测量装置的成本增高。为了获得最优的性能价格比，灵敏度的选择应根据被测磁场的强弱选用适当的元件，并满足信号传输的不失真或干扰影响最小的要求。

（2）空间分辨力　磁场信号是一空间域信号，测量元件的敏感区域是局部的，

---

[*] 国家自然科学基金资助项目(59705020、59990470)。

一般由元件的尺寸和性能决定。为了能测量出空间域变化频率较高的磁场信号，必须要求测量元件或单元具有相应的空间分辨能力。对应于空间域中的磁场信号，这一分辨能力可在一维、二维或三维空间中来描述。空间分辨力是反映测量元件或单元敏感区大小的指标，具有方向性，沿不同的方向，空间分辨能力会不同。

(3) 信噪比　在磁性检测中，信噪比可定义为电信号中有用信号（如裂纹检测信号）与无用信号（如测量中的电噪声和被测磁场中的磁噪声）幅度之比。在这里，幅度为一广义量值，它可以指信号幅度，也可以指测量信号中经信号处理后的相关特征的量值。一般而言，测量过程中的上述信噪比必须大于1，否则被测对象（如裂纹）将无法识别。

(4) 覆盖范围　磁场在空间上是广泛分布的，因而每一测量元件或单元均只能在有限的范围或区间上对磁信号敏感。随着测量元件或方法的不同，在与扫描方向垂直的平面上有效敏感区间也将不同。将测量元件或单元有效检测被测对象（如裂纹），即在垂直于扫描方向上信噪比大于1时，被测对象相对于测量单元中心可以变动的最大空间范围称为测量单元的覆盖范围。在检测中，如果要求一次测量较大的空间区域或防止检测时的漏检，则需要适当安置和选择多组测量单元。很明显，在某一方向上覆盖范围越大，在该方向上的空间分辨力将越差，因而，必须根据测量目的和要求，最优设计和选择测量单元。

(5) 稳定性　测量单元应具备对检测环境和状态的适应性，测量信号特征应不受环境条件影响。因此，应对测量单元结构进行考虑，减小检测过程中随机因素的影响。

(6) 可靠性　可靠性表现为多次检测时信号的重复性。由于测量信号大小与测量点同被测磁场信号源间位置远近关系密切，重复检测时上述位置关系会有所改变，测量方法选择不当时会增大几次测量信号的差异。

(7) 有效信息比　当采用多测量单元进行测量时，一次检测的信号量由多单元提供，同时检测中的有用信息量也将由它们均分。对单个单元而言，其测量的有效信息比为有用信息与总信息之比。因此，为了提高每个单元的有效信息比，对同一测量则应减少测量单元，这就要在不降低信噪比的前提下，提高每个单元的覆盖范围或对多单元信号进行适当组合处理。

(8) 性能价格比　选择检测元件和测量方法时，根据测量目的和要求设计最优性能价格比的检测探头。

2 磁场测量原理和元件

无形磁场的可视化可用不同的磁测量原理或元件。通常先将磁场转换成电信号，再作自动化处理。实际检测中，磁电转换原理和元件主要有下述几种。

2.1 感应线圈

通过线圈切割磁力线产生感应电压。感应线圈测量的是磁场的相对变化量，并对空间域上高频率磁场信号更敏感。根据测量目的的不同，感应线圈可以做成多种形式。线圈的匝数和相对运动速度决定了测量的灵敏度，线圈缠绕的几何形状和尺寸决定了测量的空间分辨力、覆盖范围及有效信息比等。

2.2 磁通门

磁通门传感器原理是建立在法拉第电磁感应定律和某些材料的磁化强度与磁场强度之间的非线性关系上。典型的磁通门一般有三个绕组，即激励绕组、输出绕组和控制绕组，磁心通常是跑道形的。这种磁通门的灵敏度很高，可以测量 $10^{-5} \sim 10^{-7}$ T 弱磁场，输出依赖于磁心的磁特性，分辨力等随磁心和线圈尺寸变化。近年来，某些学者采用非晶态合金作为磁通门的磁芯，使得磁通门的灵敏度又有大幅度的提高。

2.3 霍尔元件

霍尔元件基于霍尔效应原理工作，测量绝对磁场大小。元件的灵敏度、空间分辨力和覆盖范围等由其敏感区域的几何尺寸、形状以及晶体性质决定。由于其制造工艺成熟，稳定性和温度特性等均较好，在磁场测量中得到广泛应用。随着集成线路技术的发展，霍尔感应元件和线性集成电路相结合生产出的集成霍尔元件灵敏度有很大提高，一般在 7 V/T 左右，且具有较好的封装，因而可望得到更好的应用。

2.4 磁敏电阻

磁敏电阻的灵敏度是霍尔元件裸件的 20 倍左右，一般为 0.1 V/T，工作温度在 $-40 \sim 150$°C，具有较宽的温度使用范围。空间分辨力等与元件感应面积有关。

2.5 磁敏二极管和三极管

磁敏二极管是继磁敏电阻和霍尔元件之后发展起来的新型磁电转换器件。与后两者相比,磁敏二极管具有体积小和灵敏度高等特点。磁敏二极管加一正向电压后,其内阻随周围磁场大小和方向的变化而变化。通过磁敏二极管的电流越大,则在同样磁场下输出电压越大;而所加的电压一定时,在正向磁场的作用下,电流减小,反向磁场时电流加大。磁敏二极管工作电压和灵敏度随温度升高而下降,通常需要补偿。

磁敏三极管是对磁场敏感的半导体三极管,与磁敏二极管一样,是一种新型的磁敏传感器件。磁敏三极管可分 NPN 和 PNP 型两大类。

2.6 磁共振法

原子核磁性的直接和精密测量是利用核磁共振的方法。核磁共振是原子核磁矩系统在相互垂直的恒定(直流)磁场 B 和角频率为 ω 的交变磁场的同时作用下,满足条件 $\omega=\gamma B$ 时,原子核系统对交变磁场产生的强烈吸收(共振吸收)现象,γ 为原子核的旋磁化,即原子核的磁矩与角动量之比。可知,在精密测量出核磁共振的频率和磁场,并知道核的角动量或核自旋后,便可精密测定原子核磁矩。

2.7 磁光克尔效应[3]

当线偏振光被磁化了的铁磁体表面反射时,反射光将是椭圆偏振的,且以椭圆的长轴为标志的偏振面相对于入射线偏振光的偏振面旋转了一定的角度,旋转角度的大小反映了铁磁体表面及内部的磁感应强度大小。磁光效应目前在磁光记录和光传输控制技术两个领域中得到广泛应用,将它用于磁性无损检测的重要意义在于:①可以远距离测量铁磁体表面或内部的磁感应强度,而其他原理或元件只是测量元件所在点或面上的磁感应强度;②由于直接测得的是铁磁性构件表面的磁场,因而无提离距离的影响,而在磁性检测中,从测量元件到被测场源间的距离直接影响待测电信号的幅度大小,对定量检测十分不利;③在高分辨力、高精度的自动测量中,可以方便地通过控制光束大小或扫描运动,实现点、线或面磁场的精确测量;④属非接触式测量且不受构件结构形状的影响。

除了上述介绍的几种方法,还有磁膜测磁法、磁致伸缩法、磁量子隧道效应法及超导效应法(如 SQUID)等方法。为了满足检测要求和达到较优的性能价格比,应该选择合适的磁敏感元件。例如:在剩余磁场检测中采用的元件的灵敏度一般需高于有源磁场检测;在主磁通检测法和磁阻检测法中敏感元件则应能准确测量磁场的绝对量,感应线圈是不合适的;在漏磁检测法中,随着被测裂纹等缺陷几何尺寸的减小,漏磁场强度急剧减小,采用的元件灵敏度也就要相应提高。从应用来看,霍尔元件,特别是集成霍尔元件,用于测量 $10^{-5} \sim 10^{-1}$ T 范围内的磁感应强度是合适的,它可用于精确测量 0.1 mm×0.1 mm×0.1 mm 微裂纹产生的漏磁场和 0.05% 的金属横截面变化产生的主磁通量变化大小等[4,5]。

3 磁场测量方法

在检测元件选定后,磁场的测量应根据被测对象特点和检测的目的选择最佳测量方法,包括元器件的布置、安装、相对运动关系及信号处理方式等,根据检测目的和要求,磁场信号测量可采用下述几种方法或其组合形式。

3.1 单元件单点测量

单元件测量的是敏感面内的平均磁感应强度,当元件的敏感面积很小时,可认为测得的是点磁场。单元件一般用在主磁通法、磁阻法和磁导法中。例如,在管棒类铁磁性构件表面裂纹的漏磁检测中,通过绕制管状感应线圈并让这类细长构件从中穿过,则可探测到构件整个外表面缺陷产生的漏磁场,而单个半导体元件将很难实现这类构件整周上漏磁场的测量。单元件测量时后续的信号处理电路和设备相对较简单,花费成本较低,检测时有效信息较大。

3.2 多元件阵列多点测量

当需要提高测量的空间分辨力、扩大覆盖范围和防止漏检时,可采用多元件阵列组合起来进行测量。在测量信号的处理上,当需要提高空间分辨力时,采用相互独立的通道处理每个元件输出,但增大了信息量输出,降低了有效信息比。为了得到灵敏度一致的输出,对每个元件和对应通道应进行严格的标定;当只需要增大检测覆盖范围时,可将多元件测量信号叠加,以单通道或小于元件数目的通道输出,通过电路上的组合,可选择到最佳分辨力、覆盖范围和灵敏度的检测探头结构。多元件测量时,要精心选择灵敏度、温度特性较一致的元件。均匀布置元件的数量应使多元件覆盖范围总和大于被测区域。

3.3 对管测量技术

在直流磁场测量中,半导体敏感元件的温漂将影响测量的精度和稳定性。为减小温漂的影响,除了在信号处理上采用低温漂放大电路外,在测量中可采用对管技术,将特性相近的两元件的感应面尽可能靠在一起,并让它们的磁敏感方向相反,形成"对管"。由于相同材料的半导体温度特性基本一致,两元件测量信号差分的结果是检测信号温漂影响减小,同时,灵敏度可提高一倍。例如,对美国Sprague 公司生产的集成霍尔元件 UGN3501T(灵敏度为 7 V/T,温漂为 0.1 mV/℃采用对管测量技术后,测量灵敏度达 14 V/T,而温漂只有 0.01 mV/℃。该公司用对管技术生产的差动元件 UGN3501M,其温漂特性更好,但体积稍大。

3.4 差动测量技术

为了排除测量过程中振动、晃动以及被测构件中非被测特征的影响,提高测量的稳定性、信噪比和抗干扰能力,检测中适当布置一对冗余测量单元,并将两单元测量信号进行差分处理,形成差动测量。当在平行于待测磁场方向的测量面上布置对该方向敏感的测量元件并差动输出时,形成差分测量,可消除测量间隙等变动带来的影响;当在测量的磁场方向上间隔布置对该方向敏感的两测量元件并差动输出时,可对磁场的梯度进行测量,形成梯度测量,可在较强的背景磁场下测量微弱的磁场变化。

3.5 聚磁检测技术[4-6]

聚磁检测采用高导磁材料(如工业纯铁、坡莫合金等)将测量磁场主动引导至测量元件中,如图 1 所示。由高导磁材料做成的聚磁器在这里起着收集、引导及均化测量磁场的作用。根据被测构件表面形状和测量要求设计聚磁器的形状、尺寸,最大限度地收集有用磁场,并可通过设计磁场通路将磁场较好地集中并引导至测量元件中。对空间上高频变化的磁场而言,聚磁器相当于一空间上的低通滤波器,因而这一测量技术的空间分辨力将差一些。

由于单片磁敏感元件所能感受的磁场面积有限,通过聚磁器可以扩大单片元件的覆盖范围,提高有效信息比,简化测量结构和降低检测成本,并且可以通过设计聚磁器,提取感兴趣的磁场分量,提高测量的灵敏度和信噪比。例如,对钢丝绳断丝的漏磁检测,当采用前述集成霍尔元件测量时,由于每片元件沿绳周向的覆盖范围约为 8 mm 弧长,沿绳轴线方向进行扫描检测时,为防止漏检,沿 $\phi 40$ mm 钢丝绳周向至少布置 14 片这样的元件。当采用半圆环状的聚磁器对收集漏磁场,并用单片同样的霍尔元件测量环形聚磁器间磁场时,不但两路测量信号中断丝信号的幅度有较大增加,而且绳股漏磁信号被消除,信噪比得到提高。

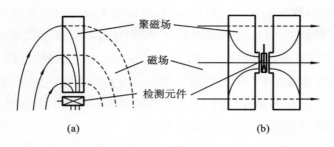

图 1 聚磁检测原理
(a) 单聚磁器；(b) 聚磁器对

聚磁检测的关键在于聚磁器的设计,当聚磁器的形状、尺寸选择不当时,测量信号质量将很差或根本得不到感兴趣的信号。因此,必须针对测量的特点设计聚磁器。在聚磁器设计中,总的原则是确保其收集磁场和引导磁场的路径畅通,也即聚磁器的每一局部都不应该有磁饱和现象发生。

3.6 磁屏蔽技术

磁场测量最易受到外界磁场的干扰,采用磁屏蔽技术后可减弱杂散磁场的影响。测量中,通常采用高导磁材料做成箱体以使腔体内的测量单元免受体外磁场的影响,一般可将体外磁场减至 $1/5\sim1/8$,好的屏蔽体可减小得更小。磁屏蔽是保证测量稳定可靠的必不可少的措施,对弱磁场测量尤为重要。

4 磁测量探头设计

与磁粉探伤相比,磁性无损检测技术除具有应用面宽、灵敏度高及可实现定量或半定量检测等特点外,还易于实现检测过程的自动化和在线或实时检测。为了适应各种状态下的检测,探头的结构就显得非常重要。在磁性检测装置或系统的设计中,为了实现探头的相对扫描运动,根据检测工况的变化,探头的结构形式可分为主动式和被动式。

4.1 主动式测量探头

当检测过程中探头与被测构件的相对运动关系可精确控制或可通过设计实现时,一般通过检测装置或系统的设计来保证单个探头的扫描覆盖范围、提离距离及敏感方向等,此时探头可以是一个独立的单元,其运动关系由检测装置控制。例如,油管自动检验线上的磁性探伤设备,其探头与油管的相对运动关系由控机构实现,有探头旋转油管直线输送、油管旋转探头直线扫描及探头固定而油管螺旋输送等多种形式。在运动机构的设计上,根据探头的覆盖范围选择自动输送的进程,保证沿钢管表面不发生漏检,并采用较小的测量间隙来减小"提离效应"影响。由于磁性测量探头一般只能测量某一方向上的磁场分量,而缺陷产生的漏磁场与其形状和走向相关,因而应让探头的敏感方向与相对扫描运动方向一致,以获得被扫描路径上较高的缺陷检测敏感性。

4.2 被动式测量探头

与主动式测量探头相反,当检测的运动关系只能由被测构件决定时,测量探头本身的结构就需要特殊考虑。例如,在钢丝绳芯输送带的检测中,输送带只能有纵向运动,为了测量每根钢丝绳的状况,需采用阵列检测方法;为了减小"提离效应"的影响,需使探头能紧贴胶带表面并跟随其上下波动。对在用抽油杆的检测,当抽油杆上下运动探头固定不动时,测量纵向裂纹将较困难,探头需在最不灵敏的方向上收集微弱的漏磁场。因此,被动式测量探头的结构随着检测要求和条件的不同而变化。

就使用工况来讲,主动式测量探头适合于检测线或检验线上批量构件的检测,探头本身较简单,互换性较强,稳定性较好,抗干扰能力强,但自适应性差。用于检测的辅助机构复杂,成本较高。被动式测量探头则更适合于现场在用构件检测,探头本身结构较复杂,通用性差,但体积小,携带安装方便,自适应力强。

参 考 文 献

[1] 杨叔子,康宜华.钢丝绳断丝定量检测原理与技术.北京:国防工业出版社,1995.

[2] 毛振珑.磁场测量.北京:原子能出版社,1980.

[3] 宛德福,马兴隆.磁性物理学.北京:电子科技大学出版社,1994.

[4] 武新军.录井钢丝裂纹及丝径检测装置的研制:[硕士学位论文].武汉:华中理工大学,1996.

[5] 金建华.录井钢丝裂纹定量检测原理及装置的研究:[硕士学位论文].武汉:华中理工大学,1997.

[6] Piekarski Kalwa E. NDT International,1987,20(5):295-301.

[7] Piekarski Kalwa E. NDT International,1987,20(6):347-353.

[8] 康宜华,薛鸿健,杨克冲,等.中国机械工程,1993,22(4):4-6.

Magnetic Signal Measurement in Magnetic Nondestructive Testing

Kang Yihua Wu Xinjun Yang Shuzi

Abstract The basic requirement for magnetic signal measurement, magnetic field measuring principle and elements, magnetic measuring method and probe design are discussed.

Keywords：Magnetic testing, Magnetic field, Probe, Magneto sensor

（原载《无损检测》1999 年第 21 卷第 8 期）

分布式网络化制造系统构想*

程 涛　胡春华　吴 波　杨叔子

提 要 从全球化趋势、市场特征和科学技术发展等方面讨论制造模式变革的必然性,提出一种面向新世纪的分布式网络化制造系统的初步构想,并就其特点及关键技术进行阐述,通过原型系统从原理上实现了这一构想。
关键词:网络化制造,制造模式,分布式系统,全球化

本文从世界经济的全球化发展趋势、新时期市场的特征与本质、科学技术的发展等方面出发,阐述变革传统制造模式的内在动力,立足于当前制造业的发展现状,借助制造技术和信息技术的最新发展,基于"敏捷制造"的基本思想,试图探索出一种适应这些形势的生产制造模式。

1　制造模式变革的内在动力

1.1　全球化趋势

世界各国之间在经济上越来越多地相互依存,不可能离开全球化的市场而独立地求得本国经济的发展,必须与国际接轨,充分利用国际市场的信息、资本、资源、技术、商品和服务,在平等竞争的基础上,互利互惠,相互合作。经济的全球化、市场的国际化、贸易和投资的自由化,以及服务的世界化,促进了竞争,提高了效率,鼓励了革新,增加了新的资本投资和加快了经济增长速度[1-3]。随着越来越多的国家接受自由市场思想,全球自由贸易体制的逐步建立和完善,世界大市场

---

* 国家自然科学基金资助项目(59705020、59990470)。

的逐步形成以及全球交通运输系统和通信网络的建立,国际间的经济贸易交往与合作更加频繁和紧密,竞争愈来愈激烈。这股全球化的洪流正推动着世界经济稳健、快速、持续地增长。这使得制造产业、制造技术和产品逐步走向国际化,导致制造业在全球范围内重新分布和组合。竞争的加剧将促使竞争对手利用一切可以利用的制造资源,主动积极地寻求市场机遇,灵捷地响应和适应客户多样化的消费需求,高质量地为全球顾客服务,从而获得规模经济,促进企业的发展与壮大。

1.2 客户化市场驱动

世界制造业正面临着一个不确定性的、急剧变化的、竞争日趋激烈的国际化市场[3]。

(1) 客户驱动市场　客户驱动着市场,市场牵动制造企业,制造业从以生产为中心转向以市场需求为中心,从以企业为主导转向以客户为主导,以满足客户需求为目标。

(2) 客户对制造产品的要求　集中表现为多品种、变批量、高性能、高质量、高可靠性、交货期短、合理的价格、完善的行销及售后服务等,而且这些要求是随着客户的经济及其他条件的变化而不断变化的,由此导致制造业面对一个产品寿命期短、更新快、动态快速多变、稳定性差和难以预测的市场。

(3) 物质产品与服务的交融　在从产品的设计与生产到行销与售后服务等过程中必须为全球范围内的客户提供优质与全方位的满意服务,这是制造企业赢得客户、开拓和占领市场的重要因素。

(4) 合作基础上的市场竞争　由于一个企业的资金、人员素质与知识和技能、设施与设备、设计与开发、制造能力、营销能力等都存在着局限性,在全球范围内日趋激烈的市场竞争中不可能取胜,因此必须进行企业间的合作,形成以竞争为基础和合作协同为主导、风险共担、利益共享、共存共荣的机制,充分实现资本、资源、技术、人才、信息和知识的交流与共享,优势互补、分工协作,才能积极有效、主动、快速响应和适应市场,夺取竞争胜利。以这种竞争与合作机制为基础,借助于计算机网络、远程通讯技术及装备,以及四通八达的交通运输网络,使在地域上分布的企业组成能适应当今和跨世纪市场需求变化的、有竞争能力的、动态可变的企业联盟,是实现在远比过去更广阔的范围内,充分有效利用比过去更广泛的资源,灵活快速响应全球范围内用户需求,提供优质服务的最佳模式。

1.3 科技进步

20世纪中叶以来,微电子、计算机、通信、网络、信息、自动化等科学技术的迅猛发展,掀起了以信息技术为核心的"第三次浪潮",正牵引着人类进入工业经济时代最鼎盛的时期,并已叩响知识经济时代的大门[4]。正是这些高新科学技术在制造领域中的广泛渗透、应用和衍生,推动着制造业的深刻变革,极大地拓展了制造活动的深度和广度,促使制造业日益向着高度自动化、智能化、集成化和网络化的方向蓬勃发展。

在全球化浪潮的冲击和高速发展的高科技的推动下,制造企业的经营、生产战略与活动应面向全球,充分合理利用以信息技术为代表的高科技,建立和实现基于分布式网络化体系结构的组织、生产和管理模式,以快速、灵活地组织和利用各种分布的、异构的制造资源,谋求企业间的相互合作、共同寻求市场机遇,在激烈的市场竞争中求得生存与发展。

2 分布式网络化制造系统的总体构想

建立和实现基于计算机网络的分布式制造系统模式,是对传统制造模式的扬弃与创新。近些年来,在理论上已初具系统,在实践中亦取得成效的先进制造生产模式主要有柔性生产、智能制造、精益生产和敏捷制造等,在总结这些成果的基础上,结合当前国内外制造业的基本形势,本文提出了分布式网络化制造系统(distributed networked-manufacturing system,DNMS)的初步构想(见图1),分布式网络化制造系统是一种由多种、异构、分布式的制造资源,以一定互联方式,利用计算机网络组成的、开放式的、多平台的、相互协作的、能及时灵活地响应客户需求变化的制造系统,是一种面向群体协同工作并支持开放集成性的系统。其基本目标是将现有的各种在地理位置上或逻辑上分布的异构制造系统/企业,通过其Agent连接到计算机网络中去,以提高各个制造系统/企业间的信息交流与合作能力,进而实现制造资源的共享,为寻求市场机遇,及时、快速地响应和适应市场需求变化,赢得竞争优势,求得生存与发展奠定了坚实的基础,从而也为真正实现制造企业研究与开发、生产、营销、组织管理及服务的全球化开辟了道路。

分布式网络化制造系统具有如下特点。

(1) 层次结构的相似性 分布式网络化制造系统具有一定的层次性,从下到上一般可分为单元级、车间级、系统/企业级等不同的层次,而且在不同的层次上

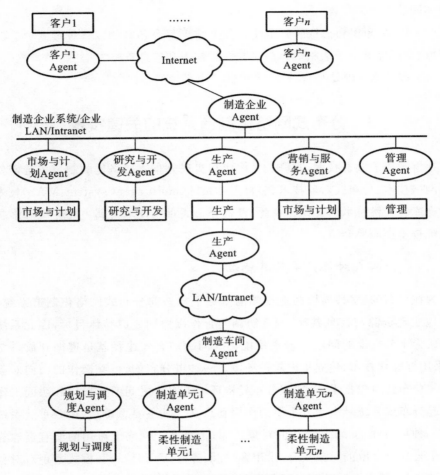

图 1　分布式网络化制造系统模式基本构想

都具有相似的体系结构,即同一层次中各结点都通过其 Agent 利用计算机网络相互连接,而处于低层次的系统,则作为一个整体通过其在上一网络层次中的代理连接到上一层次的网络中去。

(2) 分布式、开放的体系结构　在分布式网络化制造系统中,处于同一网络层次中的各个结点在逻辑结构上或地理位置上是分布的,无主从之分,能独立地、自主地完成各自的子任务,但为完成系统的整体任务,彼此间还需进行大量的交互活动,包括信息、数据的交流与共享,相互协商、协调与合作以协同完成任务。

(3) 良好的容错能力、可扩展和可重组性。

(4) 互联性　各结点通过计算机网络,以一定逻辑互联方式实现连接和通信。

(5) 互操作性　各结点或应用系统间能够交互作用、相互协调与合作以协同完成共同的任务。

（6）数据、知识和信息的分布性　各个结点都有各自的以各种形式（如文件、数据库、知识库和电子表格等）存在的数据、知识和信息资源。

（7）多样化　硬件平台、操作系统和应用平台的多样化。

3　分布式网络化制造系统的关键技术

在继承当前制造技术的基础上，构建和实现分布式网络化的制造系统需要计算机网络技术、分布式对象技术、多自主体系统（multi-agent system，MAS）技术以及数据库等关键技术的支撑。在此，仅就分布式对象技术及其主要标准和多自主体系统技术作简单介绍。

3.1　分布式对象技术及其标准

为在分布的、多种异构制造资源的基础上构造起分布式网络化制造系统，以有效地实现资源与信息共享、相互协调与合作以协同完成整体目标，因此系统集成就成为十分突出的问题。解决系统集成问题的有效途径就是遵循开放系统原则，采用标准化技术，建立集成软件环境。一种可分布的、可互操作的面向对象机制——分布式对象技术，对实现分布异构环境下对象之间的互操作和协同工作以构建起分布式系统具有十分重要的作用和意义。其主要思想是，在分布式系统中引入一种可分布的、可互操作的对象机制，把分布于网络上可用的所有资源封装成各个公共可存取的对象集合，采用客户/服务器（C/S）结构和模式实现对对象的管理和交互，使得不同的面向对象和非面向对象的应用可以集成在一起[5]。

许多计算机厂商、标准化组织等纷纷制定了分布式对象技术的相关标准[5,6]。其中，国际对象管理组织 OMG 发布的公共对象请求代理结构（common object broker architecture，CORBA），为分布异构环境下各类应用系统的集成，实现应用系统之间的信息互访、知识共享和协同工作提供了良好的可遵循的规范、技术标准和强有力的支持，它通过客户/服务器对象间的交互而实现资源的实时共享。CORBA 具有软硬件的独立性、分布透明性、语言的中立性，以及面向对象的数据管理等优点，从而成为当前十分有效的一种集成机制。因此得到包括 IBM、HP、DEC、Microsoft 等在内的计算机与软件厂商和 X/open、OSF 以及 COSE Alliance 等国际联盟的积极支持和采纳，已有几个遵循此标准的产品问世[5,6]。

基于 CORBA 标准实现的系统集成和应用开发环境在企业中将会有潜在的

巨大的应用前景,在逐步实现企业生产和管理的自动化与智能化,提高生产率,增强和提高企业及时快速响应和适应市场的能力等方面都将起到积极的推动作用。基于 CORBA 标准的系统是一个能跨越不同地理位置、穿越不同网络系统、屏蔽实现细节、实现透明传输、集成不同用户特长的基于 C/S 模式、面向对象、开放的分布式计算集成环境。

3.2 多自主体系统技术

多自主体系统理论与技术在制造领域中的应用与实现,将给制造系统/企业带来巨大变化[7]。制造系统是由若干完成不同制造子任务的环节组成的,如订货、设计、生产、销售等,各个环节上的各功能子系统既相互独立,又相互协同,以提高产品的市场竞争力和企业的经济效益为目标,共同完成制造任务。因此可以说整个制造过程是一种典型的多自主体问题求解过程,系统/企业中的每一部门(或环节)相当于该过程中的一个自主体(Agent)。制造系统/企业中的每一子任务、功能、问题或单元设备等都可由单个 Agent 或组织良好的 Agent 群来代理或实现,并通过它们的交互和相互协商、协调与合作,来共同完成制造任务。将制造系统/企业模拟成多自主体系统可以使系统易于设计、实现与维护,降低系统的复杂性,增强系统的可重组性、可扩展性和可靠性,以及提高系统的柔性、适应性和敏捷性等。

4 原型系统

基于这一构想,我们利用 StarBus22,开发了一个分布式网络化制造原型系统,见图 2。该原型系统由系统经理、任务规划、设计和生产者 4 个结点组成。

4.1 系统组成

(1) 系统经理　系统经理包括数据库服务器和系统 Agent。前者提供一个全局数据库,它可供原型系统中获得权限的结点进行数据的查询、读取、存储和检索等操作;后者则负责该原型系统在网络上与外部的交互,通过 Web 服务器在 Internet 上发布该原型系统的主页,网上用户可通过访问该主页获得该系统的有关信息,并决定是否由该原型系统来满足自己的需求,这可通过填写和提交该主页向用户提供的订单登记表来向该原型系统发出订单,系统 Agent 决定是否接受这些订单,如接受,就将其存入数据库服务器的全局数据库中;另一方面,系统

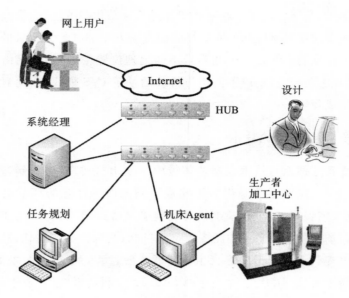

图 2　分布式网络化制造原型系统

Agent 还负责监视该原型系统上各个结点间的交互活动,如纪录和实时显示结点间发送和接收消息的情况、任务的执行情况等。

(2) 任务规划　任务规划结点的主要功能是对订单进行规划、分解成若干子任务,并通过招标-投标的方式将这些子任务分配给各个结点。该结点由一个任务经理和它的代理——任务经理 Agent 组成。

任务经理的主要功能如下:①访问全局数据库获取网络用户的订单;②根据订单进行任务登记;③将任务分解成若干子任务;④查询其他结点的类型、能力和其他信息;⑤基于价格机制,通过谈判进程以招标—投标的方式将这些子任务分配给网上其他结点。

任务经理 Agent 主要负责如下事务:①网络注册,以使该结点能加入到原型系统中,获得相应的权限;②与系统中其他结点进行交互;③管理本地数据库,包括数据查询、修改、添加、删除等;④取消注册,以使该结点退出本系统。

(3) 设计　设计结点是一个计算机辅助设计系统。该结点是由一个 CAD 工具及其代理——CAD Agent 组成的。CAD 工具是一个工具软件包,用于帮助设计人员根据用户要求进行产品设计;而 CAD Agent 则负责网络注册、取消注册、数据库管理、与其他结点的交互、决定是否接受设计任务和向任务发放者提交任务等事务。

(4) 生产者　生产者包括一台加工中心和它的 Agent——机床 Agent。该加工中心配置有我们在华中Ⅰ型数控系统的基础上开发的智能自适应数控系统[8]、刀具状况监控模块以及自诊断和自修复模块,可以提高加工效率、降低成本、保障设备运行的可靠性和安全性。此外,还具有与外部环境进行交互的能力。其主要任务是:①利用串口或网络适配器与其机床 Agent 进行信息交流,包括从机床 Agent 接受控制命令、数据和相关文档,将有关机床的运行状态和结果传送给机床 Agent;②根据从机床 Agent 接受的控制命令、数据和相关文档完成加工任务。

机床 Agent 的主要功能如下:①与网络上其他结点的 Agent 进行交互,包括谈判、协商和信息与数据交流等;②进行推理并作出决策以决定是否接受子任务,如果不接受,则寻求其他能接受该子任务的结点,并进行任务迁移;如果接受,则对该子任务进行规划以确定其行为活动;③利用串口或网络适配器与其加工中心进行信息交流,即将控制命令、数据和相关文档传送给加工中心,并接受加工中心传送来的有关机床运行状态和结果的信息;④对机床 Agent 自身和加工中心的运行状态进行监控。

4.2 系统运作

该系统中的每个结点必须通过网络注册,才能成为该原型系统的正式成员以获得相应的权限,才能与系统中的其他结点进行协作,共同完成系统任务。整个原型系统的运作过程见图 3。

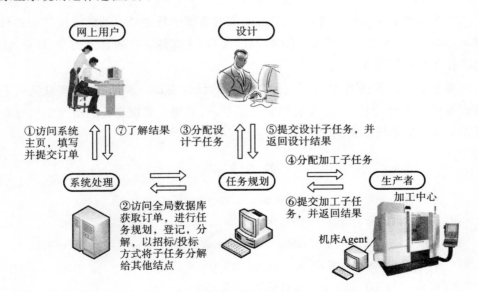

图 3　原型系统的运作过程

(1) 任一网络用户都可以通过访问该原型系统的主页获得该系统的相关信息(包括系统特征、能力和所承诺的服务等),系统主页是由 Web 服务器在 Internet 上发布的。此外,用户还可通过填写和提交系统主页所提供的用户订单登记表,向该系统发出订单。

(2) 如果接到并接受网络用户的订单,系统 Agent 就将其存入全局数据库,任务规划结点可以从全局数据库获取订单,进行任务规划,并将任务分解成若干子任务,基于招标-投标的方法将这些子任务分配给原型系统上获得权限的结点。

(3) 基于谈判方式,产品设计子任务被设计结点获得,该结点通过人机交互完成产品设计子任务,生成相应的 CAD/CAPP 数据和文档以及数控代码,并将这些数据和文档存入全局数据库。

(4) 同样方式,生产者可以获得加工子任务,一旦获得该子任务,机床 Agent 将被允许从全局数据库读取必要的数据(CAD/CAPP 数据和文档、数控代码等),并将这些数据传送给加工中心,加工中心则根据这些数据和命令完成加工子任务,并将运行状态信息传送给机床 Agent。

(5) 设计结点完成设计子任务后,向任务规划结点提交该子任务。

(6) 生产者完成加工子任务后,通过其 Agent 向任务规划结点返回结果并提交该子任务。

(7) 客户可以了解订单执行情况和结果。

在系统的整个运行期间,系统 Agent 都对系统中各个结点间的交互活动进行纪录,如消息的收发,对全局数据库数据的读写,查询各结点的名字类型、地址、能力及任务完成情况等。

基于上述过程,我们进行了一简单零件的设计与加工验证实验,实验结果表明该系统运行性能良好,能完成预期的任务,从原理上实现了我们所提出的分布式网络化制造系统模式。

参 考 文 献

[1] 张伯鹏,汪劲松,郑力,等.中国机械工程,1997,8(2):60~63.
[2] 房贵如,刘维汉.中国机械工程,1995,6(3):7~10.
[3] 罗振璧,周兆英.灵捷制造,济南:山东教育出版社,1996:9~15.
[4] 陶德言.知识经济浪潮,北京:中国城市出版社,1998:1~5.
[5] 蔡希尧,刘西洋,边定平.计算机科学,1995,22(3):9~12.
[6] 钱方,周健,邹鹏.计算机科学,1996,23(2):76~79.

[7] 段广洪,钱立,王君英等.中国机械工程,1998,9(2):23~27.
[8] 程涛,左力,刘艳明,等.中国机械工程,1999,10(1):26~31.

On a Concept of Distributed Networked-manufacturing System

Abstract　It is very necessary for manufacturing to change manufacturing mode because of the tendency of globalization, the characteristics of market place, and the rapid developments of science and technology, etc. A concept of distributed networked-manufacturing system presented in this paper is such a new idea for the next century manufacturing then its main characteristics and key techniques are introduced here. Finally, a distributed networked-manufacturing prototype system is built up on the basis of the idea.

Keywords:Globalization manufacturing mode, Distributed system, Common object request broker architecture (CORBA), Networked manufacturing

(原载《中国机械工程》1999 年第 10 卷第 11 期)

网络化制造与企业集成

杨叔子　吴波　胡春华　程涛

提　要　阐述了网络经济时代制造环境的变化与特点,指出了网络化制造模式的必然性,研究了基于 Agent 的网络化制造模式及基于利益驱动的动态重组机制。在此基础上,提出了网络环境下企业集成的基本思路及基于 Agent 的网络化企业信息模型。

关键词: 网络化制造,企业集成,系统重组,企业联盟,虚拟企业

1　网络经济与网络化制造

随着信息技术和计算机网络技术的发展,世界经济正经历着一场深刻的革命,这场革命极大地改变着世界经济面貌,塑造着一种"新世界经济"即"网络经济"[1]。其特征是信息产业将在世界范围内大大发展,以此为基础的各种服务行业将成为越来越多的国家的主导产业,它将使世界经济全球化的进程大大加速,使任何国家的市场变得过于狭小,企业跨国家和跨行业联合进一步发展,经济活动将按网络加以组织。

网络经济时代,各国的经济离不开与国际市场的信息、技术、资源和产品的交换。从这个意义上说,网络经济是跨国性的、全球性的经济,生产过程已经不再局限于一国范围内,而是形成"无国界的经济实体"在企业集团之间,竞争与合作、交流与限制并存,从而形成错综复杂的局面,引起世界经济结构与组织结构的重大变化。譬如,某种先进的计算机的设计工作可能在美国硅谷进行,芯片在韩国生产,软件在印度编制,整机在泰国组装,营销在我国香港地区进行。

网络经济使得制造环境发生了根本性的变化,见表 1。制造业面临全球性的

市场、资源、技术和人员的竞争。开放的国际市场使得消费者更具有选择性,个性化、多样化的消费需求使得市场快速多变,不可捉摸,无法预测。客户化、小批量、多品种、快速交货的生产要求不断增加。各种新技术的涌现和应用更加剧了市场的快速变化。市场的动态多变性迫使制造企业改变策略,时间因素被提高到首要地位。21世纪制造行业的竞争将是柔性和响应速度的竞争,以适应全球市场的动态变化。尽管传统的价格与质量仍然是重要的的竞争因素,但已不再是决定因素。

表 1 制造环境的变化

	传统经济时代	网络经济时代
消费者的可选择性	区域性	全球性
消费需求	物美价廉,满足基本生活需求	个性化、多样化
市场	相对稳定	快速多变、无法预测
生产需求	低成本、高质量	客户化、快速交货
生产方式	标准化、系列化、大批量	单件、小批量、多品种
技术与资源	相对集中	全球分布
竞争要素	性价比	柔性与响应速度

面对网络经济时代制造环境的变化,传统的组织结构相对固定、制造资源相对集中、以区域性经济环境为主导、以面向产品为特征的制造模式已与之不相适应,需要建立一种市场需求驱动的、具有快速响应机制的网络化制造模式,这将是当前乃至今后若干长的时期内制造业所面临的最紧迫的任务之一,是制造企业摆脱困境、赢得市场、掌握竞争主动权的关键。

网络化制造以数字化、柔性化、敏捷化为基本特征。柔性化与敏捷化是快速响应客户化需求的前提,表现为结构上的快速重组、性能上的快速响应、过程中的并行性与分布式决策。这意味着系统必须具有动态易变性,能通过快速重组,快速响应市场需求的变化。由于制造资源与市场的全球分布性,因此,这种快速重组必须建立在全球性的分布式网络化基础上。

网络环境下,制造企业的组织形态、经营模式和管理机制需要有全方位的创新,使之适应网络化制造的要求。制造企业不再是孤立的个体,而是社会化大系统中的一个成员,并作为动态的制造环境中一个可资使用的制造个体资源,以企业集成的形式,通过合作与竞争,参与动态的制造系统重组。

2 基于 Agent 的网络化制造模式

市场需求驱动的、建立在全球分布式网络基础上的网络化制造系统,其本质上是一个复杂的社会经济人文交互系统。市场需求的快速多变和不确定性决定了制造系统的暂时性,其生命周期取决于市场需求的存在,并随着市场需求的变化,快速组建与撤消,快速进入与退出市场。这就要求采用新的有效的组织形态与运行决策机制,能通过简单的控制规则来实现复杂制造系统的动态重组与运行控制。这是迄今所取得的研究成果未能解决的"瓶颈"问题之一。研究表明[2-5],基于分布式异质 Agent 协同求解的制造系统模式可望成为解决这一问题的最有效的系统组织形态,它是以个体的自律性与整体的自组织为其基本特征的分布式智能化系统模式。众多具有自律行为能力的独立的制造个体通过简单的控制规则相互作用,形成系统整体的自组织、自适应与自进化行为能力,从而实现制造系统的快速重组与行为控制。

Agent 概念最初源于分布式人工智能领域,用以表示具有推理决策与问题求解能力的智能逻辑单元。Agent 之间通过计算机网络连接,Agent 作为网络上的智能结点,构成分布式多 Agent 系统,其特点是开放性、分布式合作和适应环境变化的自组织能力。随着其应用的扩展 Agent 的含义也得到了一些扩展和延伸:在一些环境中 Agent 表示具有封闭功能、能自主决策的功能实体,称为"自主体"或"自治体";在另外一些环境中,Agent 的作用与其原来的词义相同,是功能实体利益的代表,负责代理功能实体的一切外部事务,这种 Agent 被称为"代理",本文中的 Agent 一般是这种含义。目前尚没有确切的中文术语能准确地表达 Agent 丰富的内涵,是否可以结合其音和义,将之译为"轭健体"?

从本质上讲,任何规模、任一层次的制造系统都是由若干个完成不同任务的环节组成。各环节在独立完成自身任务的同时,又相互协作,共同完成制造任务,因此,可以说制造系统实现制造资源向产品转化的整个制造过程是一个典型的多 Agent 协同求解过程。制造环境中,一个 Agent 所代表的功能实体可以是制造组织、生产单元和软件系统等。一个工厂、一个制造企业就是一个典型的功能实体,一个功能独立的车间、制造子系统、职能部门、一台或若干台制造设备均可以作为一个功能实体,这取决于 Agent 的粒度的选择。Agent 的粒度反映其功能实体的规模。功能实体的规模越小,制造环境中 Agent 的数目就越多,制造系统重组的灵活性增加,适应性增强,但系统的组织与控制的复杂程度也随之增加,并使系统的运行效率降低。

根据制造环境中各功能实体完成自身任务的方式,可以将其分为三类[5]:①运行过程中不需要人的干预的自动系统,如数控设备等;②通过人机交互进行工作的人机系统,如 CAD 系统等;③制造中的人或组织,包括完全由人来操作的简单制造设备。相应地,可以用三种具有不同功能与结构的 Agent 分别作为这三类功能实体的代理,即自动系统 Agent、人机系统 Agent 和自然 Agent,见图 1~图 3。

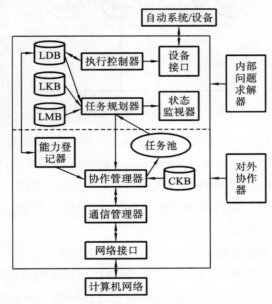

图 1　自动系统 Agent 结构

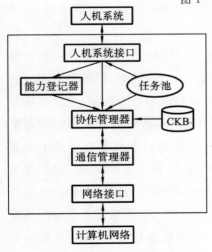

图 2　人机系统 Agent 结构

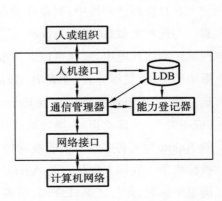

图 3　自然 Agent 结构

制造环境中的各 Agent 通过计算机网络连接起来,构成基于 Agent 分布式网络化制造环境,见图 4。一般而言,每个 Agent 都是自律的和彼此独立的组件,但也可以拥有若干 Agent 作为自己的"友元"建立一种相对紧密的联系。

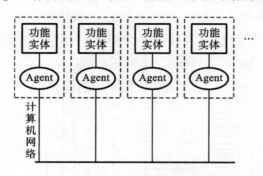

图 4　基于 Agent 网络化制造环境

为了便于系统的组织与控制,可以依照层次设计的思想规定 Agent 的粒度,形成多层次的制造环境。譬如:如果规定各 Agent 所代表的功能实体是独立的制造企业,则形成一种企业联盟网络环境,实现企业之间的集成;如果规定各 Agent 所代表的功能实体是企业内部的功能子系统,则形成一种企业内部网络环境,实现企业内部的集成。不同层次具有同构特征。

图 4 所示的网络化制造环境仅仅是一种静态的组织结构形态。为了适应不确定的快速多变的市场需求,必须建立有效的快速动态响应机制,达到系统整体的柔性与敏捷性。其中,复杂性问题和可操作性问题将变得十分突出。Agent 所具有的特性与基于利益驱动的动态重组机制相结合,将使问题相对简单化,并增强其可操作性。

在社会经济环境中,利益驱动是引起市场竞争、导致社会分化组合的核心因素。与此相类似,由于各 Agent 是其功能实体利益的代理,因而,在基于 Agent 的网络化制造环境中,同样可以引入市场竞争机制,以经济利益为纽带,通过市场竞争中的招标/投标方式,驱动各 Agent 之间的动态组合,达到制造系统的动态重组。譬如,当出现市场机遇或某个 Agent 拥有一份生产订单而其自身又不可能(或不必要)全部承担时,则向制造环境中的其他 Agent 发出任务标书。收到标书的 Agent 根据将要获得的收益与付出的代价的权衡,决定是否或以何种价格参与投标竞争。发标方收到有关 Agent 的投标后,从自身是否能获取最大收益的角度确定中标者,并与之签订合同,明确双方的权益,发出任务订单。其中,与发标方具有更紧密联系的"友元"Agent 将得到优先考虑。中标者得到任务订单后,还可以以同样的方式向其他 Agent 招标,最终建立一个动态的合同网。如此,通过合同网由制造环境中相关的 Agent 形成一种暂时的层次组织结构,达到动态的和暂时的制造系统重组。随着订单的完成,合同即告中止,这种暂时的系统组织也就随之消亡。当出现新的市场机遇时,这种重组过程重新开始。

3 网络环境下的企业集成

基于 Agent 的网络化制造模式为企业集成的实现提供了一种开放式框架。企业集成的关键是提高对环境的快速适应能力,即通过使用现代信息技术,建立灵活机动的、高效率的信息处理和反馈系统,提高企业适应市场变化的能力,使企业能够在复杂多变的市场环境中生存,并且不断发展。企业集成包括系统级集成(企业内部的集成)和企业级集成(企业之间的集成)两个层次,但其重心已经由系统级集成转向企业级集成。企业级集成的实施目标是建立增值伙伴网络或分布式合作网络。

增值伙伴网络由具有相关利益的独立企业组成,为组建一个临时的、集合性的企业提供支持。成员间共享信息、共同响应市场、共同创造商机相互协作、共同受益。

分布式合作网络是传统企业的组织结构与功能的裂化、分布化和虚拟化而形成的企业网络。面向单一场所的生产计划和控制概念(如 MRP、MRPII 等)转变为多场所协调与自主决策。同时由于外组产品的增加,严格的固定不变的层次组织结构转变为根据目标和环境的变化动态组合的临时网络结构。其运行与管理不仅需要注重本地与自身的活动,同时必须考虑和管理所有网络上的相关过程。在自治的虚拟单元之间不仅仅要支持一般应用之间的通信,如电子数据交换(EDI),还要支持分布式合作。

企业集成的总体组织原则是虚拟企业或动态企业联盟。这首先需要建立企业联盟网络环境。企业联盟网络的建立不仅需要制定联盟的行为准则,更需要先进的信息技术的支持。联盟内部的各项活动都依赖于企业集成基础设施所提供的一系列服务,这些服务的实现则依赖企业联盟的信息技术。它们之间的关系见图 5。

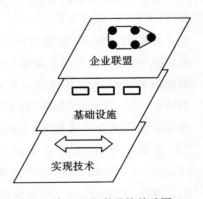

图 5 建立企业联盟的基础图

在企业联盟网络中,每个独立的企业都作为网络上的一个成员企业结点。成员企业结点一方面继续保持其自治运转的特点,另一方面与网络上的其他结点合作。联盟网络上的企业在功能上包括企业实体和企业 Agent。企业实体包含企业所有的自治功能和完整的信息结构以及企业内部的决策和内部活动等。企业 Agent 是企业外部特性的体现,代表企业利益,负责与其他结点的交互与合作。其功能:一是企业内部功能与内部行为的抽象和映射,包括企业描述信息、公共信息、质量信息、合作信息、安全信息等;二是代表企业执行商务活动。成员企业结点可以是一个真正意义上的企业,也可以是能提供某类服务的非完整的企业,这将为中小企业带来好处。

为了维护企业联盟的正常运行,联盟网络上的结点还应包含协调结点和公共信息服务结点。协调结点主要是提供一些基本的服务,如企业入盟申请的审批、入盟注册、数据访问权限的授予等。公共信息服务结点为联盟成员提供一些公共信息的存储和查询服务,如联盟成员数目、最新消息、联盟组织原则等。企业联盟网络见图 6。

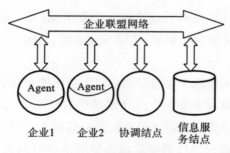

图 6　企业联盟网络

企业集成的基础设施需要提供对通信基础设施、信息交换与共享、合作应用和过程管理机制等方面的支持。其中,合作应用支持包括对地理上分布的多地点生产环境的支持、对软硬件平台和网络异构性的支持、对不同企业内部自治系统的集成的支持,以及对数据安全性的支持等。过程管理机制支持包括对集成环境中的商务过程和联合管理的支持,管理联盟网络中的信息流和物料流,使得联盟中的各项活动如订单管理、任务分配等能顺利进行。企业集成的基础设施最终表现为一系列的分布式服务,如商务服务、信息服务和表示服务等。

企业联盟网络建立的首要条件是通信网络,联盟网络可以直接建立在 Internet 上,也可以建立在专用网上,建立在专用网上的联盟是相对固定的联盟组织。其次是制定联盟的操作规则,包括企业的入盟标准和过程、权限划分、谈判规则、利益分配原则等,从信息处理的角度就是要定义联盟中所进行的活动、活动所触发的信息的流动、活动之间的触发顺序等。

入盟企业必须按照企业联盟网络的组织原则和协议规范对企业信息系统进行包装,建立网络化企业信息模型。图 7 表示了基于 Agent 的网络化企业信息模型。

企业联盟网络由市场驱动,通过基于商业利益的市场竞争机制,驱动企业联盟网络的成员企业之间的动态组合,形成临时性的动态企业联盟。一旦任务完成,利益关系终止,动态联盟解散。当出现新的市场机遇时,这种重组过程重新开始。

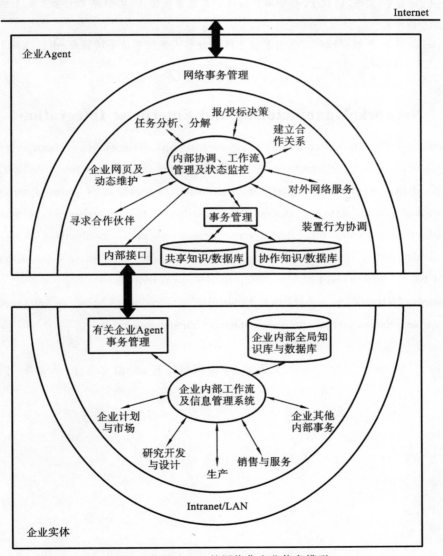

图 7　基于 Agent 的网络化企业信息模型

参 考 文 献

[1] 陶德言. 知识经济浪潮. 北京：中国城市出版社，1998：33～138.

[2] 程涛,胡春华,吴波,等. 分布式网络化制造系统构想. 中国机械工程,1999,10(11)：1234～1238.

[3] 杨叔子,吴波. 依托基金项目开展创新研究. 中国机械工程,1999,10(9)：987～990.

[4] 胡春华,吴波,杨叔子. 基于多自治体的制造系统集成. 华中理工大学学报,1996,24(9)：25～27.

[5] 胡春华,吴波,杨叔子. 基于多自主体的分布式智能制造系统研究. 中国机械工程,1998,9(7)：54～57.

Network Manufacturing and Enterprise Integration

Abstract   The manufacturing environment on networked economy era is discussed, which is characterized by fierce competition due to globalization, technical progress, and demand for customized product. It is pointed out that networked manufacturing model is coming. An agent-based networked manufacturing model and the dynamic reconfiguration mechanism based on benefit-driving are proposed. Finally, the basic idea for enterprise integration on agent-based networked manufacturing model is presented.

Keywords：Network manufacturing, Enterprise integration, System reconfiguration, Enterprise alliance, Virtual enterprisee alliance, Virtual enterprise

（原载《中国机械工程》2000 年第 11 卷第 1—2 期）

磁性无损检测技术中的信号处理技术

康宜华　武新军　杨叔子

提　要　论述磁性无损检测新技术中磁电信号的放大、时空域滤波等模拟信号处理技术和等空间间隔、等时间间隔以及时空域混合采样方法。介绍数字信号的预处理方法、信号特征量和信号的定量解释方法。

关键词：磁性检验，信号处理，模拟信号，数字信号

信号处理技术决定检测设备的总体性能，同样也决定磁性无损检测设备的技术指标。随着电子技术，特别是计算机技术的发展，检测设备在性能上得到很大程度的提高，并不断向计算机化、定量化和智能化发展。信号处理的目的，是为了达到各种检测性能的要求，将由探头输出的检测信号不失真地进行放大、滤波等处理，提高检测信号的信噪比和抗干扰能力，进一步进行信号的识别、分析、诊断、显示、存储、打印和纪录等，以显示出最佳的信号特征和检测结果。为了对信号进行计算机处理，应将模拟的磁电信号转化为数字信号，然后通过算法或程序对数字信号进行分析、评判以及显示、存储、打印和控制等。随着信号性质和检测目的的变化，信号处理的方法及策略也将不同。本文仅对磁性检测中单点图象处理技术术进行论述。

1　模拟信号的放大处理

磁场测量探头输出的信号一般较微弱，必须经放大后才能进一步处理。放大器首先应根据测量信号的性质来选择或设计。在这里，磁场信号一般有以下两类。一类是在空间局部区域内突变的磁场，如用漏磁场法检测时，铁磁性材料中的裂纹、孔洞、锈蚀斑点和气隙等产生的扩散型漏磁场，随着磁场测量方式的变

化,这一局部磁场信号产生的电信号特征也将变化。当测量漏磁磁感应强度沿磁化方向的分量时,电信号将是不过零点的叠加在背景信号上的单向脉冲信号,而测量垂直于磁化方向的分量时,则将是过零点呈对称性的脉冲信号。测量过程中,探头相对于被测磁场的运动速度变化时,电信号在时间域上的信号波形(或频率成分)将发生改变,加速时,裂纹等产生的信号的中心频率上升,减速时,中心频率下降。另一类磁场信号是在长的空间位置内变化缓慢的一类信号,如用主磁通法测量铁磁性构件长度方向上的磨损、锈蚀、厚度及直径时的主磁通变化信号。

在处理检测电信号时,局部变化的信号可采用交流放大技术,通过耦合或偏置调整消除信号中的低频或直流分量,一般来讲,这类放大电路结构较简单。缓慢变化的信号则需要采用直流放大技术或调制解调技术来处理,处理过程中的调零、温度补偿等步骤将增加电路的复杂性。检测信号放大电路的设计,应根据磁敏测量元件特性(如感应线圈测量时的速度补偿等)、测量信号特点以及检测要求来选择处理方法和元器件。

2 模拟信号的滤波处理

在磁性检测中,信号的滤波从两方面进行:一方面是对磁场信号的滤波处理,信号工作在空间域上,采用空间滤波方法;另一方面是对磁电信号的滤波处理,信号工作在时间域上,采用时域滤波方法。在检测中当探头与被测磁场间的相对运动速度 v 恒定时,空间域上的磁场信号的频率成分 f_s(单位:m^{-1})与时间域上的电信号的频率成分 f_t(单位:Hz)之间存在着对应关系

$$f_t = f_s v$$

当速度 $v(t)$ 变化时,空间域和时间域上的信号 $x(s)$ 与 $y(t)$ 间的频率对应关系为

$$y(t) = x(s) \cdot v(t)$$
$$F[y(t)] = F[x(s)] \cdot F[v(t)]$$

所以,对于测量速度恒定的磁场,时间域和空间域上的滤波处理是相互对应的,且可以替换实现。

2.1 空间域滤波

磁场信号在空间域上的滤波处理通过空间滤波器实现,其基本原理是通过导磁性能优良的材料主动引导空间分布的磁场,实现不同空间频率成分的磁场的分

流,从而有选择性地获得测量回路上的磁场信号。空间滤波器属于结构型功能构件,应根据检测对象、条件和目的进行设计。另外,磁场信号是三维矢量信号,因此,滤波器不但具有频率选择性,而且具有方向性。图 1 给出了几种典型空间滤波器的结构形状。检测中可以通过设计和组合得到所需频段的磁场信号。

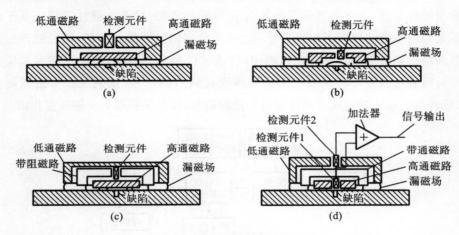

图 1 几种典型空间滤波器结构
(a)低通;(b)高通;(c)带通;(d)带阻

2.2 时间域滤波

当测量速度恒定不变时,可根据空间域滤波和时间域滤波的要求设计磁电信号滤波器,并根据速度的变动,调整滤波器的截止频率。需注意的是,放大电路和测量通道自身会产生噪声,为提高检测电信号的信噪比必须将这部分噪声信号有效滤除,因此,在确定测量速度时,应选择适当的速度范围,使得测量的有用磁场信号对应的电信号的频率与电路噪声信号频率相距较大,同时,应避免它出现在 50 Hz 的工频干扰附近。

2.3 时空域混合滤波

当测量速度波动时,也可采用时域滤波的方法来实现空间域滤波。这就要求时域滤波的特征频率随探头扫描运动速度波动而变化。较有效的方法是采用开关电容来设计滤波器(即开关电容滤波器),通过一位移测量装置测定相对移动的位置,并对它进行编码,每隔一定空间间隔发出一脉冲,脉冲的疏密对应着运动速度的快慢,用它来控制开关电容的动作,改变电容的大小,进而改变滤波器的特征频率,实现速度跟踪滤波。实际装置中通常将时空域滤波方法结合应用,一方面确保对磁场信号的选择性,另一方面排除或减小处理电路的电噪声。

计算机及人工智能等技术的发展促进无损检测技术向计算机化、定量化和智能化发展。对信号进行计算机处理,首先应将模拟的磁电信号转化为数字信号,然后通过算法或程序对数字信号进行分析、评判以及显示、存储、打印和控制等。前者由硬件完成,主要实现信号的不失真传输和变换,后者由软件完成。基于计算机处理的系统结构框图如图2所示。根据数字信号性质的不同,信号处理的方法及策略也相应不同。在磁性检测中,将数字信号分为两类,即突变信号和缓变信号。前者指在局部时间或空间区域内变化剧烈的信号,如裂纹产生的漏磁信号;后者指在较长时间或空间区域内变化的信号,如磨损引起的主磁通变化信号。

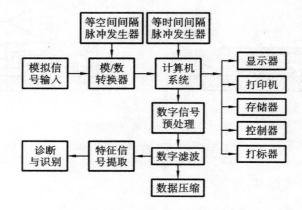

图2 数字信号处理系统框图

3 磁电信号的时空域采样[1]

磁场信号为空间域上的连续信号,经传感器测量后的磁电信号则为时间域上的连续信号。实际检测中,对检测信号的采样是在时间域中进行的。

当传感器相对于被测对象作匀速扫描运动时,空间域离散检测信号 $x_s = x_s(i \cdot \Delta s)$ 可以通过等时间间隔采样后的离散时间域信号 $x_t(i \cdot \Delta t)$ 求得

$$x_s = x_s(i \cdot \Delta s) = x_t(i \cdot \Delta t), \quad i = 0,1,2,\cdots$$

式中:Δs——空间域采样间隔;

Δt——时间域采样间隔;

$\Delta s = v_0 \Delta t$;

v_0——匀速扫描的速度。

对空间域信号的采样,首先应满足信号在空间域的采样定理,又由于采样一般是在时间域内进行的,时间域采样间隔 Δt 又应满足时域采样定理。当空间域采样间隔 Δs 确定后,时域采样间隔 Δt 必须满足

$$\Delta t \leqslant \frac{\Delta s}{v}$$

式中,v——扫描速度。

可见,时域采样间隔 Δt 直接与扫描运动的速度相关,匀速时,通过对空间域信号的空间域频谱分析,确定 Δs,进而由 v_0 确定 Δt;变速时

$$\Delta t = \frac{\Delta s}{v_c}$$

式中,v_c——扫描运动的最高速度。

因此,磁电信号的采样可通过下述三种方式实现。

3.1 等时间间隔采样

根据时空采样定理确定空间域采样间隔 Δs,考虑扫描运行的速度 v 后,确定出时域采样间隔 Δt 由时钟脉冲触发采样。

3.2 等空间间隔采样

根据空间域采样定理确定出 Δs 后,设计空间位置测量方法和脉冲编码器,如采用滚轮随探头同步进行扫描运动,用光电编码器对滚轮的转动进行编码,让滚轮每运行一小段直线位移后发出一触发脉冲,通过等空间间隔的触发脉冲序列控制 A/D 转换的进程,实现磁电信号按照空间位置进行采集。

3.3 时空混合采样[2]

当空间位置脉冲序列的间隔较大时,在空间间隔脉冲对应的时间历程内按照等时间间隔进行采集,同时,纪录空间脉冲出现的时刻。在每个空间间隔对应的时间间隔内,扫描运动的速度假设为匀速,将等时间间隔采样的信号序列映射到更小的等空间间隔点上,从而获得较小采样间隔的空间域信号序列。当精确的空间位置测量和脉冲编码器成本较高或难以实现时,可采用此种采样方法,以粗的位置测量和细的时域采样间隔相结合,实现高频信号的空间域采样。

4 数据的预处理[3]

信号采集系统在检测过程中会受到外界的各种干扰和噪声的影响,为准确判别数据,预处理的目的在于剔除数据中可能出现的短促干扰脉冲信号和数据中无意义的孤立野点,以及不感兴趣的杂散信号。实际应用中考虑到实时处理的要求,算法一般由滑动中值平滑器、汉宁滤波器等单独或组合构成。

设空间域信号序列为 $x(m), m = 0, 1, 2, \cdots$ 中值平滑器的输出为

$$y(m) = \mathrm{Median}\{x(m-1), x(m), x(m+1)\}, \quad m = 1, 2, \cdots$$

式中,Median——中值函数。

汉宁滤波器的输出为

$$s(m) = \sum_n y(n) h(m-n)$$

式中:$s(m)$——汉宁滤波器的输出;

$h(m-n)$——汉宁窗函数。

$$h(p) = \begin{cases} \dfrac{1}{4}, & p = 0 \\ \dfrac{1}{2}, & p = 1 \\ \dfrac{1}{4}, & p = 2 \\ 0, & p \text{ 为其他值} \end{cases}$$

由于磁电信号是连续信号,具有连续函数的性质,因此对磁电离散信号序列,通常可以采用下面更为简单的方法处理:

$$x(m) = \begin{cases} \dfrac{x(m-1) + x(m+1)}{2}, & \left| x(m) - \dfrac{x(m-1) + x(m+1)}{2} \right| \geqslant T \\ x(m), & \text{其他值时} \end{cases}$$

式中,T 由信号变化的幅度确定。

上述处理可消除等空间间隔脉冲误差、系统干扰和随机干扰等因素的影响。对于实时性要求不高的检测场合,可设计多种数字滤波器,采用时间序列分析和小波分析等处理方法。

5 信号特征量

对磁电信号进行定性、定量解释的关键在于信号特征量或向量的提取。在裂纹类缺陷检测中，信号特征往往是局部空间或时间轴上的异常信号，信号特征主要有以下几个参量。

5.1 信号绝对峰值 P_0

绝对峰值 P_0 的定义如图 3 所示。通常缺陷检测信号的峰值高于正常区域上检测信号的峰值，通过设置适当的门限值 D_0 可对缺陷存在的有无作出二值化处理。门限值 D_0 的选择由信号背景噪声和灵敏度大小等因素决定。其运算表示为一种非线性变换

$$c(m) = c[x(m)]$$

式中，$c[\cdot]$——门限函数。

$$c(u) = \begin{cases} 1, & u \geqslant D_0 \\ 0, & u < D_0 \end{cases}$$

对信号绝对峰值进行门限处理的算法简单，可以在定性检测仪、报警仪等低成本仪器中使用。但是，当信号基线出现波动时，将影响信号绝对值的幅度，可能引起误判。

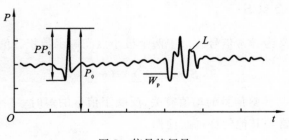

图 3 信号特征量

5.2 信号峰-峰值 PP_0

如图 3 所示，峰-峰值定义为局部异常信号的峰与谷间幅值之差的绝对值。计算时首先要找到一个峰点和两个与之相邻的谷点，计算峰谷之间的幅值差，选取其中绝对值大的一个。这一特征量排除了信号基线波动的影响，可提高异常信号识别的准确性和可靠性。

5.3 相邻信号差分值 D_1

对于信号中的局部变化异常区,其相邻离散信号的差分绝对值一般远大于非异常区的差分绝对值,因而,差分值 D_1 可用来判别正常或异常区信号,定义为

$$D_1 = |x(m) - x(m-1)|, \quad m = 1,2,3,\cdots$$

上式实际上是向后一步的差分值,当对向后 k 步信号进行差分运算时,可得向后 k 步差分值 D_{-k}。

$$D_{-k} = |x(m) - x(m-k-1)|, \quad m = k+1, k+2, \cdots$$

同时也可计算得到向前一步差分值 D'_1 和向前 k 步差分值 D'_{-k}。

$$D'_1 = |x(m) - x(m+1)|, \quad m = 0,1,2,\cdots$$

$$D'_{-k} = |x(m) - x(m+k+1)|, \quad m = 0,1,2,\cdots$$

信号的差分值在某些方面描述了相邻或相近信号之间的相关性。

5.4 波宽 W

当考虑信号在时间或空间分布的情况时,波宽是最简单的参数,常用的波宽指标有 W_{90}、W_{75}、W_{50}、W_{25} 和 W_{10} 等,如图 3 所示。

$$W_p = \sum_{m}^{m+N} c[x(m)]$$

式中:N——波形信号中超门限的采样点数;

p——门限 D_0 与峰值 P_0 的百分数。

5.5 波形面积 S

波形面积定义为异常信号波形中两个极小点间的函数积分值。

$$S(m) = \sum_{m}^{m+N} x(m)$$

波形面积 S 与异常信号的均值有关,反映了信号的短时一阶原点矩。在缺陷的定量识别中通常与其他特征一起使用。

5.6 信号周长 L

信号周长 L 指信号在幅值和时间(空间)二维空间上的路径,如图 3 所示。

5.7 短时能量 E_s

$$E_s(m) = \sum_{m}^{m+N} x(m)^2$$

上述只是局部异常信号的一些基本特征量,将这些特征组合后可得一些新的特征。从统计模式识别来看,不同的特征组合形式将形成多种特征矢量,更为重要的是,将信号进行频谱分析时,可以在频域(时频域或空频域)内获得更多的特征,通过时间序列分析方法进行建模分析时,可以获得模型特征,通过小波分析进行时-频域分析时,在不同的变换基上可以获得各自的特征等等。

6　信号的定量解释

无损检测与评价的最终难点是对检测信号的定量解释。在磁性无损检测中,由磁力线走向的不定性和很强的空间敏感性,以及检测过程中测量间隙等波动现象造成检测信号的重复性、稳定性较差。从磁场分布特征来看,不同几何形状尺寸的缺陷可能产生相似的磁场分布图形,因而,从磁场分布图形反演几何尺寸时,反演运算并非是唯一的,存在着不定性。以上这些要素对磁信号的定量解释造成了极大的障碍。从信号解释的方法学来看,对于不同性质的信号可采用不同的方法进行识别。

6.1　基于统计模式的定量识别方法[4]

基于统计模式的定量识别方法的基本原理如图 4 所示。首先从数据中分离出异常信号,即完成判别有无缺陷的定性识别,再从模式中抽取能够定量反映信号状态的特征量或特征向量,最后利用数值分类器对特征进行分类与判别。这一识别方法的准确性由训练模式样本与未知模式样本间的一致性、特征抽取准确性和分类器精度等多方面因素决定,实现的关键在于特征量(或矢量)的选取以及分类器的训练。

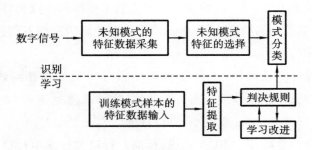

图 4　统计模式识别方法

6.2 基于模型的定量识别方法

对复杂的检测对象,仅仅采用单参量或简单的组合参量来对检测结果作出综合评价是不够的。随着计算机技术的发展,人们试图采用人工智能技术解决复杂信号的定量解释问题,数学模型则成为有力工具。基于模型的检测(MBT)方法是将检测看作一个过程,将检测仪器看作一个系统,根据被测对象和检测仪器本身特定的信息(即先验知识)和大量实验得到的数据(即后验知识),建立检测系统的数学模型,再由相应的算法对数据进行处理,从而获得信号的定量解释。如图5所示,MBT法的关键在于相关的先验知识和后验知识的获取,这也是人工智能技术的"瓶颈"问题,但如果建立了正确的模型后,将获得准确、稳定的分析处理结果。

另外还有结构模式识别、模糊模式识别、基于神经网络和神经元的识别等方法。

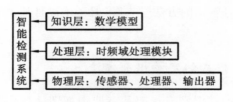

图 5　MBT 法

6.3 定量检测精度

对于一定功能的磁性检测系统,其标定一般采用已知缺陷的标准试件进行,检测信号的定量精度也只能是在一定条件下标准试件重复试验的统计分析结果。正如其他无损检测方法一样,磁性无损检测方法不可能对在役构件作出绝对准确的判别,因而定量检测的精度,实际上是检测中对缺陷识别的置信度,与参数测量的精度有着本质的区别。

在数字信号处理系统中,还存在 A/D 卡及计算机的选择问题,主要表现在 A/D 卡的精度(8、12 和 16 位或更高)、通道数(单通道、8 通道和 16 通道或更多)、转换速度(100、25、5 和 1 μs 或更小)、量程(± 5 V,± 2.5 V,0～5 V 和 0～25 V 等)、接口型式(PC 总线、PCMICA 总线、标准并行口和标准串行口等),计算机是台式机、工控机、笔记本计算机或单片机等,这些都需要根据系统的性能要求经济地选择和配置。

参 考 文 献

[1] 康宜华,杨叔子,卢文祥,等.空间域信号的采样方法.华中理工大学学报,1992(增刊)183-188.

[2] 金建华.录井钢丝裂纹定量检测原理及装置的研究:[硕士学位论文].武汉:华中理工大学,1997.

[3] 杨叔子,康宜华.钢丝绳断丝定量检测原理与技术.北京:国防工业出版社,1995.

[4] 周冠雄.计算机模式识别(统计方法).武汉:华中理工大学出版社,1987.

[5] 赵新民,陈海军.模型化测量的发展现状与展望.计量学报,1992,13(4):314-319.

Signal Processing Technology for Magnetic Nondestructive Testing

Kang Yihua   Wu Xinjun   Yang Shuzi

Abstract   The analog signal processing methods which include magneto-electric signal amplifying, filtering in time and special domains, sampling with equivalent special interval or equivalent time interval or inequivalent time interval which correlates to equivalent special intervals are discussed. Digital signal preprocessing methods and the characteristic values and quantitative interpretation of signals are introduced.

Keywords:Magnetic testing,Signal processing,Analog signal,Digital signal

(原载《无损检测》2000 年第 22 卷第 6 期)

Intelligent Machine Tools in a Distributed Network Manufacturing Mode Environment

Cheng Tao   Zhang Jie   Hu Chunhua   Wu Bo   Yang Shuzi

Abstract   Manufacturing enterprises are facing serious challenges and pressures from the growing globalisation of the economy and the market as well as from the rapid developments of science and technology. Small and medium-sized enterprises (SMEs), especially, have to reform their traditional manufacturing methods by using advanced technologies, particularly by applying information technology (IT) to succeed in the increasingly intense competition for markets. Thus, it is of great importance for them to accept the concept of agile manufacturing. For this purpose, the concept of distributed network manufacturing mode (DNMM) is outlined in this paper. In DNMM, research is concentrated on enhancing the intelligence of conventional numerical control (NC) machine tools and their ability to communicate and coordinate with the outside world. The experimental results of the distributed network manufacturing prototype system (DNMPS) show that the concept of the DNMM is correct and feasible. Moreover, the intelligent CNC system developed enhances the ability of the conventional NC milling machine to improve machining efficiency and quality, and protect the cutting tool. The capability for communicating and collaborating are improved for system integration, resource-sharing and cooperation.

Keywords: Agile manufacturing, Distributed network manufacturing, Fuzzy control, Intelligent control, Intelligent machine tool, Manufacturing mode, SMEs, Virtual enterprise

1　Introduction

Manufacturing enterprises are now facing serious challenges and pressures from the tide of globalisation (involving marketplace, economy, trade, investment, finance, technology, and service as well) which is sweeping across the world. Collaboration among enterprises distributed worldwide is growing frequent. The global market, which is being built up, is changing from a seller's market to buyer's market, by customers demand. Moreover, competition for markets is more and more unrelenting for meeting consumers' diverse demands for manufacturing products that must be supplied at lower prices, with shorter delivery dates, higher quality, better performances, and more satisfactory service[1-3]. Therefore, for a single enterprise, particularly for small and medium-sized enterprises, it is increasingly difficult to occupy a powerful position in the competition for markets[4,5].

What strategy can be adopted by a manufacturing enterprise for surviving and developing? Collaboration between enterprises can help manufacturing firms to hunt for opportunities to prosper in the face of intense competition. It is thus necessary to establish a fair and mutually beneficial collaboration mechanism for sharing out the work and helping one another, and learning from each other's strong points to offset one's weakness. Because small and medium-sized enterprises normally do not possess all the essential resources for competition for markets, improving their competitiveness is extremely important, as it is to large enterprises.

Science and technology have been advancing rapidly in recent decades. In particular, information technology (IT), including micro-electronics, computer, network and telecommunications technology, has made rapid progress and applies to many domains, which deeply affect and change production, daily work, learning, and life[6,7]. IT has become a major strategic challenge for manufacturing enterprises compelled to alter their production, management, marketing, and merchandising[1,3].

A manufacturing enterprise's strategy should be to organise flexibly and agilely its resources (including capital, techniques, talents, information, facilities, and preferential policies, etc.) as fast as possible so as to respond rapidly and adapt to the varieties of market demands. Each manufacturing enterprise does not need to adopt self-contained functions, but based on the collaboration mechanism, it should seek copartners to constitute "virtual enterprise" which are temporary networks of independent companies - suppliers, customers, even rivals- linked by IT to share skills, costs, and access to one another's markets[8]. This strategy embodies the basic concept of agile manufacturing[1-3,8,9] that focuses on rapidly setting up the whole organisation for producing different products to meet customers' demands. There must be many data, documents, sounds, images and other information exchanged among the members of a virtual enterprise. All members of a virtual enterprise must intercommunicate, interact and cooperate using IT in order to share common market opportunities. This strategy is universally applicable to small and medium-sized enterprises (SMEs) and seems to be the only way to survive in this intensely competitive era.

Full use should be made of the different manufacturing enterprises' resources in order to respond quickly to changes in the marketplace. Enterprises ought to improve their traditional organisation structure, management means and production technologies by applying IT to make the most of their own potentials and resources, and strengthen their ability to communicate and cooperate with other enterprises. The concept of a distributed network manufacturing mode (DNMM) presented in this research is a new paradigm for sharing resources and collaborating among manufacturing enterprises. In this paper, the emphasis is on how to enhance the potential of enterprises (especially SMEs) and promote their capability of interaction and collaboration in the DNMM. We attach importance to enhancing the conventional NC machine tools' intelligence and interactions with outside environments. For this purpose, a distributed network manufacturing prototype system (DNMPS) is developed. We hope this work will help to popularise agile manufacturing among enterprises (especially SMEs).

2 Distributed Network Manufacturing Mode

2.1 Basic Concept

Fig. 1 shows the basic architecture of the distributed network manufacturing mode (DNMM). Under DNMM, the enterprises, which all have the intention to collaborate with one another, may form dynamic alliances using IT for some market opportunities. Furthermore, each member of the alliances should be restructured and reengineered using IT to organise its designs, production, management, and marketing and services. Manufacturing enterprise under DNMM should have these important characteristics:

They will be market-driven and value focused.

They will be made up of relatively simple, distributed, autonomous but cooperating units or subsystems, working together to agreed targets in flattened, network-like organisations. The autonomy of the work units or subsystems will extend to each having profit and loss responsibilities, so it can be considered as a multi-agent system[10-13].

Each work unit or subsystem must use its agent, which is an intelligent entity capable of independently regulating, reasoning, and decision-making, to carry out some actions, and be in charge of the entity's interactions with outside environments and with humans[10] and be able to communicate with other work units or subsystems, that is, its agent is in charge of its interactions with others using a computer network.

There is a similar hierarchy in manufacturing cells, workshops, and enterprise layers. Each node in the same layer is linked into this layer's computer network (Internet, Intranet or LAN) via its agent, and a lower layer is a node of its upper layer and is linked into its upper layer's computer network via its agent.

The global economy and the technologies for telecollaboration will both require and enable work units or subsystems to be distributed globally and to cooperate with each other for the common objectives.

They have an open, multiplatform, and integrated manufacturing system linked by a computer network for resource-sharing and cooperation, so as to meet customers' urgent requirements quickly and agilely.

They will replace the rigid, static, hierarchical manufacturing enterprises because of its great flexibility, agility, and adaptability to the rapidly changing market.

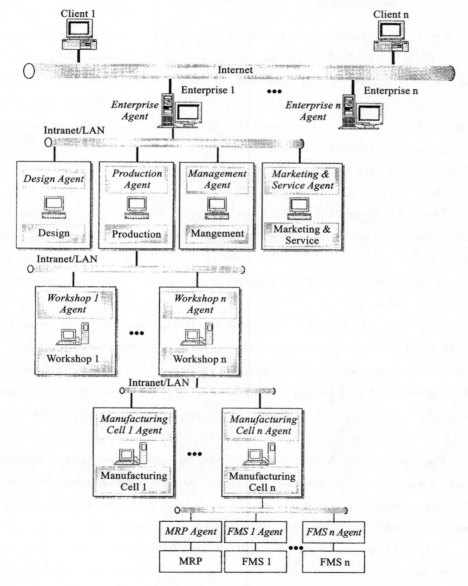

Fig. 1 The concept of the distributed network manufacturing mode

It is vital to spread DNMM among manufacturing enterprises, especially SMEs. Their abilities to communicate and collaborate can be greatly improved and their resources can be shared. The DNMM provides a mechanism for seeking market opportunities together, and jointly work to provide a rapid response to urgent changes, and help enterprises improve their competitiveness and accelerate their development.

2.2 Distributed Network Manufacturing Prototype System

A distributed network manufacturing prototype system (DNMPS) was developed for achieving our concept of DNMM. The prototype system(illustrated as Fig. 2) is composed of the following nodes: system manager, planner, designer, and manufacturing cell.

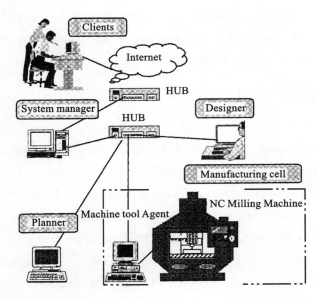

Fig. 2 The architecture of the distributed network manufacturing prototype system

1. System Manager. The system manager includes two parts: a database server and a system agent. The database server is in charge of the global database in which all nodes may query, search and store their data with each other; The system agent provides a Web server to interact with the Internet clients, including publishing information about the prototype system on the Internet with homepages, offering order register tables to Internet clients. It is also in charge of

displaying and recording the information about interactions among the nodes.

2. Planner. The planner's main function is to plan and decompose the system tasks obtained from the server's database into subtasks and assign them to the other agents by a bid construction method based on a detailed pricing mechanism. This node consists of an order manager and its agent called "order manager agent".

3. Designer. The designer is a computer-aided design system that provides humans with a good man-machine interface and a tool to design products. This node is composed of a CAD tool and its agent called "CAD agent".

4. Manufacturing Cell. The manufacturing cell has only one intelligent machine tool which consists of an NC milling machine and its agent called "machine tool agent". The NC milling machine is equipped with an intelligent adaptive CNC system developed by ourselves, and is responsible for exchanging timely information with the machine tool agent and accomplishing the machining tasks accepted by its agent. The machine tool agent completes interactions with other agents and monitors the running status of both itself and the NC milling machine tool.

2.3 Configurations

We use a local area network (LAN) to link the system agent, the server agent, the order manager agent, the CAD agent and the machine tool agent with one another with an eight-port HUB and RJ45 twisted wires, to constitute this prototype system network which is linked into the China Education and Research Network (CERN) and then is linked by the system agent into the Internet. The agents may intercommunicate by using TCP/IP network protocols. Furthermore, we chose StarBus 2.2 (a CORBA2.0-compliant[14,15] integrated developing environment developed by National Defense University of Science and Technology in China) as the basis for agent communication.

In this prototype system, the server agent which is implemented in Visual C++ 5.0 is based on a PC operating in a Windows NT 4.0 environment with a database in Microsoft SQL 6.5. The other agents which are implemented in Visual C++ 5.0 are based upon PCs running on Windows 95 and have their own local databases in Microsoft Access 97.

2.4 Operations

First of all, each node in this prototype system must enroll in the system manager so as to work cooperatively with other nodes, that is, its name, type, skills, network address and other basic information about itself should be registered in the server's database. The operation processes of the prototype system are shown as Fig. 3.

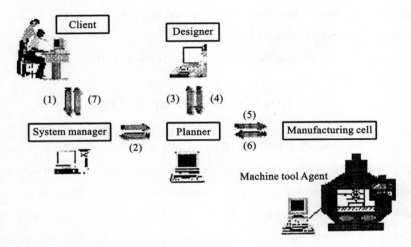

Fig. 3 The operation process of the prototype system

a. Any client may be allowed to use the Internet to visit the prototype system's homepage to obtain information about the prototype system, and then decide whether to make and submit an order to the prototype system. On receiving an order from a client, the system manager will put the order in the server's database from where the planner is capable of accessing the order.

b. The planner obtains an order from the server's database, then plans and decomposes the order into several subtasks, such as a machining task and a design task.

c. The planner assigns a design task to the designer based upon a contract net approach using a negotiation procedure.

d. In the same way, a machining task will be assigned to the intelligent machine tool by the planner.

e. If the design task is accepted, then the designer will accomplish it with the participation of design engineers, and output the corresponding CAD/CAPP data, documents, and NC codes which are stored in the server's database after submitting the subtask to the planner.

f. If the machine tool agent decides to accept the machining task, it will be allowed to read the data(CAD/CAPP data, documents, and NC codes) from the server's database, and transmit the data to the NC milling machine. Once the machining command is received from its agent, the NC milling machine accomplishes the machining task and reports its running status to its agent. Finally, the results are returned to the planner.

g. Lastly, the client may be informed about the results.

3 Intelligent Machine Tool

To adopt and install a DNMM, first of all, manufacturing enterprises ought to apply IT and other advanced technologies for reconstructing their organisation, management methods, production facilities, and training facilities. In this work, we focus on rebuilding a conventional numerical control (NC) machine tool to enhance its intelligence in machining and its ability to communicate and collaborate with the outside. The main purpose of improving its intelligence is to increase its machining efficiency, protect cutting tools, improve product quality, and reduce costs. When integrating a NC machine tool into a manufacturing cell to work jointly with other manufacturing equipment, it is necessary to reinforce the NC machine tool's capability of interaction with its outside environments.

The Huazhong I CNC system which was developed by the Huazhong University of Science and Technology in China, is an intelligent adaptive CNC system(see Fig. 4), and is built to improve its intelligence by adding some functional modules (such as an intelligent control module, a tool monitoring module, and a self-diagnosis and error-recovery module) into the existing CNC system, and to improve its ability to communicate by appending an agent.

The intelligent adaptive CNC system consists of the NC unit and the functional modules. The NC unit, which is composed of the interpolation calculating module, the input/output processing module and the user operating

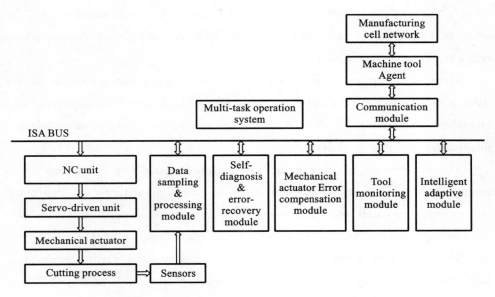

Fig. 4 The architecture of the intelligent CNC system

module, is responsible for servo control, executing the NC program for finish machining, and exchanging information with the functional modules (See Fig. 5). The functional modules complete their specific tasks.

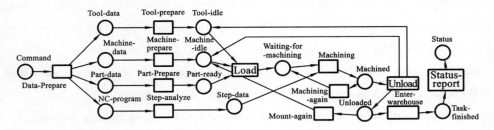

Fig. 5 The Petri net of the machine tool agent for monitoring the status of the NC machine while running

The self-diagnosis and error-recovery module, which is designed for making the NC machine tool run reliably and improving its self-adjustment, employs information, not only from sensor fusion for process monitoring, but also from feature description about the machining processes. According to the feature description, the module, based on a knowledge base, selects the option parameters for monitoring and controlling, which are best fit to the given feature machining. Moreover, once problems in machining are detected (such as tool breakage and wear), the module automatically suspends machining and searches for the protection and remedy strategies, which are represented as series atomic action codes stored in a strategy repertory[16].

The tool monitoring module is used to detect cutting tool condition on-line using a multisensor fusion technique based on artificial neural network and fuzzy identification methods[17,18], for protecting the cutting tool and providing relevant information to other modules.

The mechanical actuator error compensation module's main task is to compensate for mechanical actuator error by some algorithms for improving machining precision. Research on the above three functional modules is underway at present. Thus, the details about them will not be discussed in this paper.

The intelligent control module optimises and updates cutting parameters according to information about machining status using diversified intelligent controllers.

The communication module accomplishes communication with the machine tool agent which interacts with the other nodes in the manufacturing cell network. An agent, which is an independent operable program, can be operating on a different hardware platform, and can be written in a different programming language, and can execute under different operating systems. The machine tool agent is constructed in the model (see Fig. 6) as an autonomous software component that encapsulates a service (knowledge base, applications, and software tool). This agent's main functions are as follows.

Interacts with other agents across the computer network by passing messages.

Negotiates, coordinates and cooperates with other agents.

Decides whether or not to accept a subtask received from other nodes, if not, then searches other agents for cooperation or to transfer the subtask to them.

If accepted, then plans how to perform its activities and control the machine tool.

Communicates with the machine tool's communication module to exchange information using a serial port or a network adapter, including transporting the control command, data and documents to the machine tool, and receiving information about results and status from the machine tool.

Monitors the running status of the machine tool using a Petri net[19] (shown as Fig. 5).

The hardware layout of the intelligent CNC system is shown in Fig. 6. The CPU board is a PCA6136 all-in-one board whose CPU is an 80386DX which processes the multitask operation system which is the operating platform which manages the NC unit and the functional modules. The servo driver for each axis unit has an 80C196 microcontroller that controls its servo motor.

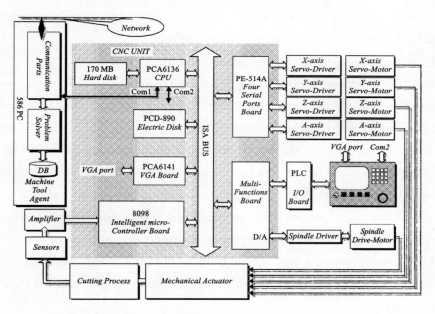

Fig. 6 The hardware layout of the intelligent adaptive CNC system

4 Intelligent Control Module

As is well known, the use of computer numerical control (CNC) machine tools in manufacturing has expanded rapidly since the middle of the twentieth century. However, the conventional CNC machine tool necessitates skilled and qualified programmers, who usually select and set operating parameters off-line based on their partial experience and knowledge or use machining databases or handbooks. In order to prevent the cutting tool from excessive wear, deflection, or even breaking, and to avoid the problems caused by non-uniform machining, faults in the material, machine tool chatter or other factors, programmers have a tendency to use extremely conservative machining parameters. Consequently, CNC machine tools are inefficient and run under poor machining conditions[20-25].

For this reason, adaptive control techniques, which provide effective ways to adjust machining parameters on-line, for overcoming the aforementioned drawbacks in the machining process, are becoming the research targets and objectives which are pursued by most commercial CNC machine tool manufacturers[20-24]. Adaptive control systems can be classified into two types[26]:

a. Adaptive control with optimisation (ACO);

b. Adaptive control with constraints (ACC).

In ACC systems applied in a machining process, one or more output parameters (typically cutting force or cutting power) are regulated to their limit values by updating the machining parameters on-line.

Since cutting force is the key factor affecting machining efficiency, cutting tool life, surface quality, and even machining cost, it should be controlled in the cutting process so as to:

a. Increase metal removal rate by regulating the cutting force to the desired value, indirectly leading to improvement of the machining efficiency;

b. Prevent premature tool failure, or damage to the workpiece, and minimise tool deflection and tool vibration to ensure product quality and decrease the cutting cost;

c. Achieve on-line adjustments of the operating parameters, so that the programmer's skill requirements can be minimised and the operators' interference in the cutting process can also be greatly reduced.

Thus, it can be seen that optimising and adjusting the machining parameters on-line is of great significance in performing successful and efficient machining operations, and it plays a key role in competitiveness in the market.

In this work, the intelligent control module, which adjusts feedrate on-line for regulating the cutting force to the desired values, can collect the cutting force signals, process the sampled data, optimise the feedrate using an intelligent controller, and exchange data with the CNC unit. The intelligent microcontroller board developed by ourselves is the principal part of the intelligent control module which is a function module of the CNC machine tool.

4.1 Intelligent Microcontroller Board

The intelligent microcontroller board (shown in Fig.7) consists of an 8098 microchip system, an E^2PROM, a double-port RAM, a voltage regulator, a low-pass filter and a series of communication circuits. Its double-port RAM is used to exchange information with its upper CPU system without refereeing by BUS. To coordinate the intelligent control module with the upper CPU or other modules conveniently, its HSO port is allowed to send an interrupt requisition signal to IRQ_5 of the ISA BUS. In addition, the board can be used in other control systems, because of its high computing speed, large memory, compatibility, and low cost.

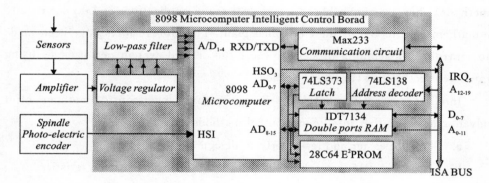

Fig. 7　The hardware block diagram of the intelligent microcontroller board

As the core of the intelligent control module, the board is in charge of sampling, signal processing, optimising machining parameters (typically spindle speed or feedrate), and exchanging data with its upper CPU or other modules. That is to say, the board is used to sample x-, y-, and z-components of the cutting force signals, calculate their resultant force, find the maximum resultant force during one spindle period, optimise feedrate with the intelligent adaptive controller, and then tell the NC unit to read the results from the double-port RAM and obtain the actual spindle speed and feedrate from the NC unit.

4.2 The Adaptive Fuzzy Controller Based on a Neuron

Since machining is a time-varying, indeterminate, dynamic, and highly nonlinear complex process, an accurate mathematical model of it is very difficult to obtain, so traditional adaptive control methods, which depend on the machining process' mathematical model, cannot control the machining process satisfactorily. From the review of the published literature it can be improved[20-25].

However, intelligent control methods (including fuzzy logic control[27-29], neural network control[30-32], and expert control[33-35]), which embody and simulate a human's capabilities of self-learning, self-organising and self-adapting in control activities, have been used in numerous industrial processes successfully. These intelligent control methods provide alternatives to the traditional control methods. They are applicable to a large variety of applications and do not require a mathematical model of the controlled plant. Therefore, efforts have been made to control the machining process to meet the desired system performance using intelligent techniques, in recent decades[27-35].

Intelligent adaptive controllers based on fuzzy logic control methods, offer advantages over other intelligent control methods[27-29].

a. They are easier and less costly to design and maintain.

b. They have stronger robustness and anti-interference characteristics.

c. They can control such highly nonlinear and complex process as machining efficiently.

Consequently, a fuzzy control strategy is applied for the end-milling operation in this research.

Generally, the fuzzy control rules of a conventional fuzzy controller[27] are established off-line by summarising operators' or experts' control strategies, experience and knowledge, and these fuzzy control rules are not changed during the control process. As a result, it is necessary to adjust the fuzzy control rules on-line for improving the control system's performance. Because of its ability of self-learning and self-adapting, an artificial neuron is used for modifying on-line the fuzzy control query table of a conventional fuzzy controller, which not only adjusts the fuzzy control rules indirectly, but also compensates for the lost information due to fuzzification. The adaptive fuzzy controller based on a neuron is illustrated as Fig. 8.

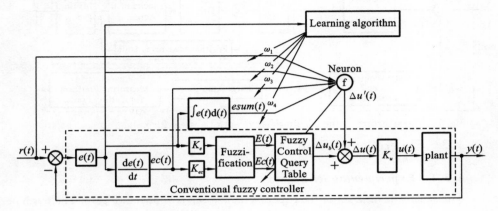

Fig. 8 The block diagram of the adaptive fuzzy controller

The fuzzy control query table can be modified as follows:
$$\Delta u_b(t) = \Delta u(t) = \Delta u'(t) + \Delta u_b(t) \tag{1}$$
the sign " = " indicates that the value on the righthand side is assigned to the variable on the lefthand side.

Here, the gradient-descent method[36,37] is selected as the neuron's learning algorithm.

The adaptive fuzzy controller is used to update the feedrate for maintaining the milling force to the desired value, so as to increase the metal removal rate in rough machining. Accordingly, the CNC milling machine's intelligence is improved to some extent, and also, its machining potential is developed as far as possible.

The research results in the literature[17] show that, milling cutter blades are likely to break at the maximum peak of the resultant force, because of the occurrence of unequal milling forces on the blades owing to their different geometry and parameters. Therefore, the maximum resultant milling force during one spindle revolution(marked by f_{max}) is chosen to be controlled in this work, which is more appropriate than using the resultant milling force averaged over one spindle rotation. The adaptive fuzzy control system for milling force regulation is shown in Fig. 9.

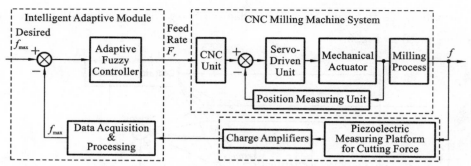

Fig. 9 The structure of the adaptive fuzzy control system for milling force regulation

4.3 Experimental Studies

4.3.1 *Experimental Set-up*

The schematic diagram of the experimental set-up is shown in Fig.10. The machine tool is a vertical milling machine, designed and manufactured by the Huazhong University of Science and Technology in China. The x-, y-, and z-components of the cutting force are measured by a Kister Model 9257A three-component measuring platform together with three Model YE5850 charge amplifiers made by Yangzhou 2nd Radio Factory in China. The experimental results in these studies are for machining HT20 cast iron with a 10.0 mm diameter 3-flute high-speed steel end mill. The machine tool agent implemented in Visual C++ 5.0 is based upon a PC running on Windows 98 and with its own local databases achieved in Microsoft Access 97. The communication between the machine tool agent and the milling machine is performed by the RS232 serial port.

Once a machining task is accepted, the machine tool agent will obtain the essential data and documents from the planner and transfer them to the NC milling machine, and then inform the NC milling machine to start machining. While executing the machining task, the NC milling machine will feed its status information back to the machine tool agent.

The milling force signals are measured by the piezoelectric three-component measuring platform, and then converted into the corresponding voltage signals and amplified by the three charge amplifiers, respectively. After being offset, amplified and filtered by the voltage regulator on the intelligent microcontroller board, the signals are acquired by the board's 10-bit A/D converter at a 1000 Hz sampling frequency. At the same time, the data processing program deals with the sampled data, calculates the resultant milling force and finds the maximum resultant milling force (called the measured f_{max}) during one spindle revolution.

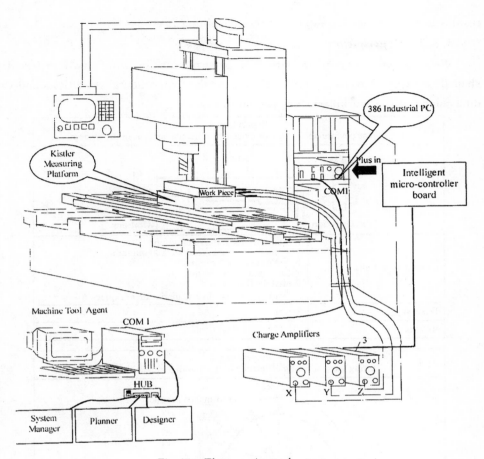

Fig. 10 The experimental set-up

Then, an intelligent controller (such as the adaptive fuzzy controller proposed in this research or another), obtains the actual spindle speed and feedrate while machining from 4010 H and 4012 H of the double-port RAM, and then optimises the feedrate based-on these cutting parameters, and the errors between the measured f_{max} and desired f_{max}, for the purpose of making the measured f_{max} approach the desired f_{max}. The control period is equal to the spindle rotation period. The intelligent control module puts the optimised feedrate in 4080 H of the double-port RAM. During 16 ms interrupt period, the CNC unit reads the optimised feedrate and puts the actual spindle speed and feedrate in 4010 H and 4012 H of the double-port RAM. Finally, the feedrate commands are transferred to the x-, y-, and z-axis servo drivers by the CNC unit. Thus, real-time control using intelligent control strategies in the milling process is achieved.

When the machining is completed, the machine tool agent will return the results to the planner for reporting the machining task.

4.3.2 *Experiments*

We conducted two main series of experiments, in which two differently shaped workpieces were machined. Details of the experimental conditions and the dimensions of the workpieces are shown in Fig.11.

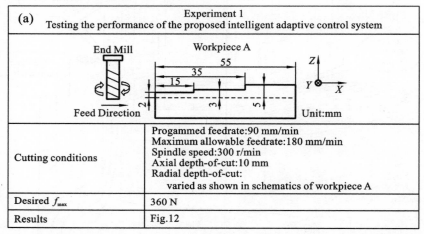

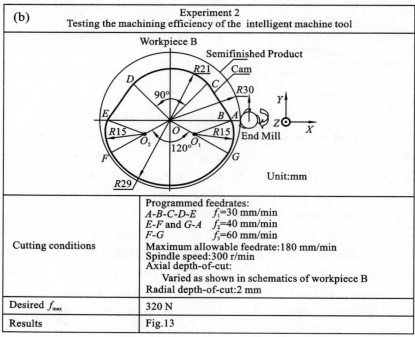

Fig. 11 Cutting conditions for (a) workpiece A, and (b) workpiece B

The first experiment is for testing the performance of the adaptive fuzzy controller based on a neuron, in which the conventional fuzzy controller and the adaptive fuzzy controller were used to control the milling process under varying radial depths-of-cut(see Fig.11(a)).

To demonstrate that the intelligent machine tool does improve cutting efficiency, in the second experiment, both a conventional milling machine without the intelligent control module, and the intelligent machine tool, which is a milling machine with the intelligent control module and other function modules, were applied respectively in the milling operations to machining the same workpieces (see Fig.11(b)).

4.3.3 Results and Discussion

Fig. 12 (a) and Fig. 12 (b), respectively, show experimental results of the conventional fuzzy controller and the adaptive fuzzy controller based on a neuron for machining the three steps of workpiece A. Although the radial depth of cut varied from 2 to 5 mm, both of them still could regulate f_{max} to 360 N or so by timely adjustment of the feedrate. Furthermore, their performance indices were all satisfactory. However, the general performance of the former is superior to that of the latter, because:

a. The overshoots and steady-state errors of the former are smaller than those of the latter while machining the first and third steps.

b. The f_{max} response curve of the former is smoother than that of the latter while machining the third step.

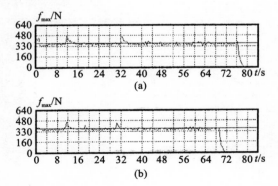

Fig. 12 The results of experiment 1. The f_{max} response curve of (a) the conventional fuzzy controller, and (b) the adaptive fuzzy controller based on a neuron

Fig. 13 (a) shows the experimental results of the conventional milling machine. The milling operation is completed for machining a cam (see Fig. 11

(b)) without the intelligent control module. From the results, we can estimate that the total cutting time is approximately 297 s. The experimental results of the intelligent milling machine with the intelligent control module, illustrated in Fig. 13(b), show that the measured f_{max} is regulated to the desired f_{max} 320 N, regardless of the variation in axial depth of cut while machining a cam(see Fig. 11 (b)), and the cutting time is about 167 s. Consequently, the intelligent machine tool is capable of increasing machining efficiency by 43.77%, from the results of the 2 series of experiments. Thus, we can conclude that the performance of the intelligent milling machine is excellent and improves the machining efficiency greatly.

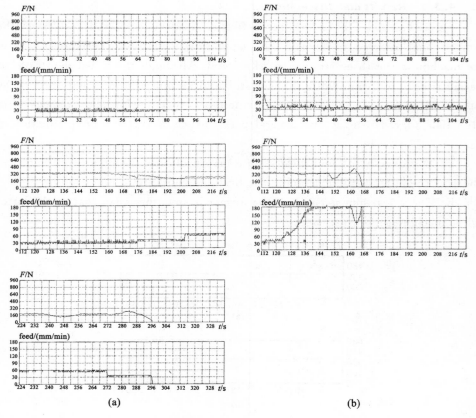

Fig. 13　The results of experiment 2. The f_{max} and feedrate response curves(a)without the intelligent control module, and (b) with the intelligent control module

5 Conclusions

This paper presents the idea of a distributed network manufacturing mode (DNMM) as a critical enabler for implementing agile manufacturing. Validation requirements are discussed in the executing mode in manufacturing enterprises, in particular in SMEs. The research is focused on how to strengthen an NC machine tool's intelligence and ability to interact with the outside world under the DNMM. An intelligent CNC system, which applies an intelligent control module and other functional modules for improving intelligence and uses the machine tool agent for reinforcing intercommunication capability, has been developed and installed in a conventional NC milling machine tool. In line with the DNMM, a distributed network manufacturing prototype system (DNMPS) has been developed for demonstrating this concept.

The experimental results show that the DNMPS architecture is achievable and its performance is good, and the concept of the DNMM is realised in the DNMPS. The intelligent machine tool was successful because an intelligent CNC system effectively enhances a conventional NC machine tool's intelligence and capability of interaction and collaboration. Owing to the open architecture of the intelligent CNC system, many functional modules may be conveniently plugged into the CNC system for extending its functions and improving its performance. Additionally, the intelligent control module which is based on an adaptive fuzzy controller, has the advantages of simple structure and algorithms, and has an excellent performance. It provides an effective and feasible means of optimising and updating cutting parameters in machining for increasing the machining efficiency and quality, protecting the cutting tool, and reducing cost.

In summary, our research should advance the acceptances of agile manufacturing by manufacturing enterprises, and offer methods for collaboration and resource sharing. It can also improve a manufacturing firm's agility, flexibility, rapid response and adaptability to the market. Based on our work, an enterprise's machining potential can be developed greatly, which should bring substantial economic benefits.

There is much to do on the DNMM in future work. Ongoing studies will be focused on the following areas.

Perfect the performance of the functional modules, particularly the self-diagnostic and error-recovery module, the tool monitoring module, and the mechanical actuator error compensation module.

Develop a commercial force sensor, or apply an indirect approach to measure cutting force, such as measuring the servo motor current and spindle motor power.

Carry out research on the ACO system in the cutting process aiming at minimum production cost[38], minimum production time[39], maximum profit rate[40], and maximum metal removal rate[41].

Strengthen the machine tool agent's autonomy, improve the mechanism of communication and coordination.

At present, the above problems are being studied. We believe that, with the solving of these problems and the developments of other key technologies concerned in the DNMM, it should be possible to spread the DNMM throughout manufacturing enterprises.

Acknowledgements

The authors gratefully acknowledge that this research is supported by the NNSFC (National Natural Science Foundation of China) under Grant no. 59705020 and 59990470.

We would particularly like to thank Professors Youlun Xiong, Bin Li, and Rensheng Du for their contributions to this project. We also thank Drs Xin Luo, Xiwen Li, and Zailin Guan for their helpful discussions and ideas.

References

1 R. N. Nagel and P. Dove. "21st century manufacturing enterprise strategy - an industry led view". Iacocca Institute, Lehigh University, Bethlehem vol. 1, 1991.

2 P. T. Kidd. "Agile manufacturing: forging new frontiers". Addison-Wesley, 1994.

3 Yiming Rong. "Agile manufacturing". Manufacturing System Program, College of Engineering, Southern Illinois University at Carbondale, December 1993.

4 H. B. Marri, A. Gunasekaran and R. J. Grieve. "An investigation into the implementation of computer integrated manufacturing in small and medium enterprises". International Journal of Advanced Manufacturing Technology, 14, pp. 935-942, 1998.

5 T. M. A. Ari Samadhi and K. Hoang. "Partners selection in a shared-CIM system". International Journal of Computer Integrated Manufacturing, 11(2), pp. 173-182, 1998.

6 Uri Klement. "A global network for plant design". Mechanical Engineering, 12, pp. 52-55, 1996.

7 Andrew S. Tanenbaum. Computer Networks, 3rd edn, Prentice-Hall, 1996.

8 J. Walton and L. Whicker. "Virtual enterprise: myth and reality". Journal of Control, pp. 22-25, 1996.

9 L. M. Camarinha-Matos, H. Afsarmanesh, C. Garita and C. Lima. "Towards an architecture for virtual enterprises". Journal of Intelligent Manufacturing, 9, pp. 189-199, 1998.

10 D. T. Ndumu and H. S. Nwana. "Research and development challenges for agent-based systems". IEE Proceedings- Software Engineering, 144(1), pp. 2-10, 1997.

11 S. C. Laufmann. "Toward agent-based software engineering for information-dependent enterprise applications". IEE Proceedings- Software Engineering, 144(1), pp. 38-50, 1997.

12 Riyaz Silora and Michael J. Shaw. "Coordination mechanisms for multi-agent manufacturing systems: applications to integrated manufacturing scheduling". IEEE Transactions on Engineering Management, 44(2), pp. 175-187, 1997.

13 Nicholas V. Findler. "Multiagent coordination and cooperation in a distributed dynamic environment with limited resources". Artificial Intelligence in Engineering, 9, pp. 229-238, 1995.

14 Http://www.omg.org.

15 OMG and X/Open. "The common object request broker: architecture and specification". Revision 2.0, July 1995, Updated July 1996.

16 Chanan Syan and Yousef Mostefai. "Status monitoring and error recovery in flexible manufacturing systems". International Journal of Integrated Manufacturing Systems, 6(4), pp. 43-8, 1995.

17 J. M. Lee and D. Choi. "Real-time tool breakage monitoring for NC milling Process". Annals CIRP Manufacturing Technology, 44(1), pp. 59-62, 1995.

18 G. Byrne, D. Dornfeld, I. Inasaki, G. Ketteler, W. Kong and R. Teti. "Tool condition monitoring(TCM)- the status of research and industrial application". Annals CIRP, 44(2), pp. 541-567, 1995.

19 Yang Wenlong, Yao Shuzhen and Xie Peijun. "Petri net theory and application". vols. 1 and 2, Department of Computer Science and Engineering, Beijing University of Aeronautics and Astronautics, 1992.

20 Tae-Yong Kim and Jongwon Kim. "Adaptive cutting force control for a machining center by using indirect cutting force measurements". International Journal of Machine Tools and Manufacture, 36(8), pp. 925-937, 1996.

21 M. Tolouei-Rad and I. M. Bidhendi. "On the optimization of machining parameters for milling operations". International Journal of Machine Tools and Manufacture, 37(1), pp. 1-16, 1997.

22 Shiuh-Tarng Chiang, Ding-I Liu, An-Chen Lee and Wei-Hua Chieng. "Adaptive control optimization in end milling using nerual networks". International Journal of Machine Tools and Manufacture, 34(5), pp. 637-660, 1995.

23 Tao Luo, Wen Lu, K. Krishnamurthy and Bruce McMillin. "An neural network approach for force and contour error control in multi-dimensional end milling operations". International Journal of Machine Tools and Manufacture, 38, pp. 1343-1359, 1998.

24 Y. S. Tarng and S. T. Cheng. "Fuzzy control of feed rate in end milling operations". International Journal of Machine Tools and Manufacture, 33(4), pp. 643-650, 1993.

25 A. Galip Ulsoy and Y. Koren. "Control of machining processes". Journal of Dynamic

Systems, Measurement, and Control, 115, pp. 301-308, 1993.

26 Y. Koren. *Computer Control of Manufacturing System*. McGraw- Hill, New York, 1983.

27 C. C. Lee. "Fuzzy logic in control systems: fuzzy logic controller, part 1, part 2". IEEE Transactions, Systems, Man and Cybernetics, 20(2), pp. 404-34, 1990.

28 R. M. A. Tong. "Retrospective view of fuzzy control systems". Fuzzy Sets and Systems, 14(3), pp. 199-210, 1984.

29 M. K. Kim, M. W. Cho and K Kim. "Application of the fuzzy control strategy to adaptive force of non-minimum phase end milling operations". International Journal of Machine Tools and Manufacture, 34(5), pp. 677-696, 1994.

30 D. Psaltis, A. Sideris and A. Yamamura. "A multilayered neural network controller". IEEE Control Systems Magazine, 8, pp. 1721, 1988.

31 IEEE. *Special issue on nerual networks*, IEEE Control Systems Magazine, 10(3), 1990.

32 K. J. Hunt, D. Sbarbaro, R. Zbikowski and P. J. "Neural networks for control systems - a survey". Automatica, 28(6), pp. 10831112, 1992.

33 Samir B. Billatos and Pai-Chung Tseng. "Knowledge-based optimization for intelligent machining". Journal of Manufacturing Systems, 10(6), pp. 464-475, 1991.

34 Panos J. Antsaklis. "Intelligent learning control". IEEE Control Systems, 7, pp. 5-7, 1995.

35 Panos Antsaklis. "Defining intelligent control". IEEE Control Systems, 6, pp. 4-5, 58-66, 1994.

36 J. J. Hopfield. "Neural networks and physical systems with emergent collective computational abilities". Proceedings of the National Academy of Sciences, 79, pp. 2557-2558, 1982.

37 J. J. Hopfield. "Neurons with graded response have collective computational properties like those of two-state neurons". Proceedings of the National Academy of Sciences USA, 81, pp. 30883092, 1884.

38 J. Wang. "Multiple-objective optimisation of machining operations based on neural networks". International Journal of Advanced Manufacturing Technology, 8, pp. 235-243, 1993.

39 M. S. Chua, M. Rahman, Y. S. Wong and H. T. Loh. "Determination of optimal cutting conditions using design of experiments and optimisation techniques". International Journal of Machine Tools and Manufacture, 33(2), pp. 297-305, 1993.

40 B. White and A. Houshyar. "Quality and optimum parameter selection in metal cutting". Computers in Industry, 20, pp. 87-98, 1992.

41 S. B. Billatos and P. C. Tesng. "Knowledge-based optimisation for intelligent manufacturing". Journal of Manufacturing Systems, 10(6), pp. 464-75, 1991.

(原载 The International Journal of Advanced Manufacturing Technology, Volume 17, Number 3, 2001)

AR 模型参数的 Bootstrap 方差估计[*]

轩建平 史铁林 杨叔子

提 要 针对故障诊断中 AR 模型参数判别门限值的确定需要大量样本和重复多次试验的问题,采用小样本统计的 Bootstrap 方法对车削颤振 AR 模型参数的方差进行了估计。结果表明,该方法具有较好的估计结果和工程应用价值。

关键词:车削颤振,AR 参数,Bootstrap,方差估计

金属切削过程颤振的监控是提高加工质量和生产效率的重要问题。在车削长轴时,在切削过程由平稳走向颤振的过渡过程中,机床振动加速度的特征变化是信号能量增加及主频带下移。对切削振动过程进行 AR 建模,可以反映这种特征变化。在这类问题中,人们更关心的是各种情况下模型参数的范围,换句话说,是模型参数的方差估计。在统计仿真上,传统的做法是,先假设信号的噪声满足一定的概率分布,如高斯分布,然后通过 MONTE-CARLO 仿真试验,得到参数估计的标准差。在工程中,多次对信号进行采样,每次对信号进行参数估计,由多个参数估计值,可得到模型参数的均值及均方误差。然而,由于工程数据特别是故障数据较难取得,或者重复试验代价太高等原因,这种方法可能行不通。本文采用 Bootstrap[1,2] 方法估计模型参数的方差。

1 Bootstrap 方法基本原理

如果总体 T 及其分布 F 已知,θ 是分布 F 未知特征量,如均值或方差,可由分析方法或 MONTE-CARLO 仿真计算总体均值和总体方差。然而,在实际问题中,总体 T 及其分布 F 未知。Bootstrap 法是下面简单思想的产物:既然人们没有总体知识,那么就充分利用已得到的观测样本 $X=\{X_1,X_2,\cdots,X_N\}$。换句话说,

---

[*] 国家"九五"攀登项目(PD9521908),国家"973"项目(998020320)。

若观测样本真实地反映了总体,则用 Bootstrap 法处理观测样本。按这种方式,从总体抽取 B 个独立同分布(i.i.d)样本的 MONTE-CARLO 方法修改如下:

从由观测样本 $\{X_1, X_2, \cdots, X_N\}$ 组成的样本总体中抽取 B 个(i.i.d)样本 $X^* = \{X^{*(1)}, X^{*(2)}, \cdots, X^{*(B)}\}$。样本总体具有经验分布,定义如下:

$$F(x) = \frac{1}{N}\sum_{i=1}^{N} 1(X_i \leqslant x) = \frac{1}{N}(\# X_i \leqslant x) \tag{1}$$

式中,$\# X_i \leqslant x$ 表示 $X_i(i=1,2,\cdots,N)$ 小于和等于实数 x 的个数之和。

$X^* = \{X^{*(1)}, X^{*(2)}, \cdots, X^{*(B)}\}$ 称为 Bootstrap 样本,从 X^* 得到的 θ^* 的分布逼近 θ 的分布。

2 AR 模型参数 Bootstrap 方差估计方法

采用修正协方差

$$x(k,l) = \sum_{n=p}^{N-1}[x(n-1)x^*(n-k) + x(n-p+l)x^*(x-p+k)] \tag{2}$$

来估计 AR 模型参数。为了保证模型是稳定的,采用 Burg 方法,该方法只对反射系数进行最小化处理。有两种 Bootstrap 方法对 AR 模型参数的方差进行估计:一种是相关数据的 Bootstrap 方法;另一种是相关数据模型的独立数据 Bootstrap 方法。本文采用后一种方法,该方法的基本原理是从相关观测样本中采用 Burg 方法估计出 AR 模型参数 $a_1, a_2, \cdots, a_p$,代入

$$x(n) = -\sum_{k=1}^{p} a_k x(n-k) + w(n) \tag{3}$$

得到噪声 $w(n)$,作为独立观测样本,即样本总体。

该方法步骤如下[3]。

a. 实验。从 AR(p), $X(n)$ 中做实验或直接由实际试验,取得 N 个样本 $x(n)$。

b. 计算残差。采用 Burg 方法估计出 AR 模型参数 $a_1, a_2, \cdots, a_p$,得到残差 $w(n)(n=1,2,\cdots,N-1)$。

c. 重采样。从残差 $w(1), w(2), \cdots, w(N-1)$,得到 Bootstrap 样本 $w(1)^*, w(2)^*, \cdots, w(N-1)^*$。

通过 $w(1)^*, w(2)^*, \cdots, w(N-1)^*$,令

$$x(0)^* = x(0)$$

$$x(n)^* = -\sum_{k=1}^{p} a_k x(n-k)^* + w(n)^* \quad (n=1,2,\cdots,N)$$

得到

$$x(0)^*, x(1)^*, \cdots, x(N-1)^*$$

d. Bootstrap 估计的计算。利用去掉均值的 $x(0)^*, x(1)^*, x(2)^*, \cdots,$ $x(N-1)^*$，采用 Burg 方法估计出 AR 模型参数
$$a^* = [a_1^*, a_2^*, \cdots, a_p^*]$$

e. 重复。重复步骤 c、d 若干次，如 $L = 1000$ 次，得到 $a(1)^*, a(2)^*, \cdots, a(L)^*$。

f. 方差估计。从 $a(1)^*, a(2)^*, \cdots, a(L)^*$ ($a(L)^*$ 表示第 L 次的 AR 模型参数 $a^* = [a_1^*, a_2^*, \cdots, a_p^*]$)，求得 a 的 Bootstrap 方差估计为

$$\sigma^{*2} = \frac{1}{L-1} \sum_{n=1}^{L} \left[a(n)^* - \frac{1}{L} \sum_{n=1}^{L} a(n)^* \right]^2 \tag{4}$$

3 AR 模型参数 Bootstrap 方差估计方法仿真结果

设式(3)中 AR 模型阶数 $p=2$，参数 $a = [a_1, a_2] = [0.6, -0.9]$；数据个数为 128。设噪声 $w(n)$ 服从均值为 0、方差为 1 的高斯分布。表 1 为未知分布的 Bootstrap 方法和已知分布的 MONTE-CARLO 法比较。Bootstrap 方法仿真过程是：先由 AR 模型参数 a 及噪声 $w(n)$ 生成一组仿真样本，再由上面介绍的方法（其中 $L = 1000$）求得 a 的 Bootstrap 方差估计。已知分布的 MONTE-CARLO 法仿真过程是：先由 AR 模型参数 a 及噪声 $w(n)$ 生成 1000 组仿真样本，对每组仿真样本，采用 Burg 方法估计出 AR 模型参数 a_1, a_2，再对得出 1000 组 AR 模型参数 a_1, a_2 求方差。共作了 3 次仿真。从表 1 可看到，第 1 次和第 3 次 Bootstrap 方法结果与 MONTE-CARLO 法仿真结果比较接近，而第 2 次 Bootstrap 方法误差较大，这是由于仿真本身造成的，因为样本总体最初是由 AR 模型参数 a 及噪声 $w(n)$ 生成一组仿真样本，且数据个数为 128，由该样本得到 AR 模型参数 a 的一次估计产生了较大误差。在此模型参数 a 下得到的残差分布偏离了真实分布，因而 Bootstrap 方法逼近该分布时产生了较大误差。

表 1 未知分布 Bootstrap(Ⅰ)方法和已知分布的 MONTE-CARLO(Ⅱ)法比较

方法	参数	1		2		3	
		均值	方差	均值	方差	均值	方差
Ⅰ	a_1	0.598	0.002	0.546	0.005	0.581	0.002
	a_2	−0.896	0.002	−0.621	0.005	−0.867	0.002
Ⅱ	a_1	0.592	0.002	0.595	0.002	0.594	0.002
	a_2	−0.880	0.002	−0.881	0.002	−0.879	0.002

4 车削颤振 AR 模型参数的 Bootstrap 方差估计

图 1 为车削状态下尾顶尖垂直方向的振动加速度所对应电压实测信号图。图 2 为三种工况下对应的信号直方图。图 2(a)为稳定状态,图 2(b)为过渡状态(信号近似满足高斯分布),图 2(c)为颤振状态(信号呈现一定的周期性)。图中 f 为横坐标中数据出现的频次。

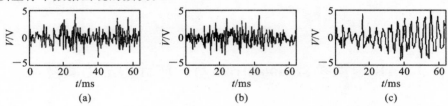

图 1 三种工况下信号图
(a)稳定状态;(b)过渡状态;(c)颤振状态

表 2 为对应三种工况下的 AR 模型参数的 Bootstrap 方差估计。其中模型阶数为 11 阶,表 2 列出前 6 个参数和方差估计。结果表明,该方法给出的方差估计具有较好的估计精度。其主要原因在于,该方法充分利用原始数据提供的信息,反复多次地抽样。然而在某些情况下,该方法可能出现较大误差。下面为影响该方法的一些因素分析。

残差的获得。在从实测数据中进行的第一次 AR 建模时,AR 参数应尽可能准确,否则进行反卷积运算所求得的残差不可靠,影响后续 Bootstrap 方法的应用。

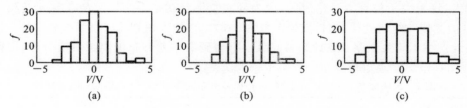

图 2 三种工况下信号直方图
(a)稳定状态;(b)过渡状态;(c)颤振状态

表 2 三种工况 AR 模型参数的 Bootstrap 方差估计

工况	统计	a_1	a_2	a_3	a_4	a_5	a_6
稳定状态	均值	0.458 2	0.211 2	−0.034 8	−0.035 6	0.213 7	0.038 5
	方差	0.007 8	0.009 0	0.009 8	0.009 6	0.009 7	0.010 2
过渡状态	均值	0.394 7	0.108 6	0.021 7	0.057 8	0.225 5	0.013 1
	方差	0.007 8	0.009 2	0.008 9	0.009 4	0.008 7	0.009 0
颤振状态	均值	0.262 3	0.168 4	0.288 6	0.326 4	0.385 3	0.395 5
	方差	0.008 8	0.008 9	0.009 6	0.008 6	0.009 9	0.009 4

数据野点的影响。主要指两方面的影响：a. 对 AR 参数估计的影响；b. 对残差的影响。数据野点将使结果难以预料。

Bootstrap 方法不一致性。AR 模型，统计量和重采样方法可能使得样本数无论多大 Bootstrap 方法不能逼近真实分布。

本文介绍的方法，在车削颤振 AR 模型参数的方差估计中得到了应用。应该指出的是，Bootstrap 方法并不是在一切场合下都适用。

参 考 文 献

[1] Efron B. *Bootstrap Methods：Another Look at the Jackknife*. The Ann. of Statistics, 1979, 7(1): 2～6.

[2] Politis D N. *Computer-intensive Method in Statistical Analysis*. IEEE Signal Processing Magazine, 1998, 15(1): 39～55.

[3] Zoubir A M, Bo as hash B. *The Bootstrap and Its Application in Signal Processing*. IEEE Signal Processing Magazine, 1998, 15(1): 56～76.

Bootstrap Variance Estimation for AR Parameters of Cutting Chatter

Xuan Jianping Shi Tielin Yang Shuzi

Abstract Estimation for variances of discriminating parameters obtained by AR modelling need a large number of samples or the experiment shall be repeated many times. In this paper, to solve those problems in variance estimation for AR parameters of cutting chatter, a Bootstrap method is introduced. The results show the method can give a good variance estimation.

Keywords: Cutting chatter, AR parameters, Bootstrap, Variance estimation

（原载《华中科技大学学报（自然科学版）》2001 年第 29 卷第 9 期）

虚拟制造系统分布式应用研究[*]

周杰韩 熊光楞 吴波 杨叔子

提 要 面对制造辅助工具异构集成、代码重用等问题,制造业进一步投资需要的是一种网络层的、可扩展的框架结构。该文分析了制造应用软件社会化开发的特点,提出了分布式制造应用的概念和以框架为中心的制造应用分布式开发方法。建立了虚拟制造系统分布式应用开发的层次结构。总结了制造组件的三类模型。以柔性制造单元分布式布局应用为例,研究了最大生成树布局算法组件的建模策略,开发了分布式柔性制造单元布局系统。

关键词:虚拟制造系统,分布式制造应用,并行工程,代理对象,分布式计算

软件界和工业界在 1990 年代末先后制定了多种基于软件重用的分布式组件对象标准,如 CORBA、DCOM、RMI 和标准支持工具。其中微软的 VC 产品全面支持 DCOM 标准。Borland 公司的 Dephi 产品则全面支持 CORBA、DCOM 两种标准。SUN 公司 Hot Java 产品则支持 RMI 标准等。

正是这些标准与工具的出现,制造业进一步投资需要的将不再是具有单一功能、难以扩展的桌面工具,而是一种网络层的、可扩展的框架结构。虚拟制造系统正是支持分布式制造应用开发的一种框架结构[1]。本文把基于标准组件开发的制造应用系统统称为分布式制造应用。

1 分布式制造应用特点

所谓分布式制造应用是指组成制造应用的资源组件分布在互连网络上,各部分以"客户请求-服务器应答"方式协同实现制造应用。这里的资源组件是指分布

---

[*] 国家自然科学基金重大项目(59990470),国家自然科学基金资助项目(69884002)。

在网络上的、按照某种标准建立的数据、算法、应用程序、车间硬件设备等。分布式制造应用开发符合制造应用自动化、社会化生产的发展方向,其前提是制造组件走标准化生产道路。所谓制造应用自动化生产是指装配符合某种标准的制造组件开发制造应用软件的一种生产方式。所谓制造应用社会化生产是指组成制造应用软件的部分组件是从市场上直接购买得到的。制造应用自动化生产、制造应用社会化生产和分布式应用开发都是基于标准的制造应用开发方式。分布式应用开发已在机床控制器、通信系统、信息处理系统等领域中得到了广泛的研究[2]。制造应用社会化生产模式如图1所示。分布式制造应用具有如下几个特征。

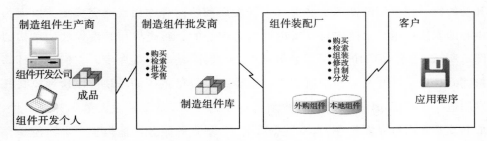

图 1 基于组件的制造应用社会化生产模式

(1)组成制造应用的对象(制造组件对象)逻辑上具有对称性。这主要体现为分布在网络上的对象无层次性。它们在地理位置上是分散的,功能上互为客户与服务器。

(2)对象具有可扩展性是指对象功能完善性上的扩展,重组性是指对象可以是其他对象的结合体;即插即用性是指制造系统有目标地选择对象,选中对象的功能对当前目标是暴露的。

(3)对象间互操作透明性。对象既有专门服务和专门请求的对象,也有互为请求服务的对象。对象间通过发送请求、提供服务进行协同工作。在协作过程中,对象的访问不需要指明具体的目标,将由称作代理的一种机制自动完成寻找工作。

(4)对象存在形式多样性。每个对象可以有各种不同的存在形式,如表格、文件、数据库、窗体、机床硬件设备等。

(5)对象实现具有语言、平台独立性。可在不同厂商制造的机型上,使用不同的语言及开发环境来创建对象。

(6)系统成本低、可靠性高、响应快捷。

2 分布式制造应用开发模式

一般来讲,以编程语言与开发工具为中心的应用开发是软件市场发展的初级阶段。其发展正逐步向以特定问题域对象框架为中心的应用开发转移。

所谓框架可以简单地认为集合了构筑特定领域系统或应用的软件模块。以框架为中心的分布式应用开发模式克服了初级阶段的作坊式和地域封闭式等开发缺点,它是一种基于框架的、集成全球化标准组件的应用开发模式。该模式下开发的每一样应用软件都是经由多家开发厂商协同合作而最终完成的。以库为中心的制造应用开发模式与以通用框架为中心的分布式应用开发模式的结构如图 2 所示。图 2(a)中的库可以是数据库、函数库、动态连接库等。它们是在计算机辅助制造工具开发过程中所积累的极少一部分可重用的模块。以这种模式开发制造应用系统,开发人员在用户新的应用需求下,将重新做大量的开发工作。如开发基本用户界面管理、基本数据库管理、基本的服务功能模块等。即使是存在少量的、可借用的数据库、函数库等,其借用也是有一定限制的。例如要求前后开发的平台一致,要求熟悉前后开发工作等。这种开发模式是目前较为普遍的一种形式。图 2(b)是以框架为中心的分布式应用开发模式。这种模式将克服上述重用量小、受开发平台限制等局限。制造通用型框架已集成了常规的、大量的用户界面管理、数据库管理、基本服务功能等模块。面对用户特定的应用需求,开发人员往往只需要开发少量的符合标准的特定制造组件。如针对用户车间布局应用需求,制造框架将提供基本的车间布局管理服务。开发人员将仅需要开发车间布局算法功能组件,并将该布局组件集成到通用框架中,即可满足用户的布局需求。

可见,分布式应用开发模式在开发效率、开发灵活性、代码重用率等方面都将明显优于传统的、以库为中心的开发模式。

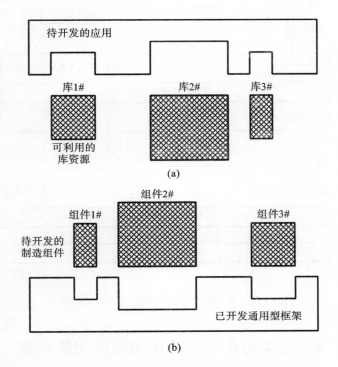

图 2 两种制造应用开发模型

(a)以库为中心的应用开发模型;(b)以框架为中心的分布式应用开发模型

3 分布式应用开发层次结构

虚拟制造系统分布式应用开发层次结构如图 3 所示。层次结构的应用层提供分布式制造应用开发定的用户界面,包括应用选择、组件选择、组件扩展与组件连接等功能。分布式制造应用有集成本地标准组件开发和集成远程标准组件开发两种方式。

应用框架层设计了若干组件管理模块和本地组件。组件管理模块与组件基于同一标准开发。组件管理模块负责接入、接出组件。信息技术支持层包括操作系统和分布式标准支持工具。

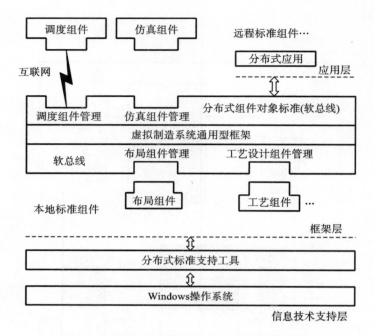

图 3 虚拟制造系统分布式应用开发的层次结构

4 三类分布式制造应用组件对象模型

为简化制造系统分布式对象建模研究,将组成完整系统的功能全部看作对象,并根据它们之间的协作关系分为三类对象,即客户对象、服务器对象和客户服务器对象。典型的分布式制造应用网络拓扑结构如图 4 所示。该图是一个有向图,共有 6 个结点,并分为三类。第一类为只有出去的箭头没有进去的箭头结点,如结点 1。第二类为既有出去的箭头又有进去的箭头结点,如结点 2,3,4,5。第三类为只有进去的箭头没有出去的箭头的结点,如结点 6。

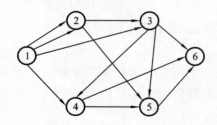

图 4 分布式制造应用网络拓扑结构图

分布式制造应用中的客户服务器对象分别对应于这三类结点。其中仅向对方提出请求服务的对象对应于第一类结点,称之为客户对象。既向对方提供服务,同时又向第三方提出请求服务的对象对应于第二类结点,称之为客户服务器对象。仅向对方提供服务的对象对应于第三类结点,称之为服务器对象。正是这三类结点的请求/服务功能体现了分布式制造应用重组协作、事务分布处理等特点。

虚拟制造系统分布式应用开发中很大一部分工作就是要将现有的、独立工作的计算机辅助制造工具中的客户对象、服务器对象以及客户服务器对象区别开来,为虚拟制造系统的框架建设与大量的制造组件库建设提供资源。下面以柔性加工单元(FMC)分布式布局为例,研究分布式制造组件建模策略。

5 FMC最大生成树布局算法组件对象建模

5.1 FMC最大生成树布局算法

车间布局是在柔性加工车间投产之前必须重点考虑的一个步骤。优化的车间布局方案将提高生产率、降低生产成本。车间布局按生产产品的分类和数量,加工设备的配置有多种方案。本文讨论的是基于物料传输路线布局类型中的直线布局,即单行机床布局。车间送料设备为自动导引小车,单行机床布局问题模型如图5[3]所示。

$$\min \sum_{i=1}^{n-1} \sum_{j=i+1}^{n} c_{ij} f_{ij} |x_i - x_j| \tag{1}$$

$$|x_i - x_j| \geqslant 1/2(l_i + l_j) + d_{ij}, \quad i = 1,2,\cdots,n-1; j = i+1,2,\cdots,n \tag{2}$$

$$x_i \geqslant 0, \quad i = 1,2,\cdots,n \tag{3}$$

式(1)是使完成机床间所需往返的总成本最小的目标函数。c_{ij}为在机床i和j之间送料,每单位距离的搬运成本;f_{ij}为在机床i和j之间往返的频率(次数);d_{ij}为机床i和j在布局中隔开的最小距离;n为机床数;l_i为机床i的长度;x_i为机床i的中心与垂直参考线vrl间的距离。式(2)保证布局中任意两台机床不重叠。式(3)保证机床位置的非负性。

针对单行布局问题提出了多种计算模型,其中最大生成树算法(MST)具有简单高效的特点。MST算法以小车在机床之间往返频率矩阵和每单位距离送料成本矩阵作为输入数据,以机床在单行布局中的排列顺序作为输出数据。算法首先

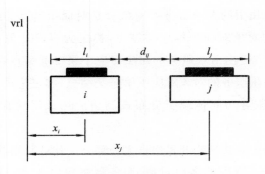

图 5 单行机床布局问题模型

实现往返频率矩阵与成本矩阵的乘法运算,求出二维流量成本矩阵。然后在流量成本矩阵中,运用最大生成树算法求解机床的排列顺序。

5.2 MST 算法组件对象设计策略

现有的计算机辅助车间布局工具往往是将面向用户的客户端与面向功能实现的服务端糅合在一起的。为了建立 MST 布局算法组件模型,首先必须将过程化语言实现的应用程序代码分成客户端与服务器端程序。其次考虑将服务器端程序设计成基于组件标准的对象。计算机辅助布局应用程序由以下三大功能模块组成。

(1) 用户界面服务。利用人机交互界面,用户提供车间布局基本数据,如小车往返频率矩阵等。

(2) 算法实现服务,即 FMC 优化布局的求解过程。

(3) FMC 车间布局服务,即输出车间布局视图。

在 MST 布局算法组件改造过程中,根据实际情况可考虑两种功能分配方案。一种是将算法服务与布局功能合在一起做成组件服务器。由于服务器将可能在远程服务,这种方案会影响到与客户交互的速度。另一种是将流量矩阵计算与布局功能放在客户方完成,从而减轻服务器在网络传输上的负担。实践中采用了第二种方案。

6 小 结

虚拟制造系统是计算机辅助制造技术发展的高级阶段。虚拟制造系统具有支持分布式应用开发的特征。作为进一步研究,本文分析了分布式制造应用开发特征,研究了虚拟制造系统分布式应用开发的层次结构。总结了制造组件的三类模型。并以 FMC 分布式布局应用为实例,研究 MST 布局算法组件的建模策略与 FMC 分布式布局应用的设计方案。实践结果表明,分布式应用开发方法为解决企业集成和软件重用等问题提供了一条途径。虚拟制造系统分布式应用开发的层次结构是合理可行的。

参 考 文 献

[1] 周杰韩,吴波,杨叔子. 虚拟制造系统综述[J]. 中国科学基金,2000,14(5):279-283. ZHOU Jiehan, WU Bo, YANG Shuzi. *Overview of virtual manufacturing system* (VMS)[J]. Science Foundation in China,2000,4(5):279-283(in Chinese).

[2] 福田好郎. 生产システムのオᴗ能プン化の动向[J]. 精密工学会志,1997,63(5):613-616. Fukuda Yoshiro. *Open system for manufacturing*[J]. Journal of the Society of Precision Engineering,1997,63(5):613-616(in Japanese).

[3] Andrew Kusiak. 智能制造系统[M]. 杨静宇,陆际联,译. 北京:清华大学出版社,1993. Kusiak Andrew. *Intelligent Manufacturing System*[M]. YANG Jingyu, LU Jilian. Peking: Tsinghua University Press,1993.

Distributed Virtual Manufacturing System

Zhou Jiehan Xiong Guangleng Wu Bo Yang Shuzi

Abstract Computer-aided manufacturing system design is an excellent area for applying software component techniques. This paper introduces a distributed manufacturing and open virtual manufacturing system which includes a common manufacturing framework, manufacturing component managers and manufacturing components developed using standard software architecture. The distributed manufacturing system is then applied to workshop arrangement. In the case of single-line machine arrangement problem solving, the paper discusses the modelling scheme of distributed arrangement components. The distributed flexible manufacturing cell workshop arrangement system is developed.

Keywords: Virtual manufacturing system, Distributed manufacturing application, Concurrent engineering, Agent, Distributed computing

(原载《清华大学学报(自然科学版)》2002 年第 42 卷第 9 期)

Wigner-Ville 时频分布研究及其在齿轮故障诊断中的应用[*]

来五星 轩建平 史铁林 杨叔子

提　要　虽然 Wigner-Ville 分布具备时-频分布的优点,但 Wigner-Ville 分布的一个致命缺点就是交叉干扰项的存在,并影响其广泛应用。本文介绍了 Wigner 双谱、Wigner 三谱,以及如何选用合适的核函数消除交叉干扰项的影响。并以齿轮为实验对象,成功应用上述方法消除交叉干扰项,达到诊断齿轮故障的目的。

关键词:故障诊断,信号处理,Wigner-Ville,状态监测

在线性系统中广泛应用的工具 Fourier 变换(FT)将信号分解为频率成分,但其功率谱或能量谱仅给出了信号数据的频域分量信息,却没给出这些分量对应的时间位置。与之相反,原始时域信号有很好的时间分辨率,却没给出任何频域信息[1]。尽管功率谱、双谱、三谱对平稳过程分析很好,但不适宜于非平稳过程或暂态信息。高阶统计量理论认为,对非平稳过程适宜采用时变累积量和高阶谱(与之相对应,二阶情况时,协方差称为时变协方差,谱称为功率谱)。举一个高阶矩应用的简单例子,考虑信号 $y(t) = s(t) + g(t)$,其中 $s(t)$ 是可分辨信号,$g(t)$ 是平稳零均值噪声,$y(t)$ 是非平稳过程,它的均值是 $s(t)$。当 $k>1$,k 阶累积量 $c_{ky} = c_{kg}$,因此只有一阶累积量携带信号 $s(t)$ 的信息。相反,$y(t)$ 的高阶矩依赖着信号 $s(t)$ 和 $g(t)$,因此在这种情形下,适宜采用高阶矩[2]。于是,描述谱或多谱(矩谱)的暂态演变的时频分布是分析这种信号的有用工具。

在时频分析工具中,常用的是短时 Fourier 变换,Gabor 展开和小波变换[3,4]。虽然线性性质用在线性系统中更为合理,但二次(双线性)时频分布还是作为时变

---

[*]　攀登预选资助项目(PD9521908)。

功率谱被提出、分析和解释。在 Cohen 类时频移不变分布中包括 Spectrogram，Rihaczek，Page，Wigner-Ville 分布和 Choi-Williams 等分布[5,6]。

Wigner-Ville 分布的一个重要的特征就是 Cohen 每一个性质均以 Wigner-Ville 分布的二维滤波形式得到[7]。三阶 Wigner-Ville 分布在文献[8]中得到介绍，它的产生和性质在文献[7,9,10-12]中得到研究。高阶 Wigner-Ville 分布描述了信号的高阶矩谱随时间的演变。

1 Wigner 谱及其高阶谱

1.1 Wigner 谱

设两个信号 $x(t)$ 和 $y(t)$，它们的 Wigner 互谱定义式为

$$W_{xy}(t,f) = \int x(t+\tau/2) y^*(t-\tau/2) e^{-j2\pi f\tau} d\tau$$
$$= \frac{1}{2\pi} \int X(f+\xi/2) Y^*(f-\xi/2) e^{j\xi t} d\xi$$

其中 $X(f)$ 和 $x(t)$ 是一对 Fourier 变换对。当 $x(t) = y(t)$，得到信号 $x(t)$ 的自相关 WS。在声纳中广泛应用的模糊度函数为 Wigner 谱的二维 Fourier 变换，定义式为

$$AF(\theta,\tau) = \int x^*(t-\tau/2) x(t+\tau/2) e^{-j2\pi\theta t} dt$$

模糊度函数可解释为联合时频分布相关函数。模糊度函数满足时频分布中的"相关化边缘特性"。它在时频分布中的 Cohen 类定义式为

$$W_c(t,f) = \iint \Phi(\theta,\tau) AF(\theta,\tau) e^{-j2\pi t\theta - j2\pi f\tau} d\tau d\theta$$

其中核函数 $\Phi(\theta,\tau)$ 取决于特定分布。

设 $x(n) = y(n) + z(n)$，它的 Wigner-Ville 分布遵从

$$W_{xx}(f,n) = W_{yy}(f,n) + W_{zz}(f,n) + W_{yz}(f,n) + W_{zy}(f,n)$$

由于 WS 是二次变换，两个信号的 Wigner 谱不是各自 Wigner 谱的和，而存在着交叉项。交叉项的存在使得 Wigner 谱的求解很困难。交叉项的抑制可通过滤波达到目的，在模糊度函数空间就是通过适当选择核函数。为了减少交叉项的影响，Choi 和 Williams 提出了利用核函数[13]

$$\Phi(\theta,\tau) = e^{-\theta^2\tau^2/\sigma^2}$$

其中参数 σ 控制交叉项的衰减量(交叉项的幅值正比于参数 σ),但是,随着交叉项抑制程度的加深,时频空间内的自相关项会发生污染或分辨率降低。

实际上,信号在时域上采样时,作者在采样频率分度上计算了 FT,为了满足连续时间 WS 在时间和频率的辨别,要求原始信号以 2 倍的 Nyquist 频率采样[10]。

离散时间算法在文献[5-6]中给出,瞬时互相关定义式为

$$r_{xy}(m,n) = x^*(n-m)y(n+m)$$

其中 n 为时间,m 为延迟,则互相关谱 WCS 的定义式为

$$W_{xy}(f,n) = \sum_m r_{xy}(m,n) e^{-j2\pi fm}$$

当 $x(n) = y(n)$ 时,得到 WS。为避免频率混叠,原始信号以 2 倍 Nyquist 或更快的速率采样。实际上频率变量 f 也是可分辨的,$f = k/K$,其中 K 控制着频域的分辨率。上式中 WCS 也可通过两次 FFT 算法得以实现[13]。两种方法的计算量和存储复杂度相同。

模糊度函数 AF 可通过下式计算:

$$AF(m,\theta) = \sum_n r_{xy}(m,n) e^{-2j\pi n\theta}$$

模糊度函数 AF 经 Choi-Williams 滤波器,

$$w(m,\theta) = e^{-(m\theta/\sigma)^2}$$

其二维 FT 产生滤波后的 WS,通过这种方法可有效抑制交叉项,但随之而来的是频率分辨率的降低,θ 是可辨别的频率分度。

1.2 Wigner 双谱

Wigner 谱的概念延伸至瞬态量的三次和四次乘积的 FT 就产生了 Wigner 双谱(WB)和 Wigner 三谱(WT),定义瞬态量的三次乘积为

$$r_3(t,\tau_1,\tau_2) = x^*(t-\alpha\tau_1-\alpha_2\tau_2)x(t+\beta\tau_1-\alpha\tau_2)x(t-\alpha\tau_1+\beta\tau_2)$$

式中,$\alpha = 1/2; \beta = 2/3$。WB 由下式给出:

$$W(t,f_1,f_2) = \iint r_3(t,\tau_1,\tau_2) e^{-j2\pi(f_1\tau_1+f_2\tau_2)} d\tau_1 d\tau_2$$

实际算法仅计算片 $f_1 = f_2$。很明显,频率轴尺度为原来的 2/3。像 WS 一样的,WB 也受到交叉项的困扰。通过适当的滤波尝试抑制交叉项。定义平滑核函数为

$$\Phi(\theta,\tau_1,\tau_2) = e^{-\theta^2(\tau_1+\tau_2)^2/\sigma}$$

滤波之后的 WB 定义为

$$W(t,f_1,f_2) = \iiiint r_3(t,\tau_1,\tau_2)\Phi(t,\tau_1,\tau_2) \\ \times e^{-j2\pi(f_1\tau_1+f_2\tau_2)} e^{-j2\pi t\theta} e^{-j2\pi u\theta} d\theta d\tau_1 d\tau_2 du$$

值得提出的是 Choi-Williams 滤波在 WB 中的应用不能保证自相关项完整，因此使用时要谨慎从事。

1.3 Wigner 三谱

定义瞬态量的四阶乘积为

$$r_4(t,\tau_1,\tau_2,\tau_3) = x^*(t-\tau)x(t-\tau+\tau_1)x(t-\tau+\tau_2)x(t-\tau+\tau_3)$$

其中
$$\tau = (\tau_1+\tau_2+\tau_3)/4$$

WT 的定义式是

$$W(t,f_1,f_2,f_3) = \iiint r_4(t,\tau_1,\tau_2,\tau_3) e^{-j2\pi(f_1\tau_1+f_2\tau_2-f_3\tau_3)} d\tau_1 d\tau_2 d\tau_3$$

注意到两项是同根的，同时注意信号的 $f_3\tau_3$ 项。实际上算法中计算片为 $f_1=f_2=-f_3=f$，也就是计算

$$W(t,f) = \iiint r_4(t,\tau_1,\tau_2,\tau_3) e^{-j2\Delta f(\tau_1+\tau_2+\tau_3)} d\tau_1 d\tau_2 d\tau_3$$

频率轴的尺度是原来的 1/2，WT 也同样受到交叉项的困扰。通过滤波抑制交叉项，平滑核函数定义式为

$$\theta(t,\tau_1,\tau_2,\tau_3) = e^{-\theta^2(\tau_1^2+\tau_2^2+\tau_3^2)/\alpha}$$

滤波后的 WT 定义式为

$$W(t,f_1,f_2,f_3) = \iiiint r_4(t,\tau_1,\tau_2,\tau_3)\theta(t,\tau_1,\tau_2,\tau_3) \\ \times e^{-j2\pi(f_1\tau_1+f_2\tau_2+f_3\tau_3)} e^{-j2\pi t\theta} e^{j2\pi u\theta} d\theta d\tau_1 d\tau_2 d\tau_3 du$$

实际上也仅计算片 $f_1=f_2=-f_3$，取片后 WT 的结果为实值。像 WS 交叉项的处理一样通过滤波加以抑制，但同时分辨率降低，但与 WB 不同的是，滤波没有引起自相关项的畸变[8,9]。

以信号

$$x(n) = \begin{cases} 5\sin(200\pi n/512), & n=1,2,\cdots,300 \\ 5\sin(200\pi n/512)+10\sin(800\pi nT/512), & n=301,302,\cdots,512 \end{cases}$$

为例说明 Wigner 三谱交叉项的抑制。$x(n)$ 是一个典型的非平稳信号，100 Hz 的频率成分贯穿整个周期，而 400 Hz 的频率成分则仅在每个周期的后半部分出现。

图 1 是直接三谱,出现了频率为 250 Hz 的干扰频率,图 2 经过滤波抑制交叉项的三谱真实地再现了信号的时频特性。总之,Choi-Williams 滤波器有助于 WS、WB 和 WT 中交叉项的抑制。但要注意在实际使用时这种方法会破坏 WB 中信号的自相关项。

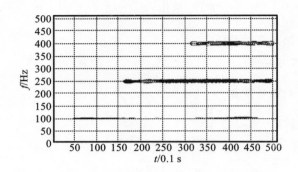

图 1 信号 $x(n)$ 的含交叉项的 Wigner 三谱

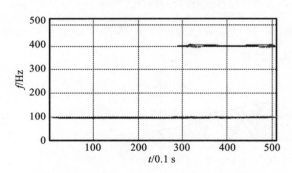

图 2 信号 $x(n)$ 的滤波后的 Wigner 三谱

2 实　　验

实验对象为 41∶37 的螺旋齿轮副,模数为 5 mm,箱盖顶端的加速率传感器用来测量齿轮传动箱声音响应,两个鉴相信号和两个红外线成影传感器,保证数据采集开始在同一对啮合齿准确啮合在同一阶段。实验的五对齿轮副为有缺损的 41∶37 螺旋齿轮副(模数为 5 mm)。信号均做了周期平均预处理。

五对齿轮副为:①低碳钢齿轮轮廓上有小剥落;②低碳钢齿轮轮廓上有严重剥落;③整个轮廓上均有剥落,但接触面上没有受损;④41 齿齿轮轮廓破损,但破损仅局限于齿顶部位,轮廓上整个接触线没有被削弱;⑤配对 37 齿的齿轮连续两个齿破损,41 齿齿轮轮廓也有轻微剥落。

限于篇幅,仅以 Wigner 三谱进行实验结果分析:上述五对齿轮副以转速 7 Hz、10 Hz 和 15 Hz 分别在无负载、负载为 40 N·m 和 80 N·m 情况下做九组实验。图 3～图 7 是 41 齿齿轮在 10 Hz 转速,80 N·m 负载情况下齿轮副 1～5 的 Wigner 三谱图。从图 3～图 7 可以看出:轻微剥落时,三谱分布主要为二倍频成分(图 3);而严重剥落时,三谱主频为三倍频,伴随频率有二倍频和介于三倍频与四倍频之间的干扰频率(图 4);齿轮剥落虽然发生在整个轮廓,但发生的部位在接触线下,于是三谱分布就类似于轻微剥落的频率,以二倍频为主,伴随频率有三倍频(图 5);而破损情况,齿轮的三谱分布主要表现在五倍频,伴随频率为二倍频(图 6);最后一例 41 齿齿轮轻微剥落,其三谱分布应该主要集中于二倍频,由于相配合的 37 齿齿轮出现严重破损,引起 41 齿齿轮三谱的变化:出现了基频、四倍频的伴随频率,以及三倍频的干扰频率(图 7)。

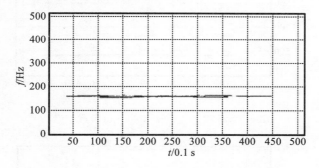

图 3　齿轮副 1,10 Hz,80 N·m 负载三谱

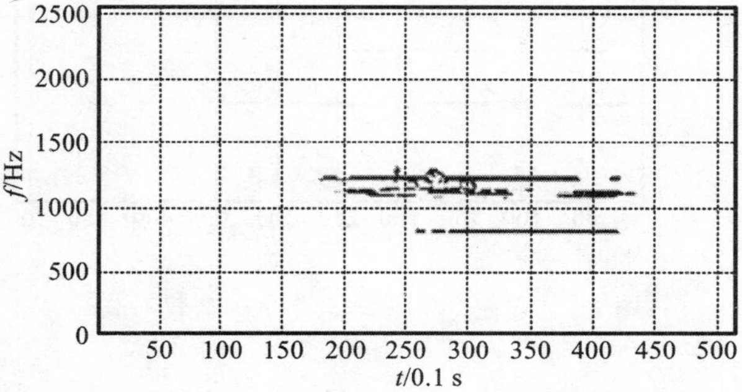

图 4　齿轮副 2,10 Hz,80 N·m 负载三谱

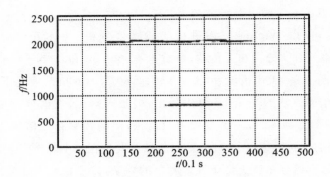

图 5　齿轮副 3,10 Hz,80 N·m 负载三谱

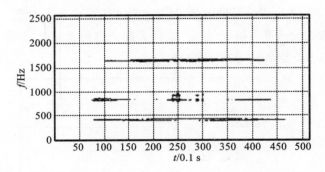

图 6　齿轮副 4,10 Hz,80 N·m 负载三谱

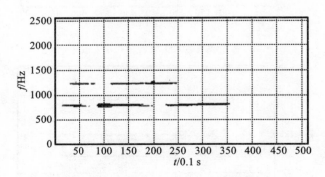

图 7　齿轮副 5,10 Hz,80 N·m 负载三谱

3　结　　论

本文介绍了 Wigner 谱及其高阶谱,并通过选择合适的核函数和 Choi-Williams 滤波器进行交叉项的抑制,达到滤波的目的,消除了非平稳过程中不同

频率混叠影响。最后在五对齿轮副上进行实验，以 Wigner 三谱为例，提取了信号的时频特征，很好地解释了齿轮副故障现象。抑制交叉项影响后的 Wigner 谱及其高阶谱可从各个角度对非平稳信号进行分析，真实再现了信号的频率成分随时间的演变过程，为特征提取、故障诊断提供了新的工具。

参 考 文 献

[1] 张贤达. 现代信号处理. 北京:清华大学出版社,1999:442-462.

[2] Swami A. *Some new results in higher-order statistics*. Proc Intl. Signal. Processing Workshop on Higher-Order Statistics,Chamrousse,France,1991;7:135-138.

[3] Mallat S. *A theory for multi-re solution signal representation:the wavelet transform*. IEEE Trans. PAMI,1989;11(7):674-693.

[4] Marple Jr. *A new autoregressive spectrum analysis algorithm*. IEEE Trans, ASSP,1990;8(28):441-450.

[5] Cohen L. *Time-frequency distributions:A review*. Proc. IEEE,1989;7:941-981.

[6] Hlaswatch F, Boudreaux-bartels G F. *Linear and quadratic time-frequency representations*. IEEE Signal Processing Magazine,1992;4:21-67.

[7] Fonollosa J R, Nikias C L. *Wigner higher-order moment spectra:definitions, properties,computation and application to transient signal detection*. IEEE Trans,SP,1993;7:842-853.

[8] Gerr N L. *Introducing a third_order Wigner distribution*. Proc IEEE,1988;3:290-292.

[9] Fonollosa J R, ikias C L. *Analysis of finite _ energy signals using higher-order moments and spectra-based time-frequency distributions*. Signal Processing, 1994; (363):315-328.

[10] Swami A. *System identification using cumalants:PhD dissertation*. University of Southern California,1988:107-108.

[11] Swami A. *Higher-order Wigner distributions*. In Proc. SPIE-92, Session on Higher-order and Time-Varying Spectral Analysis,1992;7(1770):290-301.

[12] Swami A. *Third order Wigner distributions*. Proc. I-CASSP-91,1991;3081-3084.

[13] Choi H, Williams W J. *In proved time-frequency repre-sentation of multicomponent signals using exponential kernels*. IEEE Trans ASSP,1989;7:862-871.

[14] Perry in F, Frost R. *A unified definition for the discrete-time, discrete-frequency wigner distribution*. IEEE Trans. ASSP,1986;8(34):858-867.

[15] Classen T A C M, Mecklenbra Ker W F G. *The aliasing problem in discrete-time wigner distributions*. IEEE Trans ASSP,1983;10:1067-1072.

Research of Wigner-ville Time Frequency and Application in Detecting Gear Pinion Fault

Lai Wuxing   Xuan Jianping   Shi Tielin   Yang Shuzi

Abstract   Because the Wigner-ville distribution has the advantages of so many time-frequency distributions, it attracts wide attention from engineering. But its shortcoming, cross-term, limits its' wildly application in industry. This paper introduces Wigner Bispectrum, Wigner Trispectrum and the how of choosing a kernel function to remove the affection of cross-term. Finally, an experiment on gears is described, which is successful in removing the cross-term with the above method for the fault diagnosis for gears.

Keywords: Fault diagnosis, Signal processing, Wigner-ville, Condition monitoring

(原载《振动工程学报》2003 年第 16 卷第 2 期)

先进制造技术及其发展趋势

杨叔子 吴波

提　要　分析了制造业（特别是装备制造业）在工业与国民经济中所占的重要地位，指出发展先进制造技术是我国目前紧迫的重大任务。指出现代制造业市场的特征、制造企业的特征和机械制造业的特征。从八个方面重点分析了先进制造技术的发展趋势和特色："数"是核心，"精"是关键，"极"是焦点，"自"是条件，"集"是方法，"网"是道路，"智"是前景，"绿"是必然。强调了这八个方面彼此渗透，相互支持，形成整体，并且扎根在"机械"与"制造"的基础上，服务于制造业的发展。

关键词： 制造，先进制造技术，装备，机械，信息化，数字化

0　前　言

"问渠哪得清如许？为有源头活水来。"一般认为，人类文明有三大物质支柱：材料、能源与信息。其实，应是四而非三，"制造"也应是一大支柱。可以说，没有"制造"，就没有人类。恩格斯在《自然辩证法》中讲得对："直立和劳动创造了人类，而劳动是从制造工具开始的。"的确，可形象地讲，人类是从制造第一把石刀开始的。

应该说，制造业是"永远不落的太阳"，是现代文明的支柱之一；它既占有基础地位，又处于前沿关键，既古老，又年轻；它是工业的主体，是国民经济持续发展的基础；它是生产工具、生活资料、科技手段、国防装备等及其进步的依托，是现代化的动力源之一。

值得特别提出的是机械制造业，尤其是装备制造业。马克思在《资本论》有一

---

* 国家自然科学基金中港联合科协资助项目（7001161949）。

段名言,至今仍熠熠生辉:"大工业必须掌握这特有的生产资料,即机器的本身,必须用机器生产机器。这样,大工业才建立起与自己相应的技术基础,才得以自立。"生产机器的机器,我国称为机床;英文叫 Machinetool,机器工具,有道理;德文叫 Workzeugmaschine,工具机器,更有道理。我们完全赞成这一讲法,机床制造业是装备制造业的心脏。可以说,没有制造业,就没有工业;而没有机械制造业,就没有独立的工业,即使制造业再大再多再好,也受制于人。也可以说,我国这么一个大国,如果没有强大的装备制造业,特别是同高科技相应的机床制造业,我国就不可能有独立自主的制造业与工业。例如,在信息化日益发展的今天,计算机、微电子产品在信息化中起着特别重大的作用,芯片的重要性不言而喻,而我国目前还不能生产芯片;预计 2005 年我国需要芯片 365 亿块,目前自给率小于20%,自主开发率约 5%,关键就是我国制造不出生产芯片的装备,形势是严峻的[8]。正因为制造业,特别是装备制造业如此重要,所以在党的十六大报告中,数次强调了制造业的发展,特别是装备制造业的发展。

制造业,装备制造业,绝不是"夕阳产业",但是,制造技术中确有"夕阳技术",这是同信息化大潮格格不入的技术,这是同高科技发展不相应的技术,这是缺乏市场竞争力的技术,甚至还可能是危害可持续发展的技术。我们所谓的"先进制造技术",其实就是"制造技术"加"信息技术"加"管理科学",再加上有关的科学技术交融而形成的制造技术。"先进制造技术"就是这么一个交融的技术,它生气勃勃地适应与占领现代制造业市场。

1 现代制造业的基本特征

现代的制造业市场大致有如下五个特征。第一是买方市场。这是科学技术与生产力发展的必然结果,"卖方市场"已成为过去。第二是多变性市场。由于科技发展快,技术更新快,产品换代快,如微电子产品半年到两年就得更新,从而产品非大量化、分散化、个性化的生产越来越强,市场越来越大,竞争日趋激烈,不确定因素猛增,市场变化很快。第三是国际化市场。市场打破国界,走向区域化,走向国际化,WTO 与各种区域经济组织应运而生而兴。第四是新兴产品市场。这不仅涉及对传统产品用高新技术加以改造与发展而成的产品,而且更涉及前所未有的新类型的"产品",从而导致如技术、软件、环保等产业的出现。特别在第三产业中更是如此。第五是虚拟市场。信息化的进一步是网络化。网上的产品广告、商品展示、商品交易、客户关系和代理制等均属于虚拟市场。

与此市场相应,制造业企业大致有如下六大特征。第一是满足"客户化"要求。这是最根本的,这是"买方市场"必然导致的结果,"顾客就是上帝",企业服务客户。第二是对市场的快速响应,对生产的快速重组,从而要求生产模式必须有高度柔性,有足够敏捷性,这是"客户化"必然导致的结果,而信息技术与管理科学为此提供了重要的保证。第三是既竞争又合作地参与市场,走向"双赢"、"多赢"。这是"纳什方程"给出的结果,而不一定是"鱼死"或"网破"或"两败俱伤"。网络化为此提供了更有利的条件。第四是本土化与国际化交互。走向全球化,既竞争又合作,自然导致朝这一方向发展。第五是应用虚拟技术。利用虚拟技术以加快企业有关活动的节奏,提高产品质量,节约成本,及时适应客观变化,这是实现以上各点之所需。第六是"以人为本",加强企业人文文化建设。应该说,这是现代企业成败要害之所在。在科技高度发达与快速发展的今天,如果只见"物"不见"人",只见"技术",不见"文化",不见"精神",必将导致企业走入"误区",遭到严重挫折乃至失败。

对机械制造业,特别是对装备制造业而言,除了上述六大特征外,还有以下四大变化。第一是产品本身的变化,"质"与"量"均如此。机械产品在性质上不仅取代、加强或延伸人的体力劳动,而且首先由于信息化,还有了一定"智能",即信息感知功能、信息处理功能、信息存储与显示功能以及功能整体的整合。机械产品在数量上,种类与品种日益繁多,可以说是"无所不包,无孔不入"。第二是增产方式的变化。过去以加大资金投入,加大资源消费,加大人力使用这种"粗放"方式实现增产,现在主要以开发"知识"资源这种"集约"方式作为主要增产方式来增产。第三是对产品要求的变化。开始是"物美"、"价廉",后来加上了"交货期短"、"服务好",现在还要加上"文化含量高",产品不仅是一个工业产品,还应是一个艺术产品,经得起"看"与"想"。第四是学科基础的变化。过去的基础,在理论上是靠力学,在实践上是凭经验,而现在是以多科学、新成就作为基础,而且正努力将制造技术发展与上升为制造科学,以利于制造技术进一步高质、高速、高效地发展[1]。

2 先进制造技术发展趋势

与科学技术和市场经济的发展相应,先进制造技术,特别是先进机械制造技术有如下八个方面的发展趋势与特色:"数"是核心,"精"是关键,"极"是焦点,"自"是条件,"集"是方法,"网"是道路,"智"是前景,"绿"是必然。现分述如下。

(1) "数"是发展的核心

"数"就是"数字化"。"数字化",数字地球,数字城市,数字工厂,数字制造,数字装备……狂澜巨浪,势不可当。"数字化"不仅是"信息化"发展的核心,而且也是先进制造技术发展的核心。信息的"数字化"处理同"模拟化"处理相比,有着三个不可比拟的优点:信息精确,信息安全,信息容量大[3]。

数字化制造就是指制造领域的数字化。它是制造技术、计算机技术、网络技术与管理科学的交叉、融合、发展与应用的结果,也是制造企业、制造系统与生产过程、生产系统不断实现数字化的必然趋势,它包含了三大部分:以设计为中心的数字制造,以控制为中心的数字制造和以管理为中心的数字制造。对制造设备而言,其控制参数均为数字化信号。对制造企业而言,各种信息(如图形、数据、知识和技能等)均以数字形式,通过网络,在企业内传递,以便根据市场信息迅速收集资料信息,在虚拟现实、快速原型、数据库和多媒体等多种数字化技术的支持下,对产品信息、工艺信息与资源信息进行分析、规划与重组,实现对产品设计和产品功能的仿真,对加工过程与生产组织过程的仿真,或完成原型制造,从而实现生产过程的快速重组与对市场的快速响应,以满足客户化要求。对全球制造业而言,用户借助网络发布信息,各类企业通过网络,根据需求,应用电子商务,实现优势互补,形成动态联盟,迅速协同设计并制造出相应的产品。这样,在数字制造环境下,在广泛领域乃至跨地区、跨国界形成一个数字化组成的网,企业、车间、设备、员工、经销商乃至有关市场均可成为网上的一个"结点";在研究、设计、制造、销售和服务的过程中,彼此交互,围绕产品所赋予的数字信息,成为驱动制造业活动的最活跃的因素。在此还应着重指出,制造知识(包括技能、经验)的获取、表达、存储、推理乃至系统化、公理化等,这是使制造技术发展到制造科学的关键[1],而这又与数字化不可分开。

(2) "精"是发展的关键

"精"是"精密化"。它一方面是指对产品、零件的精度要求越来越高,一方面是指对产品、零件的加工精度要求越来越高。显然,这两方面是一回事;有了前者,才要求有后者;有了后者,才促使前者得以发展。"精"可以说是指加工精度及其发展,精密加工、细微加工、纳米加工等。20 世纪初,超精密加工的误差是 10 μm,30 年代达 1 μm,50 年代达 0.1 μm,70 至 80 年代达 0.01 μm,至今达 0.001 μm,即 1 nm。再由一组数据,可以看到微电子产品对加工精度的依赖程度,电子元件制造误差为,一般晶体管 50 μm,一般磁盘 5 μm,一般磁头磁鼓 0.5 μm,集成电路 0.05 μm,超大型集成电路达 0.005 μm,而合成半导体为 1 nm。

在现代超精密机械中,对精度要求极高,如人造卫星的仪表轴承,其圆度、圆柱度、表面粗糙度等均达纳米级;基因操作机械,其移动距离为纳米级,移动精度为 0.1 nm。细微加工、纳米加工技术可达纳米以下的要求,如离子束加工可达纳米级,借助于扫描隧道显微镜(STM)与原子力显微镜的加工,则可达 0.1 nm[4,5]。实际上,纳米级的加工就是移动原子级的加工。

至于微电子芯片的制造,有所谓的"三超"[5]:①超净,加工车间尘埃颗粒直径小于 1 μm,颗粒数少于每 6.45 cm^2 0.1 个。②超纯,芯片材料中的有害杂质的质量分数小于 10^{-10},即十亿分之一。③超精,加工精度达纳米级。显然,没有先进制造技术,就没有先进电子技术装备;当然,没有先进电子技术与信息技术,也就没有先进制造装备。先进制造技术与先进信息技术是相互渗透、相互支持、紧密结合的。

(3)"极"是发展的焦点

"极"就是极端条件,就是指在极端条件下工作的或者有极端要求的产品,从而也是指这类产品的制造技术有"极"的要求。在高温、高压、高湿、强磁场和强腐蚀等条件下工作的,或有高硬度、大弹性等要求的,或在几何形体上极大、极小、极厚、极薄、奇形怪状的。显然,这些产品都是科技前沿的产品。其中之一就是"微机电系统(MEMS)",这是工业发达国家与有关国家所高度关注的一项前沿科技。甚至可以说,"极"是前沿科技或前沿科技产品发展的一个焦点。例如:在信息领域中,分子存储器、原子存储器、量子阱光电子器件、芯片加工设备;在生命领域中,克隆技术、基因操作系统、蛋白质追踪系统、小生理器官处理技术、分子组件装配技术;在军事武器中,精确制导技术、精确打击技术、微型惯性平台、微光学设备;在航空航天领域中,微型飞机、微型卫星、"纳米"卫星(0.1 kg 以内);在微型机器人领域中,脑科手术、清除脑血栓、管道内操作、窃听与收集情报、发现并杀死癌细胞;微型测试仪器中的微传感器、微显微镜、微温度计、微仪器等。MEMS 可以完成特种动作与实现特种功能,乃至可以沟通微观世界与宏观世界,其深远意义难于估量。2002 年,美国伯克利大学不仅制造了直径为 300 μm 的镜头,配以微米级探针的微米级显微镜,可深入植物细胞内观察,而且正在开发镜头直径为 500 nm 的纳米级显微镜。2002 年美国康纳尔大学还宣布研制出原子级纳米"晶体管",可以说,由单个原子输送电流的"晶体管"还是首次,这项成果被我国科技专家评为2002年世界十大科技新闻之一。

(4)"自"是发展的条件

"自"就是自动化。它就是减轻人的劳动,强化、延伸、取代人的有关劳动的技术或手段。显然,自动化是重要的,自动化总是伴随有关机械或工具来实现的。可以说,机械是一切技术的载体,也是自动化技术的载体。第一次工业革命,以机械化这种形式的自动化来减轻、延伸或取代人的有关体力劳动。第二次工业革命,电气化进一步促进了自动化的发展。据统计,从1870年至1980年,加工过程的效率提高为20倍,即体力劳动得到了有效的解放;但管理效率只提高了1.8~2.2倍,设计效率也只提高了1.2倍,这表明自动化为体力劳动所带来的效果是非常明显的。即使在美国,1984年,CAD在福特公司的应用只占40%,通用公司占34%,Chrysler为67%[2],此后,CAD发展极为迅速。今天在我国,CAD已十分普及。信息化、计算机化与网络化,不但极大地解放了人的体力劳动,而且更为关键的是有效地提高了脑力劳动自动化的水平,解放了人的部分脑力劳动。

"自动化"从自动控制、自动调节、自动补偿和自动辨识等发展到自学习、自组织、自维护和自修复等更高的自动化水平,而且今天自动控制的内涵与水平已是今非昔比,从控制理论(如多Agent系统的理论与方法、基于网络的控制理论、复杂系统的控制理论……)、控制技术(如智能化检测、多媒体信息检测……)、系统(如网络控制系统、复杂系统……)、控制元件(如具有生物特征的传感元件……)[9],都有着极大的发展。自动化是先进制造技术发展的前提条件。

(5)"集"是发展的方法

"集"就是集成化。它有三个方面:技术的集成,管理的集成,技术与管理的集成。归根结底,其本质就是知识的集成,当然亦即知识表现形式的集成。已如前述,先进制造技术就是制造技术、信息技术、管理科学与有关科学技术的集成。"集成"就是"交叉",就是"杂交",就是取人之长,补己之短,这是发展的一大方法。目前,"集"主要指以下三个方面。①现代技术的集成,机电一体化是个典型,它是高技术装备的基础,如微电子制造装备,信息化、网络化产品及配套设备,仪器、仪表、医疗、生物和环保等高技术设备。显然,在机电一体化技术中,关键往往是:(a)检测传感技术,(b)信息处理技术,(c)自动控制技术,(d)伺服传动技术,(e)精密机械技术,(f)系统总体技术。而这些技术又同许多学科有关,又是一个"集"。②加工技术的集成,特种加工技术及其装备是个典型,如增材制造(即快速原型)、激光加工、高能束加工和电加工等;当然,加工技术的集成只是现代技术集成的一个特殊部分。③企业集成,即管理的集成,包括生产信息、功能、过程的集成;包括生产过程的集成、全寿命周期过程的集成;也包括企业内部的集成,企业外部的集成。如并行工程、敏捷制造、精益生产和CIMS等都是"集"的典型表现。当然,管理的集成不可能不包含管理与技术的集成。

从长远看,还有一点很值得注意,即由生物技术与制造技术集成的"微制造的生物方法",或所谓的"生物制造";即依据生物是由内部生长而成的"器件",而非同一般制造技术那样由外加作用以增减材料而成"器件"。可以预期,这是一个崭新的充满着活力的领域,作用难以估计,道路也将是漫长的。

(6) "网"是发展的道路

"网"就是网络化。应该讲,制造技术的网络化是先进制造技术发展的必由之路,使制造业走向整体化、有序化,这同人类社会发展是同步的。制造技术的网络化是由两个因素决定的:一是生产组织变革的需要;二是生产技术发展的可能。这是因为制造业在市场竞争中,面临多方的压力,如采购成本不断提高、产品更新速度加快、市场需求不断变化、客户订单生产方式迅速发展、全球制造所带来的冲击日益加强等;企业要避免传统生产组织所带来的一系列问题,必须在生产组织上实行某种深刻的变革。这个变革体现在以下两方面。一方面利用网络,在产品设计、制造与生产管理等活动乃至企业整个业务流程中充分享用有关资源,即快速调集、有机整合与高效利用有关制造资源。另一方面,与此同时,这必然导致制造过程与组织的分散化、网络化,企业要抛弃传统的"小而全"与"大而全"这类"夕阳技术",而集中力量在自己最有竞争力的核心业务上。一个企业有无自己最有竞争力的核心业务,这是关键,"山不在高,有仙则名;水不在深,有龙则灵"。而科学技术特别是计算机技术、网络技术的发展,使得生产技术发展到可以使这种变革的需要成为可能[6]的地步。

在制造技术网络化中,值得关注的是电子商务的应用。电子商务是将业务数据数字化,并将数字信息的使用和计算机的业务处理同Internet进行集成,成为一种全新的业务操作模式。在电子商务的网络化制造中,供应链管理、客户关系管理、产品生命周期管理共同构成了制造的增值链。它具有两大优点:商务的直接化与透明化,这对降低成本、加快流通、提高效率、增加商业机会大有好处,从而对企业内部重组、经营战略与竞争模式有着深刻影响。但是,中国科学院2002年5月《发展我国电子商务的对策研究》咨询报告中指出,我国在电子商务的应用上,还存在一系列问题,使其没有得到应有的应用。这些问题大致是:在宏观层面上,不够统一;在企业层面上,"用""体"分离;在社会服务体系上,服务滞后;在软环境上,商务活动缺法又乏诚;在商业模式上,电子商务规模大于其效益,形成泡沫;而在基础设施上,又十分缺乏,形成瓶颈。

制造技术的网络化不可阻挡,它的发展会导致一种新的制造模式,即虚拟制造组织,这是由地理上异地分布的、组织上平等独立的多个企业,在谈判协商的基础上,建立密切合作关系,形成动态的"虚拟企业"或动态的"企业联盟"。此时,各企业致力于自己的核心业务,实现优势互补,实现资源优化动态组合与共享。正因为如此,对我国而言,大力发展"中场产业",使之具有精湛的最强有力的核心业务,不失为发展机械制造业的重要战略之一。

(7) "智"是发展的前景

"智"就是智能化。制造技术的智能化是制造技术发展的前景。近20年来,制造系统正在由原先的能量驱动型转变为信息驱动型,这就要求制造系统不但要具备柔性,而且还要表现出某种智能,以便应对大量复杂信息的处理、瞬息万变的市场需求和激烈竞争的复杂环境,因此,智能制造越来越受到高度的重视[3]。

智能化制造模式的基础是智能制造系统,智能制造系统既是智能和技术的集成而形成的应用环境,也是智能制造模式的载体。与传统的制造相比,智能制造系统具有以下特点:①人机一体化,②自律能力,③自组织与超柔性,④学习能力与自我维护能力,⑤在未来,具有更高级的类人思维的能力。由此出发,可以说智能制造作为一种模式,是集自动化、集成化和智能化于一身,并具有不断向纵深发展的高技术含量和高技术水平的先进制造系统,也是一种由智能机器和人类专家共同组成的人机一体化系统。它突出了在制造诸环节中以一种高度柔性与集成的方式,借助计算机模拟的人类专家的智能活动,进行分析、判断、推理、构思和决策,取代或延伸制造环境中人的部分脑力劳动。同时,收集、存储、处理、完善、共享、继承和发展人类专家的制造智能;当然,目前还只能算初步,但潜力极大,前景广阔。

随着知识经济时代的初露端倪,知识将作为发展生产力主要的源泉,并导致以知识生产率取代劳动生产率,从而智能化制造的价值日益攀升。目前,特别是分布式数据库技术、智能代理技术和网络技术等发展,将突出知识在制造活动中的价值地位。知识经济是继工业经济后的主体经济形式。尽管智能化制造道路还很漫长,但是必将成为未来制造业的主要生产模式之一。

(8)"绿"是发展的必然

"绿"就是"绿色","绿色"是从环境保护领域中引用来的。

人类社会的发展必将走向人类社会与自然界的和谐,就是走向"天人合一"。人与人类社会本质上也是自然世界的一个部分,部分不能脱离整体,更不能对抗与破环整体。《老子》讲的"无为"就是这个意思,即不去"为"违背客观规律之"为"。人类必须从各方面促使人与人类社会同自然界和谐一致,制造技术也不能例外。江泽民同志讲得好:保护环境,就是保护生产力;改善环境,就是发展生产力。

制造业的产品从构思开始,到设计阶段、制造阶段、销售阶段、使用与维修阶段,直到回收阶段、再制造各阶段,都必须充分计及环境保护。所谓环境保护是广义的,不仅要保护自然环境,还要保护社会环境、生产环境,还要保护生产者的身心健康。在此前提与内涵下,还必须制造出价廉、物美、供货期短、售后服务好的产品。作为"绿色"制造,产品还必须在一定程度上是艺术品,以与用户的生产、工作、生活环境相适应,给人以高尚的精神享受,体现着物质文明、精神文明与环境文明的高度交融。每发展与采用一项新技术时,应站在哲学高度,慎思"塞翁得马,安知非祸",即必须充分考虑可持续发展,计及环境文明。制造必然要走向"绿色"制造。

3 结 论

综上所述,数、精、极、自、集、网、智、绿这八个方面反映了先进制造技术发展的基本特点,从制造业、特别是机械制造业的发展来看,这八个方面应该是彼此渗透,相互依赖,相互促进,形成一个整体。而且,它们是服务于制造技术的,此即,"机械"是基,"制造"是础,这八个方面是一定要扎报在"机械"和"制造"这个基础上;这就是说,要研究与发展"机械"本身与"制造"本身的理论与机理,而且这八个方面的技术要以此理论与机理为基础来研究、开发、发展,要与此基础相辅相成,最终是要服务于制造业的发展。离开"机械"与"制造"的本身去研究、开发、发展这八个方面的技术,都是迷失了方向的。

同时,还值得高度重视的是,在科学技术高度发达与高速发展的今天,在先进制造技术迅速发展的今天,应深深了解"先进制造技术"如同一切先进技术一样,是不可能不"以人为本"的,不能见"物"不见"人",见"技术"不见"文化"、不见"精神";离开人,离开人文文化,离开人的精神,先进技术就失去了"灵魂",只是一个空躯壳,甚至造祸于民。我们应记住江泽民同志在"七一"讲话中所作出的深刻论断:"人是生产力中最具有决定性的力量。"进一步而言,要"以人为本",就必须"教育先导",就必须通过各种形式的教育,培养出合乎时代潮流与我国国情的制造业的科技人才与管理人才。人才是根本,教育是基础。根本如无,树凋木枯;基础不牢,地动山摇;根本深固,树荣木绿;基础坚牢,大厦凌霄。总之,要从根本、从长远着想。

在此应感谢南京航空航天大学朱剑英教授,文中某些资料是参考或采用了他在 2000 年"高等学校机械工程教学指导委员会会议"(昆明)上所做的专题报告。

参 考 文 献

[1] 杨叔子,熊有伦.重视制造科学的研究.科学时报.1999,7,14.

[2] 杨叔子.知识经济·高新科技·历史责任.中国机械工程,1999,10(3):241~246.

[3] 杨叔子,熊有伦,管在林,等.信息时代和网络条件下的制造业发展前景.湖北省 2001 年科学论坛论文集,湖北省科协,2001:5~9.

[4] 朱剑英.机械工程科学前沿与发展的思考(2).机械制造与自动化,2001.2:1~3.

[5] 朱剑英.机械工程科学前沿与发展的思考(4).机械制造与自动化.2001,4:1~7.

[6] 杨叔子,吴波,程涛.网络经济时代的制造企业策略.技术科学发展与展望——院士论技术科学(2002 年卷),山东教育出版社,2002:391~398.

[7] 宋健.制造业与现代化.机械工程学报,2002,38(12):1~9.

[8] 雷源忠,雒建斌,丁汉,等.先进电子制造中的重要科学问题.中国科学基金,2002,16(4):204~709.

[9] 王成红.关于自动化领域中若干基础科学问题的思考.中国科学基金,2002,16(4):227~230.

Trends in the Development of Advanced Manufacturing Technology

Yang Shuzi　Wu Bo

Abstract　With some analysis on the fundamental and important position that manufacturing industry (especially the equipment manufacturing industry) holds in the whole of the nation's industrial and economic lives, the present compelling importance is pointed out for the development of (AMTs). The trends in the market of modern manufacturing enterprises are then described, together with some challenges to manufacturing enterprises themselves as well as changes in the whole machine-building industry. Following these, efforts are put on the addressing of the new trends and characteristics in the development of AMTs, which asserts in the following 8 aspects that for the development of AMTs, digitalization is the core, precision holds the key, extremes in functions become the focus, automatization keeps the precondition, integration provides the technique, networking paves the pathway, intelligence forms the prospects, and sustainability stems a necessity. It is finally emphasized that all the 8 factors above overlap and support each other to form the whole, and they take roots on the basis of "machine" and "manufacturing" so as for serving the development of manufacturing industry.

Keywords: Manufacturing, Advanced manufacturing technology, Equipment machine, Informationization digitalization

(原载《机械工程学报》2003 年 10 期)

Feature Extraction and Classification of Gear Faults Using Principal Component Analysis

Li Weihua   Shi Tielin   Liao Guanglan   Yang Shuzi

Abstract   Feature extraction is a key issue to machine condition monitoring and fault diagnosis. The features must contain the necessary discriminative information for the fault classifier to have any chance of accurate classification. This paper presents a study that uses principal component analysis to reduce dimensionality of the feature space and to get an optimal subspace for machine fault classification. Industrial gearbox vibration signals measured from different operating conditions are analyzed using the above method. The experimental results indicate that the method extracts diagnostic information effectively for gear fault classification and has a good potential for application in practice.

Keywords: Fault analysis, Classification, Machinery, Analytical methods, Gears, Condition monitoring

Practical Implications

Condition-based maintenance is driven by the technical condition of the equipment. Under this strategy, all major parameters are considered in order to determine the technical condition with maximized accuracy. For this reason detailed information via diagnostic methods should be available. The principal component analysis (PCA) is a technique of linear statistical predictors that has been applied in various fields of science. It can be used as a basis for classification of variables, outlier detection, and early indication of abnormality in data structure. This paper shows how PCA can be used for building a model to identify machine failure and monitor the variation of machine working conditions.

Introduction

Fault diagnosis has received intensive study for several decades. Various approaches have been taken. In cases where the process can be represented by a suitable dynamics model either in the form of state-space, model-based signal detection and estimation approach can be applied (Isermann, 1984; Liu and Si, 1997). Besides, non-model-based methods such as fault tree, knowledge-based system, pattern recognition, and artificial neural networks are also used. (McKeever et al., 1986; Gu and Ni, 1996; McCormick and Nandi, 1997). Among these non-model-based fault diagnosis methods, pattern recognition method provides a systematic approach to acquiring knowledge from fault samples. In fact, mechanical fault diagnosis is essentially a problem of pattern recognition, in which feature extraction plays an important role (Koller and Sahami, 1996; Strackeljan, 1999), as shown in Fig. 1.

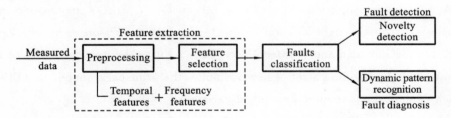

Fig. 1 Key techniques in machine condition monitoring

Features are any parameters extracted from the measurements through signal processing in order to enhance damage detection. However, the measured industrial signals are always complex because the random noises in an industrial environment degrade the signal-to-noise ratio greatly. Feature selection involves a trade-off between the computational feasibility associated with low-level features and extensive pre-processing required for high-level features. Feature extraction includes either signature or advanced signature analysis. Signature analysis employs simple feature extraction methods, based on data reduction procedures, which lead to scalar representations.

There are many techniques for extracting indicators of a machine's condition from the vibration signals. In order to capture the diagnostic information, features were computed, as many as possible. But the choice of features is often arbitrary, which will lead to situations where several features provide the same information, as well as some features providing no useful information at all. The additional burden of computing these features may increase the learning cost and affect real-time application of the condition monitoring system. So feature selection is helpful to reduce dimensionality, discard deceptive features and extract an optimal subspace from the raw feature space. It is critical to the success of fault recognition and classification.

One way of automatically extracting features is to use the well-known technique of PCA. PCA is a technique to approximate original data with lower dimensional feature vector, which is also known as Karhunen-Loeve transform in pattern recognition. It was first proposed in 1933 by Hotelling(Jackson, 1991)in order to solve the problem of decorrelating the statistical dependency between variables in multivariate statistical data derived from exam scores. Since then, PCA has found applications in statistical analysis, process monitoring and diagnosis, pattern recognition, and so on(Jackson, 1991; Wilson et al. , 1997; Jia et al. ,1998;Goodman and Hunter,1999). PCA reduces the dimensionality of the original data by defining a set of new variables, the principal components (PCs), which explain the maximum amount of variability in the data.

In this paper, PCA is used as not only a tool for feature space dimensionality reduction but also a classifier to distinguish one gearbox running condition from another. In addition, PCA is also useful as a visualization tool for monitoring the variation of gearbox operating conditions. This paper is organized as follows: first, a brief overview of PCA is given. Second, the gear failure experiment is described and experiment investigation of PCA-based fault classification is presented. Finally, the analysis results, discussion and conclusions are drawn.

PCA

PCA is a method for simple identification and classification that makes it possible to deal with data sets made by a large number of noisy and highly correlated process measurements. The basic concept behind PCA is to project the dataset onto a subspace of lower dimensionality. In this reduced space, the data are represented with removed or greatly reduced collinearity. PCA achieves this objective by explaining the variance of the original data matrix $X_{m \times n}$ that contains m observations of n variables ($m > n$) in terms of a new set of independent variables: the PCs. In this paper we will not discuss the mathematical details of PCA. A complete treatment is given by Morison(1976), whereas introduction and review of applications of PCA in process systems engineering are given by Kresta et al. (1991). These references also contain details about the PCA algorithm (Jackson, 1991). Here it suffices to say that after the PCA is carried out, the original data set X is finally expressed as a linear combination of orthogonal vectors along the directions of the PCs:

$$X = TP^T + E = t_1 p_1^T + t_2 p_2^T + \cdots + t_a p_a^T + E \tag{1}$$

Where T is the matrix of the principal component scores, P is the loading matrix and E is the residual matrix. Ideally the dimension $a \ll n$ is chosen such that there is no significant process information left in E. The matrix E represents random error and adding an extra PC would only fit the noise, consequently increasing the prediction error.

The loading vectors are orthogonal each other and subject to $|p_i| = 1$, which means that $p_i^T p_j = 0$ ($i \neq j$), $p_i^T p_j = 1$ ($i = j$). The PCs define a new orthogonal basis for the observation space of X. Each observation x_i is located on the PC subspace by score vectors t_i ($i = 1, 2, \cdots, a$). The elements of a score vector are the distances from the origin of the subspace along each PC. The principal component scores t_i are calculated as the product of the loading vector p_i and the actual observation, and can be expressed as follows:

$$t_i = Xp_i \quad (i = 1, 2, \cdots, a) \tag{2}$$

The first principal component defines the maximum variance in the data, with subsequent components explaining decreasing levels of variability. The assumption is made that large variances relate to some kind of structure in the feature space and the remaining components can be discarded with little loss of accuracy. The number of PCs is required to account for most of the variation (>85 per cent) in the data, and it can be assessed using a number of techniques (Jackson, 1991). Typically, cross-validation is employed (Wold, 1991). In practice two or three PCs are often sufficient to explain most of the predictable variations in the process.

One of the most popular approaches for calculating PCs is singular value decomposition(SVD) (Jackson, 1991). In fact, it can be performed easily by solving an eigenvalue problem of covariance matrix:

$$C = \frac{1}{m} \sum_{i=1}^{m} (x_i - \overline{x})(x_i - \overline{x})^T \tag{3}$$

Where $\overline{x}$ is the mean value of x_i.

To do this, the eigenvalue equation should be solved:

$$\lambda v = Cv \tag{4}$$

Where $\lambda \geq 0$ is eigenvalue of covariance matrix C, and v eigenvector correspondingly. The new coordinates in the eigenvector basis, i. e. the orthogonal projections onto the eigenvectors, are PCs, which are named "new features" in this study.

In this work, PC modeling is used to summarize the correlation structure in the feature datasets, both in relation to time and among the features. Two and three dimensional plots, "windows into the multidimensional feature space", can be constructed and used for condition monitoring. PC models can be used for online to visualize the current state of the process in relation to a background of known process states. The primary requirement for the development of a PCA model of a gearbox is the acquisition of training data when the gearbox is running under normal operating conditions. Based on historical data collected when the gearbox was running under normal conditions, a PCA model can be established, and then future behavior can be referenced against this monitoring model. Raw feature sets of different conditions can be projected onto the plane defined by the PCA loading vectors to obtain new features. When a fault is developing, a significant deviation from this reference model will occur. According to these new features, we can monitor the variation of gearbox operating conditions.

Case study - gear fault classification

Experiments setup

The experiments were carried out on an automobile gearbox(model: 6J90T) made by Shannxi Vehicle Gear Manufacturing Plant. The simplified gearbox transmission graph is shown in Fig. 2, in which the power transmission path is indicated by the dashed line. The input torque was 882 N · m, the input power 193.96 kW, and input rotating speed was 1270 r/min. Some working parameters of the gearbox are listed in Tab. 1. An acceleration sensor(model: YD42) was mounted externally on the gearbox bearing case for monitoring the operating conditions of the gearbox. During the experiment, the measured signals were first amplified and filtered by a B&K amplifier(model: 2626), then transmitted to a pocket computer through an analog-to-digital converter. The sampling rate was 12.5 kHz.

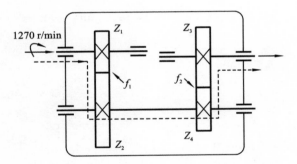

Fig. 2 Simplified diagram of gearbox transmission

Tab. 1 **Gearbox parameters**

	First stage	Second stage
Number of teeth	$Z_1/Z_2 = 26/43$	$Z_3/Z_4 = 36/24$
Module	5 mm	4.25 mm
Pitch angle	18°	14°
Meshing frequency	$f_1 = 550.33$ Hz	$f_2 = 307.16$ Hz

The test lasted about 182 hours until a tooth broke in the gearbox. During the testing process, the gearbox's running condition undergo three different stages gradually. First, the gearbox was running under normal conditions. Then a crack occurred in one tooth root and propagated gradually. At last, the tooth of the meshing gear with 24 teeth was broken when the gearbox was stripped. For more details of the experiment one may refer to Zhang(1993).

Gearbox operating conditions feature space composition

Many feature parameters have been defined in the pattern recognition field. Here, ten of them, which are usually used for the failure diagnosis of plant machinery, are denoted as follows:

$$p_1 = \sigma = \sqrt{\frac{1}{N}\sum_{i=1}^{N}(x_i - \overline{x})^2} \quad \text{(standard variation)} \tag{5}$$

Here, $x_i(i = 2, N)$ is a raw vibration signal and $\overline{x} = \frac{1}{N}\sum_{i=1}^{N} x_i$ (mean value).

$$p_2 = \sum_{i=1}^{N}(x_i - \overline{x})^3/(N\sigma^3) \quad \text{(skewness)} \tag{6}$$

$$p_3 = \sum_{i=1}^{N}(x_i - \overline{x})^4/(N\sigma^4) \quad \text{(kurtosis)} \tag{7}$$

$$p_4 = \hat{x} = \max(x_i) \quad \text{(the maximum peak value)} \tag{8}$$

$$p_5 = \overline{x}_{abs} = \frac{1}{N}\sum_{i=1}^{N}|x_i| \quad \text{(absolute mean value)} \tag{9}$$

$$p_6 = x_{rms} = \sqrt{\frac{1}{N}\sum_{i=1}^{N}x_i^2} \quad \text{(root mean square value)} \tag{10}$$

$$p_7 = C = \hat{x}/x_{rms} \quad \text{(crest factor)} \tag{11}$$

$$p_8 = W = x_{rms}/\overline{x}_{abs} \quad \text{(shape factor)} \tag{12}$$

$$p_9 = I = \hat{x}/\overline{x}_{abs} \quad \text{(impulse factor)} \tag{13}$$

$$p_{10} = L = \hat{x}/\left(\frac{1}{N}\sum_{i=1}^{N}\sqrt{|x_i|}\right)^2 \quad \text{(clearance factor)} \tag{14}$$

These parameters are non-dimensional and are used to construct "feature space" in this work. The example in the next section shows that the proposed PCA-based approach is capable of finding or generating the most discriminatory features from feature space and effectively identifying failure states of the gearbox in all cases.

Experimental analysis

Typical raw signals of different conditions were measured online from outside of the gearbox at a constant sampling rate 12.5 kHz (normal condition, tooth cracked condition, and tooth broken condition), giving 28 datasets per condition. Raw vibration signals are displayed in Fig. 3. Obviously there are distinct differences between the vibration signal of the gearbox with a broken tooth and those under the other two operating conditions, but it is very difficult to separate the signal under normal conditions from that under tooth cracked conditions.

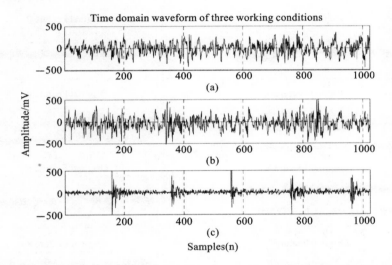

Fig. 3 Gear vibration signals of three conditions: (a) normal; (b) tooth cracked; and (c) tooth broken

Ten feature parameters of each raw signal were computed, and then 84 feature sets were divided into two groups, one group of 23 feature sets per condition is used for training, and the other one of five feature sets per condition is used for testing. None of ten features can be used to distinguish the tooth cracked condition from the normal one, as shown in Fig. 4. Then the normal condition features were normalized and trained to set up the PC monitoring model, the principal component subspace, which can detect the abnormal disturbance.

In this work, the PCA used the SVD to determine the PCs. Initially, a model was developed from normal condition feature set using k PCs and then the feature set was decomposed as:

$$X = TP^{\mathrm{T}} \tag{15}$$

where X is the raw feature sets, T is score matrix and P is the loading matrix which are the PCA model parameters. The loading matrix for normal conditions plays an important role in detecting abnormalities or faults. By projecting the feature set of different conditions onto the loading vectors, new features can be obtained and the detection of the faults is achieved. Contribution of four PCs is shown in Fig. 5, variation tendency of these new features is shown in Fig. 6, and gear fault classification is shown in Fig. 7.

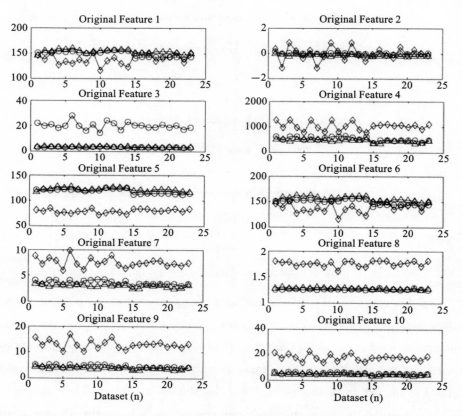

Fig. 4　Variation tendency of original features in three working modes, the "circle" represents normal conditions, the "triangle" tooth cracked and the "diamond" tooth broken

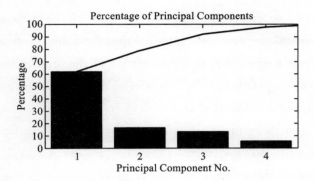

Fig. 5 Contributions of principal components

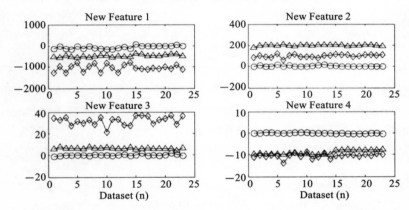

Fig. 6 Variation tendency of new features in three working modes, the "circle" represents normal conditions, the "triangle" tooth cracked and the "diamond" tooth broken

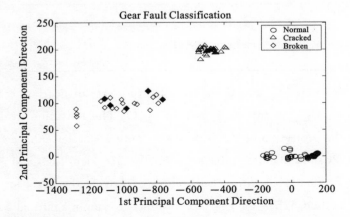

Fig. 7 Gear fault classification, the solid points are projection of test datasets

From Fig. 5, we can easily find that the sum of the variances of the first three PCs exceeds 85 per cent, which means that the subspace composed of those three PCs contains enough variation information of the original features, and this PC subspace can be regarded as the classifier to detect the damage of gear during working and separate one kind of fault from another.

From the first three principal component vectors, we can find which of the raw features contributed most to the PCs. They are listed in Tab. 2 and ordered by their weights in the relative PC. The largest weights in the first principal component are features p_4, p_1, corresponding to maximum peak value, standard variation, and features p_6, p_5, p_1 for the second PC, p_5, p_6, p_1 for the third PC, where p_5 is absolute mean value and p_6 is root mean square value. All the elements of the principal component have the same sign, making it a weighted average of all the features. Although these original features contribute most to the PCs, they are not discriminatory and efficient for recognizing gear tooth cracked running condition(see Fig. 4), whereas the three new features acquired by PCA are capable of identifying this condition(see Fig. 6). The fourth new feature is unable to separate tooth cracked condition from normal condition, which proved that adding an extra PC would only fit the noise.

Tab. 2 Features contributing most to the first three PCs

Principal component	Contributed features
1	p_4, p_1
2	p_1, p_6, p_5
3	p_5, p_6, p_1

In the two dimensional PC subspace, feature sets of gearbox conditions are classified into three clusters as shown in Fig. 7. The reduction in dimension is obtained by the projection of the feature sets onto a smaller subspace defined by the selected PCs. In the PC subspace, clusters are identified, and each cluster represents a particular condition. Such clusters provide the basis for a condition recognition method. When the gearbox continues to run in a normal condition, the feature points are bounded around the origin with little fluctuation as shown in Fig. 7. However, when a local crack is beginning to develop in a single tooth, the value of two new features increases. The feature points appear to be consistent in the same region with a growing crack. With the crack propagating, the value of the first new feature increases extremely and spreads to a much wider area, which indicates that the tooth is broken. The solid points in Fig. 7 are projection of test feature sets onto the PC subspace, and it is successful to classify three different working modes. The new features can also be used to input to a classifier, such as neural networks, for setting up an automatic condition monitoring system for machinery. When information about various faults is available, it is possible to develop the model for fault diagnosis.

Discussion

It needs to be pointed out that PCA has the advantages of having a closed-form solution (allowing an algebraic solution) and of automatically ranking the importance of the features in the projection space (PCs subspace). Over other methods, such as projection pursuit (Huber, 1985), and independent component analysis (Hyvarinen, 1999), which require exploratory analysis or nonlinear optimization and may be inexact or extremely computational intense, PCA has been chosen as the representative method here due to its straightforward interpretation and its relatively moderate computational burden for machine condition real time monitoring. As compared to the time synchronous average technique, which can reveal fault conditions only when the faults are severe, such as a tooth broken or wear-out failure, the proposed PCA-based approach can detect early fault conditions, such as tooth crack in this study, which are usually undetectable in the averaging technique alone. And besides, the PCA-based approach can be used to determine how early the fault can be detected and how the severity of the fault can be monitored, where other commonly used methods cannot, such as power spectrum and time frequency analysis etc..

It is well known that modulation is very common in the gear vibration-acoustic signal. When a fault occurs, modulation is introduced to the vibration-acoustic signal, such as frequency modulation, amplitude modulation and phase modulation. As with any empirical modeling method, this approach captures only the underlying correlations which exist in the raw features, and not necessarily the causal relationships that exist between the vibration signal and the fault.

Because vibration signals from the machinery are extremely complex, and depend greatly on the machine assembly and the working environment, it is impossible to predict the vibration spectrum in detail for any particular type of machine using only theoretical calculation. Therefore, calibration during some period is the key program for any successful application. However, it could be used to correlate any variation with assignable cause in machine condition monitoring.

Conclusion

The criterion used PCA to remove the collinearity and to reduce the dimensionality of the original feature space, at the same time different faults were classified effectively. Experimental results indicate that the proposed method is sensitive to different conditions of gearbox working, and it is capable of identifying industrial gearbox defects which occur naturally. By calibration, this model can be applied to novelty detection. Based on the current work, a fault diagnosis system will be established which can be extended during operation. To improve the accuracy of the PCA model, some designed experiments will be performed in future. In addition, the frequency features of gearbox vibration will also be studied in order to analyze failure cause and improve the performance of the model.

References

1 Gallagher, N. B. and Wise, B. M. (1996). "Application of multiway principal components analysis to nuclear waste storage tank monitoring". Computers Chem. Eng. , Vol. 22, supplement, pp. s739-744.

2 Goodman, S. and Hunter, A. (1999). "Feature extraction algorithms for pattern classification". IEE Conference Publication on Artificial Neural Network, pp. 738-742.

3 Gu, S. and Ni, J. (1996). "Multi-spindle drilling process condition monitoring and fault diagnosis". Proceedings of the 1996 ASME International Mechanical Engineering Congress and Exposition, Atlanta, GA, pp. 555-562.

4 Huber, P. J. (1985). "Projection pursuit". The Annals of Statistics, Vol. 13 No. 2, pp. 435-475.

5 Hyvarinen, A. (1999). "Survey on independent component analysis". Neural Comput. Sur. , Vol. 1 No. 2, pp. 94-128.

6 Isermann, R. (1984). "Process fault detection based on modeling and estimation methods - a survey". Automatica, Vol. 20 No. 4, pp. 387-404.

7 Jackson, J. E. (1991). "A User's Guide to Principal Components". John Wiley and Sons, New York, NY, pp. 1-25.

8 Jia, F., Martin, E. B. and Morris, A. J. (1998). "Non-linear principal components analysis for process fault detection". Computers Chem. Engng., Vol. 22, Supplement, pp. 5851-5854.

9 Koller, D. and Sahami, M. (1996). "Toward optimal feature selection". In ICML-96: Proceedings of the Thirteenth International Conference on Machine Learning, Morgan Kaufmann, San Francisco, CA, pp. 284-292.

10 Kresta, J. V., MacGregor J. F. and Marlin, T. E. (1991). "Multivariate statistical monitoring of process operating performance". Can. Jour. Chem. Eng., Vol. 69, February, pp. 35-47.

11 Liu, B. and Si, J. (1997), "Fault isolation filter design for linear time-invariant systems". IEEE Transaction onAutomatic Control, Vol. 42 No. 5, pp. 524-528.

12 McCormick, A. C. and Nandi, A. K. (1997). "Real time classification of rotating shaft loading conditions using artificial neural networks". IEEE Transactions on NeuralNetworks, Vol. 8 No. 3, pp. 745-757.

13 McKeever, B., Graham, S. and Blundell, J. K. (1986). "Faults - an expert systems environment for fault detection and diagnosis". Proceedings of the Winter Annual Meeting of ASME: Knowledge-based Expert Systems for Manufacturing, PED, Vol. 24, pp. 85-96.

14 Morison, D. F. (1976). "Multivariate Statistical Methods". McGraw-Hill, New York, NY.

15 Renwick, J. T. (1985). "Vibration analysis - a proven technique as a predictive maintenance tool". IEEE Trans. Industry Applications, Vol. 21, pp. 324-332.

16 Strackeljan, J. (1999). "Feature selection methods for softcomputing classification". Proceedings of ESIT, 99— The European Symposium on Intelligent Techniques, June.

17 Wilson, D. J. H., Irwin, G. W. and Lightbody, G. (1997). "Neural networks and multivariate SPC". IEE Colloquium on Fault Diagnosis in Process Systems, pp. 1-5.

18 Wold, S. (1978). "Cross-validatory estimation of the number of components in factor and principal component model". Technometrics, Vol. 20 No. 4, pp. 397-405.

19 Zhang, G. (1993). "Gear vibration signature analysis and fault diagnosis". Master's thesis, Xi'an Jiaotong University.

Further reading

Zhang, J., Morris, A. J. and Martin, E. B. (1996). "Fault detection and diagnosis using multivariate statistical techniques". Chem. Eng. Research and Design, Vol. 74 No. 1, pp. 89-96.

(原载 Journal of Quality in Maintenance Engineering, Volume 9, Issue 3, 2003)

制造系统分布式柔性可重组状态监测与诊断技术研究[*]

胡友民　杨叔子　杜润生

提　要　随着制造系统的分散化、柔性化与快速重组技术的发展,研究与之相适应的具有分步式柔性可重组特性的状态监测与诊断技术具有重要意义。研究如何运用"柔性可重组"思想构建分布式柔性可重组状态监测与诊断系统,建立了系统的硬件及软件模型,重点分析了模型中各部分的功能与结构。

关键词:柔性可重组,状态监测,故障诊断,先进制造系统

0　前　言

21世纪人类社会进入了知识经济时代,知识经济对制造技术乃至整个制造业产生了重大影响,其中最显著的表现是制造过程的数字化、智能化、分散化与全球化。为了适应这种新的变化,近十多年来相应地出现了许多先进的制造方式,如智能制造、柔性制造、网络制造、精良制造以及计算机集成制造等等[1,2]。

不论采用何种制造方式,状态监测技术都是保障制造过程顺利完成的重要手段。随着制造业的发展,传统意义上的那种集中式的状态监测技术已经难以适应现代制造过程的要求,迫切需要研究一些新的状态监测技术与方法以适应制造技术的数字化、智能化、分散化与全球化发展趋势。近年来出现的智能化状态监测技术、分布式状态监测技术,以及基于Internet/Intranet的网络化状态监测技术等等都是这方面的一些典型代表[3-11],较之传统的状态监测技术有了很大进步。

但是由于现代企业的生产制造模式是一种客户驱动的按需生产模式,生产品

---

[*] 国家自然科学基金重点资助项目(59990470-1),教育部智能制造技术重点实验室资助项目。

种多、个性化程度高，变化快，而批量往往不大，常常需要对制造资源进行动态重组与配置，与传统的低成本、大批量、制造资源配置基本固定不变的生产模式有了很大区别，这就造成了目前的状态监测技术和方法很难适应这种快速的动态重组要求。因为目前常用到的状态监测技术和方法，不论是智能化、分布式还是网络化等都还是主要从监测的实时性、正确性、可信性以及实现的方便性考虑，本质上讲都还是一种针对固定对象的监测模式与方法，缺乏一定的"柔性"，不便于进行动态重组和配置，在对制造资源和系统进行重组和重新配置以后，原来的监测系统就很难满足新系统的需求，也造成了监测资源的浪费。

为了克服这些缺点，适应现代制造方式的快速重组与动态配置要求，研究提出了一种柔性可重组的分布式状态监测技术和方法。在"智能化、分布式与网络化"的基础上，增加"柔性与重组"的思想与理念，以工业制造过程为研究对象，依据现场状态监测的实际情况，可以实现状态监测系统的硬件与软件资源的重组和再配置，满足现代制造系统的快速重组与动态配置需求。

1　柔性化与可重组思想

"柔"本义系指"柔软的或易弯的性质"（高级汉语大词典）；在英文中为"Flexible"，本义指"可重复弯曲而不损坏（Capable of being bent repeatedly without injury or damage）"（美国传统辞典）。本文所指的"柔性化"系采用其引申含义，即指"系统结构易于变化和改变的性能"。

柔性概念最早应用于制造技术中是 20 世纪 60 年代，英国工程师 David Williamson 将他使用的一套制造装置称为 Flexible machining system，该系统用计算机控制机床的多种操作，无须人力介入。随着计算机控制设备的发展以及在金属成形和装配方面的广泛应用，柔性加工系统（Flexibel machining system）逐渐发展为柔性制造系统 FMS（Flexible manufacturing system），美国 Kearney Trecher 公司首先使用这个名称来命名可以完成多品种、中小批量制造加工任务，并由计算机控制的自动加工线。

美国国家标准局把柔性制造系统定义为："由一个传输系统联系起来的一些设备，传输装置把工件放在其他联结装置上送到各加工设备使工件加工准确、迅速和自动化。中央计算机控制机床和传输系统，柔性制造系统有时可同时加工几种不同的零件。"

国际生产工程研究协会指出："柔性制造系统是一个自动化的生产制造系统，

在最少人的干预下,能够生产任何范围的产品族,系统的柔性通常受到系统设计时所考虑的产品族的限制。"

实际上现代的先进制造系统除了能完成多品种加工任务外,还具有根据加工的品种和任务的不同,在中央控制系统(计算机控制系统)的协调下,自动完成现场制造加工单元的重新组合与配置能力,适应不同加工品种的需要。

由此可以看出制造系统中的柔性主要是指系统能在计算机控制下完成或同时完成多品种、中小批量的任务以及对制造资源的重新组织与配置。制造系统的所有这些特点都是为适应当代制造业面临着巨大的压力,必须以低的生产成本、短的生产周期开发更多的新产品,以适应市场的多样化需求和全球范围内的激烈竞争而产生的。

在制造系统中,状态监测系统是制造系统能够实现制造资源重组与配置以完成各种加工任务的重要保障。因此从本质上讲,应用于制造系统的状态监测系统都应该是柔性与可重组的。也就是说,构建制造系统的状态监测系统时,监测系统硬件资源应该能够灵活地最大限度地进行重新组合与配置以适应制造资源的不同配置要求,而软件资源也能够方便快捷地进行重新组合以满足不同品种与批量的制造要求等等。

2 分布式柔性可重组状态监测系统结构

2.1 系统总体结构

为了便于实现"柔性可重组"目的,在进行状态监测系统总体结构设计时,遵循以下原则。

(1) 采用"模块化"设计方法,即将完成某种功能的一组硬件或软件看作一个功能模块,在实际构建系统时根据需要选用相应模块即可。

(2) 按功能不同在系统总体结构上将系统划分为四个功能层,即管理层、监测诊断层、数据采集层和设备层,如图 1 所示。

(3) 各功能层之间用以太网连接,除设备层外各层内部模块也采用网络连接方式,即整个系统构成一个监测诊断局域网,整个网络拓扑结构如图 2 所示。网络根结点设在管理层,每一层为网络中的一个结点,一个结点可以是单台主机,也可以是由一台服务器和多台主机构成的一个子网。为了提高系统安全性,子网内的主机只能对相邻层内主机的访问,且必须通过本层及被访问层服务器代理。图

图 1 状态监测系统分层结构

2 中下部一个圆形区域可以代表一个生产车间或一条生产线或一个较大型的独立生产单元如加工中心等等(为叙述方便以下均称之为车间结点),它们均独立形成一个结点并直接与管理层相连,企业规模越大则车间结点就越多,网络规模也越大,反之亦然。从网络观点看,企业增大或减小生产规模,相应监测与诊断系统只要增加或减少网络中的车间结点即可,因此在设备层以上可以很方便地实现"柔性可重组"。

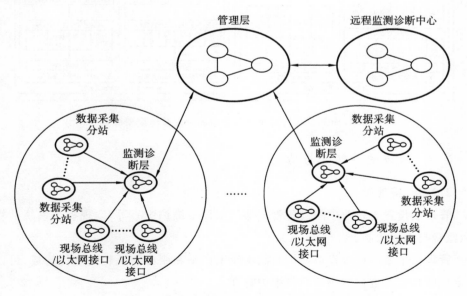

图 2 分布式柔性可重组状态监测系统网络拓扑结构

(4) 设备层最难以实现"柔性可重组",为了实现设备层的"柔性可重组",按照设备层中布置的传感器类型,划分为模拟量采集模块、数字量采集模块、基于现场总线模拟量采集模块和基于现场总线数字量采集模块等等,参见图 3。实际构建系统或要对系统进行重组时则可以根据需要灵活选用和更改这些配置方式,实现"柔性可重组"目的。

按照上述原则建立的系统总体模型框架,如图3所示,图中从上至下依次为管理层、监测诊断层、数据采集层和设备层,各层的构成方法与功能分述见第2.2节。

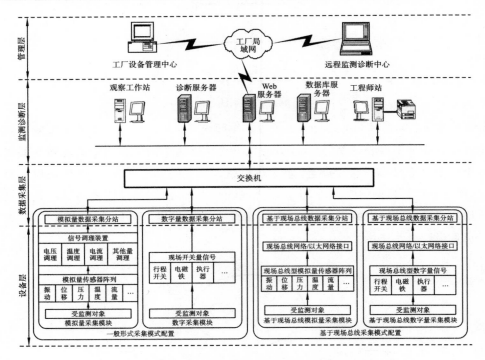

图3 分布式柔性可重组状态监测系统总体模型框架

2.2 系统硬件配置与功能

2.2.1 管理层

管理层设置于企业中的设备管理部门如设备维护中心等,实现整个系统的管理、监督与维护功能。

管理层主要由厂级设备维护监测诊断工作站和服务器及相关软件构成,可以对全厂所有设备进行监测诊断与维护管理。

管理层服务器直接接到厂级局域网上,可以根据需要将设备状态信息等在厂内局域网上发布,不同职能部门可以根据不同授权级别浏览与其相关的信息内容,另外还可以通过厂信息中心与外部远程监测诊断中心相连,实现远程监测诊断专家资源厂内共享和/或实现远程状态监测与诊断。

2.2.2 监测诊断层

监测诊断层是整个系统的核心,其主要功能如下。

(1)完成车间级监测诊断硬件设备组态。

(2)接受数据采集层上传的设备与工艺过程状态信息并将其保存在数据服务器的历史数据库中。

(3)提取设备与工艺过程状态信息的特征信息并保存在数据服务器的特征数据库中。

(4)根据特征信息对设备与工艺过程运行状态进行评估,预测设备运行趋势,对故障进行诊断并给出处理策略建议,将无法确诊的故障上传给管理层寻求其他途径诊断。

监测层主要由 Web 服务器、数据库服务器、诊断服务器、观察站和工程师站组成。

Web 服务器兼作本层子网服务器,管理层和数据采集层对本层的访问必须通过 Web 服务器代理,以保证系统运行的安全性。

数据库服务器主要用于保存数据采集层上传的设备与工艺过程状态信息、诊断服务器的诊断结果及其他系统运行的必需数据如专家知识与设备组态信息等。为此在数据服务器上装有专门开发的历史数据库、特征信息数据库、专家知识库以及设备组态信息库等。

诊断服务器主要完成设备与工艺过程实时运行状态监测、状态特征信息提取、设备与工艺过程运行故障分析、诊断与趋势预测、故障处理策略建议等等。为此诊断服务器上主要安装的软件模块有:状态监测模块、特征信息提取模块、分析诊断模块、趋势预测模块及专家系统等。

观察站用于系统的实时监测与故障分析处理。

工程师站主要供工程技术人员日常处理有关数据与资料,如数据备份、监测诊断结果打印输出及车间级结点网络维护等等。

2.2.3 数据采集层

数据采集层主要实现对受监测对象的实时运行状态信息的采集并将采集到的信息上传到监测诊断层中的数据库服务器中保存。

按照实际布置在受监测对象上的传感器类型不同,有四种数据采集模式:一般形式模拟量采集模式、一般形式数字量采集模式、基于现场总线的模拟量采集模式和基于现场总线的数字量采集模式。

各种模式的数据采集模式分别配置有相应的数据采集分站,各采集分站中装有与不同类型的数据采样模块相对应的数据采集软件。数据采集分站选用高性能的工业级 PC 机实现。

一般形式模拟量和数字量采集分别由模拟量采集分站和数字量采集分站实现,而基于现场总线的模拟量和数字量则经过现场总线/以太网转换接口转换后由基于现场总线的数据采集分站上传到监测诊断层中的数据库服务器中保存。

2.2.4 设备层

设备层主要由布置在受监测对象上的各种类型传感器及信号调理与转换接口装置构成。

由于实际受监测对象千差万别,因此这一部分最难以实现"柔性可重组"。为了尽可能最大限度实现"柔性可重组",根据传感器类型不同,将设备层分为四种信号检测方式:模拟量与数字量传感器及信号调理;基于现场总线模拟量与数字量传感器及现场总线/以太网转换装置。信号调理模块又可以分为四种基本类型模块:电压信号调理模块、电流信号调理模块、振动信号调理模块和温度信号调理模块。

在构建设备层时,可以根据所选择的传感器类型、受监测对象等因素综合考虑实际系统的设备层的信号检测方式和选用相应的信号调理模块。

3 系统软件结构原理

系统软件是整个系统功能实现的重要保证,也是"柔性可重组"思想的直接体现之一。对于本文所讨论的状态监测与诊断系统,软件功能如图 4 所示。软件由六部分组成,即系统组态模块、数据采集模块、状态监测模块、诊断分析模块、数据库管理模块和 Web 主页等。

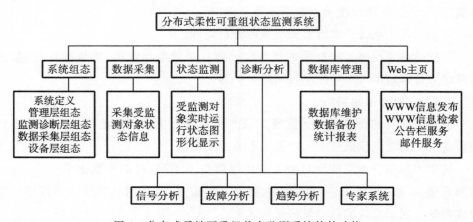

图 4 分布式柔性可重组状态监测系统软件功能

系统组态模块完成系统定义及系统四层网络模型中的硬件设备配置。

数据采集模块主要用于收集受监测对象的实时运行状态参数。

状态监测模块主要对受监测对象的实时运行状态进行图形化显示。

诊断分析模块包括了信号分析、故障分析、趋势分析和专家系统等四个子模块,主要对采集的受监测对象的状态信息进行分析,提取状态信息中的特征参数,分析运行中的(潜在)故障及故障发生原因,给出故障处理策略。

数据库管理模块主要是对数据库进行维护,定期备份数据,生成相关统计报表。

Web 主页实现系统网络监测与诊断目的,它可以将受监测对象的运行状态信息在企业内部局域网上发布,也可以在授权情况下发布在 Internet 上,另外还可以实现其他一些功能,如公告栏与邮件服务等等。

数据采集模块安装在数据采集分站,系统组态、状态监测及诊断分析模块安装在诊断分析服务器上,数据库管理模块安装在数据库服务器上,Web 主页在诊断分析服务器、数据库服务器和管理层服务器上均有。系统软件结构如图 5 所示。

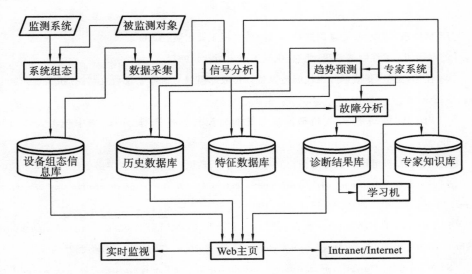

图 5 分布式柔性可重组状态监测系统系统软件结构

系统开发语言及工具选用 Java、VC++、SQL 和 Matlab 等。Java 主要用于开发与网络相关的软件功能如 Web 主页等;VC++主要用于开发与分析推理方面的软件功能如故障诊断、专家系统及学习机等;SQL 主要用于各种数据库开发;借助 Matlab 强大的信号分析工具箱则可以开发软件信号分析、趋势预测等功能。

4 结 论

为适应制造系统日益分散化、柔性化和快速重组的技术发展趋势,研究提出了一种分布式柔性可重组状态监测与诊断系统模型。基于该模型可以根据制造系统配置的变化情况灵活地变更监测诊断系统的实际配置,以最大限度满足制造系统需求,实现了状态监测与诊断系统的柔性可重组。以此为基础我们研究了一套具有柔性可重组特性的分布式状态监测与诊断原型系统,并成功运用于某厂化工生产过程与设备的监测与控制中[12-14]。

柔性可重组思想在工程技术领域越来越受到重视,相信在未来的状态监测与诊断技术研究中也会受到广泛关注,本研究项目只是作了初步尝试,还有很多问题值得进一步深入研究,例如,在设备层如何更好地适应监测环境变化实现传感器阵列的快速重组与配置、适应柔性系统快速重组特性的新型检测元件研究、柔性可重组环境下的监测与诊断软件体系结构与开发技术等等。

参 考 文 献

[1] 盛晓敏,邓朝晖.先进制造技术.北京:机械工业出版社,2000.

[2] 张培忠.柔性制造系统.北京:机械工业出版社,1997.

[3] 任伟,王坚,张浩,等.现代制造设备远程故障监测与诊断系统研究.计算机工程,2000,26(9):46~48.

[4] 周祖德,陈幼平.现代机械制造系统的监控与故障诊断.武汉:华中理工大学出版社,1999.

[5] 张安华.机电设备状态监测与故障诊断技术.西安:西北工业大学出版社,1995.

[6] Booth C. *The use of artificial neural networks for condition monitoring of electrical plant*. Neurocomput.,1998,23(1-3):97~109.

[7] Mangina E E,Mcarthur S D J,Mcdonald J R. COMMAS(Condition monitoring multi-agent system), *Autonomous Agents and Multi-Agent System*. 2001,(4):279~282.

[8] Malin J T. *Multi-agent diagnosis and control of an air revitalization system for life support in space*. In:2000 IEEE Aerospace Conference Proceedings,MT,USA,Big Sky,2000,6:309~326.

[9] PlOsh R,Weinreich R. *An agent-based environment for remote diagnosis,supervision,and control*. In:Proceedings of the international computer science conference,Hong Kong,1999:13~15.

[10] Somnath D, Sudipto G, Venkata N M, et al. *Tele-diagnosis: remote monitoring of large-scale systems*. IEEE:0-7803-5846-5/00,2000.

[11] Wijata Y I, Niehaus D, Frost V S. *A scalable agent-based network measurement infrastructure*. IEEE Communications Magazine,2000,38(9):74~83.

[12] Hu Y M, Du R S, Yang S Z. *A framework of agent-based data acquisition technology for manufacturing system*. In: Imre Hovath eds, Proceedings of TMCE2002, The Forth International Sympotium on Tools and Methods of competitive Engineering, Wuhan, P. R. China, 2002:439~450.

[13] 胡友民,杜润生,杨叔子. 液压系统运行状态监测. 液压与气动,2002,(5):35~37.

[14] 胡友民,杜润生,杨叔子. 基于PLC的高可靠性工业过程远程监控系统. 华中科技大学学报(自然科学版),2002,30(4):13~15.

[15] 胡友民,杜润生,杨叔子. 智能化状态监测技术研究. 中国机械工程,2003,14(11):946~949.

Distributed Flexible Reconfigurable Condition Monotoring and Diagnosis Technology

Hu Youmin　　Yang Shuzi　　Du Runsheng

Abstract　Accompanying the development of manufacturing technology with decentralization and flexibility and fast reconfiguration features, it is significant to researching condition monitoring and diagnosis technology with the same characteristics. The distributed condition monitoring and diagnosis technology based on "flexible and reconfigurable" concept is studied. And more, the system models are constructed. The components and their functions of models are discussed in detail.

Keyword：Flexible reconfigurable, Condition monitoring, Fault diagnosis, Advanced manufacturing system

(原载《机械工程学报》2004年第40卷第3期)

基于 Markov 模型的分布式监测系统可靠性研究*

易朋兴　杜润生　杨叔子　刘世元

提　要　监测系统作为一种保证复杂系统正常工作与提高其运行可靠性的重要手段被广泛应用。监测可靠度指监测系统能成功地对被控对象进行监测的概率，它是评价监测系统性能的一个重要尺度。以可靠性理论为指导，结合集中分布式监测系统的结构特性，对该类型监测系统的监测可靠性进行深入研究，提出一种基于 Markov 模型的监测可靠性分析方法。随后，以捏合机监测系统为例进行验证与分析。分析表明该方法对监测系统设计及其维护有一定的应用价值。

关键词：集中分布式监测系统，监测可靠度，Markov 链，可用度

0　前　言

随着科学技术的进步，人类开发了许多新的大型复杂设备和系统，例如自动生产线系统、计算机网络系统、大型现代制造系统等等。这些系统是复杂的人-硬件-软件系统，它们一旦发生故障，将会对社会、经济、环境等方面造成不同程度的损害。因此对系统效能中最重要的指标之一——可靠性就应予以严肃认真的考虑[1,2]。

目前，监测系统作为一种保证复杂系统正常工作与提高其运行可靠性的重要手段被广泛应用[3-6]。同时，在生产过程中人们对监测系统的依赖性也越来越大。由此带来一些不容忽视的问题——二次诊断与可靠性问题，即监测系统自身的可靠性问题和怎样设计满足要求的具有高度可靠性的监测系统问题等。

---

* 国家自然科学基金(50205009)和教育部智能制造技术重点实验室资助项目。

胡友民、杜润生等[4,7]针对制造系统的特殊性提出一种面向制造系统的监测系统模型并对其可靠性问题进行探讨。在其研究的基础上结合某一类分布式监测系统——集中分布式监测系统(CDMS)的结构特性,探讨其监测可靠性问题,提出一种基于 Markov 模型的监测可靠性分析方法,为监测系统的设计、日常管理及提高其工作可靠性提供一定依据。

1 集中分布式监测系统(CDMS)

多数的分布式监测系统可以定义为 CDMS(Concentrated-distributed monitoring system),这一类的监测系统由一些异构或者同构的监测子系统组成,每个子系统的功能不一样,这些子系统获得的各种监测信息传送到监控中心进行分析与处理[7,8]。该类监测系统具体结构见图 1。

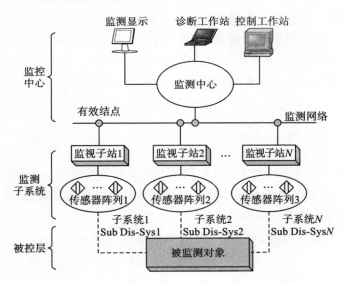

图 1 集中分布式监测系统结构

监控中心包括监测中心和一系列工作站。监测中心能有效存储、管理和分析各个监测子系统通过监测网络传过来的实时监测数据,有效结点指一些能够执行各种功能的单元如 CPU、内存以及监测中心所带的其他器件等[9]。监测中心实时分析结果在监测显示界面上显示或传到控制站,为操作人员提供操作依据。同样,诊断工程师可以依据监测中心的各种结果和数据对被控系统的状况进行分析,对系统故障进行预测与诊断。

监测子系统(Sub Dis-Sys1,…,N)由监视子站和传感器阵列等单元组成。监视子站主要完成信息的预处理,如完成监测信号的 A/D 转换、滤波等。传感器阵列对被控对象进行监测,获得监测信息。根据被控对象生产环境和要求的不同,每个子系统的拓扑结构和传感器类型亦有所不同。子系统与监测中心通过监测网络进行数据交换。

在被控系统工作过程中,监测是不可间断的,监测系统可靠性的高低直接影响到整个被控系统的正常运行。因此,系统的监测可靠性分析极为重要。

2 CDMS 监测可靠性分析

在对 CDMS 监测可靠性进行分析之前,下面作一些基于统计独立性的假设[10]:

(1)子系统工作状态及可靠度相互独立;
(2)监测中心提供的各种服务、存储的各种文件和程序相互独立;
(3)各工作单元、传递通道可靠性相互独立;
(4)每个单元处于什么状态是相互独立的;
(5)子系统的各单元正常工作的概率为常数。

2.1 CDMS 可靠性评价指标

在可靠性分析中,系统分为可修复型和不可修复型两类,CDMS 是由多数可修复的器件模块和部分不可修复的传感器阵列组成的可修复系统,其可靠性大小可以用可靠度、可用度等参数来描述[4,11]。

定义 1 单元可靠度:单元正常工作事件发生的概率

$$R_i = P\{E(i)\} \tag{1}$$

式中:R_i——单元可靠度;

P——正常工作时事件发生的概率;

$E(i)$——第 i 个单元正常工作事件。

定义 2 子系统可靠度:子系统可靠度的物理意义为监测子系统能正确工作的概率大小,即

$$RSS_i = P\{E(subsys_i)\} \tag{2}$$

定义 3 可用度:对于给定的一个随机时间 t,在时刻 t 系统处于可用状态的概率。可用度是给定随机时间 t 的函数,也叫瞬时可用度,记为 $A(t)$。

定义 4 监测中心可用度:给定时刻 t 监测中心处于可用状态的概率,记为 $A_m(t)$。

定义 5 监测可靠度:监测可靠度的物理意义为监测系统能正常对被控对象进行监测的概率,即

$$RS = P\{E(sys)\} \qquad (3)$$

2.2 CDMS 监测可靠性通用模型

CDMS 的工作过程包括信号的正常采集、数据的网络传输、监视子站完成数据的预处理并将数据通过网络传到监测中心、监测中心分析传过来的数据并形成数据分布曲线或者数据表等,每一个环节对其监测可靠性都有一定的影响。

CDMS 正常工作时,各数据采集单元、传输网络、监视子站以及监测中心的运行都应该是正常的。同时监测中心既能随时为监控工程师提供可用的数据或者指令,又能正常地接收与分析监视子站上传的数据,即监测中心是随时可用的。因此,监测中心可用度分析极为重要。

监测中心作为 CDMS 中的一个核心模块,配有专门的维修人员,二者构成一可修复系统。它一旦出现故障就不能正常工作,修复后重新开始工作。可以用取值为 1、2 的二值随机函数 $X(t)$ 来描述时刻 t 监测中心的状态,即

$$X(t) = \begin{cases} 1, & 时刻\ t\ 监测中心工作 \\ 2, & 时刻\ t\ 监测中心故障 \end{cases} \qquad (4)$$

时刻 t 监测中心的可用度 $A_m(t)$ 可以用 $X(t)$ 等于 1 的概率表示,即

$$A_m(t) = P\{X(t) = 1\}, \quad t \geqslant 0 \qquad (5)$$

根据前面的假设和分析,$X(t)$ 是一个齐次 Markov 链[1,10,11]。下面按照 Markov 过程分析监测中心的可用度。

CDMS 监测中心是一个完整的计算机系统,工程实践[3,4,7,9]表明其故障通常都是由软件故障引起的,其可用度主要取决于所带软件系统的可靠性。软件可靠性模型有很多种,其中 Goel 和 Okumoto 提出的 GO 模型应用较多[8,12]。设监测中心的失效率函数为 $\lambda(t)$,维修时间服从修复率为 μ 的指数分布。根据 GO 模型,监测中心失效率函数为

$$\lambda(t) = ab\exp(-bt) \qquad (6)$$

式中:a——预期的能最终被发现的软件故障数;

b——软件故障率。

设监测中心在时刻 t 开始工作的概率为 $P_1(t)$,$P_2(t)$ 为在时刻 t 发生故障的

概率,则相应的 Kolomogorov 方程[1,8]为

$$\begin{cases} P_1(t) = \mu P_2(t) - \lambda P_1(t) \\ P_2(t) = 1 - P_1(t) \end{cases} \tag{7}$$

设边界条件 $P_1(0)=1, P_2(0)=0$,可得到监测中心时刻 t 开始工作的概率

$$P_1(t) = \left[\int_0^t \mu\exp(\mu x - a\exp(-bx))\mathrm{d}x + \exp(-a)\right]\exp(-\mu t + a\exp(-bt)) \tag{8}$$

这也是 t 时刻 CDMS 监测中心可用度 $A_m(t)$ 的计算公式。

从图 1 知道,子系统中监测信号从检测单元到监测中心要经过多个环节,每个环节对其可靠性都有一定的影响,影响大小主要取决于子系统的复杂程度。但是,当系统结构确定后,不管一个子系统有多么复杂,其可靠度都可以用所有单元可靠度构成的可靠度函数表示[11,13]。如果简化后 CDMS 的第 i 个子系统有 M 个单元,那么它的可靠度为

$$\mathrm{RSS}_i = f(R_{i1}, R_{i2}, \cdots, R_{iM}) = \sum_{h=1}^{m}\left[a_h\prod_{j=1}^{M}R_{ij}^{k_{hj}}\right] \tag{9}$$

式中:R_{ij}——第 i 个子系统中 j 单元的可靠度,$j=1,2,\cdots,M$;

m——可靠度函数中乘积式的项数;

a_h——可靠度函数中第 h 项乘积式的系数,$h=1,2,\cdots,m$;

k_{hj}——第 h 项乘积式中 j 单元可靠度的幂次。

一个完整的 CDMS 中包括 N 个子系统和一个监测中心,它们当中任何一个出现故障就认为系统不能正常工作,则其可靠性模型是一个有 $N+1$ 个单元的串联系统。按照前面的假设,在时刻 t 它的监测可靠度为

$$\mathrm{RS}(t) = A_m(t)\prod_{i=1}^{N}\mathrm{RSS}_i \tag{10}$$

3 应用实例与分析

捏合机监测系统是保障某厂正常生产的关键,该系统包括一个监测中心和三个监测子系统,是一个典型的 CDMS 系统。下面就用前述 CDMS 可靠性模型对其进行分析。

3.1 捏合机 CDMS 结构

捏合机监测系统的拓扑结构见图 2。图中 MS(Monitoring Station)为监测中

心，与其相连的是三个子系统 $DS_i(1,2,3)$，MSS_i 为各子系统对应监测子站。温度监测子系统 DS_1 包含两个不同的温度传感器阵列 A_1 和 A_2，其中 $S_1(a_1)$，$S_1(a_2)$，$S_1(a_3)$ 是同类型传感器，三者中任意一个能正常工作就认为 A_1 工作正常；$S_1(b_1)$，$\cdots$，$S_1(b_4)$ 是同类传感器，四者中任意两个能够正常工作就认为 A_2 工作正常；当 A_1，A_2 中的任意一个、MSS_1 以及各传输链路正常工作时就认为温度监视子系统工作正常。压力、液位监测子系统 DS_2 只有一个同类传感器阵列（S_{21}，$\cdots$，S_{23}），任何一个传感器通道失常就认为它有故障。位置子系统 DS_3 也只有一个同类传感器阵列，它的四个传感器（S_{31}，$\cdots$，S_{34}）通道任意一个能正常工作就认为该子系统工作正常。图 2 中方括号内的数字是为对系统结构进一步分析而对每一个单元设置的结点号。

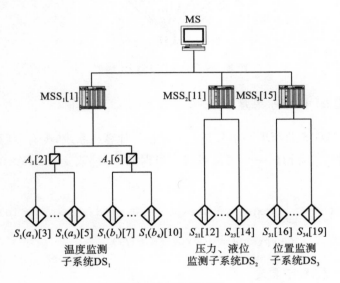

图 2　捏合机 CDMS

按照胡友民、杜润生等[3,7,9]的方法将图 2 所示系统中的每个工作单元用一个结点 i 表示，两个工作单元之间的相应信号传递通道用一条有向边 e_i 表示，得到捏合机监测系统的网络模型图，如图 3 所示。其中 S 代表被监测的捏合机，MS 表示监测中心。图中其余每个结点表示原系统中相应工作单元，每条边表示原系统中相应工作单元之间的信号传递通道，边上的箭头表示相应两个工作单元间的信号传递方向。

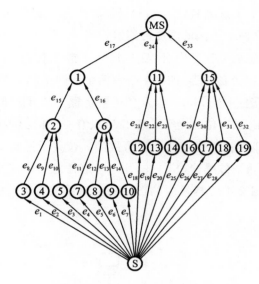

图 3 捏合机 CDMS 网络模型

3.2 监测中心可用度函数

捏合机 CDMS 监测中心本身是一个完整的计算机系统,配有专门的工作人员对其进行维护,二者构成一个可修系统。根据文献[7]取 $a=10,b=0.01$,则监测中心可靠度

$$A_m(t) = \left[\int_0^t \mu\exp(\mu x - 10\exp(-0.01x))\mathrm{d}x + \exp^{10}(-10)\right] \times \exp(-\mu t + 10\exp(-0.01t)) \tag{11}$$

3.3 子系统可靠度

图 3 中从左至右共有 3 个子系统,分别用 DS_1、DS_2、DS_3 表示,它们各自的结构如图 4 所示。图 4 中 $MS_i(1,2,3)$ 表示监测中心负责处理第 i 个子系统信息的硬件模块,在实际工作中,监测中心的硬件和被监测对象一般不会出故障,MS_i 和 S 可以看作是绝对可靠的[7]。各传感器阵列、数据采集单元、监视子站以及传递通道如果出现故障,一般难以修复,通常是以新器件予以替换,按照不可修复系统对子系统的可靠性进行分析[9,11]。

从图 4 可知,结点 1~10、传递通道 e_1~e_{17}、被监测对象 S 和监测中心 MS_1 构成 DS_1,按照图 4 系统的构成情况和假设条件,DS_1 是一个复杂的不可修复系统,其可靠性框图见图 5。

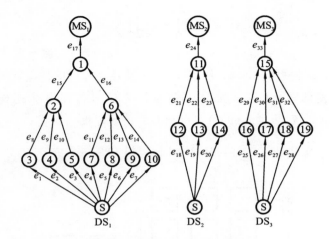

图 4　子系统各自网络拓扑图

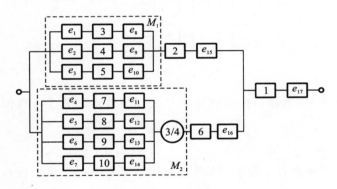

图 5　DS_1 可靠性框图

图 5 表明 DS_1 可靠性模型是一个由串联模块、并联模块和 3/4(G)模块构成的复杂混联系统[10]。结点 3~5 及各自前后传递通道构成串—并联系统 M_1；结点 7~10 及其前后传递通道构成 3/4(G)表决系统 M_2。M_1 与结点 2 及传递通道 e_{15} 成为串联旁路，M_2 与结点 3 以及传递通道 e_{16} 构成另外一个串联旁路，这两个旁路形成一个并联系统，该并联系统与结点 1 以及传递通道 e_{17} 构成一个串联系统。先单独计算各模块和旁路的可靠度，并将每个模块或旁路当作一个环节，层层计算得到 DS_1 的可靠度

$$RSS_1 = R_1 R_{e_{17}}[1-(1-R_{M_1}R_2R_{e_{15}})(1-R_{M_2}R_3R_{e_{16}})] \tag{12}$$

结点 11~14、传递通道 e_{18}~e_{24} 与监测中心 MS_2 及被监测对象构成监测子系统 DS_2。根据捏合机 CDMS 的结构特点，任何一个环节故障则 DS_2 故障。可见，DS_2 的可靠性模型为一个所有单元串在一起构成的串联系统，其可靠度

$$\mathrm{RSS}_2 = \prod_{i=0}^{3} R_{11+i} \prod_{j=0}^{6} R_{e_{18}+j} \tag{13}$$

结点 15～19、传递通道 $e_{25} \sim e_{33}$ 与监测中心 MS_3 及被监测对象 S 构成监测子系统 DS_3，其可靠性框图见图 6。结点 16～19 及各自前后的传递通道构成一个具有 4 个串联支路的串—并联系统 M_3，M_3 与节点 15 及传递通道 e_{33} 构成一个具有 3 个环节的串联系统，根据式(9)可以得到 DS_3 的可靠度

$$\mathrm{RSS}_3 = R_{15} R_{e_{33}} R_{M_3} = R_{15} R_{e_{33}} \left[1 - \prod_{i=0}^{3}(1 - R_{16+i} R_{e_{25+i}} R_{e_{29+i}}) \right] \tag{14}$$

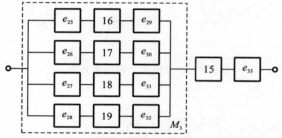

图 6　DS_3 可靠性框图

实际情况下，传感器与被监测对象 S 之间的传递通道可以看作是绝对可靠的[3,7]，即

$$R_{e_i} = P[E(e_i)] = 1, \quad i = 1, 2, \cdots, 7, 18, \cdots, 20, 25, \cdots, 28 \tag{15}$$

若各子系统中所有工作单元和传递通道的可靠度为 0.995，按照上述计算过程得到表 1 中的结果。

表 1　子系统可靠度计算

子系统	DSR_1	DSR_2	DSR_3
可靠度	0.989 9	0.965 5	0.990 0

3.4　CDMS 监测可靠度

根据式(7)～(11)和表 1 的计算结果可以得到 t 时刻捏合机 CDMS 可靠度

$$\begin{aligned}
\mathrm{RS}(t) &= A_m(t) \prod_{i=1}^{3} \mathrm{RSS}_i \\
&= 0.9462 \times \left[\int_0^t \mu \exp(\mu x - 10 \exp(-0.01x)) \mathrm{d}x + \exp(-10) \right] \\
&\quad \times \exp(-\mu t + 10 \exp(-0.01t))
\end{aligned} \tag{16}$$

当 $\mu=0.5$ 或 $\mu=0.8$ 时,捏合机 CDMS 监测可靠度 $RS(t)$ 可以用图 7 表示。根据图 7 所示监测可靠度变化曲线得出一些结论。

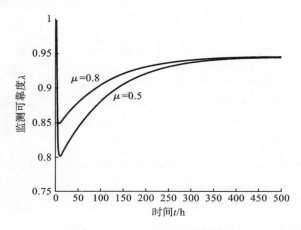

图 7　CDMS 监测可靠度时变曲线

(1) CDMS 的最小监测可靠度并没有出现在时刻 $t=0$,而是在其后某一点。原因在于:①给定边界条件是 $P_1(0)=1$,即 CDMS 监测中心在 $t=0$ 处可用度为 1;②最初监测系统本身不完善,随着分析与服务任务加重,各种故障迅速出现,导致系统可靠度急剧下降。

(2) 当 t 大于系统最小可靠度对应的时刻后,监测可靠度上升。原因在于:随着维护工程师对整个系统越来越熟悉,故障被逐渐排除,故障率下降,监测可靠度不断提高,并向各子系统可靠度的乘积 0.946 2 靠近。

(3) 监测中心修复率越高,系统监测可靠度越高。因此,给监测中心配备熟练的维护人员和模块化部件对提高 CDMS 的监测可靠度有重要意义。

(4) 监测中心是 CDMS 的核心,它的可用度大小直接决定着 CDMS 监测可靠度的高低。因此,在生产中对监测中心的维护至关重要。

4　结　　论

通过对集中分布式监测系统的结构特性进行分析,提出一种通用的分布式监测系统监测可靠性分析方法,并通过一个实例进行分析和验证。例证表明该方法对于监测系统的设计、日常维护及其故障诊断与预测具有一定的指导意义。

二次诊断与可靠性问题的研究是一个较新的研究领域,本研究只是结合集中分布式监测系统的特殊性对其进行整体可靠性建模与分析,还有很多问题有待进一步研究。如各因素（软件故障率、子系统各单元的修复特性以及子系统维护特性等）对系统监测可靠性的影响问题，多设备、多工序生产过程监测系统的可靠性分析与分配问题等等都是下一步需要研究的问题。

参 考 文 献

[1] 程侃.寿命分布类与可靠性数学理论.北京:科学出版社,1999.

[2] 张峥嵘,袁清珂.21世纪制造业的特点及其关键技术.机械工程师,1999,(1):1~3.

[3] Chou J H. *Study of condition monitoring of bridges using genetic algorithms*:[*Doctoral Dissertation*]. USA:Urbana,U of Illinois,2000.

[4] 胡友民,杜润生,杨叔子.冗余式分层监测系统可靠性分析.机械工程学报,2003,39(8):110~115.

[5] Thomas R B,Jane M B,Keller-McNulty S,et al. *Testing the Untestable*:*Reliability in the 21st Century*. IEEE Trans on Reliab,2003,52(1):118~124.

[6] Endrenyi J,Aboresheid S,Allan R N,et al. *The present status of maintenance strategies and the impact of maintenance on reliability*. IEEE Transactions on power systems,2001,16(4):638~646.

[7] 胡友民.状态监测系统可靠性研究:[博士学位论文].武汉:华中科技大学,2003.

[8] Dai Y S,Xie M,Poh K L,et al. *A study of service reliability and availability for distributed systems*. Reliability Engineering and System Safety,2003,79:103~112.

[9] 胡友民,杜润生,杨叔子.集中式监控系统可靠性分析.振动、测试与诊断,2003,23(1):6~9.

[10] 陆传赉.工程系统中的随机过程.北京:电子工业出版社,2000.

[11] 高社生,张玲霞.可靠性理论与工程应用.北京:国防工业出版社,2002.

[12] Okumoto K,Goel A L. *Optimum release time for software systems*. IEEE Trans on Reliab,1979,CH1515-6:500~503.

[13] 郭余庆,王岩.系统可靠性理论与应用.北京:煤炭工业出版社,1991.

Study of Markov Model Based Reliability for Distributed Monitoring System

Yi Pengxing　Du Runsheng　Yang Shuzi　Liu Shiyuan

Abstract　As an important means of improving the reliability of complex systems and ensuring that they can function better, distributed monitoring systems are usually designed and used in many areas. A centralized distributed monitoring system, which is usually used in monitoring system design, is presented. Based on reliability theory and the characteristics of this type of systems, monitoring reliability defined as the probability of successfully measuring and monitoring the object which is under control, an important performance measure for the distributed monitoring systems, is investigated. Furthermore, a case of application is used to illustrate the procedure. Analysis results show that the method proposed here is practical and valuable for designing monitoring systems and their maintenance.

Keywords: Centralized distributed monitoring system, Monitoring reliability, Markov chain, Availability

(原载《机械工程学报》2005 年第 41 卷第 6 期)

再论先进制造技术及其发展趋势

杨叔子 吴波 李斌

提 要 此文是《先进制造技术及其发展趋势》的深化与补充。首先,指出制造业在国家发展中所占的战略地位。接着,从三个方面十二点详细地分析了先进制造技术发展的特色与趋势:产品本身,"精"、"极"、"文";制造过程,"绿"、"快"、"省"、"效";制造方法,"数"、"自"、"集"、"网"、"智",并强调上述方面均应基于"制造"与"机械"这两个基点之上。最后,提出了先进制造技术发展的指导思想与我国应该优先发展的方向,强调必须自主创新,必须"以人为本"。

关键词:先进制造技术,数字化制造,集成制造,绿色制造,精密制造,网络化制造

0 前 言

此文是《先进制造技术及其发展趋势》[1]一文的深化与补充。

制造业在国民经济建设、社会进步、科技发展与国家安全中占有重要战略地位。概括讲,这体现在以下五个方面。

(1) 物质财富是人类社会生存与发展的基础,而"制造"是人类创造物质财富最基本的手段。人类社会发展所依赖的有四大物质文明支柱:材料、能源、信息和制造。"制造"是"永远不落的太阳"。

(2) 制造业是全面建设小康社会,加速实现现代化的第一位支柱产业。制造业是工业的主体,当今中国的制造业直接创造国民生产总值的 1/3,占整个工业生产的 4/5,为国家财政提供 1/3 以上的收入,贡献出口总额的 90%,制造业从业人员占全国工业从业人员总数的 90%[2-3]。我国要实现新型工业化,核心是要实现制造业先进化。

(3) 在制造业中,机械制造业尤其是装备制造业担负着为各行业各部门提供

装备的重要任务,是国民经济发展的基础;而其中,机床制造业又是装备制造业的心脏,是装备的"母鸡"。我们一再引用马克思在《资本论》中的一段不朽的名言:"大工业必须掌握这特有的生产资料,即机器的本身,必须用机器生产机器。这样,大工业才建立起与自己相应的技术基础,才得以自立。"

(4) 制造业是高技术产业的基础,制造业产品是高技术的载体。没有制造业,就没有高技术。信息技术、微电子技术、光电子技术、纳米技术、核技术、空间技术和生命技术等莫不与制造业有关。譬如,电子制造中所要求的高精度(控制精度趋于纳米级、加工精度趋于亚纳米级)、超微细(芯片线宽向 100 nm、运动副间隙向 12 nm 以下发展)、高加速度(芯片封装运动系统加速度向 12g 以上发展)和高可靠性(芯片千小时失效率要求小于 $1/10^9$)[4],如果没有尖端的制造装备及相应的技术,就无法达到如此高的技术水平。我国光电子制造的关键装备几乎全靠进口,尖端制造装备及技术正是西方对我国技术封锁的重点所在。

(5) 现代尖端军事装备及国防安全技术更是先进制造技术的载体,从根本上讲,这必须依靠自己,绝对不能寄希望于国外。现代战争已进入"高技术战争"的时代,武器装备的较量在相当意义上就是制造技术和高技术水平的较量。没有精良的装备,没有强大的装备制造业,一个国家既没有军事和政治上的安全,经济和文化上的安全也将受到巨大威胁。

由上可知,制造业是国家的基础性、前沿性、支柱性与战略性的产业。高度发达的制造业,对内,是实现新型工业化、加速实现现代化的必备条件;对外,是衡量国家竞争力的重要标志,是决定一个国家在经济全球化进程中国际分工地位的关键因素。可以说,没有制造,就没有一切。

因此,先进制造技术是工业发达国家的国家级关键技术与优先发展领域,我国也已将制造业科技发展中问题的研究列入我国科技发展中长期规划,这是极为正确的。我国目前在迅速和平崛起中,经济正在展翅高飞,创造奇迹。然而,就制造业而言,众所周知,我国是制造大国,不是制造强国;严格讲,是加工大国,不是加工强国,更不是制造强国,因为制造包含设计、加工、管理等方面的内容。

同科学技术与市场经济的发展相适应,作者在文献[1]中指出,先进制造技术,特别是先进机械制造技术有如下的发展趋势与特色:"数"是核心,"精"是关键,"极"是焦点,"自"是条件,"集"是方法,"网"是道路,"智"是前景,"绿"是必然。这八点实质上是指的三个不同方面:"精"的精密化,"极"的极端条件,是指产品本身而言;"数"的数字化,"自"的自动化,"集"的集成化,"网"的网络化,"智"的智能化,是指所采用的制造方法而言;而"绿"的绿色,是指制造过程而言。这三个方面,彼此根本不同,但又相互密切联系。朱熹有一名言:"问渠哪得清如许?为有源头活水来。"如果将产品比做渠,制造过程比做水,制造方法比做源,那么,好的产品就是清渠,好的制造过程就是活水,好的制造方法就是高品位的富源。高品位的富源涌出活水,活水流成清渠。正因为如此,我们所要求的是适应时代需要的好的产品,这一好的产品须由适应时代要求的好的制造过程产出,而这一好的制造过程又须依靠适应时代发展的好的制造方法实现。但是,与源、水、渠这三者的关系有所不同:第一,它们三者还存在着互动的关系,任何一方面的重大变化或发展,将可导致其他两方面的重大变化或发展;第二,还有其他因素特别是管理因素同这三者也有互动关系,尤其是管理因素同制造过程的关系很大。然而,从整体上看,对产品本身的要求毕竟处于主导地位。此文就是在前文的基础上对这三个方面来作深化与补充。

1 先进制造技术的发展趋势

1.1 先进制造产品的发展趋势

对产品本身而言,应包括"精"、"极"、"文"这三点。

(1)"精"应理解为"精确化"。制造就是为了生产出产品,产品必须满足要求。对产品的要求,可归纳为对产品几何方面与物理方面的要求。产品本身的几何形体(包括几何尺寸、几何形状和表面形貌等)就是要精密,而且越来越高,即所谓几何"精度"越来越高;产品的性能(包括物理性能、力学性能、化学性能乃至生化性能等)就是要精确,而且越来越严,这可称为物理"精度"越来越高。正因为产品的"精度"越来越高,所以对制造过程、制造方法才提出相应的要求,并推动它们的发展,以保证产品的"精度"得以实现。显然,产品的"精度"要得以实现,检测原理与技术占有极为重要的地位。"精"可以讲是先进制造技术发展的"关键"。

(2)"极"指"极端化"。"极端"或"极"在此是借用的,应理解为"苛刻化"。产品本身不但往往要"精",而且同时往往要"极";在几何形体上,极大、极小、极厚、极薄、极平、极柔、极圆、极方以及奇怪形状;在物理性能上,极高硬度、极高塑性、极大弹性、极大脆性、极强磁性、极强辐射性、极强腐蚀性、奇性怪能;有时还得在极端条件下制造。显然,这些产品往往就是科技前沿的产品,例如,"微机电系统(MEMS)"的产品,不但要求"精",而且要求"极",在制作中是关键的关键。所以,"极"可以讲是先进制造技术发展的"焦点"。

(3)"文"应理解为"人文化"。社会进步到今天,产品不仅是一个工业产品,只解决"实用"的问题,满足物质层面上的需要;还应该是一个艺术产品,文化含量高,特别是人文文化含量高,真正解决"物美"问题,满足精神层面上的需要,能同环境协调,能供欣赏,能悦人心,经得起"看",经得起"想"。"工业设计"等学科即由此而生。当然,"文"还可扩大到制造过程、制造方法,即"文明化"生产。所以,"文"可以讲是先进制造技术发展的"新义"。

1.2 先进制造过程的发展趋势

对制造过程而言,应包括"绿"、"快"、"省"、"效"这四点,而不仅仅是"绿"。

(1)"绿"是指"绿色","绿色"是从环境保领域中引用来的。工业的发展,特别是制造业的发展,众所周知,导致生态失衡、环境污染、资源枯竭,难于持续发展,甚至可能招来社会大灾难。人类社会的发展必将走向人类社会与自然界的和谐,走向我国传统一贯主张的"天人合一"。制造业的产品从构思开始,到设计、制造、销售、使用、维修,直到回收,再制造,都必须考虑环境保护。正如文献[1]中所述,所谓环境保护是广义的,不仅要保护自然环境,还要保护社会环境、生产环境,保护劳动者的身心健康。显然,制造过程的"绿"又是重中之重。"绿色"制造在我国科技发展中长期规划中也得到了高度的重视。所以,"绿"可以讲是先进制造技术发展的历史"必然"。

在这里应指出,所谓"绿色"产品是指无危害的产品,这是社会十分关注的。其实,此文上面所述产品本身的"精"与"文",已经包括了"绿色"的含义,这是十分明显的。

(2)"快"是指"快速化",即指对市场的快速响应,对生产的快速重组,这两个快速必然要求生产模式有高度柔性与高度敏捷性[5]。这一点是市场经济走向"买方市场"、"多变市场"、"顾客是上帝"、企业为了客户、满足"客户化"的必然结果。正是这一"快"的结果,强有力地推动着制造技术的进步与制造方法的发展。所以,"快"可以讲是先进制造技术发展的"动力"。

(3)"省"是指"节省",即指制造过程必须节省、节约、节俭,这是市场经济必然的要求。任何一个经济行为,都不同程度地讲节省,讲成本;市场经济,尤其是我国这么一个并不富裕的大国的市场经济,制造过程就更是不能不讲节省,不能不讲成本,不能不讲资源的优化配置,不能不讲制造过程各有关环节的优化配置。我国正用极大力气来倡导与推动建设节约型社会,就是一个极具战略性的举措。我国有句很好的谚语"一分钱办两分钱的事"就是这个意思。节约是我国一贯的传统美德。《老子》讲:"治人事天莫如啬。"这个啬,实质上就是要节约生产,节约使用。毛泽东同志在 20 世纪 50 年代中期一再强调,要勤俭办一切事业,"什么事情都应当执行勤俭的原则,这就是节约的原则"。所以,"省"可以讲是先进制造技术发展的"原则"。

(4)"效"指"高效",主要指"高生产率",即指单位时间内生产的产品数量多。固然市场经济与科技的发展,导致不确定性的因素猛增,市场的需求变化加快,使得产品非大量化、分散化、个性化的生产越来越强,但决不意味着单位时间内产品生产数量减少,相反,还应增加。这是市场快速响应的应有推论与含义。社会的进步由生产力的发展所推动,而生产力的发展则由生产率的提高所推动。高效低耗无污染应是生产过程所追求的。所以"效"可以看做是先进制造技术发展的"追求"。

1.3 先进制造方法的发展趋势

对制造方法而言,应包括"数"、"自"、"集"、"网"和"智"这五点,这同参考文献[1]中的论述没有太大的变化。

(1)"数"即指"数字化"。"工业化"发展的核心是"信息化","信息化"的核心是"数字化","数字化"也是先进制造技术发展的核心。数字化制造包括以设计为中心的、以控制为中心的与以管理为中心的数字制造。"数字化"贯穿与渗透制造全过程所用的方法中;并且发展得更广泛、更深刻,乃至导致生产方法与过程的重大变革。这就是说,"数字化"技术的发展必将导致制造技术的重大发展与变革[6]。所以,"数"是先进制造技术发展的"核心"。

(2)"自"即指"自动化"。"自动化"就是减轻人的劳动,就是强化、延伸、取代人的有关技术或手段,甚至是完成人无法完成的有关工作的技术或手段,是制造业发展必需的前提。"自动化"已从一般的自动控制、自动调节、自动补偿等发展到自学习、自组织、自维护、自修复等更高的水平,自动控制理论、技术、系统、元件在制造技术领域中有着极大的发展。所以,"自"是先进制造技术发展的"条件"。

(3)"集"即指"集成化"。"集成化"是技术的集成,是管理的集成,是技术与管理的集成。现代的制造技术就是集成的技术,先进制造技术就是制造技术、信息技术、管理科学与有关科学技术的集成。集成,就是交叉,就是融合,就是"博采百家";甚至可以讲,一个高新技术的出现,极少不是"兼收并蓄"其他有关学科或领域的理论与技术的。所以,"集"是先进制造技术发展的"方法"。

(4)"网"即指"网络化"。"网络化"是信息技术(特别是数字技术)与通信技术的交融。网络化既是制造业信息化、集成化必需的基础，又是制造业信息化、集成化的进一步发展的需要，这将使制造企业走向全球化、整体化、有序化，资源互补共享。在"网络化"中，企业有无自己具有竞争力的核心业务，是企业能否成功走向网络化的关键所在。所以，"网"是先进制造技术发展的"道路"。

（5）"智"即指"智能化"。"智能化"将使制造过程具有处理大量信息与不完整信息、错误信息的能力，具有强大的自诊断、自修复和自组织能力，主动协调与协同能力以及非逻辑处理能力。它将是数字化、自动化、集成化、网络化的有机交融与高度发展。但是，它决不意味完全取代人，而是"人"与"机"的高度协调与有机统一。所以，"智"是先进制造技术发展的"前景"。

2　先进制造技术发展的指导思想与优先发展方向

正如前文所指出，先进制造技术有两个基点，一是"机械"，一是"制造"。上述三点加四点加五点的三个方面十二点，必须扎根在这两个基点上，此即必须研究与发展有关这两个基点的理论与机理，以上十二点必须在这两个基点上来研究、来开发、来发展。

为了进一步发展先进制造技术，应确立如下四点指导思想。

（1）加强基础，立足应用。技术就是为了应用，面向实际。为了应用，就是立足之点。但是，为了技术有足够的发展后劲，有原创性的源头，能自主创新，就必须加强基础工作，包括基础研究。

（2）关注全面，突破重点。有所不为，才有所为；没有重点，就没有政策。因此必须抓住重点，但是必须关注全面。因为：第一，重点不是孤立的，没有"面"的支撑，也难于有"点"的良好的发展；第二，"面"上可能出现新的重"点"，有了"面"，就可能不会丢失这些新出现的重"点"。

（3）扎根实际，务求超越。我们发展先进制造技术，必须面对与扎根我国的实际，实事求是。脱离实际，违背实际，定遭失败。但是，我们的扎根实际，就是为了下一步的务求超越。如果不在若干重要方面超越，那就只能长期跟踪，永远落后，这决非我们的期望，也决不许可。

（4）实干巧干，出奇制胜。为了既能扎根实际，又能务求超越，那就只能脚踏实地，苦干实干；同时，开动脑筋，寻求捷径，巧干妙干，包括利用"后发"的有关优势，以奇取胜。我们应坚信：真理是不可穷尽的，捷径是存在的；只要在战略上敢于藐视困难，而在战术上高度重视困难，下定决心，团结一致，排除万难，一定能求得捷径,自主创新,实现超越。

为此,在我国可考虑将以下十点作为优先发展方向。这十点的前八点,同制造方法有关;这八点中,前五点同制造本身直接有关,后三点同学科交叉更有关;而第九点同制造过程有关,第十点同产品本身有关。

(1) 机电产品创新设计、优化设计的理论与方法。

(2) 网络协同制造的策略理论与关键技术。

(3) 新型成形制造原理与技术。

(4) 数字装备制造理论与技术。

(5) 量值溯源与新型测量的原理与技术。

(6) 纳米制造科学与技术。

(7) 生物制造与仿生机械的理论与技术。

(8) 微系统与新一代电子制造理论与关键技术。

(9) 绿色制造科学与技术。

(10) 面向国家重大工程与国家安全的制造科学与技术。

3 结 论

如同文献[1]一样,在此文结束时,应强调指出:在科学技术高度发达与高速发展的今天,要更高度重视人的作用,就必须"以人为本";有了人,有了人才,有了拔尖创新的人才,才可能拥有真正的实力,才可能真正实现自主创新;没有了人,丢失了人才,就会丧失一切,"人是生产力中最具有决定性的力量"。国力的竞争,归根结底在于人才特别是具有高素质的高级人才的竞争,制造业自不例外。

参 考 文 献

[1] 杨叔子,吴波.先进制造技术及其发展趋势[J].机械工程学报,2003,39(10):73-78.

[2] 路甬祥.团结奋斗 开拓创新 建设制造强国[J].制造技术与机床,2003,1:6-13.

[3] 宋健.制造业与现代化[J].机械工程学报,2002,38(12):1-9.

[4] 雷源忠,雒建斌,丁汉,等.先进电子制造中的重要科学问题[J].中国科学基金,2002,16(4):204-209.

[5] 杨叔子,吴波,胡春华,等.网络化制造与企业集成[J].中国机械工程,2000,11(1-2):45-48.

[6] FREEDMAN S.. *Overview of fully integrated digital manufacturing technology*[C]//. Proceedings of IEEE Winter Simulation Conference,cl999:281-285.

Further Discussion on Trends in the Development of Advanced Manufacturing Technology

Yang Shuzi　Wu Bo　Li Bin

Abstract　This one is the deepening and supplement of "Trends in the development of advanced manufacturing technology". The strategic status of manufacturing industry in state development is pointed out firstly. Then, a detailed analysis is given on the features and trends in the development of advanced manufacturing technology in altogether 12 points from three aspects: "precision", "extreme" and "culture" from the aspect of the product itself; "green", "rapidness", "saving" and "efficiency" during manufacturing process; "digit", "auto", "integration", "networking" and "intelligence" in view of manufacturing method. In addition, it is emphasized that all the above aspects should be based on the two base points of "manufacture" and "machinery". Finally, that the guiding ideologies for the development of advanced manufacturing technology and the aspects China should give priority to in development is put forth, while the principles of independence, innovation and "human-orientation" are stressed.

Keywords: Advanced manufacturing technology, Digitalization manufacture, Integration manufacture, Green manufacture, Precision manufacture, Network manufacture

(原载《机械工程学报》2006 年第 42 卷第 1 期)

基于神经网络信息融合的铣刀磨损状态监测[*]

李锡文　杨明金　谢守勇　杨叔子

提　要　为了获得铣削加工过程中铣刀后刀面磨损的全面评价,用铣刀后刀面磨损带面积作为衡量刀具磨损量的一个评价指标。提取和精选了 8 个对铣刀后刀面磨损状态敏感的无量纲特征参数并经归一化处理后,作为基于神经网络信息融合的铣刀磨损状态监测系统的输入信号。采用 3 层 BP 神经网络模型,利用其多传感器信息融合功能在线监测了铣刀后刀面磨损带宽度和磨损带面积。监测系统的输出结果与实际测量结果基本吻合。

关键词:铣刀,磨损,状态监测,神经网络,信息融合

0　引　言

刀具的磨损程度常用后刀面磨损带中间部分平均宽度(常以 L_{VB} 表示)来评价,ISO 规定的外圆车刀刀具磨钝标准只适用于车削实验[1-2]。铣削是断续切削,其磨损比车刀的磨损复杂,不同的切削条件下产生不同的磨损方式和磨损带形状,因此 ISO 制定的刀具磨钝标准很难在实际生产中推广。

铣刀磨损状态可由切削力、扭矩、主轴端径向振动位移和主轴功率等多个物理量来感知,单一感知参数不能全面反映铣削加工中刀具磨损状态的变化,因而需使用多传感器信息融合方法。基于神经网络的信息融合方法具有很强的鲁棒性、容错性和自学习能力[3-5],本文基于此,应用神经网络信息融合法,对铣刀磨损

---

[*] 国家"973"重点基础研究发展计划资助项目(项目编号:2005CB724101)和国家自然科学基金资助项目(项目编号:50575087)。

状态监测进行研究。

1 铣刀磨损量的评价指标

影响刀具后刀面磨损带的因素很多,不同加工条件、工件材料、刀具几何角度等都可能造成不同的磨损带,仅仅用刀具后刀面磨损带中间部分的带宽很难真实、准确和完整地评价刀具的磨损状态。多次实验研究表明[3],可用后刀面磨损带面积(以 A_{VB} 表示)作为衡量刀具磨损程度的另一个指标。其原因如下。

(1) 铣刀的磨损比车刀的磨损复杂,L_{VB} 是磨损带的一维度量,只能评价磨损带宽度变化,不能评价磨损带长度的变化。铣削加工时,轴向切削深度的变化可能造成磨损带宽度不变,而磨损带长度变化。

(2) A_{VB} 是磨损带的二维度量,能评价磨损带长度、宽度两个方向的变化,能更准确地反映真实的磨损带变化。

图 1 为 HSS ϕ10 mm 三刃立铣刀渐进磨损过程中后刀面磨损情况(加工条件:主轴转速 360 r/min,切削宽度 9 mm,切削深度 2 mm,每齿进给量为 0.083 3 mm/齿。工件材料为铸铁,顺铣加工)。图中表明:立铣刀渐进磨损过程中每刀齿 L_{VB} 和 A_{VB} 的变化曲线与车刀的磨损曲线相似,即存在初期磨损、正常磨损和急剧磨损三个阶段。但在急剧磨损阶段,A_{VB} 比 L_{VB} 能更真实、准确地评价刀具的磨损状态。

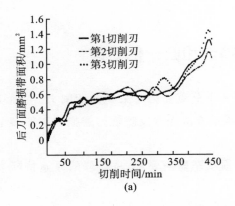
(a)

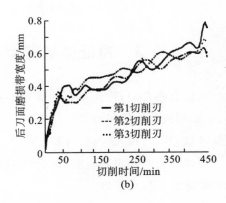
(b)

图 1 刀具后刀面磨损带参数变化示意图

2　神经网络信息融合基本原理

神经网络是由许多神经元按照一定的拓扑结构相互连接的网络结构[5-6]。根据 Kolmogorov 定理，3 层 BP 神经网络可以逼近任一连续函数，因而能实现任意复杂非线性映射问题[7-8]。采用 3 层 BP 神经网络模型来实现铣刀磨损状态监测多传感器信息融合。

设输入向量为 $\boldsymbol{x}=(x_1,x_2,\cdots,x_N)$，输出向量为 $\boldsymbol{y}=(y_1,y_2,\cdots,y_M)$，则 $\boldsymbol{x}$ 和 $\boldsymbol{y}$ 之间的映射关系满足

$$y_i = \sum_{j=1}^{H} W_{ji}^{(o)} S_j \left(\sum_{k=1}^{N} W_{kj}^{(i)} x_k + \Phi_j \right) \quad (i=1,2,\cdots,M) \tag{1}$$

式中：H——隐层节点数；

$W_{ji}^{(o)}$——隐层节点 j 到输出层节点 i 的输出权值；

$W_{kj}^{(i)}$——输入层节点 k 到隐层节点 j 的输入权值；

Φ_j——隐层节点 j 的门槛值；

S_j——作用函数，取 Sigmoid 函数：$f(x)=(1+\mathrm{e}^{-x})^{-1}$，$f(x)$ 取值范围 $[0,1]$。

由式(1)可知，当网络的拓扑结构固定后，输出向量和输入向量之间的映射关系只与输入层和输出层的权值有关。因此，可以给定一组学习样本集，对网络进行反复训练，通过调整权值，直至达到期望的性能为止。经过学习和训练后的神经网络，可以对输入向量进行网络联想，获得输出。

3　特征参数提取、精选和归一化处理

在信息融合中，特征参数的提取和精选十分重要。对采集的切削力、主轴功率、主轴端振动位移和加速度信号按图 2 所示方法提取反映铣刀磨损状态的多个特征参数，并进行特征再提取和特征精选。

根据已有的知识和经验，结合铣刀磨损状态监测要求，从提取的原始特征参数中精选如下特征参数。

(1) 切削力

在时域分析中，X 方向的切削力每转均值 $\overline{F}_x$ 和方差 σ_X 对刀具磨损敏感，把这两个特征参数变换为切削力变动系数 C_1，则

$$C_1 = \sigma_X / \overline{F}_x \tag{2}$$

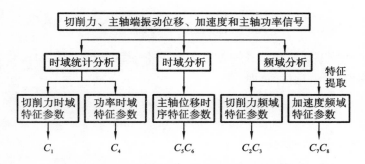

图 2　监测信号特征参数提取和精选框图

在频域分析中,切削频率处的功率谱值对刀具磨损敏感。设锋利刀具切削时 X、Y 方向切削力切削频率处的功率谱值为 G_x、G_y,磨损刀具切削时 X、Y 方向切削力切削频率处的功率谱值为 G'_x、G'_y,则功率谱值的变化系数 C_2、C_3 为

$$\begin{cases} C_2 = G'_x/G_x \\ C_3 = G'_y/G_y \end{cases} \tag{3}$$

(2) 主轴功率

时域分析中主轴功率每转均值 $\overline{P}$ 和方差 σ_P 对刀具磨损敏感,把这两个特征参数变换为主轴功率的变动系数 C_4,则

$$C_4 = \sigma_P/\overline{P} \tag{4}$$

(3) 主轴端振动位移

通过对 Y 方向振动位移信号进行时域、频域和时序分析,提取的多种特征参数对刀具磨损敏感,对这些特征参数进行分析和精选,选择自回归 AR(p) 时序模型系数作为主轴端振动位移信号原始特征参数。设锋利铣刀切削时主轴端振动位移信号的模型系数为 φ_1、φ_2,磨损刀具切削时主轴端振动位移信号的模型系数为 φ'_1、φ'_2,则模型系数的变动系数 C_5、C_6 为

$$\begin{cases} C_5 = \varphi'_1/\varphi_1 \\ C_6 = \varphi'_2/\varphi_2 \end{cases} \tag{5}$$

(4) 主轴端振动加速度

X、Y 方向振动加速度频谱高频处的功率谱值对刀具磨损敏感,选择(0.5～2.5) kHz 间的频段能量作为反映铣刀磨损的原始特征参数。设锋利铣刀切削时 X、Y 方向振动加速度在(0.5～2.5) kHz 间的频段能量为 G_{ax}、G_{ay},磨损刀具切削时 X、Y 方向振动加速度在(0.5～2.5) kHz 间的频段能量为 G'_{ax}、G'_{ay},则频段能量变动系数 C_7、C_8 为

$$\begin{cases} C_7 = G'_{ax}/G_{ax} \\ C_8 = G'_{ay}/G_{ay} \end{cases} \tag{6}$$

式(2)~(6)中所选择的各特征参数幅值不一致,对其进行归一化处理,使特征参数被压缩到[0,1]区间,即

$$q' = (q - q_{\min})/(q_{\max} - q_{\min}) \tag{7}$$

且
$$q_{\max} = \max(q), \quad q_{\min} = \min(q)$$

式中:q——归一化前的特征参数;

q'——归一化后的特征参数。

式(2)~(6)中定义的8个特征参数都为无量纲参数,对铣刀磨损状态敏感,能有效反映铣削加工过程中铣刀的磨损状态。

4 实 验

4.1 实验方案

(1) 实验条件

机床:XHK 5140 立式加工中心;刀具:高速钢直柄立铣刀 $\phi 10$,螺旋角 θ_h 为 30°,刀齿数 N_f 为 3;工件材料:铸铁(190 mm×90 mm×50 mm);铣削方式:顺铣,不使用切削液;切削参数:主轴转速 360 r/min,切削宽度 9 mm,切削深度 2 mm,铣刀每齿进给量 0.083 3 mm。

(2) 测试仪器

Kistler 9257A 型三向测力平台,测量工件所受的 X、Y、Z 方向切削力分量;AVX 92B 型振动电涡流位移传感器,测量主轴端 X、Y 方向振动位移;B & K 4369 型压电式加速度传感器,测量主轴端 X、Y 方向振动加速度;JC 10 型读数显微镜,测量铣刀刀齿后刀面磨损带尺寸;研华 PCL 818HG 型高速多通道数据采集板,对输入的模拟信号进行 A/D 转换、脉冲信号计数和计时;主轴功率信号从主轴驱动单元直接引出,用 PC 工控机完成数据采集和信号处理。数据采样频率 2 000点/转。

4.2 信息融合过程

用测试仪器监测铣刀加工过程中后刀面磨损状态(L_{VB} 和 A_{VB})。采集多路传感器的输出信号,进行预处理和特征提取;根据预处理结果对特征信号进行特征

精选；将精选的特征参数进行归一化处理，为 3 层 BP 神经网络模型的输入和模式识别提供标准数据形式。归一化后的特征参数如表 1 所示。

表 1 归一化后的特征参数

磨损带宽度/mm	磨损带面积/mm²	C_1	C_2	C_3	C_4	C_5	C_6	C_7	C_8
0	0	1	0.583 2	0	0	0.805 5	0.199 5	0.595 0	0.176 1
0.100	0.063	0.585 8	0.491 2	0.013 1	0	0.561 9	0	0.037 5	0.040 1
0.192	0.192	0.269 8	1	0.258 5	0.133 5	1	0.198 8	0.177 9	0.160 0
0.325	0.363	0.226 6	0.292 4	0.487 7	0.355 8	0.735 9	0.100 2	0	0
0.403	0.533	0.135 1	0.625 3	0.353 8	0.236 7	0.332 0	0.901 8	0.805 8	0.988 2
0.520	0.573	0.062 4	0	0.539 5	0.263 9	0	0.833	0.733 7	0.676 9
0.590	0.875	0.016 9	0.011 1	0.806 4	0.589 1	0.026 4	0.614 7	1	1
0.690	1.207	0	0.414 9	1	1	0.217 2	1	0.395 4	0.412 9

将归一化处理后的特征参数和已知的铣刀后刀面磨损值（离线测量）作为神经网络训练样本集，反复训练，直至达到期望的性能。其中，信息融合模型网络输入层节点数为 8，与特征参数个数相等；网络隐层节点数为 8（根据经验确定）；网络输出层节点数为 2，分别对应铣刀后刀面磨损带宽度和磨损带面积。提取的特征参数为主轴转动一周或几周的平均切削情况，未考虑铣刀单齿磨损量。网络学习采用误差逆传播算法，训练样本数不小于 8。

将训练好的神经网络应用于铣刀磨损状态在线监测，经神经网络联想，获得加工过程中当前铣刀后刀面磨损值。待检验的归一化后的特征参数如表 2 所示。

表 2 待检验的特征参数

检验序号	磨损带宽度/mm	磨损带面积/mm²	C_1	C_2	C_3	C_4	C_5	C_6	C_7	C_8
1	0.213	0.213	0.347 8	0.465 5	0.315 5	0.269 1	0.974 6	0.188 5	0.492 8	0.091 9
2	0.393	0.505	0.249 2	0.000 2	0.509 1	0.443 3	0	0.530 4	0.385 1	0.603 8
3	0.447	0.599	0.090 5	0.131 1	0.479 2	0.279 0	0.099 1	0.944 0	0.374 3	0.858 7
4	0.617	1.012	0.109 5	1	0.948 2	0.934 7	0.228 7	1	0.607 3	0.239 4

4.3 实验结果及分析

经过上述神经网络信息融合过程，对铣刀后刀面的磨损状态进行了在线监测,结果如图3所示。从图中可以看出,铣刀后刀面磨损状态监测系统输出的磨损带宽度和磨损带面积与实际测量结果基本吻合,表明本研究所建立的基于神经网络融合法的铣刀磨损状态监测模型是合理的、有效的。

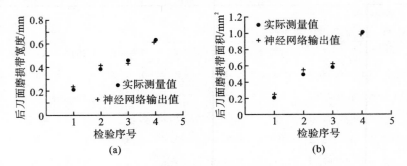

图3 神经网络信息融合法监测结果
(a)后刀面磨损带宽度;(b)后刀面磨损带面积

5 结 束 语

在对反映铣刀后刀面磨损状态的特征参数进行提取和精选的基础上,选择归一化处理后的8个无量纲参数作为基于神经网络的铣刀磨损状态监测系统的输入信号,通过训练好的神经网络的信息融合功能在线监测了铣刀后刀面磨损宽度和磨损面积,获得了较好的监测效果。

基于神经网络信息融合的铣刀磨损状态监测系统,充分利用多特征参数的信息互补性,对来自多传感器的原始信息进行智能综合,从中获取被监测对象的一致性评价,有效地提高了铣刀磨损状态监测的可靠性,能满足铣刀磨损状态在线监测要求。

参 考 文 献

[1] 孟少龙.机械加工工艺手册:第一卷[M].北京:机械工业出版社,1992.

[2] 陈日曜.金属切削原理[M].北京:机械工业出版社,1985.

[3] 李锡文.螺旋立铣刀状态监测的关键技术研究[D].武汉:华中理工大学,1998.

[4] 马平,吕锋,杜海莲,等.多传感器信息融合基本原理及应用[J].控制工程,2006,13(1):48~51,77.

[5] 虞和济,陈长征,张省,等.基于神经网络的智能诊断[M].北京:冶金工业出版社,2000.

[6] 李强,杨晓京,魏岚. 基于神经网络信息融合的智能机器人[J]. 机电工程技术,2006,35(6):72~74.

[7] 丛爽. 神经网络、模糊系统及其在运动控制中的应用[M]. 合肥:中国科学技术大学出版社,2001.

[8] Lippmann R P. *An introduction to computing with neural nets*[J]. IEEE ASSP Magazine,1987,4(2):4~22.

Wear Condition Monitoring of Helical Cutters Based on Neural Network Information Infusion Method

Li Xiwen　Yang Mingjin　Xie Shouyong　Yang Shuzi

Abstract　To obtain comprehensive evaluations of major flank wear of a helical cutter in the milling process, the wear land area was proposed as an index for estimating the wear out of milling cutters. In the research, 8 dimensionless characteristic parameters, which are sensitive to major flank wear condition of the cutter, were extracted, selected and normalized as input signals of the wear condition monitoring system based on neural network information infusion method. By three-layer back propagation neural network model, with its capability of multi-sensor information infusion, major flank wear land width and wear land area of the helical cutter were monitored online. The output results of the monitoring system were consistent with the tested data.

Keywords:Helical cutter, Wear, Condition monitoring, Neural network, Information infusion

(原载《农业机械学报》2007 年第 38 卷第 7 期)

以人为本——树立制造业发展的新观念

杨叔子　史铁林

提　要　首先从历史角度论述了制造业对人类社会发展的重大作用；其次，论述了如同所有技术一样制造技术对社会正反两方面的作用；再次，引用了"国际工业设计联合会"关于工业设计的定义，论述了创新技术要人性化；最后，论述了"以人为本"的制造业发展的内涵。

关键词：以人为本，制造，机械制造，装备制造

0　前　言

路甬祥同志2007年11月4日，在中国机械工程学会年会上所作的《坚持科学发展，推进制造业的历史性跨越》[1]主题报告，是一个极为重要的指导性文件。他在报告中明确指出："我们必须准确把握时代特征，深刻认识我国国情，树立新的发展观念，以科学发展观为指导，促进制造业和制造技术的发展和创新，推动并加快实现我国由制造大国向制造强国、创造大国的跨越。"的确，"这既是我们面临的挑战，也是我们难得的历史机会"。本文试图论述一些我们对这一报告的认识与体会及有关想法。

新世纪的我国制造业必须进一步树立新的发展观念。这个新的发展观念，毫无疑问，应是科学发展观在制造业的贯彻与实现，其核心就是"以人为本"。科学发展观指导着我国各个领域、各个部门、各个行业及各个企业的发展，显然，也指导着我国经济部门、工业、制造业、机械制造业、装备制造业的发展。仔细考察与分析一下，可以确认一个事实，随着人类文明的进步，世界经济、科技、工业、制造业等的发展，也正在不以人的意志为转移，或直或曲或快或慢地转移到以人为本的轨道上来。

1　人类的创造与工具的制造

人类文明的基础是物质文明，人类物质文明的基础应该说是从制造开始建造，并由制造、材料、能源、信息四者作为支柱所构成的。恩格斯在《自然辩证法》中深刻指出："直立和劳动创造了人类，而劳动是从制造工具开始的。"他接着形象地讲，而制造工具是从制造第一把石刀开始的。到今天，工具发展成为机械、装备乃至系统，但本质上仍是工具；当然，最简单的机械仍是工具。制造、工具制造创造了人类，人类为了自己的需要，也创造了制造、工具，更用制造、工具制造创造了科学技术、物质文明、人类文明，严格讲，人类以制造、工具制造为科学技术、物质文明，乃至人类文明的创造与发展不断地奠定基础，同时，在这一创造、发展与奠定过程中，制造、工具制造业不断发展着自己。

人类的发展、人类需要的发展、人类文明的发展，同制造的发展、工具制造的发展是分不开的；而且，人类一旦诞生，人为了自己的需要，就在此发展中始终处于主导地位，但这绝不是讲，制造、工具制造对人的发展、人的需要、人类文明没有极为巨大的作用。可以说，没有制造，就没有工具制造；没有金属发现，就没有金属农具制造；没有热力学定律发现，就没有蒸汽机制造；没有电磁现象发现，就没有电机与电器制造；没有半导体发现，就没有芯片与计算机制造；从而就没有农业革命，就没有第一次工业革命，就没有第二次工业革命，就没有计算机与信息革命等，就不能满足人类不断发展的需要。

可以肯定，先进制造、光电子制造、纳米制造、生物制造等，必将导致人类生产、生活、思维、社会更深刻的革命与变化。从这个意义上讲，没有制造，没有工具制造，没有装备制造，没有机床制造，就没有人类的过去、今天与未来。马克思在《资本论》中讲得极为正确，一点也没有过时："大工业必须掌握这一特有的生产资料，即机器的本身，必须用机器生产机器，这样，大工业才能建立起自己相应的技术基础，才得以自立。"机器的本身，实质上就是工具；"用机器生产机器"，实质上就是工具制造。

"制造"不只是"加工"。从制造第一把石刀开始，"制造"就包含了三点内涵：构思、加工与使用，而对加工对象的检验也自然蕴涵于其中。到今天，即便是制造最复杂的装备，这三点内涵也未变，只不过是内容或环节更加发展、更加明确、更加丰富、更加系统、更加深刻罢了！构思与设计（即加工前），加工（含装配、包装，即加工中），营销、服务与使用（即加工后），当然，扩大点讲，乃至回收与再制造，仍是制造的内涵。至于检验与测量，还有管理与经营也自然蕴涵于其中。

我们常讲，我国是"制造大国"，不是"制造强国"，严格地讲，不但不是"制造强国"也不是严格意义上的"制造大国"，只是"加工大国"而已。众所周知，"加工"并非"制造"的全部。制造中，构思与设计是前提，是先天，此点失误，一切皆空；加工是关键，是后天，此点失误，先天作废；营销、服务与使用是兑现，是根本，此点失误，前功尽弃；而作为质量保证手段的检测，作为现代企业中枢的管理，是制造生命之所系，更是先进制造所不能须臾离开的灵魂。在上述所有环节中，都存在着以这种或那种方式的怎么创新或改进、怎么操作或使用，而怎么创新或改进、怎么操作或使用，这都是制造过程中或者准确讲都是产品整个生命周期中有关人员的怎么创新或改进、怎么操作或使用。总之，上述所有环节都以这种或那种方式离不开人。

2　科技的双刃剑与人的价值观取向

制造、科学技术，为人所创造（创新，包括发现、发明、改进），为人所使用（包括操作），最终会以这种或那种方式作用（包括服务）于人。所谓这种或那种方式的"作用于"可以作用于人与自然环境的关系，或作用于社会内部的人际关系，或作用于制造过程中的有关人员，或作用于用户。为什么创，为什么用，怎么创，怎么用，创得怎样，用得怎样，往往不取决于制造、科学技术本身；制造、科学技术的最终价值如何，往往不取决于制造、科学技术本身，这一切往往取决于能否合宜"创"，恰当"用"并使之能否正确服务于人的有关人员。

武器能否成为武器，能否发挥最大效力，往往不在武器本身，而在于使武器能否合宜地根据客观规律与需要创制出来、能否按其原理恰当使用、能否在使用中有利于人类社会的人，总之，关键在于策划用武器来干什么的人。武器可以用来抵御猛兽枭禽、坏人的侵害，也可以用作杀戮无辜的手段。某种化工装备能不能有益于人类社会，关键不在于能否按科学原理来创制与使用，能否在当前发挥巨大的经济效益，而在于从长远看，有无相应的措施能保证不破坏自然环境与人的和谐，而能有益于人类社会的可持续发展。"人无远虑，必有近忧。"这是真理！目前，造成环境严重污染、生态严重失衡、气候严重恶化等的主要源头莫不同工业中的化学过程与化工产品有关，固然首先是化工工业，但也包括冶金工业、电子工业等。这些工业都是制造业。我国 2005 年 GDP 的 33.3% 由制造业所创，工业中的产值 80% 由制造业提供，工业中的从业人员 90% 由制造业占有。至于资源危机，特别是能源危机,也莫不同制造业密切有关。这些均严重导致社会不可持续发展。

科学技术的迅猛发展,工业革命的伟大胜利,人类以为可以依靠科学技术,凭着制造手段,创造各种工具与方法,战胜自然,征服自然,驾驭自然,奴役自然,满足人类无止增长的欲望,成为自然界的主宰。然而,"福兮,祸之所倚",物极必反,战胜堪忧,征服未成,自然界正在严重报复与沉重惩罚人类,人类正在自食恶果,人类社会面临着难于持续发展,甚至面临着严重灾难。

早在1992年,国际上1 575位科学家就联合发表了一个宣言,叫做《世界科学家对人类的警告》。宣言一开始就尖锐地指出,人类与自然正走上一条相互抵触的道路。诚然,人类大不同于其他生物,创造出自然界中从未有过的万事万物,认识出自然界中千奥万妙,展示着"人为万物之灵"的人类智慧的伟大与奇迹,人类绝不是自然界可以任意奴役的奴隶。正如《老子》所讲:"道大,天大,地大,人亦大,域中有四大,而人居其一焉。"但是,人类更绝对不能成为自然界的主宰,任意"改造"、驱使、奴役、宰割自然界,绝不能把文艺上精神世界作用的夸张当成现实中可以实现的事实,人类只能是也只应是自然界、生物界中具有自觉的主动性与高度的创造性的有机组成部分。宋代程颐讲得多么深刻:"安知有天道而不知有人道者乎?道,一也!岂人道自是一道,天道自是一道?"朱熹讲得更透彻:"天即人,人即天。人之始生,得之于天,即生此人,则天又有人矣!"人与自然不可分割。人能不断深刻认识自然,积极主动顺应自然,合宜恰当利用自然,能全面协调按客观世界规律"制天命而用之",适度地改造相应的自然,促使人与自然更加有机融合,更加和谐共生。这就是我国一贯主张的"天人合一"。

2001年诺贝尔化学奖得主日本科学家野依良治2007年3月在北京的学术研讨会上明确指出:科技给人类带来了巨大的利益,但也带来了巨大的伤害。他告诫说:人们的价值观不改变,就将面临灾难。他认为,科学与人文以及社会科学应该成为一个统一体系,才可摆脱此一困境。他讲的是人的价值观的问题,是需要与培养什么人的问题。众所周知,我国制造业的发展给我国社会进步创立了强大的物质基础,给我国人民带来了巨大的财富,但与此同时也付出了惨重的环境代价与资源代价!2007年中国科学技术协会年会上有专家讲得多么透彻:节能减排,与其讲关键在科技,不如讲根本在制定与执行有关政策与措施的人。一切的要害就在于有关人员要真正贯彻与落实科学发展观。

3 创新技术人性化

"解铃还须系铃人。"制造、工业、科技带来了社会不可持续发展的严重危机，解决这一严重危机的办法不是倒退，倒退到工业革命以前去，倒退是没有出路的，还是要靠发展，靠深入技术革命、推进科技发展、实现科学发展的人，靠人以相应的更好的制造、工业、科技解决这一危机，并推动社会可持续向前发展。读一读"国际工业设计联合会"(International Councel of Societies of Industrial Design, ICSID)对工业设计的定义，很有启发，很有裨益。ICSID对工业设计的定义由目的与任务两部分组成。

目的是："设计是一种创造性的活动，其目的是为物品、过程、服务以及它们在整个生命周期中构成的系统建立起多方面的品质。因此，设计既是创新技术人性化的重要因素，也是经济文化交流的关键因素。"设计是经济文化交流的关键因素，其义易明；而设计是创新技术人性化的重要因素，其义极要。工业、工业生产、工业产品的发展要求相应的技术不断创新，而创新技术应该而且必须人性化。此即应在创新技术中不断融入人文思想，融入人文关怀，融入各有关方面的和谐。

任务是："设计致力发现和评估与下列项目在结构、组织、功能、表现和经济上的关系：①增强全球可持续性发展和环境保护（全球道德规范）；②给全人类社会、个人和集体带来利益和自由；③最终用户、制造者和经营者互利共赢（社会道德规范）；④在世界全球化的背景下支持文化的多样性（文化道德规范）；⑤赋予产品、服务和系统以表现性的形式（语义学）并与它们的内容相协调。"显然，这所要发现与评估的五条关系，讲的就是道德、道德规范，也就是经由道德、道德规范以达到创新技术人性化的目的这一整体思想所彰显的和谐。所以，可以认为，第一条是讲人类社会与自然环境关系的和谐，第二条是讲人类社会内部关系的和谐，第三条是讲人类社会中工业商业活动有关方面关系的和谐，第四条是讲跨地区间多种文化间关系的和谐，第五条是讲产品、服务与系统整个过程有关方面的形式与内涵间关系的和谐。当然，第五条的关系紧密地同前四条相联互融。

我们可以概括地认为，ICSID对工业设计的定义给我们的启发是，时代发展到今天，科学发展到今天，工业生产、制造业生产，不仅涉及产品本身，不仅涉及产品生产过程与生产方法，而且涉及产品本身、生产过程与生产方法同有关环境的关系，涉及全局与长远。不仅涉及物质层面、操作层面、财富层面，还涉及精神层面、制度层面、文化层面。因此，对工业、制造业的发展，应通过以技术为手段的"实"，来创造以科学为基础的"真"与以人文为内涵的"善",来体现以艺术为形式的"美",从而达到真善美相互和谐统一的且能不断满足人的需要的"新"。

科学求"真",即接触、研究、认识、把握客观世界实际及其规律,力求所作所为符合客观实际,办事能够成功。

人文务"善",即研究、了解、认识、体悟主观的精神世界,力求满足精神世界需要,实现人文关怀。

艺术完"美",即以相应悦人目、乐人耳、赏人心、耐人品的和谐形式,力求体现形式所包含的真与善的内涵。

技术致"实",即创造合乎科学的恰当的方法或手段,去达到能够体现真与善内涵所需的美的形式。

简而言之,以人性化的技术创造科学、人文与艺术相互和谐统一的新。我们还可以讲,学术(包括科学与人文)是觅源,寻觅客观与主观世界源头的真相,是"发现";艺术完美,反映源头真相,是"表现"。技术致实,实现与达到真善美的统一,是创新性的活动,是"实现"。工业、工业生产、工业产品,毫无疑问,首先包括制造,就应做到这三"现"相统一的"新"。这种"新",可以讲首先就应是 ICSID 所讲的创新技术人性化。因此,创新技术既应有科学的真的严谨,又应有人文的善(爱)的深沉,也应有艺术的美的浪漫,从而就应有三者相统一的技术所应体现人性化的和谐。在《再论先进制造技术的发展及其趋势》[2] 的 12 个字的趋势中,就或隐或现或直接或间接或多或少体现着先进制造技术的人性化趋势。对产品本身,"精"、"极"、"文",即产品本身的精确化是关键,主要就是保证产品质量,极端化条件下的产品质量是焦点,是关键的关键,这是为了人的需要;产品应有人文文化含量,是新义,是人性化的直接体现。对生产过程,"绿"、"快"、"省"、"效",绿色化,快速化,节约化,高效化,归结起来,就是无污染,高效率,低消耗,这正是"以人为本"的减排节能的,建设环境友好型、资源节约型社会的要求,这就是"绿色制造",这也是制造历史发展的必然。对生产方法,"数"、"自"、"集"、"网"、"智",即数字化、自动化、集成化、网络化、智能化,这正是实现上述生产过程、保证产品质量的方法,也是解放人的体力劳动与脑力劳动的、发挥人的聪明与智慧的、为了人与依靠人的方法,都与"以人为本"不可分割[3]。

历史与现实告诉我们:工业是重要的,制造业是重要的,制造业是国家的战略性产业。在今天,更可以说,没有强大的工业,没有先进的制造业,就没有民族真正的独立,就没有国家真正的富强,就没有人民真正的幸福,就没有社会真正的进步,就谈不上"以人为本"。在我国科学技术中长期发展规划中,制造业不仅直接占有极为重要的地位,而且非制造业几乎处处都间接地或本质地同制造业紧密相关。仔细分析一下,规划中所要解决的制造业的问题大致可归纳为四个方面:①先进制造的基本工艺、基本材料、基本元件、基本组件、基本部件与基本技术、基本理论;②绿色制造;③关键的高性能成套装备制造;④关键的高性能工作母机制造。

我们不仅要努力促进信息技术同制造技术融合,也要大力加速前沿技术同制造技术的结合,淘汰落后生产力,推动制造技术走向综合化、科学化、人性化,以实现我国由制造大国走向制造强国、创造强国的跨越。这一切都要靠从事制造业的有关的"人"。

4 结 论

可以认为,在制造业中树立"以人为本"的新的发展观念,就是要调动人在构思与设计、加工(含装配)、营销与使用,以及服务、检测与管理,乃至在制造各方面各环节上的聪明与智慧、主动性与创造性,积极创造人性化的技术,合理使用这种人性化的技术,使之能够正确全局地长远地处理好同各有关方面的关系,真正服务于人类,谋求社会的可持续的发展。是的,"以人为本,不仅是发展为了人,而且发展也必须依靠人"。毫无疑问,教育、培训占有极为重要的地位。事实已清楚表明:在科技高度发达并迅猛发展的今天,企业的竞争,制造业的竞争,关键在科学技术,基础在企业人文文化,焦点在人才,在人才的教育,在于能否营造一种文化环境、一种制度环境与一种政策环境,尊重人,关心人,爱护人,吸引人,团结人,培养人。充分激励、开拓与发挥人的巨大创造性的潜力,使得能拥有大批这样的高素质人才,他们能树立远大理想,饱含人文关怀,深具开阔视野,富有创新激情,深怀忧患意识,具备实施能力,长于团结他人,善于自主决策,及时不断学习。而且能在制造相应的过程中、环节上生动活泼地发挥其聪明才能,在建设中国特色社会主义、实现制造业与制造技术的跨越发展的伟大事业中,以天下为己任,充分实现个人自我价值。我们想,可以认为,这就是制造业中的"以人为本",这就是制造业坚持科学发展新观念的核心。

本文有关工业设计的论述得到武汉理工大学陈汉青教授的帮助,特此致谢。

参 考 文 献

[1] 路甬祥. 坚持科学发展, 推进制造业的历史性跨越[J]. 机械工程学报, 2007, 43(11): 1-6.

[2] 杨叔子, 吴波, 李斌. 再论先进制造技术及其发展趋势[J]. 机械工程学报, 2006, 42(1): 1-5.

[3] 杨叔子, 吴波. 先进制造技术及其发展趋势[J]. 机械工程学报, 2003, 39(10): 73-78.

Humanism——Establishing a New Concept of Manufacture's Development

Yang Shuzi　Shi Tielin

Abstract　This paper begins with discussing manufacture's vital functions on human society's development from the historical angle. Then it expounds both the positive and negative functions of manufacture, which all the other technologies have on the society. Furthermore, by quoting the definition of the industrial design given by International Council of Societies of Industrial Design (ICSID), this paper demonstrates that bringing new ideas into technology needs humanization. Finally, it discusses the connotation of the manufacture's development with the core of humanism.

Keywords：Humanism, Manufacture, Mechanical manufacture, Equipment manufacturer

（原载《机械工程学报》2008 年第 44 卷第 7 期）

走向"制造-服务"一体化的和谐制造

杨叔子　史铁林

提　要　指出了自然、人类和制造三者之间的关系,讨论了制造及其发展的内涵、制造的效益化和人性化,介绍了"工业设计"给予的启示,最后阐述了走向制造与服务一体化的和谐制造。

关键词: 制造,服务,一体化,和谐制造

0　前　言

路甬祥同志 1997 在中国机械工程学会长沙年会上所做的《坚持科学发展,推进制造业的历史性跨越》主题报告是一个重要的指导性文件。笔者在《中国机械工程》2008 年第 7 期的论文《以人为本——树立制造业发展的新观念》中,论述了对这一报告的有关认识,根本点就是十分赞同制造业必须树立科学发展观这一新观念。

科学发展制造业就必须:要准确把握时代特征;要把握科技发展趋势;要准确把握我国国情。只有以人为本,树立新的发展观念,即"科学发展观",有力而正确地促进制造业、制造技术的科学发展与自主创新,才能走向和谐制造。本文更着重于阐述制造与服务的关系,发展和谐制造。

1　自然、人类和制造三者之间的关系

大自然创造了人类,人类创造了文化、文明。人类以自然为生存与发展的基础,以文化、文明为生存与发展的方式。人类文化、文明的基础是物质文明,人类的物质文明是从制造开始的,并由制造、材料、能源、信息四者作为支柱所构成。

制造是人类物质财富创造的基础,制造是国民经济建设中第一大产业部门,制造是为各个领域提供装备的核心部门,制造是高新技术载体的提供者,制造与国防、国家安全休戚相关,制造显然非常重要。

恩格斯在《自然辩证法》中深刻指出:"直立和劳动创造了人类,而劳动是从制造工具开始的。"并形象地描述,制造工具是从制造第一把石刀开始的。

毛泽东同志 1964 年春发表的词《贺新郎·读史》上阕是:"人猿相揖别。只几个石头磨过,小儿时节。铜铁炉中翻火焰,为问何时猜得,不过(是)几千寒热。人世难逢开口笑,上疆场彼此弯弓月。流遍了,郊原血。"词中描述了人类与猿的分道扬镳就是从制造石头工具开始的,发展到几千年前,人就懂得冶金了;懂得冶金,据称也是文明与野蛮区别的重要标志之一。

人类的发展、人类文明的发展与制造的发展、工具制造的发展是分不开的;而且人类在此发展中始终处于主导地位。自然在演化中,经由劳动、制造、工具创造了人类;人类在进步中,创造与发展了制造、工具,并以此为物质文化、文明与精神文化、文明的创造与发展不断奠定基础。与此同时,制造、工具也在不断发展。可以说,没有制造,就没有工具制造;没有金属的发现,没有金属农具制造,就没有农业革命;没有热力学定律的发现,没有蒸汽机制造,就没有第一次工业革命;没有电磁的发现,没有电机和电器制造,就没有第二次工业革命;没有半导体的发现,没有芯片与计算机制造,就没有信息革命。可以肯定,纳米科技、生命科技、光量子科技等的发展,更加发展了先进制造,它必将导致人类生产、生活、思维、社会更为深刻、更为广泛的革命。可以说,没有科技的发展,没有制造的发展,就不能持续满足人类不断发展的需要。

制造有层次,从制造、工具制造、机械制造、装备制造到机床制造。机械是复杂的工具,装备是关键的机械,机械制造的核心是装备制造,装备制造中更为核心的是机床制造——工作母机制造,机床是制造机器的工具。马克思在《资本论》中的论述极为正确:"大工业必须掌握这一特有的生产资料,即机器的本身,必须用机器生产机器,这样,大工业才能建立起自己相应的技术基础,才得以自立。""用机器生产机器",实质上就是用工具制造工具。据不久前统计,我国制造光电子装置的装备 100% 靠进口,制造 IC 的装备 80% 靠进口,高档数控 70% 以上也要靠进口,这些都受制于人。所以装备制造是制造发展的关键。国际上,把一个国家关键技术依靠进口的比例定义为技术依存度,以此作为考核一个国家科技实力一个重要的指标。在 20 世纪 90 年代中期,工业发达国家的平均技术依存度为 10%,日本 6.6%,美国 1.6%,韩国 24%,中国远在 50% 以上。我国中长期科技发展规划,争取到 2020 年将技术依存度降到 30% 以下。所以我们必须高度重视制造领域关键技术的发展,重视制造领域的知识产权。

2 制造及其发展的内涵

"制造"不等于"加工"。制造内涵有三点:前端的构思与设计,指加工前;中端的加工,指加工中;后端的营销、服务与使用,乃至回收与再制造,指加工后。检验测量及管理经营都涵于其中,其实,这就是依靠信息联系把制造全过程综合成整体。所以严格讲,目前我国仅仅是个"加工大国",还不算是"制造大国"。我们必须把"制造"的全过程都做好了,才有可能成为"制造大国",走向"制造强国"。经过努力,我们能够做到这点。制造发展过程大体如下。

a) 原始制造工艺系统(图 1)。原始的制造是人手拿了石器对加工对象进行加工,人脑按照需求指挥人体进行加工动作,眼睛观察加工对象后,获得反馈信息,人脑据此判定是否需继续加工。这里就有了需求、加工、使用这三个阶段,即前端、中端与后端。

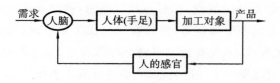

图 1　原始制造工艺系统

b) 金属工具出现后的制造工艺系统(图 2)。随着材料的发展,有了金属工具的出现,加工过程、加工强度剧增,这时人的体力不够,要通过机构机器操作工具加工。

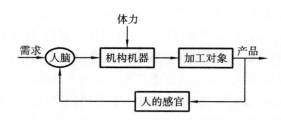

图 2　金属工具出现后的制造工艺系统

c) 加工设备出现后的制造工艺系统(图 3)。随着材料、动力等的发展,出现了蒸汽机、电动机,使用动力的加工设备替代了人力机械,检测装置替代了人的感官,有了工艺文件下达指令。制造还是三个阶段。

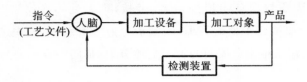

图 3　加工设备出现后的制造工艺系统

d) 信息(检测、控制)技术发展后的制造工艺系统(图 4)。随着材料、动力、信息(检测、控制)技术等的发展,出现了自动、半自动金属切削机床、织布机、印刷机等,都是用刚性控制器,诸如凸轮、行程挡块等部分替代人脑。例如 1873 年美国出现的单轴、多轴自动机床。

图 4　信息(检测、控制)技术发展后的制造工艺系统

e) 信息(计算机、控制)技术发展后出现的制造工艺系统(图 5)。随着计算机的出现,计算机可以存储加工指令及处理检测信息,实现数字控制,控制器、制造系统开始具有柔性。例如 1952 年,美国出现了数控机床。

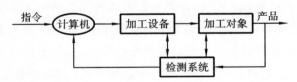

图 5　信息(计算机、控制)技术发展后出现的制造工艺系统

信息技术与计算机技术的进一步发展,实现了数字控制,实现了制造的柔性。从控制一台机床到控制一条生产线、一个车间、一个工厂、一家企业,进一步发展为跨地区的联合企业、虚拟企业、企业动态联盟,发展到了网络制造。信息技术的发展极大地提高了生产自动化水平。例如 1974 年,美国有了 CMS(计算机集成制造系统,此处 C 为 Computer);1976 年,日本有了 FMS(柔性制造系统);当代又有了当代的 CMS(当代集成制造系统,此处 C 为 Contemporary)。

不管发展到什么程度,制造的内涵本质未变,不过它的内涵更丰富、更深刻、更系统了。制造的前端包括构思、决策、规划与设计等等,是前提,是先天,此点失误,一切皆空;制造的中端包括加工、装配、包装等等,是关键,是后天,此点失误,先天作废;制造的后端包括营销、服务、咨询、维修与使用等等,是根本,此点失误,前功尽弃;而作为质量保证手段的检测、考核,则无处不在;"三分技术,七分管理",作为现代企业神经中枢的管理,是制造生命之所系,更是先进制造须臾不可离的灵魂。在上述整个生命周期的每一个环节中,都离不开人对每一个环节的不断改进与创新。

3 制造的效益化和人性化

制造发展趋势中有两个特点:

a) 效益化——制造的参与者工作效率不断提高,人尽其用,物尽其利,即制造的生产水平与生产力不断提高;

b) 人性化——制造者的劳动不断减轻,用户的需要不断满足,一切参与者的潜力开发不断改善,即对一切参与者的人文关怀不断提高。

"效益化"与"人性化"这两者是对立统一的,这也是科技双刃剑作用的重要体现。统一,相助,即两者相互促进;对立,矛盾,即两者彼此抵触。特别是"效益化"的反"人性化",最终也会导致反"效益化",此即:一是导致与自然环境不和谐(环境污染,生态失衡,资源浪费,等等);二是导致与人际关系不和谐(恶化劳动现场,强化劳动强度,加剧社会矛盾,等等)。结果是社会、制造均不可持续发展。

考虑制造的效益化和人性化,还必须要考虑人与自然的关系。人不是自然的奴隶,听凭自然驱使与宰割,但人更不是自然的主宰,能为所欲为地去奴役与宰割自然。人只能够是而且必须是自然的有机组成部分,包括人的自觉的主动性与高度的创造性。人是"万物之灵",能不断深刻认识自然,积极主动地顺应自然,恰如其分地利用自然,因势适度地改造自然,全面协调地按自然规律办事,"天人合一",和谐共生。在这方面中国人很值得骄傲。美国诺贝尔化学奖金获得者普里高津说过:"中国传统的学术思想着重研究整体性与自然性,研究协调与协同,现代科学的发展更符合于中国的哲学思想。因此,中国的思想对于那些想扩大西方科学范围与意义的哲学家与科学家来说,始终是一个启迪的源泉。"有事实为证:早在2200多年前(约公元前276—前251)李冰创建的都江堰水利工程,就是在深入调查研究、总结前人治水经验的基础上,精心选择在成都平原顶点的岷江上游干流出山口处作为工程地点,团结和组织西蜀各族人民,经过艰苦奋斗建成的。实践证明,历2000多年而效益不衰的都江堰水利工程,地理位置优越、合理,工程布置适合自然规律,分水堤、溢洪道、宝瓶口三项工程相互制约、相辅相成,综合发挥引水、分洪、排沙输沙的重要作用,旱涝都不怕,从此有了富饶美丽的成都大平原。都江堰水利工程甚至经受住了四川汶川2008年5月12日的大地震,至今仍然发挥着巨大的效益。李冰早就全面协调按照客观规律办事了。此外,诸如广西的灵渠等何尝不是如此!

随着工业革命的伟大胜利,随着科学技术迅猛发展,越来越"效益化",人间"奇迹"一个接一个创造出来,人类误以为:按照人的主观意志,依靠科学技术,凭借制造手段,创造出各种工具与方法,征服自然,人可以成为自然的主宰。美国未来学家约翰·赖斯比特1999年出版了《高科技,高思维——科技和人性意义的追寻》一书(2000年出了中文版),书中叙述了科学技术的两面性,人类科学技术发展很快,正在改变人类的环境,改变人类与环境的关系,改变得好坏尚难定论。特别是生物科学技术的发展,不但改变了人类的环境,改变了人类与环境的关系,而且还改变了人类本身。这个改变得好坏更不清楚。该书中文版序言更明确指出,科学技术正在给人类带来神奇的创新,但同时带来了潜在毁灭性的后果。有一位生物学家更指出,当人类彻底揭开生命奥秘之日,就是人类彻底灭亡之时。如果生命奥秘一旦被狂热分子掌握,灾难必然降临。所以在美国2001年9月11日恐怖事件之后,我相信了他们的看法。"祸兮福所倚,福兮祸所伏",事情一旦做过了头,必然走向反面,"物极必反"。

出路何在?"解铃还须系铃人",根本在"人"!倒退没有出路!遵循"天人合一"与"社会和谐"。一靠发展:靠科技发展;二靠人:靠人以相应的更好的制造、工业、科技,谋求"效益化"与"人性化"的有机统一,即靠科学发展观发展。2007年中国科协年会有位专家在大会报告中讲:"节能减排,与其讲关键在科技,不如讲根本在制定与执行政策与措施的人。"他的观点完全正确,科技是重要,但如何去应用更为重要。做表面文章害死人,如果只讲眼前政绩,不管后果,就要危害子孙。"为官一任,造福一方。"这个讲法,值得研究:怎么造福?眼前的、局部的?长远的、整体的?今年中国科协工作活动的主题是:"发展、责任",要发展,还要负责任,即要负责任的发展,科学发展。佛教禅宗六祖讲:"邪人用正法,正法亦邪;正人用邪法,邪法亦正。"所以关键还是在人,非常正确。各地节能减排存在的问题,往往不是科技的问题,而是干部问题,首先要解决干部问题。

4　工业设计的启示

工业设计是制造全过程中一项很重要的前端工作。根据国际工业设计联合会 ICSD(International Council of Societies of Industrial Design)2006 年发布的关于工业设计的定义,包含工业设计的目的和任务两项。

a) 工业设计的目的:"设计是一种创造性的活动,其目的是为物品、过程、服务以及它们在整个生命周期中构成的系统建立起多方面的品质。因此,设计既是创新技术人性化的核心因素,也是经济文化交流的关键因素。"

b) 工业设计的任务:"设计致力发现和评估与下列项目在结构、组织、功能、表现和经济上的关系:1)增强全球环境保护和可持续性发展(全球道德规范);2)给个人和集体、全人类社会带来利益和自由;3)最终用户、制造者和经营者互利共赢(社会道德规范);4)在世界全球化的背景下支持文化多样性(文化道德规范);5)赋予产品、服务和系统以表现性的形式(语义学),并与它们的内容相协调。"

从工业设计要创新技术人性化这个提法,我们应该获取非常重要的启示。工业设计不但是设计产品外形,还要考虑产品与周围环境的配合。好些年前,联合国有个组织评价第三世界生产的工业产品便宜、实用,但是不太美,经不起看,经不起想,经不起品赏。工业产品要有文化含量,要与周围环境构成整体,经得起看,经得起想。有人讲:"建筑是凝固的音乐,音乐是流动的建筑。"我们可以讲:"产品是凝固的音乐,音乐是流动的产品。"所以工业设计中产品创新技术人性化,就是在创新技术中不断融入人文思想,不断融入人文关怀,不断融入各方面的和谐。

工业设计的任务应该服务于目的。工业设计任务中的 5 条其实就是讲"关系和谐",对应为:1)人类社会与自然环境关系的和谐;2)人类社会内部关系的和谐;3)人类社会工业商业活动有关方面关系的和谐;4)人类社会跨地区间多种文化关系的和谐;5)产品、服务与系统整个过程有关的形式与内涵间关系的和谐。

"人性化"与"效益化"的和谐,就是对工业、制造业的生产要求以技术为手段的"实",来创造以科学为基础的"真",以人文为内涵的"善",以艺术为形式的"美",以技术作为手段,达到真善美和谐统一,能满足人所需要的"人性化"与"效益化"相融合的"新"。核心就是坚持走这种自主创新的道路,以人为本,实现创新技术人性化。

5 走向制造与服务一体化的和谐制造

5.1 服务贯穿于制造及其发展全过程

原始的制造,就是开始于服务,制作石刀,就是为了服务于"用"的。服务本质上一直贯穿于制造及其发展全过程。服务的"效益化",是直接服务于产品生产过程及其各有关环节,主要同提高制造的生产力与生产水平有关;服务的"人性化",是直接服务于与产品有关的参加者的身心需要,核心同强化人文关怀有关。两者均从属于以人为本的服务,但应以"人性化"为主,不是为生产而生产。

路甬祥认为:数百年来,以产品为中心的制造业正在向服务增值延伸,制造业的结构也从以产品为中心迈向以提供产品和增值服务为中心,这是制造业的历史性发展和进步,是制造业走向高级化的重要标志。这十分深刻。服务既包括"服务于"、"我为他服务";也包括"被服务","他为我服务"。制造后端的产品售后技术服务比较直观,实际上服务存在于:制造的前、中、后端的战略分析,概念创意,规划设计,管理决策,管理维护,软件支持,咨询服务等等之中,各个环节都要把服务深入进去。服务当然包括"效益化"服务与"人性化"服务。一些统计数据得重视,路甬祥在他的报告中举出了一些国家服务业占GDP的百分比:美国75%,德国71%,日本67%,韩国54%,印度51.2%,中国40.7%(2004年)。企业服务的收入占总收入的百分比:美国通用电气公司(GE)60%,美国的IBM公司55%。我国大企业中的海尔公司,更明确提出了:产品的成本竞争,要变成服务竞争。服务竞争做好了,产品的成本竞争自然就迎刃而解。

5.2 先进制造技术的"人性化"与"效益化"

先进制造技术的发展趋势可以用以下12个字概括。
制作产品:精,极,文。
制造过程:绿,快,省,效。
制造方法:数,自,集,网,智。

"问渠哪得清如许?为有源头活水来。"有了源头,才有活水,有了活水,才有清渠。源头是制造方法,清渠就是制造过程,渠就是产品。所不同的是,活水、清渠与源头没有相互影响,而产品、制造过程与制造方法是相互影响的。

a) 制造产品的发展趋势如下。

"精",指精确化,包括几何量、物理量的精确。这是发展的关键。

"极",指极端条件化,包括几何量、物理量的极端条件化。在国家中长期科技发展规划中,在基础研究这部分,就列有极端条件下制造的课题。极端条件化是发展的焦点,是关键的关键。

"文",指人文化,产品要经得起看,经得起欣赏。要考虑产品与周围环境的配合。产品既是实用品,也是艺术品。人文化是发展的新含义,是人性化的直接体现。

精确化、极端条件化和人文化,是在更高层次上,直接体现"保证产品质量,满足用户需要"的"人性化"趋势。

b) 制造过程的发展趋势如下。

"绿",指绿色化,是发展的必然。绿色化是广义的,不仅不能污染环境,还要保证社会环境、劳动环境以及人的身心的和谐。我国中长期科技发展规划特别要求制造过程中绿色化要过关。凡是环保不合格的项目绝对不准上。

"快",指快速化,是先进制造发展的动力。市场的变化是迅速的,为了保证能够快速响应市场需求,制造过程一定要能够快速重新组合。能够快速重新组合的生产系统,一定要是敏捷的、高度柔性的生产系统。

"省",指节省、节约化,是发展的原则。搞经济就是要讲节约,搞经济不讲节约,就不是搞经济了。节约绝对不是小气,而是为了更好地利用有限的资源,以达到更高的效益。

"效",指高效、高生产率化,是发展的追求。列宁讲,共产主义之所以能够战胜资本主义,是因为共产主义社会有更高的生产效率。

所以不污染、少耗能、高效益,正是体现着"人性化"、同时体现着"效益化"的趋势。

c) 制造方法的发展趋势如下。

"数",指数字化,是先进制造方法的核心。信息化的核心就是数字化。数字化的优点是精确、安全、容量大,这是模拟技术达不到的。所以设计、加工、管理等所有有关过程与环节都要实现数字化。数字化发展很快,1980年美国三家大汽车公司,CAD的使用大约是80%,现在我国却已普遍使用。虚拟技术是数字化发展的一个高级阶段。

"自",指自动化,是发展的条件。只有实现自动化才能解放人。

"集",指集成化,是发展的方法。集成化包括知识集成、技术集成等各个方面的集成,开放就是搞国内外集成。机电一体化技术就是传感技术、信号处理技术、随动技术、自动控制技术、精密机械技术以及总体集成技术等相关技术的综合集成。

"网",指网络化,是发展的道路。

"智",指智能化,是发展的前景。机器智能已经越来越聪明,但是人的智能特征是具有原创性,一般机器智能难以实现非线性、非逻辑控制,计算机也难于处理海量信息、错误信息、不完整信息等,人的智能却可以实现。机器人下棋挺厉害,可以战胜世界冠军,但是只要给它下一步程序中无此程序的棋,它就只能以"查无此程序"而告输。

上述制造方法的5点,即数、自、集、网、智,重中之重是网络化,因为:

(1)"网络"是现代制造企业生存的运行环境;

(2)"网络"是现代新型制造模式的基础设施;

(3)"网络"是现代制造企业发展必须走的道路。

从一定意义上讲,制造企业的网络化就是管理现代化的深刻体现。网络化导致企业间有关资源互补、共享、最佳配置;导致制造的有序化、整体化、全球化;导致"效益化"与"人性化"的一体化,达到最优服务,实现"服务-制造"一体化。所以我们必须高度重视网络化的强大功能,否则会脱离时代。

5.3 "产业集聚"问题

产业集聚,是指一个区域从实际出发,发挥区域优势整合产业,形成自身的整体优势。即以优势企业为主导,延伸其产业链,利用地域邻近、交通方便,从而联系紧密、资源易于共享、协作易于稳定,从而相应的服务易于优化,达到"制造-服务"易于一体化,既利于产品生产与销售,能够提高产品品质、响应市场,又利于缩短周期,降低成本。宁波慈溪市原来是个很穷的小县城,依山傍海,什么都不能生产,但是,从1973年到2003年的30年间,发生了翻天覆地的变化,其原因就是靠产业集聚。到2003年,这个县级市的GDP达120亿美元。其中有一个电风扇厂,除了几把起子、几个老虎钳,没有什么像样的设备,所有原料、元件等都来自于当地,靠就地取材,靠产业集聚。这个市的产品一点也不土,有人造军用纤维,高档保护漆,还有航天用的微型轴承等等。王宏甲写了一本书,书名为《贫穷致富与执政》,生动地介绍了慈溪的发展。当地干部管理政策非常宽松,使个体企业能够充分发展,能够很好组合,充分发挥了产业集聚的优势。

5.4 什么是和谐制造？

和谐制造就是以人为本，制造走向"效益化"与"人性化"的一体化和"制造与服务"一体化的和谐制造。

和谐制造的重要特色与发展趋势：

a) 它是以制造技术、信息技术、管理科学以及有关高科技成就相结合而形成的先进制造技术；

b) 它是与社会相和谐的人文制造技术；

c) 它是与环境相友好的绿色制造技术；

d) 它是立足于科学发展、学科交叉的，能不断开拓新领域的，可持续发展的先进制造技术。

上面的看法，供大家参考。请批评！请指教！李白有首诗："长风破浪会有时，直挂云帆济沧海。"我国改革开放 30 年了，天翻地覆，长风破浪今是时，可以直挂云帆济沧海了。谢谢大家！

本文根据杨叔子院士在南京举行的"江苏省科技高端论坛"的报告录音与 PPT 整理，并经作者本人审阅。

Towards Manufacture and Service Incorporated Harmonious Manufacture

Yang Shuzi   Shi Tielin

Abstract　The paper points out relation between nature, humanity and manufacture firstly. Then meaning, benefit and humanization for manufacture are discussed. An apocalypse of industry design is introduced. Finally tend towards manufacture and service incorporated harmonious manufacture is expatiated.

Keywords：Manufacture, Service, Integration, Harmonious manufacture

（原载《机械制造与自动化》2009 年第 38 卷第 1 期）

Kinematic-parameter Identification for Serial-robot Calibration Based on POE Formula

He Ruibo   Zhao Yingjun   Yang Shunian   Yang Shuzi

Abstract   This paper presents a generic error model, which is based on the product of exponentials (POEs) formula, for serial-robot calibration. The identifiability of parameters in this error model was analyzed. The analysis shows the following: 1) Errors in all joint twists are identifiable. 2) The joint zero-position errors and the initial transformation errors cannot be identified when they are involved in the same error model. With either or neither of them, three practicable error models were obtained. The joint zero-position errors are identifiable when the following condition is satisfied: Coordinates of joint twists are linearly independent. 3) The maximum number of identifiable parameters is $6n+6$ for an n-degree-of-freedom (DOF) generic serial robot. Simulation results show the following: 1) The maximum number of identifiable parameters is $6r+3t+6$, where r is the number of revolute joints, and t is the number of prismatic joints. 2) All the kinematic parameters of the selective compliant assembly robot arm (SCARA) robot and programmable universal machine for assembly (PUMA) 560 robots were identified by using the three error models, respectively. The error model based on the POE formula can be a complete, minimal, and continuous kinematic model for serial-robot calibration.

Keywords: Calibration and identification, Identifiable parameters, Product of exponentials(POEs)

Nomenclature

A_i Differential of the exponential map at $\hat{\xi}_i$.

$|A_i|$ Determinant of A_i.

A_{st} Differential of the exponential map at $\hat{\xi}_{st}$.

g Forward kinematics map of the robot.

g_a Actual end-effector pose.

g_n Nominal end-effector pose.

$\delta g g^{-1}$ Pose error at $\{T\}$ expressed in $\{S\}$.

$g_{st}(0)$ Initial transformation matrix.

J_i Jacobian matrix for the ith joint.

J_{st} Jacobian matrix of ξ_{st}.

J Identification Jacobian matrix.

n Number of DOF of the robot, which belongs to $\Re$.

q_i Joint variable of the ith joint, which belongs to $\Re$.

q Vector of all joint variables, which belongs to $\Re^n$.

δq Kinematic errors in q.

$\{S\}$ Base coordinate frame of the robot.

$\{T\}$ Tool coordinate frame of the robot.

v_i Position vector of the ith joint axis relative to $\{S\}$, which belongs to $\Re^3$.

ξ Vector of all joint twist coordinates, which belongs to $\Re^{6n}$.

ξ_i Twist coordinate of $\hat{\xi}_i$, which belongs to $\Re^6$.

ξ_{st} Twist coordinate of $\hat{\xi}_{st}$ which belongs to $\Re^6$.

$\delta \xi$ Kinematic errors in ξ.

$\delta \xi_{st}$ Kinematic errors in ξ_{st}.

$\hat{\xi}_i$ Twist of the ith joint, which belongs to $se(3)$.

$\hat{\xi}_{st}$ Twist of initial transformation, which belongs to $se(3)$.

ω_i Unit directional vector of the ith joint axis, which is expressed in $\{S\}$, belonging to $\Re^3$.

$\hat{\omega}_i$ Skew-symmetric matrices of ω_i, which belongs to $so(3)$.

I. Introduction

BECAUSE of the manufacturing and assembly tolerance, the actual kinematic parameters of a robot deviate from their nominal values, which are referred to as kinematic errors. The kinematic errors would result in the end-effector errors if the nominal kinematics were used to estimate the pose of the robot. With the restriction of cost, the kinematic calibration is an effective way to improve the absolute accuracy of robots.

The kinematic calibration is a four-step procedure[1]: kinematic modeling, pose measurement, kinematic identification, and kinematic compensation. This paper focuses on kinematic modeling and kinematic identification.

A kinematic model should meet the following three basic requirements of the kinematic-parameter identification[2,3].

1) *Completeness*: A complete model must have enough parameters to describe any possible deviation of the actual kinematic parameters from the nominal values[1].

2) *Continuity*: Small changes in the geometric structure of the robot must correspond to small changes in the kinematic parameters. In mathematics, the model is a continuous function of the kinematic parameters.

3) *Minimality*: "The kinematic model must include only a minimal number of parameters"[3]. The error model for the kinematic calibration should not have redundant parameters.

Although the standard Denavit-Hartenberg(DH)[4] convention is most often used to describe the robot kinematics at the design stage, the error models based on DH convention are not continuous when consecutive joint axes are almost parallel[5,6]. To avoid the singularity of DH convention, many authors have suggested using different models, such as the Hayati et al. models[5-7], Veitschegger and Wu's model[8], Stone and Sanderson's S-model[9], and the Zhuang et al. complete and parametrically continuous (CPC) model[10]. However, some redundant parameters are involved in the error models based on all the above conventions. These redundant parameters need to be eliminated from the error models to obtain the robustness of parameter identification[1]. There are two types of redundant parameters in robot calibration[11,12]: 1) Due to the structure of the robot and the error model used, these types of redundant parameters are involved, regardless of the joint variables, and 2) due to the choice of the measurement poses, these types of redundant parameters are associated with the joint variables.

To eliminate redundant parameters, analytical algorithms[3,11,13,14] and numerical algorithms[15-17] have been adopted. Analytical algorithms are only implemented for the first type of redundant parameters prior to identifying kinematic parameters, whereas numerical algorithms are implemented during the identification procedure for both types of redundant parameters. It is practical to process the measurement data with the numerical algorithms, but numerical algorithms cannot reveal the principles of the kinematic-parameter identification.

The analytical algorithms were implemented to analyze the identifiability of kinematic parameters in the above error models by some authors[3,10,13,14,18,19]. Inevitably, the redundant parameters are involved in these error models. Thus, Schröer et al.[3] and Santolaria et al.[20] thought that a single convention without redundancy cannot exist due to "fundamental topological reasons concerning mappings from Euclidean vector spaces to spheres"[3]. However, they neglected the mappings from Lie algebra to Lie group.

In addition, we have a suspicion that the maximum number of identifiable parameters is $4r + 2t + 6$[18,19], where r and t are the numbers of revolute and prismatic joints, respectively. Because the identifiability of the errors in joint variables, which are termed as the joint zero-position errors or joint offsets, has not been investigated clearly. There are different views on the joint zero-position errors. The joint zero-position errors needed to be identified as in Ref.[20]~[24]. Conversely, the joint zero-position errors did not have to be identified as in Ref.[2] and Ref.[3] and were regarded as redundant parameters[10], and the identifiability of joint zero-position errors in the error-model-based POE formula has not been investigated.

To avoid the problem of singularity, Park and Okamura[25] first employed the POE formula to serial-robot calibration. Since then, the POE formula[26] and local

POE formula[27-30] have been applied to robot calibration. Park and Okamura[25,26] presented a generic error model, which is based on POE formula, for serial-robot calibration. However, the expressions of their error models were not in explicit form, as the differentials of the exponential map at joint twists were presented in definite-integral form. Furthermore, they have not analyzed the identifiability of kinematic parameters in their error models. These definite integrals need to be further derived to analyze the identifiability of kinematic parameters. One of the contributions of this paper is that we propose the error model based on POE formula with explicit expressions. Chen et al.[28] thought that "the direct identification of the actual joint twists is difficult and may result in complicated computations" only the errors in the initial poses of the local frames are involved in their error model. Therefore, this error model cannot be used to analyze the identifiability of kinematic parameters due to the lack of errors of joint twists and joint variables.

Since the exponential map from a Lie algebra to a Lie group is a smooth map and its derivative at the identity is the identity map, the exponential map gives a diffeomorphism of a neighborhood of zero in Lie algebra onto a neighborhood of the identity in Lie group. Thus, the error models based on POE and local POE are singularity-free. In addition, the kinematic model based on POE formula is complete because any rigid-body motion as a screw motion corresponds to a certain twist according to Chasles's theorem. According to the requirement of parameter identification, the identifiability of kinematic parameters in the error models based on the POE formula needs to be clarified. However, the identifiability of kinematic parameters in these error models has not been analyzed in existing papers, including Ref. [25]～Ref. [30].

In this paper, the error model based on the POE formula is presented with explicit expressions, and the identifiability of kinematic parameters was analyzed. This paper is organized as follows. Section Ⅱ presents a generic error model based on the POE formula for generic serial robots. Section Ⅲ analyzes the identifiability of kinematic parameters in the error model and presents three practicable error models without redundant parameters. Section Ⅳ presents the verification of the three error models by some simulations. Sections Ⅴ and Ⅵ discusses and concludes the paper, respectively.

II. Error Model

A. Forward Kinematics Using Product-of-Exponential Formula

Brockett [31] initially described the forward kinematics of a serial robot with the POE formula, and only two coordinate frames need to be attached, as shown in Fig. 1. The base frame $\{S\}$ is attached to any position of robot and is stationary with respect to link 0. The tool frame $\{T\}$ is attached to the end-effector of the robot. The $g_{st}(0)$ is the initial transformation from $\{T\}$ to $\{S\}$ when robot is in its reference configuration, with all joint variables being equal to zero, and a twist $\hat{\xi}_i \in se(3)$ is associated with the ith joint. All twists are expressed in the base frame $\{S\}$ as follows:

$$\hat{\xi}_i = \begin{bmatrix} \hat{\omega}_i & v_i \\ 0 & 0 \end{bmatrix} \quad (1)$$

where $v_i = [v_{1i}, v_{2i}, v_{3i}]^T \in \Re^3$, and $\hat{\omega}_i \in so(3)$ is the skew-symmetric matrix of ω_i, $\omega_i = [\omega_{1i}, \omega_{2i}, \omega_{3i}]^T \in \Re^3$ and is given by

$$\hat{\omega}_i = \begin{bmatrix} 0 & -\omega_{3i} & \omega_{2i} \\ \omega_{3i} & 0 & -\omega_{1i} \\ -\omega_{2i} & \omega_{1i} & 0 \end{bmatrix} \quad (2)$$

Let $\xi_i := [\omega_i, v_i]^T \in \Re^6$ be the twist coordinate of $\hat{\xi}_i$. Let us define the operator

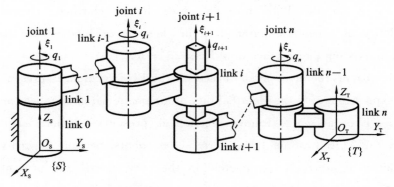

Fig. 1 Forward kinematics of an n-DOF serial robot

V, which maps $\hat{\xi}_i \in se(3)$ and $\hat{\omega}_i \in so(3)$ into $\xi_i \in \mathfrak{R}^6$ and $\omega_i \in \mathfrak{R}^3$, respectively. In addition, q_i is the joint variable. Then, a generic forward kinematics for an n-degree-of-freedom (DOF) serial robot is given by

$$g = \exp(\hat{\xi}_1 q_1)\exp(\hat{\xi}_2 q_2)\cdots\exp(\hat{\xi}_n q_n)g_{st}(0) \tag{3}$$

Eq. (3) is called the product of exponentials formula (POEs formula) for the robot forward kinematics, which is described in more detail in Ref. [32] and Ref. [33]. In Ref. [25] and Ref. [26], $g_{st}(0)$ is also given by

$$g_{st}(0) = \exp(\hat{\xi}_{st}) \tag{4}$$

Then, $\hat{\xi}_{st}$ is referred as twist of initial transformation. Therefore, Eq. (3) can be expressed as

$$g = \exp(\hat{\xi}_1 q_1)\exp(\hat{\xi}_2 q_2)\cdots\exp(\hat{\xi}_n q_n)\exp(\hat{\xi}_{st}) \tag{5}$$

B. Generic Linearized Kinematic Error Model

The error model can be obtained by linearizing the forward kinematic Eq. (5) as follows:

$$\delta g g^{-1} = \left(\frac{\partial g}{\partial \xi}\delta \xi + \frac{\partial g}{\partial q}\delta q + \frac{\partial g}{\partial \xi_{st}}\delta \xi_{st}\right)g^{-1} \tag{6}$$

where

$$\boldsymbol{\xi} = [\xi_1, \xi_2, \cdots, \xi_n]^T \in \mathfrak{R}^{6n} \tag{7}$$

$$\boldsymbol{q} = [q_1, q_2, \cdots, q_n]^T \in \mathfrak{R}^n \tag{8}$$

The right-hand side of Eq. (6) shows that the end-effector error results from the kinematic errors in ξ, ξ_{st} and $\boldsymbol{q}$, which are denoted as $\delta\xi$, $\delta\xi_{st}$ and $\delta\boldsymbol{q}$, respectively. As for the left-hand side of Eq. (6), let g_a be the actual end-effector pose obtained from the measurement data, and let g_n be the nominal end-effector pose. Then, $\delta g g^{-1} \in se(3)$ is represented as the deviation of g_a from g_n. Thus, $\delta g g^{-1}$ is also given by [25]

$$\delta g g^{-1} = (g_a - g_n)g_n^{-1} = g_a g_n^{-1} - \boldsymbol{I}_3 \tag{9}$$

If the deviation is sufficiently small, then $g_a g_n^{-1}$ is in the neighborhood of the identity in the group; thus [27], we have

$$\log(g_a g_n^{-1}) = \sum_{k=1}^{\infty}(-1)^{k-1}\frac{(g_a g_n^{-1} - \boldsymbol{I}_3)^k}{k} \tag{10}$$

With first-order approximation, Eq. (9) can be rewritten as [25]

$$\delta g g^{-1} = \log (g_a g_n^{-1}) \tag{11}$$

Hence, after completing the procedure of pose measurements, the identification of kinematic parameters is to solve the cost function

$$\text{Min} \parallel \delta g g^{-1} - (\frac{\partial g}{\partial \xi}\delta\xi + \frac{\partial g}{\partial \boldsymbol{q}}\delta\boldsymbol{q} + \frac{\partial g}{\partial \xi_{st}}\delta\xi_{st}) g^{-1} \parallel^2 \tag{12}$$

From Eq. (6), including further derivation of the definite integrals given in Ref. [25] and Ref. [26], the explicit expression is given by

$$\begin{aligned}
[\delta g g^{-1}]^V =& [(\delta\exp(\hat{\xi}_1 q_1))\exp(-\hat{\xi}_1 q_1)]^V \\
&+ [\exp(\hat{\xi}_1 q_1)(\delta\exp(\hat{\xi}_2 q_2))\exp(-\hat{\xi}_2 q_2)\exp(-\hat{\xi}_1 q_1)]^V \\
&+ \cdots + [(\prod_{i=1}^{n-1}\exp(\hat{\xi}_i q_i))(\delta\exp(\hat{\xi}_n q_n))\exp(-\hat{\xi}_n q_n) \times (\prod_{i=1}^{n-1}\exp(\hat{\xi}_i q_i))^{-1}]^V \\
&+ [(\prod_{i=1}^{n-1}\exp(\hat{\xi}_i q_i))(\delta\exp(\hat{\xi}_{st}))\exp(-\hat{\xi}_{st}) \times (\prod_{i=1}^{n}\exp(\hat{\xi}_i q_i))^{-1}]^V \\
=& [(\delta\exp(\hat{\xi}_1 q_1))\exp(-\hat{\xi}_1 q_1)]^V \\
&+ Ad(\exp(\hat{\xi}_1 q_1))[(\delta\exp(\hat{\xi}_2 q_2))\exp(-\hat{\xi}_2 q_2)]^V \\
&+ \cdots + Ad((\prod_{i=1}^{n-1}\exp(\hat{\xi}_i q_i)))[(\delta\exp(\hat{\xi}_n q_n))\exp(-\hat{\xi}_n q_n)]^V \\
&+ Ad(\prod_{i=1}^{n-1}\exp(\hat{\xi}_i q_i))[(\delta\exp(\hat{\xi}_{st}))\exp(-\hat{\xi}_{st})]^V
\end{aligned} \tag{13}$$

where

$$\begin{aligned}
&[(\delta\exp(\hat{\xi}_i q_i))\exp(-\hat{\xi}_i q_i)]^V \\
&= (q_i \boldsymbol{I}_6 + \frac{4-\theta_i \sin(\theta_i) - 4\cos(\theta_i)}{2\parallel\omega_i\parallel^2}\Omega_i + \frac{4\theta_i - 5\sin(\theta_i) + \theta_i\cos(\theta_i)}{2\parallel\omega_i\parallel^3}\Omega_i^2 \\
&+ \frac{2-\theta_i\sin(\theta_i) - 2\cos(\theta_i)}{2\parallel\omega_i\parallel^4}\Omega_i^3 + \frac{2\theta_i - 3\sin(\theta_i) + \theta_i\cos(\theta_i)}{2\parallel\omega_i\parallel^5}\Omega_i^4)\delta\xi_i + \xi_i\delta q_i \\
&= A_i\delta\xi_i + \xi_i\delta q_i
\end{aligned}$$

$$\tag{14}$$

and

$$\Omega_i = \begin{bmatrix} \hat{\omega}_i & \boldsymbol{0} \\ \hat{v}_i & \hat{\omega}_i \end{bmatrix}$$

$$\parallel\omega_i\parallel = (\omega_{1i}^2 + \omega_{2i}^2 + \omega_{3i}^2)^{1/2}$$

$$\theta_i = \parallel\omega_i\parallel q_i$$

$$[(\delta\exp(\hat{\xi}_{st}))\exp(-\hat{\xi}_{st})]^V = (\mathbf{I}_6 + \frac{4-\theta_{st}\sin(\theta_{st})-4\cos(\theta_{st})}{2\theta_{st}^2}\Omega_{st}$$

$$+ \frac{4\theta_{st}-5\sin(\theta_{st})+\theta_{st}\cos(\theta_{st})}{2\theta_{st}^3}\Omega_{st}^2$$

$$+ \frac{2-\theta_{st}\sin(\theta_{st})-2\cos(\theta_{st})}{2\theta_{st}^4}\Omega_{st}^3$$

$$+ \frac{2\theta_{st}-3\sin(\theta_{st})+\theta_{st}\cos(\theta_{st})}{2\theta_{st}^5}\Omega_{st}^4)\delta\xi_{st}$$

$$= A_{st}\delta\xi_{st} \tag{15}$$

and

$$\Omega_{st} = \begin{bmatrix} \hat{\omega}_{st} & \mathbf{0} \\ \hat{v}_{st} & \hat{\omega}_{st} \end{bmatrix}$$

$$\theta_{st} = \|w_{st}\| = (\omega_{1st}^2 + \omega_{2st}^2 + \omega_{3st}^2)^{1/2}$$

Letting

$$J_i = \begin{cases} [A_1,\xi_1], & i=1 \\ Ad(\prod_{k=1}^{i-1}\exp(\hat{\xi}_k q_k))[A_i,\xi_i], & 1<i<n+1 \\ Ad(\prod_{k=1}^{i-1}\exp(\hat{\xi}_k q_k))A_{st}, & i=n+1 \end{cases} \tag{16}$$

$$J_{st} = J_{n+1} \tag{17}$$

then Eq. (13) can be expressed as

$$y = Jx \tag{18}$$

where

$$y = [\delta g g^{-1}]^V \in \mathfrak{R}^6 \tag{19}$$

$$J = [J_1, J_2, \cdots, J_n, J_{st}] \in \mathfrak{R}^{6\times(7n+6)} \tag{20}$$

$$x = [\delta\xi_1, \delta q_1, \delta\xi_2, \delta q_2, \cdots, \delta\xi_n, \delta q_n, \delta\xi_{st}]^T \in \mathfrak{R}^{7n+6} \tag{21}$$

where J is the identification Jacobian matrix, y is the error vector of the tool-frame pose expressed in the base frame, and x is the error vector of kinematic parameters. Up to now, a generic error model for serial-robot calibration has been proposed.

III. Identifiability of Kinematic Parameters

In the identification Jacobian matrix J, if a column of J is the linear combination of other columns of J, then the kinematic parameter that corresponds to the column is unidentifiable, which is referred to as redundant parameter [13,14]. Whether any redundant parameter is involved in this generic error model, which is based on the POE formula or not is analyzed in this section.

A. Identifiability of δq and $\delta \xi_{st}$ in Generic Error Model

In this section, the identifiability of the joint zero-position error δq and the initial transformation errors $\delta \xi_{st}$ are analyzed. From Appendix II, the determinant of A_i is given by

$$|A_i| = \begin{cases} \dfrac{4q_i^2(-1+\cos(\|\omega_i\|q_i))^2}{\|\omega_i\|^4}, & i<n+1, \|\omega_i\| \neq 0 \\ q_i^6, & i<n+1, \|\omega_i\| = 0 \\ \dfrac{4(-1+\cos(\|\omega_{st}\|))^2}{\|\omega_{st}\|^4}, & i=n+1, \|\omega_{st}\| \neq 0 \\ 1, & i=n+1, \|\omega_{st}\| = 0 \end{cases} \quad (22)$$

If $\|\omega_i\| < 2$, $q_i \in [-\pi, \pi]$, then only when $q_i = 0$, $|A_i| = 0$. Thus, A_i is full rank when $q_i \neq 0$. In addition, since A_i is equal to $\mathbf{0}_{6\times 6}$ when $q_i = 0$, all columns of A_i are the functions of joint variables q_i.

When $q_i \neq 0$, the set of vectors A_i is a basis of $\Re^6$ so that the column associated with joint variable as a constant is always the linear combination of column vectors of A_i. Hence, only when all joint variables are equal to zero, δq may be identified. However, in this configuration, the end-effector error would be the linear combination of δq and $\delta \xi_{st}$, because the set of constant vectors A_{st} is a basis of $\Re^6$. Thus δq and $\delta \xi_{st}$ cannot appear in the error model at the same time; otherwise, the redundant parameters would be involved. In other words, it is impossible to distinguish the end-effector error contributed either by δq or by $\delta \xi_{st}$. Therefore, δq and $\delta \xi_{st}$ should be separately modeled, and sometimes, supposing no errors in $g_{st}(0)$ and q, then $\delta\xi_{st}$ and δq would not be involved in the error model.

Thus, Eq. (6) is modified into three practicable error models as follows:

$$\delta g g^{-1} = (\frac{\partial g}{\partial \xi}\delta\xi + \frac{\partial g}{\partial \xi_{st}}\delta\xi_{st})g^{-1} \quad (23)$$

$$\delta g g^{-1} = (\frac{\partial g}{\partial \xi}\delta\xi + \frac{\partial g}{\partial q}\delta q)g^{-1} \quad (24)$$

$$\delta g g^{-1} = (\frac{\partial g}{\partial \xi}\delta\xi)g^{-1} \quad (25)$$

and then, Eq. (16) is also modified for (23) and (25) as follows:

$$J_i = \begin{cases} A_1, & i = 1 \\ Ad(\prod_{k=1}^{i-1} \exp(\hat{\xi}_k q_k))A_i, & 1 < i < n+1 \\ Ad(\prod_{k=1}^{i-1} \exp(\hat{\xi}_k q_k))A_{st}, & i = n+1 \end{cases} \quad (26)$$

However, when only δq is involved in the error model without $\delta\xi_{st}$ as in Eq. (24), δq is unidentifiable unless the matrix formed by all coordinates of joint twists is column full rank. Hence, the condition should be satisfied: Any coordinate of joint twist cannot be the linear combination of the others.

B. Identifiability of $\delta\xi$

Since the purpose of this section is to investigate the identifiability of $\delta\xi$, the columns of J associated with the twist coordinates of any two consecutive joints are analyzed by the method similar to that given in Ref. [13]. From Eq. (26), J_{i-1} and J_i are expressed as

$$J_{i-1} = Ad(\prod_{k=1}^{i-2} \exp(\hat{\xi}_k q_k))A_{i-1}, \quad L_{i-1} := A_{i-1} \quad (27)$$

$$\left.\begin{aligned} J_i &= Ad(\prod_{k=1}^{i-1} \exp(\hat{\xi}_k q_k))A_i \\ L_i &:= Ad(\exp(\hat{\xi}_{i-1} q_{i-1}))A_i \end{aligned}\right\} \quad (28)$$

Since $Ad(\prod_{k=1}^{i-2} \exp(\hat{\xi}_k q_k))$ is not only independent of the consecutive joint parameters $\xi_{i-1}, q_{i-1}, \xi_i$ and q_i but also full rank, L_{i-1} and L_i have the same linear combinations as those of J_{i-1} and J_i. Therefore, instead of J_{i-1} and J_i, the linear combinations of L_{i-1} and L_i will be analyzed as follows.

1) *Column Vectors of L_{i-1} and L_i*: Because both $Ad(\exp(\hat{\xi}_{i-1} q_{i-1}))$ and A_i are block-triangular matrices, hence $Ad(\exp(\hat{\xi}_{i-1} q_{i-1})) A_i$ are block-triangular matrices whose diagonal blocks are formed by $R_{i-1} HH_i$. The column vectors of HH_{i-1} and $R_{i-1} HH_i$ are linearly independent (for further details, see Appendix IV) so that the columns of $Ad(\exp(\hat{\xi}_{i-1} q_{i-1})) A_i$ and A_{i-1} are also linearly independent.

Thus, the columns of J_{i-1} and J_i are linearly independent.

2) *Nonconsecutive Analysis*: As for those nonconsecutive joints, such as ith and $(i+k)$th ($k = 2, 3, \cdots, n-i$), L_i and L_{i+k} are expressed as

$$L_i = A_i \tag{29}$$

$$L_{i+k} = \prod_{j=i}^{i+k-1} Ad(\exp(\hat{\xi}_j q_j)) A_{i+k} \tag{30}$$

Since additional joint variables $\{q_{i+1}, q_{i+2}, \cdots, q_{i+k-1}\}$ are involved in L_{i+k}, any column of L_i and L_{i+k} is not the linear combination of the others according to the analysis of consecutive joints. Thus, $\delta\xi$ is identifiable.

C. Section Summary

There are no redundant parameters in error models Eq. (23) and Eq. (25). Error model Eq. (24) is also not involved in redundant parameters under the aforementioned condition. Hence, the maximum number of the identifiable parameters in the error model based on the POE formula is $6n + 6$ for an n-DOF generic serial robot as in Eq. (23).

IV. Simulation Result and Discussion

To verify the above three error models, some simulations were carried out on a 4-DOF SCARA robot and a 6-DOF PUMA 560 robot. The data used in simulations are obtained from Ref. [26] and Ref. [28].

A. Instantiation of Error Model

Up to now, the generic error model has been discussed. However, the revolute joints and the prismatic joints are the two most common types of joints in serial robots, whose error models will be derived from the generic error model as follows.

1) *Error Model for Revolute Joint*: Twist coordinate of revolute joint is defined by[32]

$$\xi_i = [\omega_i, v_i]^T = [\omega_i, p_i \times \omega_i]^T \tag{31}$$

where p_i is the coordinate of any point lying on the joint axis with respect to the base frame, and $\|\omega_i\| = 1$.

Considering revolute-joint twist ξ_i as a general twist, its derivative has the same form as that of Eq. (14). During the procedure of parameter identification, two conditions should be satisfied: (1) $\|\omega_i + \delta\omega_i\| = 1$, (2) $(\omega_i + \delta\omega_i)^T(v_i + \delta v_i) = 0$. Otherwise, the identified result would not conform to the definition of revolute joint.

2) *Error Model for Prismatic Joint*: Twist coordinate of prismatic joint is defined by [32]

$$\xi_i = [\mathbf{0}, v_i]^T \tag{32}$$

where v_i is a unit vector for translation direction, and the elements of ω_i are always equal to zero.

Hence, $[(\delta\exp(\hat{\xi}_i q_i))\exp(-\hat{\xi}_i q_i)]^V$ is derived from Eq. (14) as follows:

$$[(\delta\exp(\hat{\xi}_i q_i))\exp(-\hat{\xi}_i q_i)]^V = q_i \mathbf{I}_6 \begin{bmatrix} \mathbf{0} \\ \delta v_i \end{bmatrix} + \xi_i \delta q_i \tag{33}$$

According to the definition of prismatic joint, the condition $\|v_i + \delta v_i\| = 1$ should be satisfied during parameter identification.

B. Simulation Process

The procedure of simulation is shown in Fig. 2, which is similar to that given in Ref. [2], Ref. [26], and Ref. [28]; however, the updating kinematic parameters are conformed to the definitions of corresponding joints at each iterative step. The measurement data are obtained by calculating the actual tool-frame poses at different configurations.

As for m measured pose data, an equation can be obtained from Eq. (18) as follows:

$$\begin{bmatrix} y^1 \\ y^2 \\ \vdots \\ y^m \end{bmatrix} = \begin{bmatrix} J^1 \\ J^2 \\ \vdots \\ J^m \end{bmatrix} x \tag{34}$$

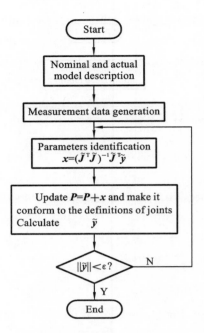

Fig. 2 Simulation process

and in compact form, it is given by

$$\tilde{y} = \tilde{J} x \quad (35)$$

As shown in Fig. 2, x is solved by the iterative least-squares algorithm. Starting with the nominal kinematic parameters, the solution of each step least squares for x is obtained by

$$x = (\tilde{J}^T \tilde{J})^{-1} \tilde{J}^T \tilde{y} \quad (36)$$

Let P be the kinematic parameters of previous solution. After the solution of current step for x has been obtained, P can be updated as

$$P = P + x \quad (37)$$

Then, P is processed in conformance with the definitions of corresponding joints. The iterative procedure is terminated when all elements of x approach zeroes and elements of P converge to some stable values.

C. Simulation I: Using Eq. (24) for SCARA Robot

The kinematic parameters of SCARA robot from Ref. [28] are presented in Tab. 1, along with the identified results with the method of Ref. [28] and of this paper, respectively. From Ref. [28], $g_{st}(0)$ is set to I_4, and it is assumed that $\delta\xi_{st}$ is not involved. In addition, the joint zero-position errors are preset, as given in Tab. 2. Therefore, Eq. (24) is taken as the error model for the kinematic identification in this simulation, which is named simulation I.

Tab. 1 SCARA-Type Robot in Simulation I and Simulation III

	Nominal values [28]	Actual values [28]	Identified result [28]	Simulation I result	Simulation III result
ξ_1	$\begin{bmatrix}0\\0\\1\\0\\0\\0\end{bmatrix}$	$\begin{bmatrix}0.01999\\0\\0.9998\\0\\0.013033\\0\end{bmatrix}$	$\begin{bmatrix}0.03999\\0\\0.9992\\0\\0.028074\\0\end{bmatrix}$	$\begin{bmatrix}0.01999\\2.3162\times10^{-13}\\0.9998\\1.3434\times10^{-16}\\0.013033\\-3.0221\times10^{-15}\end{bmatrix}$	$\begin{bmatrix}0.01999\\-1.3233\times10^{-15}\\0.9998\\-2.5845\times10^{-16}\\0.013033\\2.2418\times10^{-17}\end{bmatrix}$
ξ_2	$\begin{bmatrix}0\\0\\1\\0\\-0.25\\0\end{bmatrix}$	$\begin{bmatrix}0\\0.0004\\1\\-0.0003\\-0.25399\\0.000102\end{bmatrix}$	$\begin{bmatrix}-0.02\\0\\0.9998\\0\\-0.268918\\0\end{bmatrix}$	$\begin{bmatrix}2.1604\times10^{-13}\\0.0004\\1\\-0.0003\\-0.25399\\0.000102\end{bmatrix}$	$\begin{bmatrix}2.5348\times10^{-17}\\0.0004\\1\\-0.0003\\-0.25399\\0.000102\end{bmatrix}$
ξ_3	$\begin{bmatrix}0\\0\\0\\0\\0\\-1\end{bmatrix}$	$\begin{bmatrix}0\\0\\0\\0.02\\0.0196\\-0.99961\end{bmatrix}$	$\begin{bmatrix}0\\0\\0\\0.03999\\0.0196\\-0.99901\end{bmatrix}$	$\begin{bmatrix}0\\0\\0\\0.02\\0.0196\\-0.999608\end{bmatrix}$	$\begin{bmatrix}0\\0\\0\\0.02\\0.0196\\-0.999608\end{bmatrix}$
ξ_4	$\begin{bmatrix}0\\0\\-1\\0\\0.47\\0\end{bmatrix}$	$\begin{bmatrix}0.04077\\0.03917\\-0.9984\\-0.026683\\0.504558\\0.018706\end{bmatrix}$	$\begin{bmatrix}0.03999\\0.0192\\-0.99902\\-0.0128416\\0.504308\\0.00917818\end{bmatrix}$	$\begin{bmatrix}0.04077\\0.03917\\-0.9984\\-0.026683\\0.504558\\0.018706\end{bmatrix}$	$\begin{bmatrix}0.04077\\0.03917\\-0.9984\\-0.026683\\0.504558\\0.018706\end{bmatrix}$

The number of measured poses is set to six, which is the same as that given in Ref. [28], and these poses are randomly generated. The identified parameters are presented in Tab. 1, and the mean end-effector errors during the iterative procedure are shown in Fig. 3.

Tab. 2 Identified Joint Zero-position Errors in Simulation I

	δq_1 /rad	δq_2 /rad	δq_3 /m	δq_4 /rad
Preset values [28]	0	0.02	0.002	0.02
Initial values	0.000	0.000	0.000	0.000
First iteration	0.01070	−0.0002431	0.001685	0.01051
Second iteration	0.0002154	0.01956	0.001998	0.01978
Third iteration	1.2084×10^{-7}	0.02000	0.002000	0.02000
Fourth iteration	4.6589×10^{-14}	0.02000	0.002000	0.02000

At the end of the fourth iteration, the identified result shows that all the identified values of kinematic parameters are sufficiently close to the actual values, as shown in Tab. 1. Since the actual v_3 [28] is not a normalized vector against the definition of prismatic-joint twist, a small difference between the actual value and the identified value exists. The normalized actual v_3 is equal to the identified value as given in the following equation:

$$\frac{v_3}{\| v_3 \|} = \begin{bmatrix} 0.02 \\ 0.0196 \\ -0.999608 \end{bmatrix} \quad (38)$$

and the preset-joint zero-position errors are fully recovered, as shown in Tab. 2. However, these preset-joint zero-position errors cannot be recovered in [28], as the errors in joint twists and joint variables are replaced by the errors in the initial poses of the local frames.

From Fig. 3 and Tab. 1, comparing with the simulation result in Ref. [28], the convergence rate of parameter identification in this simulation is slower; however, the identified parameters are closer to the actual values.

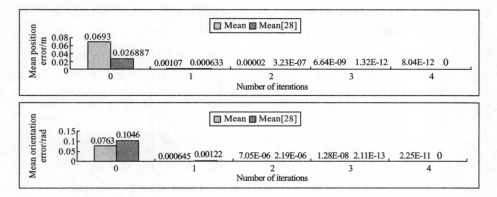

Fig. 3　Mean pose errors during iterative procedure of simulation I

D. Simulation II: Using Eq. (23) for PUMA 560 Robot

The simulation II is to identify the kinematic parameters of PUMA 560 robot [26], including ξ and ξ_{st}, so that Eq. (23) is taken as the error model. The nominal, actual kinematic parameters and the identified results are presented in Tab. 3.

From Ref. [26], 100 pose measurements are randomly generated for identification, whereas an additional 50 pose measurements are also randomly generated for verification. To check the identification effectiveness influenced by the measurement noise, uniformly distributed random noise is injected into each measurement used for kinematic identification. The uniformly distributed noise in the range $[-0.1, 0.1]$ (in millimeters) and $[-0.001, 0.001]$ (in radians) is injected to the measured end-effector position and orientation, respectively. The 50 pose measurements used for verification are not injected noise.

In addition, 40 pose measurements without noise are used for kinematic identification. In addition, an additional 50 pose measurements without noise are used for verification.

After the fourth step of iterations, the identified results are presented in Tab. 3. Since the actual values of v_3 and v_6 do not satisfy the definition of revolute-joint twist, small differences between the identified results and the actual values exist. According to the definition of revolute-joint twist, v_i is orthogonal to ω_i and $\|\omega_i\|$ is equal to 1. Thus, the Gram-Schmidt process is adopted to orthogonalize v_i and ω_i, so that v_3 and v_6 are given in Eq. (39) and Eq. (40), respectively. Therefore, the identified results are correct:

$$v_3 = v_3 - \frac{\omega_3^T v_3}{\omega_3^T \omega_3} \omega_3 = \begin{bmatrix} -0.08418 \\ 0.08741 \\ -101 \end{bmatrix} \tag{39}$$

$$v_6 = v_6 - \frac{\omega_6^T v_6}{\omega_6^T \omega_6} \omega_6 = \begin{bmatrix} -51.273 \\ 248.911 \\ 2.8596 \end{bmatrix} \tag{40}$$

Tab. 3 PUMA-560-Type Robot in Simulation II

	Nominal Values[26]	Actual Values[26]	Identified Result	
			Without Noise	With Noise
ξ_1	$\begin{bmatrix} 0 \\ 0 \\ 1 \\ 0 \\ 0 \\ 0 \end{bmatrix}$	$\begin{bmatrix} 0.04 \\ -0.02 \\ 0.999 \\ 0.02 \\ 0.04 \\ 0 \end{bmatrix}$	$\begin{bmatrix} 0.04 \\ -0.02 \\ 0.999 \\ 0.02 \\ 0.04 \\ -7.75 \times 10^{-12} \end{bmatrix}$	$\begin{bmatrix} 0.04005 \\ -0.01982 \\ 0.9990 \\ 0.02761 \\ 0.05091 \\ -0.00009711 \end{bmatrix}$
ξ_2	$\begin{bmatrix} 0 \\ -1 \\ 0 \\ 0 \\ 0 \\ 0 \end{bmatrix}$	$\begin{bmatrix} 0 \\ -1.00002 \\ 0 \\ -0.02 \\ 0 \\ 0.05 \end{bmatrix}$	$\begin{bmatrix} 5.69 \times 10^{-13} \\ -1 \\ -1.73 \times 10^{-12} \\ -0.02 \\ -9.79 \times 10^{-14} \\ 0.05 \end{bmatrix}$	$\begin{bmatrix} -0.00001486 \\ -1 \\ 0.0002063 \\ -0.02413 \\ 9.3104 \times 10^{-6} \\ 0.04338 \end{bmatrix}$
ζ_3	$\begin{bmatrix} 0 \\ -1 \\ 0 \\ 0 \\ 0 \\ -100 \end{bmatrix}$	$\begin{bmatrix} 0.178 \\ -0.984 \\ -0.001 \\ -0.07 \\ 0.009 \\ -101 \end{bmatrix}$	$\begin{bmatrix} 0.1780 \\ -0.9840 \\ -0.00100 \\ -0.08418 \\ 0.08741 \\ -101 \end{bmatrix}$	$\begin{bmatrix} 0.1780 \\ -0.9840 \\ -0.0001960 \\ -0.09455 \\ 0.002997 \\ -100.987 \end{bmatrix}$

续表

	Nominal Values[26]	Actual Values[26]	Identified Result	
			Without Noise	With Noise
ζ_4	$\begin{bmatrix} 0 \\ 0 \\ -1 \\ -50 \\ 250 \\ 0 \end{bmatrix}$	$\begin{bmatrix} 0.062 \\ 0.013 \\ -0.998 \\ -51 \\ 249 \\ 0.0752 \end{bmatrix}$	$\begin{bmatrix} 0.06200 \\ 0.01300 \\ -0.9980 \\ -51 \\ 249 \\ 0.07515 \end{bmatrix}$	$\begin{bmatrix} 0.06200 \\ 0.01273 \\ -0.9980 \\ -51.011 \\ 248.986 \\ 0.01334 \end{bmatrix}$
ζ_5	$\begin{bmatrix} 0 \\ -1 \\ 0 \\ -20 \\ 0 \\ -250 \end{bmatrix}$	$\begin{bmatrix} 0.001 \\ -1.00004 \\ 0 \\ -20.6 \\ -0.0206 \\ -249 \end{bmatrix}$	$\begin{bmatrix} 0.001000 \\ -1 \\ 2.40\times10^{-10} \\ -20.6 \\ -0.02060 \\ -249 \end{bmatrix}$	$\begin{bmatrix} 0.0009310 \\ -1.0000 \\ 0.0004551 \\ -20.6327 \\ -0.1325 \\ -249 \end{bmatrix}$
ζ_6	$\begin{bmatrix} 0 \\ 0 \\ -1 \\ -50 \\ 250 \\ 0 \end{bmatrix}$	$\begin{bmatrix} 0.095 \\ 0.031 \\ -0.995 \\ -51 \\ 249 \\ 0 \end{bmatrix}$	$\begin{bmatrix} 0.09500 \\ 0.03100 \\ -0.9950 \\ -51.273 \\ 248.911 \\ 2.8596 \end{bmatrix}$	$\begin{bmatrix} 0.09485 \\ 0.03066 \\ -0.9950 \\ -51.2993 \\ 248.932 \\ 2.7794 \end{bmatrix}$
ζ_{st}	$\begin{bmatrix} 0 \\ 0 \\ 0 \\ 250 \\ 50 \\ -20 \end{bmatrix}$	$\begin{bmatrix} 0.02 \\ -0.01 \\ 0.01 \\ 249 \\ 51 \\ -20.6 \end{bmatrix}$	$\begin{bmatrix} 0.02 \\ -0.01 \\ 0.01 \\ 249 \\ 51 \\ -20.6 \end{bmatrix}$	$\begin{bmatrix} 0.01962 \\ -0.009741 \\ 0.009918 \\ 249.012 \\ 51.0204 \\ -20.6056 \end{bmatrix}$

When the number of measured poses injected noise exceeds 40, the mean errors of calibration poses are stable and smaller than 0.1 mm and 0.001 rad, as shown in Fig. 4. Comparing with the simulation results given in Ref. [26], the convergence rate of parameter identification is faster and the calibration accuracy is higher. The mean errors of calibration poses were obtained in Ref. [26] after at least the eighth iteration, which are almost larger than those of this paper and are close to the maximum errors of calibration poses, as shown in Fig. 4. Hence, the algorithm in this paper not only needs fewer iterations but is also more robust against the measurement noise.

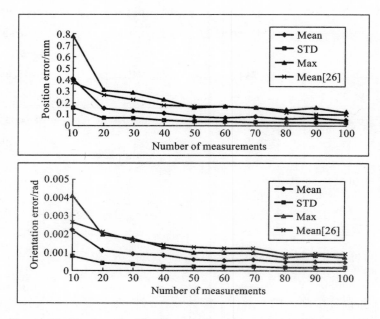

Fig. 4 Verification of calibration results against noise in simulation II.

At the end of the fourth iteration, the identified parameters are sufficiently close to the actual values by using the 40 pose measurements without noise, as shown in Tab. 3, and the errors of calibration poses are almost zero. It takes fewer iterations and fewer pose measurements than those in Ref. [26], as similar results were obtained in Ref. [26] by using 50 pose measurements without noise after the fourth or fifth iteration.

E. Simulation III: Using Eq. (25) for SCARA Robot

Considering Eq. (25) as the error model, joint zero-position errors are set to zero and the other simulation conditions are the same as those in simulation I.

The identified coordinates of joint twist are close to the actual values at the end of the fourth iteration, with higher accuracy than those given in Ref. [28], as shown in Tab. 1. From Fig. 5, the convergence rate of parameter identification in this simulation is a little faster than that in simulation I and a little slower than that in Ref. [28].

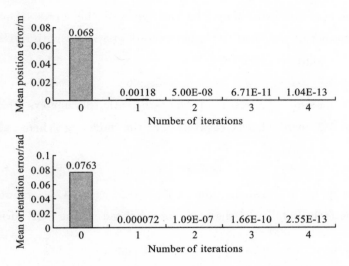

Fig. 5   Mean pose errors during iterative procedure of simulation III

F. Discussion of Simulations

Three error-modeling methods [see Eq. (23)~Eq. (25)] have been verified by the above simulations. After the fourth step of iterations, all the kinematic parameters, which are based on the POE formula, were fully recovered under the ideal experimental conditions. After the fourth step of iterations, the impacts of measurement noise were reduced. From Appendix V, the iterative algorithm for parameter identification is convergent when the deviations of the actual kinematic parameters from the nominal values are small, and the pose measurements are reasonably chosen to make the identification-Jacobian-matrix column full rank.

V. Discussion

A. Maximum Number of Identifiable Parameters

The identifiability of kinematic parameters in the three practicable error models, which are based on the POE formula, has been investigated. The maximum number of identifiable parameters is $6n + 6$ for an n-DOF generic serial robot. Since the three error models are used for generic serial robot, the error models must be instantiated by the types of the joints. Therefore, the maximum number of the identifiable parameters in a robot with r revolute joints and t prismatic joints is given by

$$c_1 = 6r + 3t + 6 \qquad (41)$$

whereas the maximum number of the kinematic parameters that must be identified for any manipulator, regardless of the modeling scheme used is given by [19]

$$c_2 = 4r + 2t + 6 \qquad (42)$$

According to the investigation of this paper, Eq. (42) cannot effectively represent the identifiability of the error model based on the POE formula.

B. Discussion of Joint Zero-position Errors

The joint zero-position errors can be identified when the initial transformation errors are not involved in the same error model and the aforementioned condition is satisfied. According to the requirement of this condition, Eq. (24) cannot be used to calibrate the serial robot whose number of DOF exceeds six. The error modeling for these applications [20-24], where the joint zero-position errors need to be identified, should meet the above requirement. Otherwise, the joint zero-position errors are regarded as redundant parameters and need to be eliminated from the error model as in Ref. [10].

C. Discussion of Redundancy

As for a serial robot, its forward kinematics is composed of a series of homogeneous transformations, which are the elements of special Euclidean group $SE(3)$. Therefore, homogeneous transformation, near the identity, can be written as the exponential of a certain twist, which is an element of $se(3)$. However, in most kinematic models, the homogeneous transformation, which is defined by consecutive link frames, is expressed as a function of the geometric parameters that describe the relationship between the locations of the links. Unfortunately, these geometric parameters do not belong to $se(3)$. Hence, kinematic-parameter identification based on such models results in redundancy and, sometimes, in singularity. On the contrary, joint twist belongs to $se(3)$ so that it can be mapped into $SE(3)$ by exponential map and is identifiable according to the above analysis in Section III. Thus, a single modeling convention without redundancy can be obtained from Eq. (23), Eq. (24), or Eq. (25), based on the POE formula.

VI. Conclusion

A generic error model, which was based on the POE formula, for serial-robot calibration was proposed in explicit form. Three practicable error models, in which all kinematic parameters are identifiable, were presented. Additional condition to identify errors in joint variables was proposed: Any coordinate of joint twist cannot be the linear combination of the others. The identifiability of kinematic parameters is clarified, and one can choose an appropriate error model for serial-robot calibration.

From the above discussion, the following conclusions can be reached.

1) The maximum number of identifiable kinematic parameters is $6n + 6$ for an n-DOF of generic serial robot. Then, for a specific serial robot with r revolute joints and t prismatic joints, its maximum number of identifiable kinematic parameters is $6r+3t+6$.

2) The redundant parameters are involved in the error model when its kinematic parameters do not belong to $se(3)$, and a complete, continuous, and minimal kinematic model for serial-robot calibration can be obtained from the POE formula.

Appendix I Math Background[32,34]

A. Exponential of $se(3)$

Given $\hat{\xi} = \begin{bmatrix} \hat{\omega} & v \\ 0 & 0 \end{bmatrix} \in se(3)$ and $q \in \Re$, the exponential of $\hat{\xi}q$ is an element of

$SE(3)$, $\|\omega\| = (\omega_1^2+\omega_2^2+\omega_3^2)^{1/2}$ and is given by

$$a = \exp\begin{bmatrix}\hat{\omega}q & vq \\ 0 & 0\end{bmatrix} = \begin{bmatrix}R & b \\ 0 & 1\end{bmatrix} \quad (43)$$

where

$$R = I_3 + \frac{\sin(\|\omega\|q)}{\|\omega\|}\hat{\omega} + \frac{1-\cos(\|\omega\|q)}{\|\omega\|^2}\hat{\omega}^2 \quad (44)$$

$$b = \left(qI_3 + \frac{1-\cos(\|\omega\|q)}{\|\omega\|^2}\hat{\omega} + \frac{\|\omega\|q - \sin(\|\omega\|q)}{\|\omega\|^3}\hat{\omega}^2\right)v \quad (45)$$

B. Logarithm of $SE(3)$

When the Lie group is in the neighborhood of the identity, and trace $(R) \neq 1$ and $1 + 2\cos(\phi) = \text{trace}(R)$, $\|\phi\| < \pi$, then the log function on $SE(3)$ is given by

$$\hat{\xi} = \log\begin{bmatrix}R & b \\ 0 & 1\end{bmatrix} = \begin{bmatrix}\hat{\omega} & A^{-1}b \\ 0 & 0\end{bmatrix} \quad (46)$$

where

$$\hat{\omega} = \log[R] = \frac{\phi(R-R^T)}{2\sin(\phi)} \quad (47)$$

$$M^{-1} = I - \frac{1}{2}\hat{\omega} + \frac{2\sin(\|\omega\|) - \|\omega\|(1+\cos(\|\omega\|))}{2\|\omega\|^2\sin(\|\omega\|)}\hat{\omega}^2 \quad (48)$$

If ϕ is sufficiently small, then $\hat{\omega} = (R-R^T)/2$.

C. Adjoint Transformation

Given $a \in SE(3)$, which maps one coordinate system into another, $Ad(a)$: $\Re^6 \to \Re^6$ is given by

$$Ad(a) = \begin{bmatrix}R & 0 \\ \hat{b}R & R\end{bmatrix} \quad (49)$$

And is invertible such that

$$Ad^{-1}(a) = \begin{bmatrix}R^T & 0 \\ -R^T\hat{b} & R^T\end{bmatrix} \quad (50)$$

D. Adjoint Representation of the Lie Algebra

Considering the derivative of $Ad(a)$ at the identity for $s \in \Re$

$$ad(\hat{\xi}) := \frac{d}{ds}Ad(\exp(\hat{\xi}s))\bigg|_{s=0} = \begin{bmatrix} \hat{\omega} & 0 \\ \hat{v} & \hat{\omega} \end{bmatrix} \tag{51}$$

then

$$\begin{aligned}
ad(\hat{\xi}_1)\xi_2 &= \frac{d}{ds}Ad(\exp(\hat{\xi}_1 s))\xi_2\bigg|_{s=0} \\
&= \frac{d}{ds}\exp(\hat{\xi}_1 s)\hat{\xi}_2 \exp(-\hat{\xi}_1 s)\bigg|_{s=0} \\
&= (\hat{\xi}_1\hat{\xi}_2 - \hat{\xi}_2\hat{\xi}_1)^V \\
&= [\hat{\xi}_1, \hat{\xi}_2]^V
\end{aligned} \tag{52}$$

E. Velocity Screw

Let $ad(\hat{\xi}) := \Omega = \begin{bmatrix} \hat{\omega} & 0 \\ \hat{v} & \hat{\omega} \end{bmatrix}$, and $\theta = (\omega_1^2 + \omega_2^2 + \omega_3^2)^{1/2}$

Letting $\hat{\xi} = \hat{\xi}(t) = \begin{bmatrix} \hat{\omega}(t) & v(t) \\ 0 & 0 \end{bmatrix} \in se(3)$ be a differentiable function of a scalar variable t, then the differentials of the exponential map are given in Eq. (53) [33]. The definite-integral expression in (53) was given by Park and Okamura [25,26], i.e.,

$$\begin{aligned}
&\left[\frac{d(\exp(\hat{\xi}))}{dt}\exp(-\hat{\xi})\right]^V \\
&= \int_0^1 \exp(\hat{\xi}s)\frac{d\hat{\xi}}{dt}\exp(-\hat{\xi}s)ds = \sum_{k=0}^{\infty}\frac{1}{(k+1)!}(ad(\hat{\xi}))^k\frac{d\xi}{dt} \\
&= (I_6 + \frac{4-\theta\sin(\theta)-4\cos(\theta)}{2\theta^2}\Omega + \frac{4\theta-5\sin(\theta)+\theta\cos(\theta)}{2\theta^3}\Omega^2 \\
&\quad + \frac{2-\theta\sin(\theta)-2\cos(\theta)}{2\theta^4}\Omega^3 + \frac{2\theta-3\sin(\theta)+\theta\cos(\theta)}{2\theta^5}\Omega^4)\frac{d\xi}{dt}
\end{aligned} \tag{53}$$

Letting $Y = \hat{\xi}q = \hat{\xi}(t)q(t) = \begin{bmatrix} \hat{\omega}(t)q(t) & v(t)q(t) \\ 0 & 0 \end{bmatrix}$, and $q(t) \in \Re$, then

$$\begin{aligned}
&\left[\frac{d\exp(Y)}{dt}\exp(-Y)\right]^V \\
&= \sum_{k=0}^{\infty}\frac{1}{(k+1)!}(ad(Y))^k\frac{dY}{dt}
\end{aligned}$$

$$= \sum_{k=0}^{\infty} \frac{1}{(k+1)!} (ad(Y))^k (q\frac{d\xi}{dt} + \xi\frac{dq}{dt})$$
$$= \sum_{k=0}^{\infty} \frac{1}{(k+1)!} (ad(Y))^k q\frac{d\xi}{dt} + \xi\frac{dq}{dt} \quad (54)$$

Because $ad(Y) = ad(\hat{\xi}q) = ad(\hat{\xi})q$, then

$$\sum_{k=0}^{\infty} \frac{1}{(k+1)!} (ad(Y))^k = \mathbf{I}_6 + \frac{4 - \theta q \sin(\theta q) - 4\cos(\theta q)}{2\theta^2 q}\Omega$$
$$+ \frac{4\theta q - 5\sin(\theta q) + \theta q \cos(\theta q)}{2\theta^3 q}\Omega^2$$
$$+ \frac{2 - \theta q \sin(\theta q) - 2\cos(\theta q)}{2\theta^4 q}\Omega^3$$
$$+ \frac{2\theta q - 3\sin(\theta q) + \theta q \cos(\theta q)}{2\theta^5 q}\Omega^4 \quad (55)$$

Then, we have

$$\left[\frac{d\exp(Y)}{dt}\exp(-Y)\right]^{\vee} = \sum_{k=0}^{\infty} \frac{1}{(k+1)!} (ad(Y))^k q\frac{d\xi}{dt} + \xi\frac{dq}{dt}$$
$$= (q\mathbf{I}_6 + \frac{4 - \theta q \sin(\theta q) - 4\cos(\theta q)}{2\theta^2}\Omega$$
$$+ \frac{4\theta q - 5\sin(\theta q) + \theta q \cos(\theta q)}{2\theta^3}\Omega^2$$
$$+ \frac{2 - \theta q \sin(\theta q) - 2\cos(\theta q)}{2\theta^4}\Omega^3$$
$$+ \frac{2\theta q - 3\sin(\theta q) + \theta q \cos(\theta q)}{2\theta^5}\Omega^4)\frac{d\xi}{dt} + \xi\frac{dq}{dt}$$
$$= A\frac{d\xi}{dt} + \xi\frac{dq}{dt}$$
(56)

Where

$$A = q\mathbf{I}_6 + \frac{4 - \theta q \sin(\theta q) - 4\cos(\theta q)}{2\theta^2}\Omega$$
$$+ \frac{4\theta q - 5\sin(\theta q) + \theta q \cos(\theta q)}{2\theta^3}\Omega^2$$
$$+ \frac{2 - \theta q \sin(\theta q) - 2\cos(\theta q)}{2\theta^4}\Omega^3$$
$$+ \frac{2\theta q - 3\sin(\theta q) + \theta q \cos(\theta q)}{2\theta^5}\Omega^4$$

Thus

$$\left[\frac{d\exp(\hat{\xi}q)}{dt}\exp(-\hat{\xi}q)\right]^V = A\frac{d\xi}{dt} + \xi\frac{dq}{dt} \qquad (57)$$

Appendix II Determinant of A in Eq. (56)

Letting $A = \begin{bmatrix} HH & 0 \\ D & HH \end{bmatrix}$, and $H = \begin{bmatrix} HH & 0 \\ 0 & HH \end{bmatrix}$, then

$$HH = q\mathbf{I}_3 + \frac{1-\cos(\|\omega\|q)}{\|\omega\|^2}\hat{\omega} + \frac{\|\omega\|q - \sin(\|\omega\|q)}{\|\omega\|^3}\hat{\omega}^2 \qquad (58)$$

The determinant of HH is given by

$$|HH| = \frac{2q(-1+\cos(\|\omega\|q))}{\|\omega\|^2} \qquad (59)$$

Since the determinant of A is equal to $|H| = |HH|^2$, then

$$|A| = \frac{4q^2(-1+\cos(\|\omega\|q))^2}{\|\omega\|^4} \qquad (60)$$

Only if $q = 0$, then $A = \mathbf{0}$; else, if $q \neq 0$, then A is full rank. And if $\omega = 0$, then $A = q\mathbf{I}_6$.

Appendix III Relationship between HH_{i-1} and R_{i-1}

According to Appendix I and Appendix II, HH_{i-1} and R_{i-1} are given by

$$\begin{aligned} HH_{i-1} &= q_{i-1}\mathbf{I}_3 + \frac{1-\cos(\|\omega_{i-1}\|q_{i-1})}{\|\omega_{i-1}\|^2}\hat{\omega}_{i-1} \\ &+ \frac{\|\omega_{i-1}\|q_{i-1} - \sin(\|\omega_{i-1}\|q_{i-1})}{\|\omega_{i-1}\|^3}\hat{\omega}_{i-1}^2 \end{aligned} \qquad (61)$$

$$\begin{aligned} R_{i-1} &= \mathbf{I}_3 + \frac{\sin(\|\omega_{i-1}\|q_{i-1})}{\|\omega_{i-1}\|}\hat{\omega}_{i-1} \\ &+ \frac{1-\cos(\|\omega_{i-1}\|q_{i-1})}{\|\omega_{i-1}\|^2}\hat{\omega}_{i-1}^2 \end{aligned} \qquad (62)$$

Then, the relationship between HH_{i-1} and R_{i-1} is given by

$$R_{i-1} = HH_{i-1}\hat{\omega}_{i-1} + \mathbf{I}_3 \qquad (63)$$

Letting $R_{i-1} = [\alpha_1, \alpha_2, \alpha_3]$, and $HH_{i-1} = [\gamma_1, \gamma_2, \gamma_3]$, then Eq. (63) becomes

$$\begin{cases} \alpha_1 = \omega_{3i-1}\gamma_2 - \omega_{2i-1}\gamma_3 + e_1 \\ \alpha_2 = -\omega_{3i-1}\gamma_1 + \omega_{1i-1}\gamma_3 + e_2 \\ \alpha_3 = \omega_{2i-1}\gamma_1 - \omega_{1i-1}\gamma_2 + e_3 \end{cases} \quad (64)$$

where

$$e_1 = \begin{bmatrix} 1 \\ 0 \\ 0 \end{bmatrix}, \quad e_2 = \begin{bmatrix} 0 \\ 1 \\ 0 \end{bmatrix} \quad \text{and} \quad e_3 = \begin{bmatrix} 0 \\ 0 \\ 1 \end{bmatrix}$$

Appendix IV Proof of all columns of $R_{i-1}HH_i$ and HH_{i-1} are linearly independent

Proof: Letting $R_{i-1} = [\alpha_1, \alpha_2, \alpha_3]$, $HH_{i-1} = [\gamma_1, \gamma_2, \gamma_3]$, and

$$HH_i = \begin{bmatrix} a_{11} & a_{12} & a_{13} \\ a_{21} & a_{22} & a_{23} \\ a_{31} & a_{32} & a_{33} \end{bmatrix}$$

then

$$R_{i-1}HH_i = [a_{11}\alpha_1 + a_{21}\alpha_2 + a_{31}\alpha_3, a_{12}\alpha_1 + a_{22}\alpha_2 + a_{32}\alpha_3, \\ a_{13}\alpha_1 + a_{23}\alpha_2 + a_{33}\alpha_3] \quad (65)$$

Suppose that c_1, c_2, c_3, c_4, c_5, and c_6 are elements of $\mathfrak{R}$ such that

$$c_1(a_{11}\alpha_1 + a_{21}\alpha_2 + a_{31}\alpha_3) + c_2(a_{12}\alpha_1 + a_{22}\alpha_2 + a_{32}\alpha_3) \\ + c_3(a_{13}\alpha_1 + a_{23}\alpha_2 + a_{33}\alpha_3) + c_4\gamma_1 + c_5\gamma_2 + c_6\gamma_3 \equiv 0 \quad (66)$$

Substituting Eq. (64) into Eq. (66), we have

$$c_1(a_{11}(\omega_{3i-1}\gamma_2 - \omega_{2i-1}\gamma_3 + e_1) + a_{21}(-\omega_{3i-1}\gamma_1 + \omega_{1i-1}\gamma_3 + e_2) \\ + a_{31}(\omega_{2i-1}\gamma_1 - \omega_{1i-1}\gamma_2 + e_3)) + c_2(a_{12}(\omega_{3i-1}\gamma_2 - \omega_{2i-1}\gamma_3 + e_1) \\ + a_{22}(-\omega_{3i-1}\gamma_1 + \omega_{1i-1}\gamma_3 + e_2) + a_{32}(\omega_{2i-1}\gamma_1 - \omega_{1i-1}\gamma_2 + e_3)) \quad (67) \\ + c_3(a_{13}(\omega_{3i-1}\gamma_2 - \omega_{2i-1}\gamma_3 + e_1) + a_{23}(-\omega_{3i-1}\gamma_1 + \omega_{1i-1}\gamma_3 + e_2) \\ + a_{33}(\omega_{2i-1}\gamma_1 - \omega_{1i-1}\gamma_2 + e_3)) + c_4\gamma_1 + c_5\gamma_2 + c_6\gamma_3 \equiv 0$$

Rearranging Eq. (67), we get

$$\begin{aligned}
&(-c_1 a_{21}\omega_{3i-1} + c_1 a_{31}\omega_{2i-1} - c_2 a_{22}\omega_{3i-1} \\
&+ c_2 a_{32}\omega_{2i-1} - c_3 a_{23}\omega_{3i-1} + c_3 a_{33}\omega_{2i-1} + c_4)\gamma_1 \\
&+ (c_1 a_{11}\omega_{3i-1} - c_1 a_{31}\omega_{1i-1} - c_2 a_{32}\omega_{1i-1} \\
&+ c_2 a_{12}\omega_{3i-1} + c_3 a_{13}\omega_{3i-1} - c_3 a_{33}\omega_{1i-1} + c_5)\gamma_2 \\
&+ (-c_1 a_{11}\omega_{2i-1} + c_1 a_{21}\omega_{1i-1} - c_2 a_{12}\omega_{2i-1} \\
&+ c_2 a_{22}\omega_{1i-1} - c_3 a_{13}\omega_{2i-1} + c_3 a_{23}\omega_{1i-1} + c_6)\gamma_3 \\
&+ (c_1 a_{11} + c_2 a_{12} + c_3 a_{13})e_1 + (c_1 a_{21} + c_2 a_{22} + c_3 a_{23})e_2 \\
&+ (c_1 a_{31} + c_2 a_{32} + c_3 a_{33})e_3 \equiv 0
\end{aligned} \tag{68}$$

And then

$$\begin{cases}
-c_1 a_{21}\omega_{3i-1} + c_1 a_{31}\omega_{2i-1} - c_2 a_{22}\omega_{3i-1} \\
+ c_2 a_{32}\omega_{2i-1} - c_3 a_{23}\omega_{3i-1} + c_3 a_{33}\omega_{2i-1} + c_4 \equiv 0 \\
c_1 a_{11}\omega_{3i-1} - c_1 a_{31}\omega_{1i-1} - c_2 a_{32}\omega_{1i-1} \\
+ c_2 a_{12}\omega_{3i-1} + c_3 a_{13}\omega_{3i-1} - c_3 a_{33}\omega_{1i-1} + c_5 \equiv 0 \\
-c_1 a_{11}\omega_{2i-1} + c_1 a_{21}\omega_{1i-1} - c_2 a_{12}\omega_{2i-1} \\
+ c_2 a_{22}\omega_{1i-1} - c_3 a_{13}\omega_{2i-1} + c_3 a_{23}\omega_{1i-1} + c_6 \equiv 0 \\
c_1 \begin{bmatrix} a_{11} \\ a_{21} \\ a_{31} \end{bmatrix} + c_2 \begin{bmatrix} a_{12} \\ a_{22} \\ a_{32} \end{bmatrix} + c_3 \begin{bmatrix} a_{13} \\ a_{23} \\ a_{33} \end{bmatrix} \equiv 0
\end{cases} \tag{69}$$

According to Appendix II, the columns of HH_i are linearly independent so that $c_1 = c_2 = c_3 = 0$, which results in $c_4 = c_5 = c_6 = 0$.

Thus, all columns of $R_{i-1} HH_i$ and HH_{i-1} are linearly independent.

Appendix V Proof of Convergence for the Iterative Least-squares Algorithm in Parameter Identification

Proof: Let P_* be the actual kinematic parameters, and let $\{P_k\}$, and $\{J_k\}$, $\{\tilde{y}_k\}$ be the sequence of iterations, where $\tilde{J}_* \neq 0$. Vector $\tilde{y}_k$ is composed of logarithm of $g_a g_{nk}^{-1}$ at different pose and $\tilde{y}_* = 0$, where g_{nk} is the forward kinematics based on the POE formula after the kth iteration. Letting

$\| (\tilde{J}_k^T \tilde{J}_k)^{-1} \tilde{J}_k^T \| = \beta$, and $\| \tilde{J}_k \| = \lambda$, then

$$\begin{aligned}
\| P_{k+1} - P_* \| &= \| P_k + (\tilde{J}_k^T \tilde{J}_k)^{-1} \tilde{J}_k^T \tilde{y}_k - P_* \| \\
&= \| (\tilde{J}_k^T \tilde{J}_k)^{-1} \tilde{J}_k^T (\tilde{J}_k P_k + \tilde{y}_k - \tilde{J}_k P_*) \| \\
&= \| (\tilde{J}_k^T \tilde{J}_k)^{-1} \tilde{J}_k^T ((\tilde{y}_k - \tilde{y}_*) + \tilde{J}_k (P_k - P_*)) \| \\
&\leqslant \| (\tilde{J}_k^T \tilde{J}_k)^{-1} \tilde{J}_k^T \| \| (\tilde{y}_k - \tilde{y}_*) + \tilde{J}_k (P_k - P_*) \| \\
&\leqslant \| (\tilde{J}_k^T \tilde{J}_k)^{-1} \tilde{J}_k^T \| (\| (\tilde{y}_k - \tilde{y}_*) \| + \| \tilde{J}_k \| \| P_k - P_* \|)
\end{aligned}$$
(70)

Since logarithm map and exponential map are Lipschitz [35], and $\tilde{y}_k$ is the composition of logarithm map and exponential map, then $\tilde{y}_k$ is Lipschitz such that $\| \tilde{y}_k - \tilde{y}_* \| \leqslant \alpha \| P_k - P_* \|$, where α is the Lipschitz constant. Then, we have

$$\begin{aligned}
\| P_{k+1} - P_* \| &\leqslant \| (\tilde{J}_k^T \tilde{J}_k)^{-1} \tilde{J}_k^T \| (\alpha \| P_k - P_* \| + \| \tilde{J}_k \| \| P_k - P_* \|) \\
&= \beta(\alpha + \lambda) \| P_k - P_* \|
\end{aligned}$$
(71)

Therefore, the iterative least-squares algorithm in parameter identification is local convergent when the deviation of the actual kinematic parameters from the nominal values is small enough to make logarithm map valid, and appropriate pose measurements are chosen to make $\tilde{J}_k$ column full rank.

Acknowledgements

The authors would like to thank F. Han, Q. Zhao, Q. Zeng, and C. Chen for giving some good writing advice.

References

1 H. Zhuang and Z. Roth. "Camera-aided Robot Calibration". CRC Press, 1996, pp. 1; pp. 63-64; pp. 77.

2 G. Gatti and G. Danieli. "A practical approach to compensate for geometric errors in measuring arms: application to a six-degree-of-freedom kinematic structure". Meas. Sci. Technol., vol. 19, no. 1, p. 015107, 2008.

3 K. Schröer, S. Albright, and M. Grethlein. "Complete, minimal and model-continuous kinematic models for robot calibration". Robot. Comput.-Integr. Manuf., vol. 13, no. 1, pp.

73-85, 1997.

4 R. Paul. "Robot Manipulators: Mathematics, Programming, and Control". Cambridge, MA: MIT Press, 1982, pp. 50-55.

5 S. A. Hayati. "Robot arm geometric link parameter estimation". in *Proc.* 22th IEEE Conf. Decision Contr., vol. 22, Dec. 1983, pp. 1477-1483.

6 S. Hayati and M. Mirmirani. "Improving the absolute positioning accuracy of robot manipulators". J. Robot. Syst., vol. 2, no. 4, pp. 397-413, 1985.

7 S. Hayati, K. Tso, and G. Roston. "Robot geometry calibration". in Proc. IEEE Conf. Robot. Autom., 24-29 April 1988, pp. 947-951.

8 W. Veitschegger and C. Wu. "Robot accuracy analysis based on kinematics". IEEE Trans. Robot. Autom., vol. 2, no. 3, pp. 171-179, 1986.

9 H. Stone and A. Sanderson. "A prototype arm signature identification system". in Proc. IEEE Conf. Robot. Autom., vol. 4, Mar. 1987, pp. 175-182.

10 H. Zhuang, Z. S. Roth, and F. Hamano, "A complete and parametrically continuous kinematic model for robot manipulators". IEEE Trans. Robot. Autom., vol. 8, no. 4, pp. 451-463, Aug. 1992.

11 L. Everett and L. Ong. "Determining essential parameters for calibration". in Proc. 1993 ASME Winter Annual Meeting, vol. 49, New Orleans, LA, Nov. 1993, pp. 295-302.

12 D. Etienne and W. Khalil. *Modeling, Performance Analysis and Control of Robot Manipulators*. Wiley Blackwell, Dec 2006, pp. 31-33.

13 M. A. Meggiolaro and S. Dubowsky, "An analytical method to eliminate the redundant parameters in robot calibration". in Proc. IEEE Conf. Robot. Autom., vol. 4, 24-28 Apr. 2000, pp. 3609-3615.

14 W. Khalil, M. Gautier, and C. Enguehard. "Identifiable parameters and optimum configurations for robots calibration". Robotica, vol. 9, pp. 63-70, 1991.

15 J. Hollerbach and C. Wampler. "The calibration index and taxonomy for robot kinematic calibration methods". Int. J. Robot. Res., vol. 15, no. 6, pp. 573-591, 1996.

16 J. Borm and C. Menq. "Experimental study of observability of parameter errors in robot calibration". in Proc. IEEE Conf. Robot. Autom., 1989, pp. 587-592.

17 K. Schröer. "Theory of kinematic modelling and numerical procedures for robot calibration". in Robot Calibration, R. Bernhardt and S. Albright, Eds. London: Chapman & Hall, 1993, ch. 9, pp. 157-196.

18 L. J. Everett and T. W. Hsu. "The theory of kinematic parameter identification for industrial robots". ASME J. Dynam. Syst. Meas. Contr., vol. 110, no. 1, pp. 96-100, 1988.

19 L. J. Everett and A. H. Suryohadiprojo. "A study of kinematic models for forward

calibration of manipulators". in Proc. IEEE Conf. Robot. Autom., 24-29 April 1988, pp. 798-800.

20 J. Santolaria, J. Aguilar, J. Yagüe, and J. Pastor. "Kinematic parameter estimation technique for calibration and repeatability improvement of articulated arm coordinate measuring machines". Prec. Eng., vol. 32, no. 4, pp. 251-268, 2008.

21 R. P. Judd and A. B. Knasinski. "A technique to calibrate industrial robots with experimental verification". IEEE Trans. Robot. Autom., vol. 6, no. 1, pp. 20-30, Feb. 1990.

22 J. M. Renders, E. Rossignol, M. Becquet, and R. Hanus. "Kinematic calibration and geometrical parameter identification for robots". IEEE Trans. Robot. Autom., vol. 7, no. 6, pp. 721-732, Dec. 1991.

23 A. Goswami, A. Quaid, and M. Peshkin. "Identifying robot parameters using partial pose information". IEEE Control Syst. Mag., vol. 13, no. 5, pp. 6-14, Oct. 1993.

24 P. Rousseau, A. Desrochers, N. Krouglicof, W. Autom, and Q. Montreal. "Machine vision system for the automatic identification of robotkinematic parameters". IEEE Trans. Robot. Autom., vol. 17, no. 6, pp. 972-978, 2001.

25 F. Park and K. Okamura. "Kinematic calibration and the product of exponential formula". in Advances in Robot Kinematics and Computational Geometry, J. Lenârciĉ and B. Ravani, Eds. Cambridge: MIT Press, 1994, pp. 119-128.

26 K. Okamura and F. Park. "Kinematic calibration using the product of exponentials formula". Robotica, vol. 14, pp. 415-421, 1996.

27 I. Chen and G. Yang. "Kinematic calibration of modular reconfigurable robots using product-of-exponentials formula". J. Robot. Syst., vol. 14, no. 11, pp. 807-821, 1997.

28 I. Chen, G. Yang, C. Tan, and S. Yeo. "Local POE model for robot kinematic calibration". Mech. and Mach. Theory, vol. 36, no. 11-12, pp. 1215-1239, 2001.

29 S.-H. Kang, M. W. Pryor, and D. Tesar. "Kinematic model and metrology system for modular robot calibration". in Proc. IEEE Conf. Robot. Autom., vol. 3, Apr 26-May 1, 2004, pp. 2894-2899.

30 S. K. Mustafa, G. Yang, S. H. Yeo, and W. Lin. "Kinematic calibration of a 7-DOF self-calibrated modular cable-driven robotic arm". in Proc. IEEE Conf. Robot. Autom., 19-23 May 2008, pp. 1288-1293.

31 R. Brockett. "Robotic manipulators and the product of exponentials formula". in Mathematical Theory of Networks and Systems, P. A. Fuhrman, Ed. New York: Springer Verlag, 1984, pp. 120-129.

32 R. Murray, Z. Li, and S. Sastry. "A Mathematical Introduction to Robotic

Manipulation". Boca Raton, FL: CRC Press, 1994.

33 J. Selig. "Geometric Fundamentals of Robotics". ser. Monographs in Computer Science. Springer Verlag, 2005.

34 F. C. Park, "Computational aspects of the product-of-exponentials formula for robot kinematics". IEEE Trans. Autom. Control, vol. 39, no. 3, pp. 643-647, Mar. 1994.

35 I. Hambleton and E. Pedersen, "Compactifying infinite group actions". in Geometry and Topology, Aarhus: Conference on Geometry and Topology, August 10-16, 1998, Aarhus University, Aarhus, Denmark, vol. 258. Amer Mathematical Society, 2000, pp. 203-211.

（原载 IEEE Transactions on Robotics, Volume 26, Issue 3, 2010）

高端制造装备关键技术的科学问题

杨叔子　丁汉　李斌

提　要　分析了高端制造装备的国家重大需求和面临的发展机遇,讨论了现阶段高端数控机床和极大规模集成电路制造装备研发的主要任务,通过分析这两类高端制造装备的研究现状与发展趋势,总结出了高性能运动控制、高速高精度驱动、装备与工艺动态交互作用、精密加工的环境控制四方面的关键科学问题。

关键词:高端制造装备,高端数控机床,极大规模集成电路制造装备,发展趋势,科学问题

1　国家需求与发展机遇

装备制造业真正体现了一个国家的实力,在产业链上具有非常强的带动性。我国虽然是世界制造大国,制造业占国家 30% 的产值,工业增加值居世界第四位。但与工业发达国家相比,我国的制造技术仍有较大差距,突出表现高端装备制造技术落后,对国家各领域发展的装备支撑能力弱,能源和资源消耗大,环境污染严重。针对这一状况,我国提出了走新型工业化道路,2020 年成为世界制造强国的战略目标。在国家 16 个重大科技专项中,核心电子器件、高端通用芯片及基础软件、极大规模集成电路制造技术及成套工艺、高端数控机床与基础制造技术、大型油气田及煤层气开发、大型先进压水堆及高温气冷堆核电站、大型飞机、高分辨率对地观测系统、载人航天与探月工程等专项均与制造技术息息相关,其成功实施都需要先进的制造装备予以保障。

我国已经将制造业列为国家中长期的重点发展领域之一[1]。胡锦涛主席在 2010 年 6 月 7 日的中国科学院和工程院院士大会上强调指出:"大力发展先进制造科学技术。"2010 年 9 月 8 日国务院决定将高端装备制造业确定为战略性新兴

产业之一,要求高端装备制造产业优先发展先进航空装备和高速铁路交通等先进运输装备、海洋工程装备、高端智能制造装备,推进空间基础设施建设,促进卫星及其应用产业发展,强化基础配套能力,积极发展以数字化、柔性化及系统集成技术为核心的智能制造装备[2]。

高端数控装备首先指技术上的高端,表现为知识、技术密集,体现多学科和多领域高、精、尖技术的集成,同时具有高附加值,是价值链的高端,处于产业链的核心部位,其发展水平决定产业链的整体竞争力[3]。具体包括超大型多轴复合机床、高速高精度制造装备、巨型重载制造装备、超精密制造装备等,其技术指标往往接近装备的物理极限[4]。限于笔者的知识和研究领域,本文重点探讨高端数控机床和极大规模集成电路制造装备。图1是大型客机机舱整体加工用超重型五轴立柱移动式镗铣床和超大规模集成电路制造用的光刻机,其中五轴镗铣床具有超大的工作范围,行程达 40 m×12 m×3 m,光刻机需要实现大尺度运动和大惯量负载下多自由度的纳米定位精度,是对运动控制的极限挑战。高端数控机床和极大规模集成电路制造装备均已列入国家16个重大科技专项,体现了其在国家需求、国家发展战略中的重要位置。

(a)   (b)

图 1   两类高端制造装备

(a)客机机舱整体加工用超重型五轴立柱移动式镗铣床;(b)极大规模集成电路制造用的光刻机

高端数控机床是支撑航空航天、船舶、汽车、发电设备等制造领域发展的核心装备。面对我国高端数控机床几乎全部依赖进口的情况,现阶段的主要任务[5]是以这些领域所需要的高端数控机床为目标,重点研发高速精密复合数控金切机床、重型数控金切机床、数控特种加工机床等主机产品,基本掌握高档数控装置、电动机及驱动装置、数控机床功能部件、关键部件等的核心技术,逐步提高我国高档数控机床与基础制造成套装备的自主开发能力,满足国内主要行业对制造装备的基本需求。

极大规模集成电路制造装备在国际上称为战略装备,是集成电路产业链中最重要的一环,不仅支撑集成电路产业的发展,同时也支撑了整个信息产业的发展。超大规模集成电路的总体战略目标[6]是重点实现 90 nm 制造装备产品化,若干关键技术和元部件国产化;研究开发 65 nm 制造装备整机产品;重点突破 45 nm 以下加工关键技术;研发出光刻技术与装备、薄膜生长工艺与装备、化学机械抛光(CMP)技术与装备、高密度后封装工艺与装备,初步建立我国集成电路制造产业创新体系。

2 研究现状与发展趋势

2.1 高档数控机床

航空航天、船舶、汽车、发电设备制造领域的需求是高档数控机床发展的主要推动力。这些领域工况渐趋于超高温、超高压、超高速等极端条件,所用零件的材料向超高强度、耐高温和超低温、耐腐蚀、轻量化方向发展,几何形状越来越复杂,生产中要求自动化程度高、多种工艺复合,对数控机床提出了极大的挑战,如飞机起落架和发动机整体叶轮的加工,材料去除率高达 90% 以上,具有型面复杂、材料难加工、加工周期长且对加工品质一致性要求高等一系列加工特点,要求多轴复合数控机床具有适应强冲击、时变工况的能力。

从行业需求和近年国际机床展的主流产品上看,大型、复合、多轴、重载、高速高精、智能控制已成为高档数控机床的发展趋势,2010 年芝加哥国际制造技术展会上,奥地利林茨的 WFL 机床可以一次装夹完成飞机起落架等大型圆柱形零件的车铣复合加工,如图 2 所示。普通数控机床加工精度 5 μm,精密级加工中心 1 μm,超精密加工已进入纳米级 0.01 μm,图 3 是美国 Lawrence Livemore 国家研究室研制的纳米级精度数控机床。高档数控机床的研究重点包括高档数控装置、机床动态特性和高速高精驱动系统。

a)高档数控装置

高档数控装置技术的研究主要包括三个方面:多轴精细数控插补、柔性复合加工技术、加工过程闭环控制等智能控制技术。

日本和德国在数控系统的精细插补方面做了大量的研究工作,日本 FANUC 数控系统提出的高速纳米插补与纳米平滑等方法可以实现微小细分线段的高速高精度加工[8](图 4),德国 SIEMENS 840D 数控系统推出的五轴 NURBS 插补功能有效实现了高精度插补和速度平滑[9]。五轴刀心点高速平滑插补(RTCP)也是

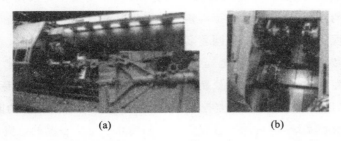

图 2　高档数控机床的最新产品

(a)奥地利林茨的 WFL 车铣复合机床；(b)多任务机床的多主轴多塔式刀架结构

图 3　美国 Lawrence Livemore 研制的纳米级精度数控机床[7]

图 4　FANUC 数控系统的高速纳米插补

重要的趋势，德国、日本的高档数控装置已经具备了该项功能，我国"高档数控机床及基础制造装备"重大科技专项已经把五轴刀心点控制和精细样条插补功能作为高档数控系统技术的重要指标[10]。

柔性复合加工技术方面，在 2009 年的北京国际机床展上，五轴车铣复合数控机床是德国 DMG、奥地利 WFL、瑞士 Mikron 和日本 Mazak 等国际知名机床品牌展出的主流产品，采用多主轴、多塔式刀架结构，可一次装夹完成复杂零件加工，具备高复合加工能力。柔性加工还体现在智能机器人与数控机床的融合上，工业机器人应用领域不但从搬运、码垛、喷漆、焊接工作范围扩展到了机床上下料、换刀、测量、切削加工、装配及抛光领域，而且从减轻劳动强度的繁重工种发展到提高劳动效率的颜色分检、视觉跟踪等领域。

加工过程闭环控制方面，德国著名的高端数控系统 Heidenhain 在 2009 年的欧洲机床展和北京国际机床展上提出了全闭环加工数控系统的概念(图 5)，有效克服机构间隙和热误差等，以保证制造精度。日本 Mazak 展出的机床具有智能化防干涉、振动防止与控制、主轴监控、车削工作台平衡失调检测等功能，将加工状态控制引入数控系统，通过加工过程闭环控制提高了复杂工况的适应性。

图 5　Heidenhain 数控系统的全闭环演示

b) 机床动态特性

数控机床动态特性的研究主要包含三个方面：数控装备状态辨识与动态行为仿真、数控装备与加工工艺的交互作用，以及基于动态特性的装备可靠性评估。

数控装备状态辨识和动态行为仿真是保证动态加工精度的关键，复合、多轴、重载、高速高精数控加工装备的动态行为分析是目前的研究趋势，德国、加拿大和美国针对机床本体结构和运动部件的动力学特性，研究了以机床整机及部件动力学特性驱动的机床设计建模方法，用于指导高档数控机床的设计。

数控机床与加工工艺的动态交互作用对加工品质有很大影响，由于许多关键零件具有加工工序复杂、制造周期长的特点，要求数控装备具有钻、镗、车、铣等一体的复合加工能力，因而机床-工艺交互作用的影响更加明显。装备与工艺交互作用也是目前的研究前沿，国际生产工程学会(CIRP)在 2008—2010 年连续以"装备-工艺交互作用"为主题组织召开国际会议，研究机床和工艺交互作用机理、动态测量、数字仿真预测等技术，如图 6 所示。

针对机床长期连续工作可靠性的问题，德国从工况信息的度量、工艺系统运行状态特征的提取和辨识对保证零件加工品质的作用研究了装备的可靠性评估

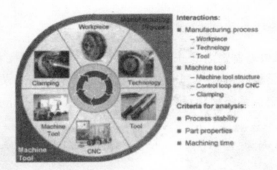

图 6　数控机床和工艺的交互作用[11]

方法。国内现有机床的可靠性差,导致加工工艺能力指数远远低于国际先进水平的能力指标,亟须开展高档数控装备动态行为演变规律和服役可靠性评估方面的研究。

c) 高速高精驱动系统

高档数控机床的高速高精驱动系统主要得益于直接驱动电动机的广泛使用,所谓直接驱动就是将直接驱动旋转电动机(DDR)或直接驱动直线电动机(DDL)直接耦合或连接到从动负载上,从而实现了与负载的直接耦合。相对于传统的旋转电动机加机械传动方式,直驱方式消除了机械传动带来的间隙、柔性及与之相关的系列问题,可以实现高速、高加速度、高刚性,提高驱动系统的动态性能,降低运行成本,而且具有紧凑的结构。2010 年芝加哥国际制造技术展会上,日本 THK 公司展出的直线电动机加速度可达到 9g,速度达到了 720 m/min。大功率、大扭矩直线电动机已经用于重载、高速数控机床,摆动与旋转运动也开始采用力矩电动机驱动,瑞士 Mikion(图 7)、日本 Mazak 和德国 DMG 等著名机床厂商都推出了使用直线电动机和力矩电动机直接驱动的多轴、复合加工机床。

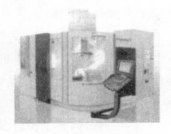 

图 7　采用直驱电机技术的 Mikron 五轴联动铣削机床

2.2 极大规模集成电路制造装备

极大规模集成电路制造装备主要的研究目标包括：光刻技术与装备、薄膜生长工艺与装备、化学机械抛光（CMP）技术与装备、高密度后封装工艺与装备。其中，光刻机的纳米定位系统和后封装用的引线键合机，如图 8 所示。

图 8 极大规模集成电路制造装备中的纳米定位系统和引线键合系统

大规模集成电路制造装备对运动工作台的速度和精度有极高要求。例如光刻机工作台是光刻机精密运动主体，负责完成光刻曝光运动，是高速、大行程、大质量、大惯性、6 自由度的纳米级超精密运动，特别是其硅片台和掩模台间的纳米级精密同步要求。要求对数十千克的运动质量，在数百毫米行程上，对 15 个运动控制轴，实现加速度高至 2g，速度 4 m/s 以上、精度为纳米级的 6 维运动。45 nm 光刻机要求定位精度达到 45 nm，同步精度移动平均差（MA）＜25 nm，移动标准差（MSD）＜5 nm，这种大尺度运动和大惯量负载下多自由度的纳米定位精度要求是对运动平台系统的极限挑战。引线键合机等后封装装配过程要求运动平台处于高加速度、高速的不间断的精密启停，运动加速度达 12～15 g，往复频率 20～25 Hz，定位精度 1～5 μm。大规模集成电路制造装备的研究重点包括：工作台结构与驱动系统、高速高加速度运动控制和运行环境控制技术。

a) 工作台结构与驱动系统

主流光刻机、引线键合机、真空中晶元传输设备均采用直驱电机，适应了芯片加工对工作台的要求。但直线电机直接驱动方式在结构上仍然是在一个方向进给轴上放置搭乘另一个方向进给轴，这种层叠式结构使得定位工作台质量难以最大限度地减小，也使得上下进给轴的惯性负载不一致，因此工作台固有频率较低、难以进一步提高响应速度。同时，工作台上下层叠式结构将引起阿贝测量系统误差、导轨直线度误差等问题。

针对芯片光刻平台的高精度要求，有研究者采用压电陶瓷电机实现直线移动,满足了超精密定位的要求[12]。压电陶瓷电机驱动的工作平台具有以下优点:1)在输入控制量为电流参数时,系统模型为线性模型;2)具有宽频带的特点,一般是在千赫兹量级,平台易于控制,精度可以达到纳米级。但缺点在于:1)行程非常小,通常在 100 μm 左右,且压电式电机的自由度只有一个;2)当控制输入量为电压时,系统会出现滞后现象,这种情况在较大范围移动和运动频率达到 100 Hz 时尤为明显;3)压电陶瓷的变形需要高压电源(1 kV),散热困难导致机械结构变形,而且压电陶瓷的抗张强度小,易折断。目前常用的做法是将其搭载在另一个粗定位工作台上,实现粗精定位分开,达到大位移和超精密定位的目的,这给高速、高加速度和超精密运动的控制带来了巨大的挑战。

为了开发适用于下一代光刻机的高速、高加速度、高精度定位工作台,从 20 世纪 90 年代开始,美、日、韩等国对悬浮式精密定位工作台开展了大量研究。这种悬浮运动工作台具有运动范围大、推重比大、精度高、能量损失少、摩擦极小、无磨粒污染等特点,是下一代光刻设备超精密工作台的基础,受到世界各 IC 芯片制造商和科研机构的密切关注。

b) 高速、高加速度和超精密运动的控制

极大规模集成电路制造装备对运动速度、加速度和精度的要求已近物理极限,仅靠提高机械部件精度无法实现,需由控制技术作软件修正。因此,运动控制技术是整机精度实现的重要支撑。

图 9 所示的引线键合机要求同时实现高加速度和高精度的运动控制。在高加速度工况下,弹性变形、摩擦、间隙、变阻尼、传感器的动态特性等因素对伺服系统定位精度的影响已不可忽略;同时,封装过程中的微力(克量级)受运动机构的弹性变形、摩擦以及空气阻力的影响极大,也对控制系统提出了苛刻的要求。高速、高加速度精密运动控制的研究分为三个方面:1)精确的高频采样定时,为了实现短行程、高加速度的快速高精度定位,位置环控制的采样频率必须等于大于 8 kHz,通常通过 DSP 和 FPGA 等硬件系统实现;2)消除或减小负载惯量和外界干扰对运动性能的影响,提高控制系统的鲁棒性,许多先进控制算法如自适应控制、鲁棒控制、滑膜控制、干扰观测器和变结构控制等都得到了应用;3)抑制高加速度引起的平台高频振动的方法,现有的算法包括使用滤波器来过滤振动信号,或者使用 FIR 来进行信号整形,但这两种方法都会引起响应的延时。重复学习控制算法(ILC)在平台运行重复的运动时具有很大的优势,不仅可以消除或减弱负载惯量和外界干扰的影响,还能有效抑制振动,且不造成响应延时,但它仅适用于重复运动。

图 9 引线键合机

c) 运行环境控制

工作环境对电气器件的制造精度有很大影响。美国麻省理工学院(MIT)的空间纳米技术实验室,采用扫描电子束干涉光刻(SBIL)方法加工纳米级基准光栅[13],对系统中的激光干涉仪测量误差的分析表明系统总的测量误差为 84.96 nm,其中由于环境因素应起的测量误差为 31.16 nm,占到总误差的 45% 左右。

微电子制造前道工艺对运行环境有极高的要求,如 CVD、PVD 和等离子刻蚀等制造阶段,对 300 mm 的晶圆的操纵环境的真空度要求达到 $5×10^{-9}$ Torr(1 Torr= 133.32 Pa),如图 10 所示。随着特征线宽从 65 nm 向 45 nm、32 nm 的延伸,对操纵环境的真空度的要求也越来越高($1×10^{-10}$ Torr)。光刻机中环境控制系统由送风单元、水冷单元、氮气净化单元、气体控制单元、热空气抽排单元和空气颗粒抽排单元组成。机内环境温度要求 $22±0.1$ ℃,相对湿度 35%～60% RH。关键部件的温度控制更加苛刻,投影物镜为 $22±0.01$ ℃。对于光学投影式光刻机而言,无论是物镜系统,还是工件台系统都要求极高的恒温精度(局部最高精度达 0.001 ℃)。需要通过空气减振器和有源减振器进行减振和隔振,自然频率 0.5～4 Hz 噪声达到 30 dB of acoustic isolation,需集成地震波侦听器等高灵敏传感器。外界的环境振动、系统中局部运动单元激励引起的振动,都对系统中的工艺单元稳定性带来极大的不良影响,对光刻的分辨率、离子注入的均匀性、检测分辨率指标等都有着不可忽视的制约。

(a) (b)

图 10 超真空环境下的制造装备

(a)超高真空 65 nm 刻蚀机;(b)超高真空环境下的晶元操作

3 关键技术科学问题

a) 高性能运动控制技术

精密复杂零件的高效加工是高端数控机床的主要挑战。高档数控装置面临的主要技术科学问题包括:机床动力学约束下刀心点高速高精度平滑插补技术、融合智能机器人和复合加工的多轴多通道及复合加工控制技术。

电子元件的微细化和高效化发展趋势,需要研究微/纳制造环境控制下的超精密定位与操纵系统的控制。核心挑战是:在多尺度(宏观、微观以及机器视觉等尺度共存)空间上,制造过程中的高加速度、高精度带来的光、机、电、控制等一体化融合的科学问题。关键问题包括:多维高精度极端运动系统动力学建模与面向控制的优化设计;超精密运动系统的精度设计理论与新型控制方法;制造过程多影响参数耦合作用模型及高精度协同控制。

b) 高速高精度驱动技术

直接驱动电动机的高精度、高动态特性、结构紧凑和低成本特性符合高端数控装备的要求,已经应用在高端数控机床和大规模集成电路制造装备中。使用中存在推力波动导致的速度波动及定位精度问题、系统温升及冷却问题和生产清洁度问题。需要研究永磁直线伺服电动机推力稳定性控制、高精度直线电动机加工工艺、电动机温升分析及冷却系统和基于推力补偿控制的高精度控制技术。

悬浮驱动是面向下一代光刻设备超精密工作台的技术,寻找合适的控制方法、并通过合理的结构设计优化驱动与悬浮磁阵列、采用智能材料增加空气轴承的刚度特性,是磁浮(气浮)的研究热点,同时解决纹波推力扰动、齿槽推力扰动、磁链谐波扰动和端部效应也是促使无摩擦驱动平台走向实用化的关键问题。

c) 装备与工艺动态交互的机理与控制

在加工型面复杂、高强度零件时,时变、强切削力的输入将激发机床和工艺系统的复杂响应,易导致加工过程失稳,因此需要研究制造装备与工艺的动态交互作用,实现高效率、高精度、批量稳定加工。主要问题包括:强激励下装备复杂响应与工艺过程的动态交互机理、多轴加工动力学系统稳定性判别方法、多工序制造过程的数字化监测技术和加工过程物理行为的主动控制方法、适应复杂时变工况的高精度全闭环智能控制技术。

d) 精密加工的环境控制技术

随着半导体工业的快速发展,对工作环境的洁净度提出了苛刻的要求,要求工作环境的洁净度大于 10 级。纳米级精度运动系统的环境控制是实现纳米级精度的前提,多轴精密、超精密加工的热环境与加工精度也有着密切的关系。需要研究对振动、洁净度、温度场与真空度等环境的控制与相应的驱动技术,制造环境的微扰动与物理参数之间的映射关系,微扰动在操纵系统作用界面上的动态演化规律。

4 结 语

高端装备制造业是为国民经济各行业提供技术装备的战略性产业,是国家安全的重要保障,也是各行业产业升级、技术进步的重要保障和国家综合实力的集中体现。国家重大科技专项、装备制造业调整和振兴规划等国家战略计划的实施为高端装备制造业发展提供了前所未有的发展机遇。本文分析了高档数控机床和极大规模集成电路制造装备的重大需求,探讨了目前的研究现状与发展趋势,总结并提出了高性能运动控制、高速高精度驱动、装备与工艺动态交互作用、精密加工的环境控制四方面的关键技术科学问题。

参 考 文 献

[1] 中华人民共和国科学技术部. 国家中长期科学和技术发展规划纲要(2006—2020 年)[R]. 2006.

[2] 中华人民共和国国务院. 国务院关于加快培育和发展战略性新兴产业的决定[R]. 2010.

[3] 陆燕荪. 高端装备制造产业是振兴装备制造业的突破口[R]. 第五届中国电工装备创新与发展论坛,2010.

[4] 国家自然科学基金委员会工程与材料学部机械学科. 机械工程学科发展战略报告(2011— 2020)[R]. 北京:科学出版社,2010.

[5] 中华人民共和国国务院. 装备制造业调整和振兴规划[R]. 2009.

[6] 中华人民共和国工业和信息化部. 国家"十一五"科学技术发展规划[R]. 2006.

[7] US Lawrence Livemore National Laboratory, https://www.llnl.gov/, 2010.

[8] FANUC Series 30i/31i/32i, ALNano CNC for High-Speed, High-Accuracy machining, 2009.

[9] SINUMERIK 840D/840Di/810D Description of functions special functions (FB3), 2008.

[10] 中华人民共和国科学技术部. "高档数控机床与基础制造装备"科技重大专项指南[R]. 2009.

[11] Brecher C., Esser M., Witt S.. *Interaction of manufacturing process and machine tool*. CIRP Annals-Manufacturing Technology, 2009, 58 (2): 588-607.

[12] 王国彪. 纳米制造前沿综述[M]. 北京:科学出版社, 2009.

[13] Chen C. G., Konkola P. T., Heimann R. K., Joo C., Schattenburg M. L.. Nanometer-accurate grating fabrication with scanning beam interference lithography Proceedings of SPIE-The International Society for Optical Engineering, 4936, pp: 126-134, 2002.

Scientific Problems Originated from Key Techniques of Advanced Manufacturing Equipments

Yang Shuzi　Ding Han　Li Bin

Abstract　In this paper, the national requirements and development opportunities of the advanced manufacturing equipments are first analyzed. The major research and development works in advanced NC (Numerical Control) machine tools and very-large-scale integrated circuit equipments are then discussed. Based on the summarizations of the current states and development tends of these two types of manufacturing equipments, four scientific problems originated from the key techniques are presented. They cover high performance motion control, high-speed and high-precision driving interaction mechanism between manufacturing process and machine tool, and environment control for high-precision manufacture.

Keywords: Advanced manufacturing equipment, Advanced NC machine tool, Very-large-scale integrated circuit equipment, Development trend, Scientific problem

(原载《机械制造与自动化》2011 年第 40 卷第 1 期)

后　记

我首先引用我在《杨叔子教育雏论选》"后记"中的第一句话："事情还得追溯到我75周岁生日那天。2008年9月我生日那天，参加祝贺的我的学生，一致商定要将我的论著结集出版，一共四种。这在本书的'序言'中讲清楚了。"

四种书，已出了两种，另一种也即将出版。这三种书都是大"文科"类的，只有这种是大"理科"类的了。我大学念的是工科机械制造专业，毕业后也一直从事机械工程特别是机械制造领域中的教学与科研。虽然自1993年我担任校长工作以来，主要精力放在高等教育有关方面去了，但是，立足的领域仍然没变，如同以往一样，一直高度关注着科学技术特别是机械工程、机械制造技术的发展及其趋势。

正因为我一贯高度关注科技整体的发展，又重点关注机械工程、机械制造技术的发展，所以，1976年后，研究范围逐渐扩大，进入80年代，可以讲，我是立足于机械工程、机械制造领域，致力于同新兴学科、基础学科的交叉。进入90年代以来，又高度关注同生命科技、纳米科技、微电子科技等的结合，特别是在先进制造技术、机电设备诊断、振动噪声、信号处理、无损检测、人工智能与网络的应用等方面，和同事、研究生一起，开展了一系列的研究工作，取得了一系列的相应成果。本文集从千余篇学术论文中，本着"序言"中所讲的区别，选取了具有代表性的48篇，其中先进制造技术（含智能制造、网络制造、虚拟制造）11篇，机电设备诊断13篇，机床刚度、振动、噪声6篇，时序分析、信号处理（包括δ函数、小波变换、倒谱分析、BP网络等）7篇，无损检测（主要是钢丝绳的）10篇，其他1篇，其中以英文发表的11篇。

这些论文之所以具有代表性，是因为合乎世界科技发展的潮流，顺乎机械工程、机械制造技术发展的趋势，切合我国工程与生产实际的要求，适应高层次人才培养的需要，而且在有关实践中得到了考验。例如：在机械制造方面，关于智能制造我们发表了国内第一篇学术论文，接着又发表了一批论文，这里选了代表作如《多施主协作设计问题的研究》，这些论文为有关学者十分关注；在网络制造方面选了代表作《网络化制造与企业集成》等论文;而在先进制造技术发展方面,我们在《机械工程学报》2003年与2004年连续发表两文,并选入了本文集。这批论文引起了国内外高度关注,点击次数与引用次数都很高,因为它们精练而又准确地分析了先进制造技术的发展及其趋势,并分析了这些发展与趋势之间的关系。又如:在机电设备诊断方面,我们努力建立设备诊断学的学科体系,本文集中选登了1989年发表在德国英文版刊物上 *Plant Condition Recognition—A Time Series Model Approach* 一文与发表在《计算机学报》1991年第3期上的《基于深知识的多故障两步诊断推理》一文,并以此两文与有关学术论文为基础,在清华大学出版社于1993年出版了《基于知识的诊断推理》一书;同时,对工程与生产实际中的关键问题展开了研究,且于1998年领头承担了国家攀登项目(B类)"大型机电系统中若干动力学关键技术的研究",取得了一系列成果。又如:在机床刚度、振动、噪声方面,最具有代表性的成果是我们同第二汽车制造厂(简称"二汽",今东风汽车公司)合作,在"二汽"的领导与多方面的支持、合作下,一举解决了"二汽"生产关键难题——MX-4曲轴连杆颈车床强烈的振动与噪声问题;这一难题连美国生产MX-4机床的厂家也无法解决,如果不解决,则可能导致曲轴停产、东风汽车停产。同时,与此成果相应,发表了一系列论文,其中,在《机械工程学报》1992年的英文版上发表了代表作 *Forced Regenerative Chatter and its Control Strategies in Machine Tools* 一文,文中提出了切削过程中强迫再生颤振这一新发现、新概念,这一论文后来还被评为《机械工程学报》创刊50周年优秀论文之一。再如,在钢丝绳断丝定量检测技术方面,80年代中期,我与师汉民教授合作的团队攻克了钢丝绳断丝定量检测这一国际难题,取得了突破性进展,在生产与工程中广泛获得了应用,目前这一成果还在继续深化与扩大应用范围。本文集选登了有关论文,具有代表性的是刊登在美国 Material Evaluation 杂志1990年3期上的 *Space-Domain Feature-Based Automated Quantitative Determination of Localized Faults in Wire Ropes*,该刊特别回函给予高度评价。又如,在信号处理,包括 δ 函数分析、时间序列分析以及有关方面,有些基础性的研究成果至今仍很有学术价值,例如,在时间序列分析方面不少成果已成为1991年出版的《时间序列分析的工程应用》一书中的内容,此书已出了第二版,至今仍为有关人员的重要参考书。

这本文集之所以能够出版,一句话,因为有实实在在的科技成果及与之有关成果。我之所以有这些成果,不能不感谢我的老师与前辈,不能不感谢我的同事、朋友与同辈,不能不感谢我的学生与晚辈,对于他们,我都应表示深挚的谢意。的的确确,我无法将他们一一列出;只能提出少数的少数,而且这还只是直接与本文集内容有关的很少数的人,但已是挂一漏万了。在此要特别声明,在本文集所附论文目录中,论文第一作者如果是我,即是我执笔的;如果不是我,则其中不少,甚至其主要工作也不是我参与的,而是第一作者(往往是研究生)、与我共同指导研究生的导师和有关人员所做的。

我应该感谢西安交通大学阳含和教授,是他 20 世纪 80 年代前期第一个将我推上了面向全国的课程教学指导组织舞台,他将他负责的全国高等学校机械工程控制课程教学指导委员会的主任工作交给了我。我应该感谢哈尔滨工业大学黄文虎教授,是他 80 年代中期第一个将我推上了全国的学术组织舞台,在他力荐下,我参加了中国振动工程学会常务领导班子,后成为副理事长。我应该感谢天津大学彭泽民教授,是他 80 年代大力支持下,我参加了高等学校机械工程这一大学科教学指导委员会的工作,而且较早成为博士生导师。我应该感谢中国科学院胡海昌教授与大连理工大学钱令希教授,是他们 80 年代有力地支持了我在学术上的攀登,使我在学术上有了重要的进展。无疑,我应该感谢我校我的专业老师,如路亚衡、陈日曜、李如沅等教授。当然,我必须再次重申,我不会忘记我的老师这老一辈,我能够作出一些贡献,都是他们奠下的基础。没有老师,就没有学生,学生永远是踩着老师的肩膀上去的,学生的生命永远是老师的生命的延续。这点在《杨叔子散文、序、函类文选》的"后记"中讲得更清楚。

我还应该感谢兄弟学校、兄弟单位的我的同行、同事、同辈们,特别是涂序彦、屈梁生、黄仁、严普强、刘又午、童忠钫、黄逸云、陈继武、徐宗俊、朱骥北、李传奇、朱剑英、朱德懋、徐善祥、徐敏等专家、教授,本文集的有关论文都同他们以这种或那种形式的支持与帮助分不开!

我还应感谢校内的同行、同事、同辈们,特别是卢文祥、杜润生、熊有伦、师汉民、胡庆超、宾鸿赞、段正澄、陈尔昌、李柱、陈志祥、费奇、黄玉盈、刘永长、马元镐、黄长艺等教授,不仅本文集的有关论文,而且我的教学、科研工作都同他们多方面的合作、支持、帮助分不开!

我还应该感谢我的弟子们,这包括在"序言"落款中署了名的弟子们,也包括在本文集论文中署了名的弟子们,这里既有署了名的,也有因论文未选入而未见名的,这里既有还在校的,还有先后离校的,既有大量在国内的,还有一批在国外的;当然,令人惋惜,有的还已去世,吴雅就是 1999 年底在美国因车祸去世的,她做了大量极为出色的工作。不言而喻,还应该感谢在我的教学、科研团队内工作过的同志,在我的成果中也凝集了他们的心血。在此,我应该特别感谢史铁林同志,不仅因为这本文集几乎是靠他亲自动手并组织有关人员特别是研究生完成的,还不仅因这四种文集的出版他付出了大量精力,而且自我担任校长工作以后,我所在团队的工作就主要由他负责,并延续至今;并且我在校长工作岗位上,在校学术委员会主任工作岗位上,以及离开学校工作岗位回到团队以后,他同有关同志一起,全力支持我的工作,照顾我的方方面面,特别这些年来,极为关注我的健康。应该说,没有史铁林同志,没有史铁林同志与有关同志的关心、帮助与照顾,许多工作、许多事情就不好去设想了。在这里,我尤应提及,在史铁林同志与有关同志努力下,我所在团队的科研工作如同团队所在的华中科技大学机械学院一样,有着巨大的发展,团队紧紧把握着科研方向,将机械工程、机械制造技术同微纳科技、生物科技、智能科技、光电科技、国防科技等紧密而有机地结合起来,使得人才培养、科学研究、社会服务充满着蓬蓬勃勃的朝气,洋溢着人与人之间的人文关怀。我喜见"长江后浪超前浪,世上新人换旧人"这一客观规律的生动体现。这正如前几年我改陈子昂《登幽州台歌》一诗为:"前既见古人,今更见来者;看大江之滔滔,喜奔腾而东下!"

当然,我应感谢为本文集出版付出了辛勤劳动的研究生,诸如王珂、张嘉琪、蒋淑兰、饶和昌、张财胜等40位同学,特别是何锐波同志这位博士后,他们在史铁林同志组织下做了大量扎扎实实的工作,"序言"中也已讲了。当然,还应感谢陈惜曦同志,配合本文集出版,她做了不少默默无闻的有关工作。可以说,没有他们,本文集就出版不了。

当然,我还一定要感谢华中科技大学出版社的领导和朋友。出版社对我一直十分信任、十分友好、十分配合。1984年、1991年分别出版了我与我的同事合作编著的本科生教材《机械工程控制基础》、研究生教材《时间序列分析的工程应用》,出版得很成功,至今仍畅销,一再获奖,赢得读者好评。现今这四种文集中,已出了两种,社会反映热烈,又获好评。本文集出版后,最后一种文集即将随之出版。在此关键之际,我谨向出版社的领导与朋友,由衷地表示感谢。

今天,恰恰是"七一",是中国共产党成立九十一周年的日子。这九十一年是星火燎原的岁月,是改天换地的岁月,是"国际悲歌歌一曲,狂飙为我从天落"的峥嵘岁月。九十一年,中国共产党由第一次代表大会十三名代表所代表的五十多名党员发展到八千二百多万名党员,靠小米加步枪打败了现代化武装到牙齿的敌人,将沦为"华人与狗,不得入内"的半殖民地的旧中国改造成在世界上影响举足轻重的新中国,经济总量居世界第二,"神舟"上天、"蛟龙"入海、科技迅速发展,靠的是什么?是人!是用马克思主义中国化的伟大理论与中华民族几千年来的优秀文化传统武装起来的人!毫无疑问,科学技术是立世之基、文明之源。历史一再证明,没有科学技术的进步,就没有生产力的进步,没有工具理性的进步,社会文明就会丧失源头。然而,人文文化是为人之本、文明之础,历史也一再证明,没有人文文化的进步,没有价值理性的进步,社会文明就会面临崩溃。因为科学技术的确是第一生产力,社会越进步,情况就越如此;然而,掌握科学技术的是人。人是物质的,更是精神的;人的躯体是载体,人的思想是灵魂;才是形而下的,德是形而上的。司马光在《资治通鉴》中讲得极深刻:"才者,德之资也;德者,才之帅也。"才,体现着、深化着德;德,统率着、激活着才。这就是20世纪五六十年代所讲的"又红又专,以红为先";也就是今天讲的"德才兼备,以德为先"。所以,我一再认为,首先必须学会做好人,同时,必须学会做好事。一个高层次人才若不能做好事,不能掌握好本专业的业务,用什么来体现做好人?!若不能用主要精力去做好事,不能用主要精力掌握好本专业的业务,用什么来体现做好人?!当然,一个人不能做好人,而很会做事,那么后果将会十分严重;富才劣德,是大灾难!方向一错,全盘全错!现在,我国正处在快速发展、改革高潮之中,困难重重,挑战频频,但挑战与机遇并存,陷阱与大道同在。我们在中国共产党领导下,把准方向,坚定信心,练好本领,扎实工作,敢于开拓,善于创新,完全可以战胜前进道路上的一切艰难,奔向本世纪中叶我国的第三步目标。愿这本文集的出版为我国科技园中增添一朵小花,并作为向党的十八大的真诚献礼!

杨叔子

2012年7月1日(农历壬辰年五月十三日)

论 文 附 录

[1] 杨叔子.机床主轴部件静刚度的分析与计算.华中工学院学报,1978(01): 41-201.
[2] 杨叔子.机床两支承主轴部件静刚度的分析与计算.机床,1979(03):1-11.
[3] 苏士铭,杨叔子.主轴部件轴向支承的角刚度及其对径向刚度的影响.机床, 1979(03):12-13.
[4] 杨叔子.三支承主轴部件静刚度的分析与讨论.机床,1979(09):11-20.
[5] 杨叔子.关于机床多支承主轴部件的支承孔不同轴时一些问题的讨论.华中工学院学报,1979(04):193-198.
[6] 杨叔子.支承反力矩对机床主轴部件静刚度的影响.华中工学院学报,1979(S1):13-25.
[7] 杨叔子,师汉民.δ 函数在机械制造中的应用.华中工学院学报,1980(04): 145-154.
[8] 杨叔子.专业课教师教些基础课,好处甚多.高等教育研究,1981(02):21-24.
[9] 杨叔子.多自由度线性阻尼系统频率特性的研究——兼论机床动态性能分析中的一些问题.华中工学院学报,1981(04):65-72.
[10] 王治藩,杨叔子.多自由度线性阻尼系统与机床主轴部件动态性能的状态空间分析.华中工学院学报,1981(05):63-72.
[11] 杨叔子,师汉民.控制理论在机械制造中的应用.湖北机械,1981(05): 10-13.
[12] 杨叔子.为培养研究生努力开出新课.高等教育研究,1981(04):37-39.
[13] Yang Shuzi. *Study of the static stiffness of machine tool spindles*. International journal of machine tool design & research,v 21,n 1,p 23-40 (1981).

[14] 王治藩,杨叔子.关于评定主轴部件刚度的几个问题.华中工学院学报,1982(01):71-80.

[15] 杨叔子.平稳时间序列的数学模型及其阶的确定的讨论.华中工学院学报,1983(05):9-14.

[16] 杨叔子.动态数据的系统处理.机械工程,1983(05):41-43.

[17] 杨叔子.动态数据的系统处理——第二讲 ARMA 模型及其特性.机械工程,1983(06):41-45.

[18] 杨叔子.时间序列分析学术讨论会在华中工学院召开.机械工程,1984(01):27.

[19] 杨叔子.动态数据的系统处理——第三讲 建模.机械工程,1984(02):38-43.

[20] 杨叔子.动态数据的系统处理——第四讲 最佳预测.机械工程,1984(03):43-45.

[21] 王治藩,杨叔子.建模、辨识、滤波与动态数据处理.振动、测试与诊断,1984(05):30-36.

[22] 杨叔子,熊有伦,师汉民,王治藩.时序建模与系统辨识.华中工学院学报,1984(06):85-92.

[23] 杨叔子,王治藩,赵星,杜润生,陈小鸥,丁洪.信号(动态数据)的微型机在线处理.华中工学院学报,1984(6):93-100.

[24] 韦庆如,傅建西,陈尔昌,杨叔子.磨床砂轮主轴在旋转状态下的主要特性.华中工学院学报,1984(06):131-138.

[25] Yang Shuzi and Wang Zhifan. *Principle and application of vibration time domain monitoring technique.* Proceedings of the International Modal Analysis Conference & Exhibit, v 1, p 379-383 (1985).

[26] 王治藩,杨叔子.ARMA 谱分析中若干问题的研究.应用力学学报,1985(01):33-47.

[27] 杨叔子.振动信号的微型机在线分析与监视.振动、测试与诊断,1985(02):1-8.

[28] 杨克冲,杨叔子.控制理论在机械制造中的应用.机床,1985(09):40-44.

[29] 杨克冲,王平明,杨叔子.系统分析·信号分析与机械工程·生物工程(上).机械工程,1985:2-4.

[30] 杨叔子,宾鸿赞,陈志祥,王治藩.时间序列方法及其在机械工业自动化中

的应用.机械工业自动化,1985(02):1-11.

[31] 杨克冲,王平明,杨叔子.系统分析·信号分析与机械工程·生物工程（下）.机械工程,1985(04):5-8.

[32] 杨叔子,杨克冲,赵星,王治藩,熊有伦.平稳时序连续模型建模的探讨.华中工学院学报,1985(05):1-10.

[33] 吴雅,熊有伦,杨叔子.二维时序模型与闭环辨识.华中工学院学报,1985(05):19-26.

[34] 吴雅,杨叔子.向量 ARMA 模型的一种建模方法.华中工学院学报,1985(05):27-33.

[35] 王治藩,杨叔子,赵星.在线信号处理系统研制的总体考虑.华中工学院学报,1985(05):35-40.

[36] 丁洪,赵星,陈小鸥,杨叔子.最大熵谱的两种快速算法及局部表示.华中工学院学报,1985(05):55-64.

[37] 杨叔子,刘经燕,师汉民,梅志坚.金属切削过程颤振预兆的特性分析.华中工学院学报,1985(05):79-86.

[38] 梅志坚,师汉民,刘经燕,杨叔子.金属切削过程颤振早期诊断的一个有效的综合判据.华中工学院学报,1985(05):87-94.

[39] 刘经燕,杨叔子,卢文祥,杜润生.刀具磨损的信号功率谱分析与监视.华中工学院学报,1985(05):95-102.

[40] 王平明,杨克冲,徐毓琪,杨叔子.人体消化道肠鸣音的时间序列分析初探.华中工学院学报,1985(05):103-108.

[41] 秦争鸣,宾鸿赞,杜润生,杨叔子,王玉璞,王晓明.滚珠丝杆副的频谱分析与在线验收.华中工学院学报,1985(05):109-116.

[42] Yang Shuzi, Shi Hanmin, Mei Zhijian and Liu Jingyan. *Computer aided early stage diagnosis of chatter in the metal-cutting process and its control*. Mechanical Engineering Publ Ltd, p 269-274 (1986).

[43] 丁洪,杨叔子.IBM-PC 上绘图软件库的建立及其在机械工程中的应用.机械工程,1986(01):32-35.

[44] 杜润生,赵卫,杨冲克,陈卓宁,宾鸿赞,杨叔子,王玉璞.单板机至 IBM-PC 机的串行数据传输.陕西科技大学学报,1986(01):13-19.

[45] 王玉璞,杨叔子,宾鸿赞.滚珠丝杠副的频谱分析与导程误差的在线测量.装备机械,1986(02):19-24.

[46] 赵卫,杜润生,杨克冲,杨叔子,王玉璞.滚珠丝杆副导程误差的微型机在线回归分析与频谱分析.华中工学院学报,1986(03):371-378.

[47] 杨叔子.时序模型的诊断方法.机械工程,1986(02):5-14.

[48] 丁洪,杨叔子.IBM-PC机上绘图软件库的设计.微计算机应用,1986(06):50-56.

[49] 赵卫,杜润生,陈卓宁,杨克冲;丁洪,宾鸿赞,杨叔子,王玉璞.滚珠丝杠副导程误差分析及微型机在线处理.机械工业自动化,1986(03):6-18.

[50] 杨叔子.机械设备诊断技术的发展与研究(上).机械工程,1987(01):18-20.

[51] 赵卫,杜润生,杨克冲,杨叔子,王玉璞.机床传动链传动误差采集及滚珠丝杠副导程误差分析.机床,1987(05):4-7.

[52] 赵云坤,赖常芹,朱莉,林武延,杨叔子,王治藩,陈小鸥,赵星.微机在眼震电图上的应用.临床耳鼻咽喉科杂志,1987(01):53-55.

[53] 刘经燕,梅志坚,郁国平,卢文祥,师汉民,杨叔子.金属切削过程振动信号的分析与监视.振动与冲击,1987(01):59-72.

[54] 赵云坤,赖常芹,朱莉,林武延,杨叔子,王沼藩,陈小鸥,赵星.微电脑在眼震电图上的应用.广后医学,1987(01):30-34.

[55] 杨叔子,师汉民,熊有伦,王治藩.机械设备诊断学的探讨.华中工学院学报,1987(02):1-8.

[56] 郑小军,杨叔子,师汉民,周安法.设备诊断专家系统的层次诊断模型.华中工学院学报,1987(02):9-14.

[57] 熊有伦,杨叔子,王跃科.复时间序列与主轴回转精度.华中工学院学报,1987(2):15-22.

[58] 郑小军,杨叔子.自回归谱估计算法的分类及应用.华中工学院学报,1987(02):23-28.

[59] 吴雅,杨叔子,杜跃芬,周有尚.多维时序模型及其在人口死亡预报中的应用.华中工学院学报,1987(02):29-34.

[60] 梅志坚,刘经燕,师汉民,杨叔子,韩敬礼,周光前.金属切削过程颤振的计算机模式识别及在线监控.华中工学院学报,1987(02):41-48.

[61] 杜润生,赵卫,杨克冲,杨叔子.传动链误差微计算机诊断系统.华中工学院学报,1987(02):73-76.

[62] 丁洪,何军,杨叔子,李家镕,李再光.激光器工作噪声的表面声强测量及频

谱分析.华中工学院学报,1987(02):89-94.

[63] 丁洪,杨叔子.信号特征的谱阵图分析方法.华中工学院学报,1987(02):101-104.

[64] 卢文祥,丁洪,杨叔子.刀具磨损时声发射信号的谱特征.华中工学院学报,1987(02):105-108.

[65] 欧阳普仁,杨克冲,杨叔子.关于"平稳时序连续模型建模的探讨"一文的几点注解.华中工学院学报,1987(02):141-143.

[66] 杨叔子.机械设备诊断技术的发展与研究(下).机械工程,1987(02):22-25.

[67] 梅志坚,杨叔子,师汉民,刘经燕.一种无颤振机床技术的研究.机床,1987(9):9-12.

[68] 杨叔子,郑小军,周安法,师汉民.人工智能在机械设备诊断中的应用.机械工程,1987(04):10-13.

[69] 杨叔子,郑小军.专家系统的原理、现状和发展趋势.水利电力机械,1987:33-37.

[70] Zhou Anfa, Shi Hanmin, Yang Shuzi and Zheng Xiaojun. *Fuzzy inductive learning based on rough sets*. Proc 1988 IEEE Int Conf Syst Man Cybern, p 525-529 (1988).

[71] Zheng Xiaojun, Yang Shuzi, Zhou Anfa and Shi Hanmin. "A knowledge-based diagnosis system for automobile engines". Proceedings of the 1988 IEEE International Conference on Systems. Man, and Cybernetics (IEEE Cat. No. 88CH2556-9), v 2, p 1042-1047 (1988).

[72] Zheng Xiaojun, Yang Shuzi, Zhou Anfa and Shi Hanmin. *A knowledge-based diagnosis system for automobile engines*. International Journal of Advanced Manufacturing Technology, v 3, n 3, p 159-169(1988).

[73] Wang Yangsheng, Shi Hanmin and Yang Shuzi. *Quantitative wire rope inspection*. NDT international, v 21, n 5, p 337-340 (1988).

[74] 杨叔子,丁洪.机械制造的发展及人工智能的应用.机械工程,1988(01):32-34.

[75] 欧阳普仁,杨叔子.一种改进的Marple算法.南京理工大学学报(自然科学版),1988(01):41-47.

[76] 吴雅,杨叔子.门限自回归模型建模的有关探讨及其在人口死亡率预报中的应用.天津大学学报,1988(01):23-29.

[77] 周光前,杨叔子,梅志坚,刘经燕,师汉民.金属切削过程颤振及在线监控的研究.长沙铁道学院学报,1988(01):68-74.

[78] 欧阳普仁,杨叔子,谢月云.模态参数识别Prony法的一种改进.振动与冲击,1988(01):68-71.

[79] 吴雅,杨叔子,杜跃芬,周有尚.人口死亡的时序模型分析及其群体预报.信号处理,1988(01):106-114.

[80] 杨叔子,欧阳普仁,郑小军,丁洪.信号的人工智能处理系统——基于知识的信号处理方法的探讨.华中理工大学学报,1988(03):1-6.

[81] 师汉民,郑小军,杨叔子,周安法,刘永长,马元镐.设备诊断专家系统的核心结构探讨.华中理工大学学报,1988(03):7-12.

[82] 郑小军,杨叔子,师汉民,周安法,马元镐,刘永长.复杂系统的诊断问题求解——一种基于知识的方法.华中理工大学学报,1988(03):13-18.

[83] 周安法,师汉民,杨叔子,郑小军,马元镐,刘永长.知识获取的多层证据网络模型.华中理工大学学报,1988(03):19-26.

[84] 吴雅,杨叔子,陶建华.灰色预测和时序预测的探讨.华中理工大学学报,1988(03):27-34.

[85] 吴雅,杨叔子,师汉民.门限自回归模型与非线性系统的极限环.华中理工大学学报,1988(03):35-42.

[86] 丁洪,杨叔子,董双文.ARMA谱估计中若干问题的研究.华中理工大学学报,1988(03):43-48.

[87] 赵星,王治藩,陈小鸥,杨叔子.在线信号处理系统改进中的若干问题的讨论.华中理工大学学报,1988(03):49-54.

[88] 王阳生,师汉民,杨叔子,李劲松,叶兆国.检测局部异常信号的一个新特征量.华中理工大学学报,1988(03):61-67.

[89] 叶兆国,王阳生,杨叔子,师汉民,韩连生,刘连顺.钢丝绳断丝定量检测中径向随机晃动误差的补尝.华中理工大学学报,1988(03):69-74.

[90] 李劲松,刘克明,卢文样,杨叔子,蔡建龙,谢德珍.钢丝绳断丝探伤传感器的研制.华中理工大学学报,1988(03):75-80.

[91] 梅志坚,师汉民,昌松,杨叔子.信号功率谱特征变化的时域快速诊断.华中理工大学学报,1988(03):85-90.

[92] 昌松,梅志坚,师汉民,杨叔子.机床颤振预兆早期诊断的频率矩心判别法.华中理工大学学报,1988(03):91-94.

[93] 欧阳普仁,杨叔子,谢月云,杨克冲.时域模态参数识别法的改进及实验研究.华中理工大学学报,1988(03):95-100.

[94] 刘克明,李劲松,卢文祥,杨叔子.钢丝绳断丝无损探伤原理及探伤系统信号预处理装置的研究.强度与环境,1988(03):17-22.

[95] 王阳生,师汉民,杨叔子,叶兆国.钢丝绳断丝信号的定量解释.振动工程学报,1988(02):9-17.

[96] 杨叔子,师汉民,郑小军,梅志坚,周安法,刘经燕.机械设备及其工作过程的计算机诊断.机械工业自动化,1988(02):2-6.

[97] 杜跃芬,周有尚,吴雅,杨叔子.时间序列分析在死亡率季节变动规律研究中的初步应用.中国卫生统计,1988(04):39-40.

[98] 梅志坚,杨叔子,师汉民.机床颤振的早期诊断与在线监控.振动工程学报,1988(03):8-17.

[99] 丁洪,杨叔子,康宜华.ARMA谱值估计的快速算法的研究.郑州工学院学报,1988(03):21-30.

[100] 梅志坚,杨叔子,师汉民,刘经燕.机床颤振的计算机控制技术的研究.工业控制计算机,1988(05):17-20.

[101] 昌松,梅志坚,师汉民,杨叔子.在线变刀具几何角度抑制机床颤振的研究.洛阳工学院学报,1988(04):19-25.

[102] Zheng Xiaojun and Yang Shuzi. *Plant condition recognition - a time series model approach*. Computers in Industry, v 11, n 4, p 333-340 (1989).

[103] 梅志坚,杨叔子,师汉民,昌松.机床颤振的一种新的控制方法.江苏工学院学报,1989(01):39-44.

[104] 李震,赵星,陈小鸥,杨叔子,肖行贯.脉搏波特征点的自动识别.信息与控制,1989(01):59-64.

[105] 吴雅,杨叔子.死因死亡率的非平稳时序模型预报及有关探讨.成都科技大学学报,1989(01):125-130.

[106] 吴雅,杨叔子.时间序列分析及其在机床工业中的应用.机床,1989(06):46-48.

[107] 康宜华,丁洪,杨叔子.最大熵谱谱峰的计算.四川工业学院学报,1989(01):1-9.

[108] 杨玉坚,梅志坚,杨叔子,师汉民.用机床颤振的非线性理论对变速切削抑制颤振的研究.四川工业学院学报,1989(01):10-15.

[109] 梅志坚,杨叔子,昌松,师汉民,陈日曜.在线调整刀具前角和后角抑制颤振的原理与试验研究.武汉工学院学报,1989(01):53-61.

[110] 欧阳普仁,杨叔子.一个智能编程系统.武汉水利电力学院学报,1989(03):76-81.

[111] 王丁,王克雄,刘仁俊,陈佩薰,谌刚,王治藩,卢文祥,杨叔子.白鳍豚声行为及听觉灵敏度的初步研究.湘潭大学自然科学学报,1989(02):116-121.

[112] 康宜华,丁洪,杨叔子.最大熵谱的一种新的快速算法.西安理工大学学报,1989(02):130-136.

[113] 周安法,师汉民,桂修文,杨叔子,郑小军.诊断问题求解的一种扩展解释模型.陕西科技大学学报,1989(02):52-62.

[114] 吴雅,杨叔子,姜莉,孙世荃.云锡矿工肺癌的时间序列预测方法.广西大学学报(自然科学版),1989(03):64-71.

[115] 王阳生,师汉民,杨叔子,李劲松,叶兆国,韩连生,刘连顺.钢丝绳断丝定量检测的原理与实现.中国科学(A辑 数学 物理学 天文学 技术科学),1989(09):993-1000.

[116] 王丁,刘仁俊,陈佩薰,王治藩,卢文祥,杨叔子.白鱀豚的发声及其与环境适应的初步研究.水生生物学报,1989(03):210-217.

[117] 杨叔子,桂修文,郑小军,陆鑫森.专家系统与振动工程.机械科学与技术,1989(05):2-5.

[118] 李震,陈小鸥,赵星,杨叔子,肖行贯,蔡用之,朱家麟,章晓东.一种脉搏波处理系统.中国医疗器械杂志,1989(06):322-326.

[119] 康宜华,李劲松,陈根林,卢文祥,杨叔子.钢丝绳断丝在线定量检测装置中的抗干扰措施.强度与环境,1989(06):1-8.

[120] 郑小军,杨叔子.集成节约覆盖模型与概率推理的新方法.华中理工大学学报,1989(04):1-8.

[121] 丁洪,杨叔子,桂修文.复杂系统诊断问题的研究.华中理工大学学报,1989(04):9-16.

[122] 桂修文,丁洪,杨叔子.用树表达法进行波形理解与释义.华中理工大学学报,1989(04):17-22.

[123] 王阳生,师汉民,杨叔子.深度图像用于机器人视觉导引.华中理工大学学报,1989(04):117-122.

[124] 史铁林,杨叔子,周继洛.新型管道有源降噪系统的试验研究.强度与环境,1990(01):7-12.

[125] 史铁林,杨叔子,周继洛.高效管道有源降噪系统——理论与试验.振动工程学报,1990(01):34-39.

[126] 昌松,梅志坚,杨叔子,师汉民.机床颤振信号互谱特性分析.山东工业大学学报,1990(03):25-31.

[127] 梅志坚,昌松,师汉民,杨叔子.实现机床颤振早期诊断的一个新的特征量.武汉工业大学学报,1990(02):69-77.

[128] 钟毓宁,杨叔子,陈继平,孙贵坤.冷连轧机工作辊轴承失效的定量分析.钢铁研究,1990(04):46-51.

[129] 康宜华,李劲松,卢文祥,杨叔子.钢丝绳断丝在线检测装置.电工技术杂志,1990(05):46-47.

[130] 史铁林,杨叔子,周继洛.一个高效的管道有源降噪系统的试验研究.噪声与振动控制,1990(05):3-7.

[131] 郑小军,杨叔子.汽车发动机诊断专家系统 AEDES.自动化学报,1990(05):393-399.

[132] 郑尚龙,桂修文,丁洪,杨叔子.规则波形的智能理解方法.南昌大学学报(工科版),1990(04):11-17,60.

[133] 杨叔子.俯首甘为孺子牛.高等工程教育研究,1990(04):47-55.

[134] 郑尚龙,丁洪,桂修文,马元镐,杨叔子.一种从状态信息中自动提取特征知识的智能方法.振动工程学报,1990(04):31-38.

[135] Wu Ya, Ke Shiqiu, Yang Shuzi, Xu Shanxiang, Li Weiguo and Jiang Qiang. *Mixed time series models for time-varying metal cutting process*. American Society of Mechanical Engineers, Design Engineering Division (Publication) DE, v 36, p 307-312 (1991).

[136] Ding Hong, Gui Xiuwen and Yang Shuzi. *An approach to state recognition and knowledge-based diagnosis for engines*. Mechanical Systems and Signal Processing, v 5, n 4, p 257-266 (1991).

[137] Wu Ya, Ke Shiqiu, Yang Shuzi, Zhang Qilin, Xu Shanxiang and Wen Yaozu. *Experimental study of cutting noise dynamics*. American Society of Mechanical Engineers, Design Engineering Division (Publication) DE, v 36, p 313-318 (1991).

[138] Ding Hong, Yang Shuzi and Gui Xiuwen. *Elicitation and management of knowledge for engine diagnosis* 91. Proceedings of the IFIP WG5.4/IFAC Workshop on Dependability of Artificial Intelligence Systems (DAISY_91), p 5-13 (1991).

[139] 李培生,陈卓宁,杨克冲,杨叔子.微机补偿磨削高精度丝杆的研究.磨床与磨削,1991(02):31-37.

[140] 徐海贤,杜润生,杨叔子,陈迈群.一种故障诊断专家系统信号接口的研制.振动、测试与诊断,1991(01):43-50.

[141] 胡庆夕,李劲松,杜润生,卢文祥,杨叔子.架空索道钢丝绳检测中遥测技术的研究.焦作矿业学院学报,1991(01):104-114.

[142] 郑小军,杨叔子,师汉民.基于深知识的多故障两步诊断推理.计算机学报,1991(03):206-212.

[143] 吴雅,柯石求,杨叔子,徐善祥,李维国,蒋其昂.时变金属切削过程颤振的线性、非线性时序模型.振动工程学报,1991(01):25-32.

[144] 吴雅,柯石求,杨叔子,徐善祥,张启林,李维国,蒋其昂,闻耀祖.曲轴车床颤振薄弱环节的识别与控制.机床,1991(07):29-35.

[145] 吴雅,杨叔子,柯石求,李维国.机床切削系统的强迫再生颤振与极限环.华中理工大学学报,1991(02):69-75.

[146] 王阳生,师汉民,杨叔子,波波,毕替.激光寻距器作可见物体测量.宇航计测技术,1991(02):1-6.

[147] 蔡建龙,李劲松,康宜华,杨叔子.矿井提升钢丝绳状态在线定量检测技术的开发及其应用.有色金属(矿山部分),1991(03):35-40.

[148] 丁洪,杨叔子,桂修文.基于知识的发动机诊断系统的研究.振动工程学报,1991(02):35-42.

[149] 康宜华,刘启茂,卢文祥,等.实时钢丝绳检测信号处理系统.华中理工大学学报,1991(04):97-102.

[150] 康宜华,杨叔子,卢文祥.钢丝绳磨损和绳径缩细无损检测的研究.强度与环境,1991(4):26-31.

[151] 梅志坚,杨叔子,师汉民,等.机床非线性颤振的描述函数分析.应用力学学报,1991(3):69-78.

[152] 戴林钧,谌刚,杨叔子.弹性空腔噪声的有限带宽预估法.振动工程学报,1991(3):27-33.

[153] 张文祖,胡瑞安,杨叔子.用向前差分基函数生成二次、三次曲线的算法.微型机与应用,1991(12):10-47.

[154] 杨叔子,丁洪,史铁林,等.机械设备诊断学的再探讨.华中理工大学学报,1991(S2):1-7.

[155] 史铁林,杨叔子,师汉民,等.机械设备诊断策略的若干问题探讨.华中理工大学学报,1991(S2):9-14.

[156] 康宜华,李劲松,卢文祥,杨叔子.钢丝绳励磁磁路及磁化段表面漏磁场的分析.华中理工大学学报,1991(S2):113-120.

[157] 丁忠平,李劲松,杨叔子,等.钢丝绳断丝定量检测装置的可靠性研究.华中理工大学学报,1991(S2):127-131.

[158] 陶涛,杨叔子,熊有伦,等.目标状态的"全面"自适应估计.华中理工大学学报,1991(S2):133-139.

[159] 吴功平,吴波,杨叔子,等.检测高压油路压力的夹持式传感器.华中理工大学学报,1991(S2):149-152.

[160] Ding Hong, Yang Shuzi and Zhu Xinbiao. *Intelligent prediction and control of manufacturing processes using neural networks*. Proceedings of the International Conference on Manufacturing Automation, p 268-276 (1992).

[161] 钟毓宁,翁平,谢月云,杨叔子.工程数据库管理系统与应用程序的接口.微计算机应用,1992(1):11-14.

[162] 马宏党,黄锐,李劲松,师汉民,卢文祥,杨叔子,蔡建龙.钢丝绳截面损耗缺陷的检测.强度与环境,1992(1):47-53.

[163] 吴功平,杨光友,杨叔子,等.动压测量装置的研究与实践——夹持式压力传感器.强度与环境,1992(1):59-63.

[164] 吴功平,吴波,杜润生,等.动压检测——夹持式压力传感器及其信号的提取方法.振动、测试与诊断,1992(1):29-33.

[165] 熊有伦,杨叔子.测量自动化、集成化和智能化,中国机械工程,1992(1):23-25.

[166] 吴波,吴功平,杨叔子,等.柴油机喷油系统压力波形的特征抽取及描述方法.振动工程学报,1992(1):34-40.

[167] 钟毓宁,谢月云,翁平,杨叔子.工程数据库管理系统中版本的动态管理与控制.武汉工学院学报,1991(01):30-36.

[168] 吴波,杨叔子,李白诚.柴油机喷油压力波形的符号描述与结构模式分类.国防科技大学学报,1992(2):12-16.

[169] 梅宏斌,阎明印,杨叔子.滚动轴承故障诊断的高频共振法.机械设计与制造,1992(2):14-16.

[170] 蒋泽汉,周述果,卢平,谌刚,杨叔子,谢月云.汽车对桥梁作用力的荷载模型分析研究.振动、测试与诊断,1992(2):24-30.

[171] 史铁林,杨叔子,师汉民,等.基于模糊理论与覆盖技术的诊断模型.计算机学报,1992(4):313-317.

[172] 杨叔子,丁洪.智能制造技术与智能制造系统的发展与研究.中国机械工程,1992(2):18-21.

[173] 丁忠平,康宜华,杨叔子,等.集成霍尔元件在钢丝绳缺陷检测中的应用.强度与环境,1992(3):59-64.

[174] 黄锐,孙迎,卢文祥,杨叔子.钢丝绳截面损耗定量检测系统的研究.机械与电子,1992(3):25-28.

[175] 吴雅,梅志坚,杨叔子.机床切削颤振的定常与时变特性.固体力学学报,1992(3):271-276.

[176] 杨叔子,史铁林,丁洪.机械设备诊断的理论、技术与方法.振动工程学报,1992(3):193-201.

[177] 徐志良,桂修文,丁洪,杨叔子.精密丝杠磨削补偿过程中校正装置的智能控制.机械工业自动化,1992(03):29-31.

[178] 钟毓宁,杨叔子,桂修文.设备诊断型专家系统的一种开发工具.自动化学报,1992(5):559-564.

[179] 黄其柏,杨叔子,师汉民.低噪声齿轮副最佳齿侧间隙的确定方法研究.噪声与振动控制,1992(5):26-28.

[180] 王雪,史铁林,阎明印,杨叔子等.大型设备监测与诊断计算机系统的选型.中国机械工程,1992(6):10-12.

[181] 谌刚,谢月云,欧阳普仁,杨叔子等.钢筋混凝土T型梁在不同荷载(裂纹)下的试验模态分析及非线性外推.应用力学学报,1992(4):84-87.

[182] 丁洪,杨叔子.制造过程描述与建模的生命期工程方法.华中理工大学学报,1992(S1):121-126.

[183] 丁洪,杨叔子.多施主协作设计问题的研究.华中理工大学学报,1992(S1):127-133.

[184] Ding Hong, Gui Xiuwen, Yang Shuzi and Zheng Shaolong. *An implementation architecture of knowledged-based system for engine diagnosis*. Applied Artificial Intelligence,v 7,n 4,p 397-417 (1993).

[185] Ding Hong, Yang Shuzi and Zhu Xinbiao. *Intelligent prediction and control of a leadscrew grinding process using neural networks*. Computers in Industry,v 23,n 3,p 169-174 (1993).

[186] 徐志良,桂修文,丁洪,朱心飚,曹伟,杨叔子,李培生,赵建东,徐才元,辛守义.3米C级精度滚珠丝杠磨削的研究.机床,1993(3):17-21.

[187] 丁洪,杨叔子.智能制造工程的研究与发展现状.计算机辅助工程,1993(1):1-7.

[188] 杨叔子,史铁林,丁洪.机械设备的诊断方法与分类.华中理工大学学报,1993(1):1-5.

[189] 史铁林,王雪,何涛,杨叔子.层次分类诊断模型.华中理工大学学报,1993(1):6-11.

[190] 史铁林,王雪,阎明印,杨叔子.一种基于因果网络模型的诊断推理模型.华中理工大学学报,1993(1):12-19.

[191] 王雪,肖人彬,史铁林,杨叔子.面向对象专家系统中知识的不确定性.华中理工大学学报,1993(1):20-24.

[192] 阎明印,史铁林,王雪,杨叔子.诊断知识获取过程的不确定性问题.华中理工大学学报,1993(1):25-29.

[193] 王雪,史铁林,杨叔子.专家面向对象诊断系统中知识表示方法.华中理工大学学报,1993(1):30-34.

[194] 吴雅,雷鸣,杨叔子,国效香.基于参数模型的智能化预测及其系统研制.华中理工大学学报,1993(1):41-46.

[195] 雷鸣,吴雅,杨叔子.非线性时间序列建模与预测的神经网络法.华中理工大学学报,1993(1):47-52.

[196] 王雪,史铁林,阎明印,杨叔子.基于Petri网的知识库检验方法.华中理工大学学报,1993(1):53-56.

[197] 王雪,史铁林,阎明印,杨叔子.监测与诊断系统中传感器信号可靠性的确认.华中理工大学学报,1993(1):57-62.

[198] 吴波,杨叔子,李白诚.基于知识的诊断系统软件开发方法学探讨.华中理工大学学报,1993(1):68-72.

[199] 何岭松,吴雅,杨叔子.设备工况监测中三维信号图形的消隐.华中理工大学学报,1993(1):73-76.

[200] 蔡志强,吴雅,周笠,杨叔子.故障诊断与切削颤振的小波分析.华中理工大学学报,1993(1):88-94.

[201] 何景光,吴雅,皮钧,杨叔子.一种包络谱细化的新方法及其在故障诊断中的应用.华中理工大学学报,1993(1):95-99.

[202] 何涛,史铁林,谢月云,杨叔子.汽轮发电机组故障诊断系统的知识库维护.华中理工大学学报,1993(1):100-104.

[203] 伍行健,史铁林,叶能安,杨叔子.汽轮发电机组故障的振动特征与自动识别.华中理工大学学报,1993(1):105-110.

[204] 伍良生,杨叔子,师汉民,林金铭.大型发电机-励磁机转子系统不平衡响应分析.华中理工大学学报,1993(1):144-148.

[205] 伍良生,师汉民,杨叔子,林金铭.大型汽轮发电机组转子系统轴承安装位置的优化.华中理工大学学报,1993(1):149-152.

[206] 黄锐,卢文祥,杨叔子.钢丝绳断丝缺陷漏磁场模型及其应用.华中理工大学学报,1993(1):153-158.

[207] 高红兵,谈兵,卢文祥,杨叔子.细丝小直径钢丝绳断丝形式与断丝信号特征.华中理工大学学报,1993(1):159-162.

[208] 王雪,杨叔子,肖人彬.机械产品智能制造中的知识获取问题.中国机械工程,1993(1):1-3.

[209] 雷鸣,朱心飚,尹申明,杨叔子.智能系统的知识自动获取.中国机械工程,1993(1):4-6.

[210] 张文祖,吴雅,桂修文,胡瑞安,杨叔子.机械产品模型的研究与评述.中国机械工程,1993(1):7-9.

[211] 张文祖,桂修文,吴雅,胡瑞安,杨叔子.一种尺寸与公差的表达模式.中国机械工程,1993(1):10-13.

[212] 雷鸣,朱心飚,尹申明,杨叔子.自构形神经网络及其在刀具智能监控系统中的应用.中国机械工程,1993(1):14-16.

[213] 雷鸣,吴雅,杨叔子.智能化预测系统的工程应用.机床,1993(1):23-27.

[214] 何岭松,吴雅,杨叔子.带通信号处理技术.数据采集与处理,1993(3):161-169.

[215] 黄其柏,师汉民,杨叔子,夏薇.计及气流和线性温度梯度的内燃机穿孔声

管排气消声器研究.内燃机学报,1993(1):77-82.

[216] 蒋泽汉,谌刚,杨叔子.汽车对桥梁作用力的荷载模型.中国公路学报,1993(1):40-46.

[217] 吴雅,雷鸣,杨叔子.基于参数模型的智能化预测系统及其应用.振动工程学报,1993(1):35-41.

[218] 张文祖,丁洪,胡瑞安,杨叔子.机械产品的可视知识建模.华中理工大学学报,1993(2):120-125.

[219] 周汉明,吴雅,周笠,杨叔子.双谱的一致估计及在颤振辨识中的应用.振动、测试与诊断,1993(2):1-5.

[220] 梅宏斌,吴雅,杨叔子,崔乐芳,吴克勤.用2032信号分析系统诊断滚动轴承损伤类故障.强度与环境,1993(2):39-44.

[221] 朱心飚,丁洪,杨叔子.人工神经网络简介.机床,1993(9):40-43.

[222] 雷明,李作清,陈志祥,吴雅,杨叔子.神经网络在预报控制中的应用.机床,1993(11):33-35.

[223] 周建国,张石柱,吴雅,杨叔子.排粉风机的振动分析与诊断.振动、测试与诊断,1993(3):37-41.

[224] 解源,郑军,康宜华,杨克冲,杨叔子.钢管管缝在线检测的研究.冶金自动化,1993(3):7-11.

[225] 康宜华,杨克冲,卢文祥,杨叔子.便携式钢丝绳断丝定量检测仪.仪器仪表学报,1993(3):269-274.

[226] 李军旗,闫明印,史铁林,杨叔子.基于神经网络的故障诊断系统研究.机械与电子,1993(3):16-18.

[227] 吴雅,师汉民,梅志坚,杨叔子.金属切削机床颤振理论与控制的新进展.中国科学基金,1993(2):99-105.

[228] 梅宏斌,吴雅,杨叔子,崔乐芳,吴克勤.用包络分析法诊断滚动轴承故障.轴承,1993(8):38-40.

[229] 黄其柏,师汉民,卢文祥,杨叔子,吴佳常.齿轮啮合噪声辐射特性研究.农业机械学报,1993(4):80-85.

[230] 康宜华,杨克冲,朱文凯,杨叔子.钢丝绳断丝断口漏磁场分析计算.中国机械工程,1993(4):1-3.

[231] 康宜华,薛鸿健,杨克冲,杨叔子.钢丝绳断丝漏磁场的聚磁检测原理.中国机械工程,1993(4):4-6.

[232] 黄锐,卢文祥,杨叔子.钢丝绳内部断丝探伤传感器的设计.中国机械工程,1993(4):7-9.

[233] 杨叔子.新中国高教事业发展的一个缩影——庆祝华中理工大学建校四十周年.华中理工大学学报,1993(5).

[234] 赵英俊,杨克冲,杨叔子.非晶态合金脉冲感应型磁场传感器.仪表技术与传感器,1993(5):4-6.

[235] 梅宏斌,何景光,杨叔子,崔乐芳,吴克勤.轴承工况监视与故障诊断.轴承,1993(12):30-34.

[236] 黄其柏,曹剑,杨叔子.齿轮齿顶修缘量与降噪量的正交回归设计.噪声与振动控制,1993(6):2-4,11.

[237] 杨叔子.努力开拓现代机械工程学研究领域.中国科学院院刊,1993(4):340-341.

[238] 何岭松,吴波,吴雅,杨叔子.两类小波函数的性质和作用.振动工程学报,1993(4):317-326.

[239] 吴波,杨叔子,杜润生,马元镐.基于波形智能识别的诊断方法.机械工程学报,1993(1):50-55.

[240] 雷鸣,朱心飚,尹申明,杨叔子.自构形神经网络及其应用.计算机科学,1993(1):52-54,74.

[241] Lei Ming, Wu Bo, Guan Zailing, Wu Ya and Yang Shuzi. *Multi-agent systems for AGV applications*. PRICAI-94. Proceedings of the 3rd Pacific Rim International Conference on Artificial Intelligence, v 2, p 1073-1078 (1994).

[242] Lei Ming, Wu Bo, Guan Zailing, Wu Ya and Yang Shuzi. *Multi-agent systems for AGV applications*. The Third International Conference on Automation, Robotics and Computer Vision. Proceedings, v 1, p 595-599 (1994).

[243] Cui Hanguo, Hu Ruian, Jin Duanfeng and Yang Shuzi. *The ViSC system of 3D computer's electromagnetic radiation field*. Proceedings of the 4th International Conference on Computer - Aided Drafting, Design and Manufacturing Technology, v1, p 60-63(1994).

[244] Lei Ming, Gui Xuwei, Shun Curen, Lian Jizhong, Wu Ya and Yang Shuzi. Intelligent scheduling editor for production management with human-

computer cooperation. International Conference on Data and Knowledge Systems for Manufacturing and Engineering. Conference Proceedings, v1,p 162-167(1994).

[245] 赵英俊,杨克冲,杨叔子,郑军,解源.无缝钢管冷轧机芯棒断裂、窜动监测原理与实现.冶金自动化,1994(1):29-31,57,59.

[246] 刘艳明,黄一夫,杨叔子.机械加工过程的离散滑模自校正控制.华中理工大学学报,1994(1):95-98.

[247] 王贤江,杜润生,杨叔子.一种新型钢丝绳缺陷检测方法的实验研究.无损探伤,1994(1):20-23.

[248] 黄锐,卢文祥,杨叔子.神经元网络钢丝绳断丝信号定量识别技术.中国机械工程,1994(1):1-3.

[249] 雷鸣,尹申明,杨叔子.神经网络自适应学习研究.系统工程与电子技术,1994(3):19-27.

[250] 康宜华,杨叔子,杨克冲,卢文祥.钢丝绳缺陷检测信号的计算机辅助定量评估.振动与冲击,1994(1):23-30.

[251] 崔汉国,金端峰,胡瑞安,杨叔子.三维数据场的可视化.海军工程学院学报,1994(1):24-31.

[252] 梁建成,李圣怡,温熙森,杨叔子.机床智能加工的体系结构.国防科技大学学报,1994(2):24-28.

[253] 徐志良,杨叔子,桂修文.人工神经网络在制造过程智能预测控制中的应用研究.制造技术与机床,1994(5):26-28.

[254] 黄锐,袁建华,卢文祥,杨叔子.钢丝绳断丝报警仪.机械与电子,1994(3):42-43.

[255] 尹申明,陆建东,雷鸣,杨叔子.自适应神经网络学习方法研究.计算机研究与发展,1994(6):24-29.

[256] 刘艳明,黄一夫,杨叔子.机械加工过程的神经网络最优自适应控制.机械工业自动化,1994(2):4-6.

[257] 刘建素,杨克冲,孙健利,杨叔子.直线滚动导轨的刚度分析.制造技术与机床,1994(6):26-28.

[258] 雷鸣,吴雅,杨叔子.用多施主求解策略集成智能制造中的车间层控制问题.制造技术与机床,1994(6):38-41.

[259] 吴昌林,刘辉,程愿应,杨叔子.全工况齿轮传动振动特性的研究.机械设

计与研究,1994(2):43-45.

[260] 朱向阳,杨叔子,万德钧,黄仁.应用最优化方法辨识指数自回归模型.振动工程学报,1994(2):112-116.

[261] 张文祖,吴雅,胡瑞安,杨叔子.智能制造系统的一种研究方法.华中理工大学学报 1994(7):1-4.

[262] 刘延林,王瑜辉,饶运清,杨叔子.智能加工中心的视觉系统.华中理工大学学报,1994(7):5-9.

[263] 李军旗,王雪,史铁林,杨叔子.层次分类诊断模型的面向对象实现.华中理工大学学报,1994(7):10-13.

[264] 王雪,李军旗,史铁林,杨叔子.基于知识的机械设备故障诊断的解释学习问题.华中理工大学学报,1994(7):14-18.

[265] 李军旗,阎明印,史铁林,杨叔子.诊断专家系统的不确定性问题研究.华中理工大学学报,1994(7):19-21.

[266] 谈兵,高红兵,杜润生,杨叔子.港口起重机钢丝绳状态监测仪的研制.华中理工大学学报,1994(7):28-31.

[267] 刁柏青,谢月云,吴雅,杨叔子.钢丝绳检验数据的参变量磨光样条插值法.华中理工大学学报,1994(7):32-35.

[268] 谈兵,杜润生,康宜华,杨叔子.大直径钢丝绳轴向励磁磁路的研究.华中理工大学学报,1994(7):36-39.

[269] 鲁宏伟,吴雅,杨叔子.快速采样数据建模的最小二乘算法.华中理工大学学报,1994(7):44-48.

[270] 罗欣,李光斌,朱涵梁,杨叔子.时间分割法正弦曲线插补算法研究.华中理工大学学报,1994(7):54-58.

[271] 崔汉国,伍晓宇,胡瑞安,杨叔子.双三次参数曲面求交的一种新方法.华中理工大学学报,1994(7):59-63.

[272] 高红兵,曾桂芳,杨叔子.运用多帧序列图像估计运动参数的一种改进算法.华中理工大学学报,1994(7):83-86.

[273] 庆华,陆枫,杨叔子.图像处理中一种 4×4 细化算法.华中理工大学学报,1994.

[274] 张松,杨叔子,虞烈,朱均.铁磁性流体润滑滑动轴承性能的计算和分析.华中理工大学学报,1994(7):102-105.

[275] 易传云,杨叔子,杨元山.非共轭齿面鼓形齿联轴器运动分析.华中理工大

学学报,1994(7):111-116.

[276] 赵英俊,杨克冲,郑军,解源,杨叔子.无缝管材高速冷轧机芯棒监测装置和方法.专利号:CN1090798.

[277] 刁柏青,钟诗胜,谢月云,杨叔子.钢丝绳安全系数可靠性评估.机械强度,1994(3):19-23.

[278] 伍世虔,余兴倬,陈尔昌,杨叔子.机床整体结构及其设计思想的发展.湖北工学院学报,1994(3):170-173.

[279] 劳俊,伍世虔,杨叔子.模块化与现代制造技术.制造技术与机床,1994(9):40-43.

[280] 王雪,吴雅,肖人彬,杨叔子.基于粒度的知识获取方式.应用力学学报,1994(9):128-132.

[281] 朱向阳,杨叔子,黄仁.非线性随机振动的时间序列模型.振动工程学报,1994(3):195-200.

[282] 贤江,杜润生,卢文祥,杨叔子.基于决策树的钢丝绳断丝定量识别.信息与控制,1994(5):299-303.

[283] 伍世虔,劳俊,孙炜,杨叔子.机床模块化发展与其内涵的演变.标准化报道,1994(5):3-4.

[284] 高宝成,何文丰,杨叔子.质量快速测量研究.电子技术应用,1994(11):17-18.

[285] 郭兴,杨叔子.无导师神经网络用于刀具状态的监测系统.武汉交通科技大学学报,1994(4):397-402.

[286] 崔汉国,余海涛,胡瑞安,杨叔子.计算机三维电磁辐射场的可视化研究.系统仿真学报,1994(4):57-64.

[287] 陈维克,张松,吴雅,杨叔子.我国水电机组振动监测与故障诊断技术的现状和发展方向.大电机技术,1994(6):1-3.

[288] 雷鸣,杨叔子.美日重视发展智能制造技术.机械与电子,1994(6):39-40.

[289] 石磊,崔汉国,杜润生,杨叔子.华中理工大学加工中心实时智能控制系统.海军工程学院学报,1994(4):25-29.

[290] 罗欣,李斌,吴雅,杨叔子.并行处理技术在智能加工中的应用.交通与计算机,1994(6):14-19.

[291] Guan Zailin, Lei Ming, Wu Bo, Wu Ya and Yang Shuzi. *Application of decentralized cooperative problem solving in dynamic flexible scheduling.*

[292] Zhang Wenzu, Wang Fengyin, Ding Hong and Yang Shuzi. *Geometry-oriented information modeling for mechanical products*. Proceedings of the 3rd International Conference, Computer Integrated Manufacturing, v 1, p 236-241 (1995).

[293] Liang Jiancheng, Wen Xisen, Li Shengyi and Yang Shuzi. *Optimum cutting parameters selection strategy based on neural network and artificial intelligence*. Proceedings of SPIE - The International Society for Optical Engineering, v 2620, p 458-462 (1995).

[294] Yang Shuzi, Lei Ming, Guan Zailin and Xiong Youlun. *Intelligent manufacturing: the challenge for manufacturing strategy in China in the 21st century-what we will do*. Proceedings of the SPIE - The International Society for Optical Engineering, v 2620, p 2-13 (1995).

[295] Xu Zhiliang, Gui Xiuwen and Yang Shuzi. *Intelligent control of precision thread grinding*. International Journal of Advanced Manufacturing Technology, v 10, n 5, p 311-316 (1995).

[296] Huang Xiuqing, Yang Shuzi and Ouyang J. Y. *Study on computer-controlled precision assembly*. Proceedings of the IEEE International Conference on Systems, Man and Cybernetics, v 2, p 1753-1758 (1995).

[297] Li Junqi, Shi Tielin, Wang Xue and Yang Shuzi. *Parallellism of fault diagnosis for large-scale mechatronic systems*. Proceedings of the SPIE - The International Society for Optical Engineering, v 2620, p 347-352 (1995).

[298] 杨叔子. 第五讲 机电一体化. 学习月刊, 1995(1): 17-18.

[299] 熊蔡华, 熊有伦, 杨叔子. 机器人多指手抓取中的规划问题. 机器人, 1994(1): 58-64.

[300] 陈维克, 张赛珍, 郭予康, 郁明山, 杨叔子. BP网络在CAPP中的应用研究. 机械科学与技术, 1995(1): 20-23.

[301] 刁柏青, 姚来昌, 杨叔子. 基于随机点过程的钢丝绳寿命预测及更换决策的研究. 煤炭学报, 1995(1): 21-24.

[302] 熊良才, 杨克冲, 杨叔子. 一种新型电涡流传感器的研制. 无损探伤, 1995

(1):34-35.

[303] 梁积中,雷鸣,吴雅,杨叔子.加工中心镗削刀具磨、破损检测的研究.组合机床与自动化加工技术,1995(2):2-5.

[304] 谈兵,杜润生,杨叔子,王贤江.钢丝绳绳径测量传感技术的研究.计量技术,1995(2):4-6.

[305] 郑定阳,周汉明,杨叔子.输送链不稳定运行的力学分析.起重运输机械,1995(3):3-5.

[306] 李军旗,史铁林,王雪,杨叔子.大型机电系统故障诊断中的并行性研究.振动工程学报,1995(1)230-36.

[307] 张文祖,吴雅,桂修文,胡瑞安,杨叔子.一种计算机表达的机械零件模型.武汉工业大学学报,1995(2):84-87.

[308] 李白诚,钟毓宁,杨叔子.设备智能诊断系统的可靠性研究.电子产品可靠性与环境试验,1995(2):18-21.

[309] 崔汉国,胡瑞安,金端峰,杨叔子.三维任意区域中点集的三角剖分算法.计算机辅助设计与图形学学报,1995(2):103-108.

[310] 卢江舟,熊有伦,杨叔子.机器人多指手的控制与传感器技术.机器人,1995(3):184-192.

[311] 杨叔子,胡以怀,史铁林.机械设备状态的计算机集成监控.机械强度,1994(2):31-37.

[312] 石磊,罗欣,杜润生,杨叔子.数控加工中心的在线通信与实时控制.工业控制计算机,1995(3):3-4.

[313] 罗欣,吴雅,李斌,杨叔子.基于 RISC 技术的智能数控系统体系结构研究.华中理工大学学报,1995(6):57-61.

[314] 石磊,杜润生,吴雅,杨叔子.面向智能结点的智能制造系统的分析与建模.华中理工大学学报,1995(6):62-65.

[315] 石磊,杨叔子,刘延林,杜润生.一种智能加工中心的设计与原型构造.华中理工大学学报,1995(6):66-69.

[316] 康宜华,何岭松,杨叔子.ARMA 谱的理论频率分辨力.华中理工大学学报,1995(6):101-104.

[317] 鲁宏伟,汤燕斌,吴雅,杨叔子.机床颤振的混沌特征.华中理工大学学报,1995(6):105-108.

[318] 鲁宏伟,吴雅,杨叔子.基于时序的李亚普诺夫指数谱的计算.华中理工大

学学报,1995(6)109-112.

[319] 管在林,雷鸣,吴雅,杨叔子.动态柔性生产调度中的分布式协同决策.华中理工大学学报,1995(6):113-116.

[320] 刁柏青,吴雅,杨叔子,张孝令.设备诊断的 Bayesian 预测及其改进的 SPRT 方法.振动工程学报,1995(2):167-171.

[321] 陈培林,史铁林,余佳兵,杨叔子.大型机电设备分布式故障诊断专家系统的求解策略.中国机械工程,1995(3):20-22.

[322] 薛鸿健,杨克冲,杨叔子.浅孔钻削加工状态监测系统.华中理工大学学报 1995(7):102-105.

[323] 易传云,欧阳渺安,杨元山,杨叔子.数控化改造滚齿机加工鼓形齿轮.机械与电子,1995(4):3-4.

[324] 鲁宏伟,汤燕斌,吴雅,杨叔子.分维数估计及其在机床颤振混沌研究中的应用.力学与实践,1995(4):46-48.

[325] 何岭松,黄载禄,杨叔子.窄带信号的压缩与重构算法.信号处理,1995(3):201-205.

[326] 薛鸿健,杨克冲,杨叔子,杨容.机械制造中的过程测量.华中理工大学学报,1995(8):81-84.

[327] 黄秀清,杨叔子,顾崇衔.机器人用于表面精加工作业的若干问题研究.华中理工大学学报,1995(8):85-89.

[328] 余佳兵,史铁林,杨叔子.可编程自补偿式倍频器的设计与实验.电子技术,1995(9):2-4.

[329] 吴雅,柯石求,杨叔子.金属切削机床切削噪声的动力学研究.机械工程学报,1995(5):76-85.

[330] 崔汉国,胡瑞安,金瑞峰,杨叔子.三维点集 Delaunay 三角剖分的自动生成与修改算法.工程图学学报,1995(2):1-7.

[331] 梅宏斌,林志航,吴雅,杨叔子.用于滚动轴承早期损伤诊断的模型及特征量.振动工程学报,1995(3):191-197.

[332] 薛鸿健,杨克冲,杨叔子,杨容.切削加工状态监测技术的现状与发展.机械与电子,1995(5):36-39.

[333] 鲁宏伟,刁柏青,吴雅,杨叔子.一种非线性系统参数辨识算法.信息与控制,1995(5):277-282.

[334] 杨叔子.继承传统,面向未来,加强人文素质教育.高等工程教育研究,

1995(4):1-6.

[335] 虎刚,杨叔子,谢友柏.弹流润滑膜高压流变性能影响因素的综合分析.华中理工大学学报,1995(10):114-119.

[336] 何岭松,杨叔子.一种最平通带 QMF 滤波器组最小二乘设计方法.华中理工大学学报,1995(10):120-123.

[337] 崔汉国,伍晓宇,胡瑞安,杨叔子.三维非规则数据场可视化中等值面构造算法.计算机辅助设计与图形学学报,1995(4):277-282.

[338] 杨叔子.设备诊断技术的现状与未来.设备管理与维修,1995(11):18-21.

[339] 黄其柏,师汉民,卢文祥,杨叔子.声软边界离心风机蜗壳高阶声学模态分析.华中理工大学学报,1995(11):41-45.

[340] 刘建素,杨克冲,孙健利,杨叔子.滚动直线导轨预过盈量的一种优化准则.华中理工大学学报,1995(11):60-63.

[341] 梁建成,李圣怡,温熙森,杨叔子.基于神经网络多传感器融合的刀具磨损定量监测的研究.机械科学与技术,1995(6):125-130.

[342] 陈维克,吴波,杨叔子,郭予康,郁明山.基于范例推理在 CAPP 中的研究与应用.机械与电子,1995(6):26-27.

[343] 孙健利,刘建素,杨克冲,杨叔子.影响滚动直线导轨副摩擦力的因素.制造技术与机床,1995(12):26-28.

[344] 刘艳明,黄一夫,杨叔子.车削加工过程的变结构控制.华中理工大学学报,1995(S2):88-91.

[345] 刁柏青,史铁林,杨叔子.自激过程与非齐次 Poisson 过程用于钢丝绳断丝计数过程的建模与预测.振动工程学报,1995(4):311-316.

[346] 易传云,肖来元,杨元山,杨叔子.非共轭齿面鼓形齿联轴器鼓形齿轮鼓度曲线的优化.中国机械工程,1995(S1):4-6.

[347] 孙楚仁,桂修文,叶文,杨叔子,李银巧.计算机辅助车间实时调度系统.中国机械工程,1995(5):68-70.

[348] 阎明印,郭宝成,杜玉波,杨叔子.基于深、浅知识集成表达的故障诊断策略.沈阳工业学院学报,1995(4):29-35.

[349] 黄其柏,卢文祥,师汉民,杨叔子.离心风机蜗舌气动噪声 D－最优声学模型的研究.振动、测试与诊断,1995(4):1-7.

[350] *Lei Ming, Guan Zailin and Yang Shuzi. Mobile robot fuzzy control optimization using genetic algorithm.* Artificial Intelligence in Engineering,v

10,n 4,p 293-298 (1996).

[351] He Hanwu, Xiong Youlun, Yang Shuzi and Wu Bo. *Virtual manufacturing systems and environment*. Proceedings of the IEEE International Conference on Industrial Technology (ICIT'96) (Cat. No. 96TH8151), p 13-24 (1996).

[352] Wang Xue and Yang Shuzi. *A parallel distributed knowledge-based system for turbine generator fault diagnosis*. Artificial Intelligence in Engineering, v 10, n 4, p 335-341 (1996).

[353] Xue Hongjian, Yang Kechong and Yang Shuzi. *Quantitative inspection of broken wire in wire ropes: Method and apparatus*. International Journal of Occupational Safety and Ergonomics, v 2, n 1, p 35-40 (1996).

[354] 徐宜桂,史铁林,杨叔子.BP网络的全局最优学习算法.计算机科学,1996(1):73-75.

[355] 薛立宏,王雪,史铁林,余佳兵,杨叔子.分布式汽轮发电机组在线监测故障诊断系统的设计与实现.电测与仪表,1996(2):41-44.

[356] 虎刚,杨叔子,谢友柏.纯滚动集中接触润滑入口区热效应的数值分析.润滑与密封,1996(1):13-17.

[357] 高宝成,杨叔子.混凝土T型简支架桥承载力动态评估方法探讨.北方工业大学学报,1996(1):92-96.

[358] 薛鸿健,杨克冲,杨叔子.小波分析在钢丝绳断丝定量检测中的应用.强度与环境,1996(1):19-24.

[359] 高宝成,张征,杨叔子,基于神经网络的结构荷载识别研究.振动、测试与诊断,1996(1):14-19.

[360] 张征,吴昌林,毛汉领,杨叔子.齿轮传动装置对流换热系数的实验研究.机械与电子,1996(6):24-26.

[361] 王雪,杨叔子.基于模糊Petri网络的分布式知识表示方法.华中理工大学学报,1996(3):13-16.

[362] 徐宜桂,史铁林,杨叔子.应用遗传优化算法枚举结构系统的主要破坏模式.海军工程学院学报,1996(1):39-43.

[363] 冯昭志,黄载禄,杨叔子.CMOS电流式符号多值PLA's.计算机学报,1996(4):313-316.

[364] 崔汉国,胡瑞安,金端峰,杨叔子.计算机三维电磁辐射场可视化系统.计

算机辅助设计与图形学学报,1996(2):121-126.

[365] 杨叔子,刘克明.中国古代工程图学的成就及其现代意义.世界科技研究与发展,1996(2):10-15.

[366] 杨叔子.访台有感.世界科技研究与发展,1996(2):92.

[367] 韩西京,史铁林,陈培林,杨叔子.故障诊断中事例推理的理论与方法.华中理工大学学报(社会科学版),1996(4):33-35.

[368] 黄秀清,杨叔子,顾崇衔.机器人力控制的稳定性研究.华中理工大学学报(社会科学版),1996(4):63-66.

[369] 廖道训,熊有伦,杨叔子.现代机电系统(设备)耦合动力学的研究现状和展望.中国机械工程,1996(2):44-46.

[370] 杨叔子.现代工程技术的发展态势与我们的对策.学习与实践,1996(5):18-21.

[371] 刘艳明,左力,程涛,杨叔子.自适应滑模跟踪控制研究及其应用.华中理工大学学报,1996(5):44-46.

[372] 叶伯生,杨叔子,彭炎午.基于IPC的开放式体系结构的CNC系统.华中理工大学学报,1996(5):47-49.

[373] 刘艳明,程涛,左力,杨叔子.机械加工中切削用量的K—L优化研究.华中理工大学学报,1996(5):50-52.

[374] 陈培林,史铁林,韩西京,杨叔子.机械设备分布式诊断专家系统的任务描述与分解.机械科学与技术,1996(3):455-458.

[375] 刁柏青,谈兵,鲁宏伟,杨叔子.轴承事故的神经网络(ANN)分析.机械强度,1996(2):21-58.

[376] 鲁宏伟,杨叔子.基于非线性模型的切削过程的混沌研究.振动工程学报,1996(2):169-172.

[377] 陈培林,史铁林,余佳兵,杨叔子.分布式故障诊断系统中控制的组织方法.华中理工大学学报,1996(6):41-43.

[378] 郭兴,杜润生,杨叔子.基于人工神经网络的铣刀破损功率监测.华中理工大学学报,1996(6):47-49.

[379] 袁建华,杨叔子,杜润生.空间等位信号采集技术.测控技术,1996(3):25-26.

[380] 高宝成,刘红霞,杨叔子.神经网络用于结构动荷载识别的研究.郑州工学院学报,1996(2):91-94.

[381] 余佳兵,史铁林,杨叔子.基于 ELAN 网的分布式数据采集系统.电子技术,1996(7):18-19.

[382] 曹伟,张鸿海,周学夫,江福祥,师汉民,杨叔子.一种宽范围扫描隧道显微镜.计量学报,1996(3):226-228.

[383] 高宝成,杨叔子.提高 A/D 采样器的频率和精度的方法.机械与电子,1996(4):13-14.

[384] 刘艳明,杨叔子.基于 Perceptron 的非线性滑模控制.信息与控制,1996(4):193-198.

[385] 郭兴,杜润生,杨叔子.铣刀破损功率监测方法的研究.武汉交通科技大学学报,1996(4):465-469.

[386] 熊有伦,罗欣,何汉武,杨叔子.先进制造技术——制造业走向 21 世纪.世界科技研究与发展,1996(Z1):31-40.

[387] 杨叔子,刘克明.取精用弘——中国机械设计方法的历史回顾与前瞻.世界科技研究与发展,1996(Z1):126-131.

[388] 冯昭志,黄载禄,杨叔子.一种新的单层神经网络学习算法分析模型.华中理工大学学报,1996(08):21-23.

[389] 陈维克,吴波,杨叔子,郭予康,褚启权.重型机械工艺设计中机床资源动态模型的研究与应用.中国机械工程,1996(04):62-64.

[390] 叶伯生,杨叔子.带 DOS 平台的 CNC 系统实时多任务的并行处理.制造业自动化,1996(03):35-40.

[391] 叶伯生,杨叔子.空间等半径过渡曲面成形刀纵向加工的算法.华中理工大学学报,1996(09):12-14.

[392] 吴波,胡春华,陈维克,杨叔子.基于多自治体的制造系统集成.华中理工大学学报,1996(09):25-27.

[393] 胡春华,吴波,刘琦,杨叔子.基于 Petri 网的离散制造过程建模工具.华中理工大学学报,1996(09):28-31.

[394] 胡春华,吴波,杨叔子.虚拟制造系统概念与结构框架.华中理工大学学报,1996(09):32-34.

[395] 周杰韩,吴波,左朝辉,杨叔子.改进决策树方法及其用于故障诊断知识获取.华中理工大学学报,1996(09):35-38.

[396] 韩西京,李录平,史铁林,杨叔子.设备监测与诊断系统构成技术的研究和实践.华中理工大学学报,1996(09):39-41.

[397] 黄其柏,师汉民,卢文祥,杨叔子.掠形叶片轴流叶轮旋转频率噪声辐射的理论研究.振动工程学报,1996(03):308-312.

[398] 杨叔子.诗词二首.高等工程教育研究,1996(4):6.

[399] 余佳兵,史铁林,杨叔子.新型可编程自适应数字倍频器.电子测量与仪器学报,1996(4):24-29.

[400] 叶伯生,杨叔子.任意三维抛物线的一种高速插补方法.华中理工大学学报,1996(11):15-17.

[401] 陈维克,吴波,杨叔子,宋汉珍.一种在CAPP系统中嵌入基于范例推理的方法.华中理工大学学报,1996(11):18-21.

[402] 程涛,左力,刘艳明,杨叔子.基于参数在线自调整的自适应模糊控制器研究.华中理工大学学报,1996(11):22-25.

[403] 康宜华,武新军,杨叔子.录井钢丝直径的恒磁测量原理.华中理工大学学报,1996(11):30-32.

[404] 左力,程涛,李锡文,刘艳明,杨叔子.连接单片机与PC机的纽带——双端口RAM的应用.微型机与应用,1996(12):20-21.

[405] 胡春华,吴波,杨叔子.先进制造系统中的智能控制器.航空制造工程,1996(12):3-5.

[406] 李东晓,刘世元,史铁林,杨叔子.面向大型成套设备的分布式监测诊断系统.计算机与通信,1996(06):44-47.

[407] 冯昭志,黄载禄,杨叔子.改进的神经网络快速学习算法.华中理工大学学报,1996(S2):28-31.

[408] 曹伟,张鸿海,师汉民,杨叔子.STM纳米级轮廓仪的研制.仪器仪表学报,1996(S1):384-387.

[409] 余佳兵,史铁林,陈培林,杨叔子.窗谱校正方法的实用峰值搜寻算法研究.振动工程学报,1996(04):378-382.

[410] Zuo Li, Cheng Tao, Zuo Jing, Liu Yanming and Yang Shuzi. *A hierarchical intelligent control system for milling machine*. 1997 IEEE International Conference on Intelligent Processing Systems (Cat. No. 97TH8335), v 1, p 813-817 (1997).

[411] Zhang Wenzu, Wang Fengyin, Wu Bo and Yang Shuzi. *Surface patches for filling arbitrary topological networks*. Proceedings of the International Conference on Manufacturing Automation, ICMA, v 2, p

680-684(1997).

[412] 冯昭志,黄载禄,杨叔子.单层神经网络的快速学习算法研究,自动化学报,1997(01):68-72.

[413] 韩西京,陈培林,史铁林,杨叔子.300MW汽轮发电机组状态监测与故障诊断专家系统——知识获取与机器学习的研究.汽轮机技术,1997(01):8-13.

[414] 刘琦,吴波,胡春华,杨叔子.智能加工系统中通信与协作规范探讨.中国机械工程,1997(01):71-76.

[415] 刘克明,杨叔子.先秦车轮制造技术与抗磨损设计.华中理工大学学报(社会科学版),1997(01):115-119.

[416] 胡阳,杨叔子.一种基于Fourier复变换的压缩比自调整ECG数据压缩算法.中国医疗器械杂志,1997(02):75-78.

[417] 韩西京,史铁林,陈培林,杨叔子.事例库的管理与维护系统的研究.模式识别与人工智能,1997(01):55-60.

[418] 刘克明,杨叔子.《周礼》中的图学记载及其有关问题的探讨.地图,1997(01):42-46.

[419] 张中民,卢文祥,杨叔子,张英堂.滚动轴承故障振动模型及其应用研究.华中理工大学学报,1997(03):50-53.

[420] 胡以怀,刘永长,杨叔子.柴油机活塞-缸套磨损故障振动诊断机理的研究.上海海运学院学报,1997(01):1-8.

[421] 徐宜桂,史铁林,杨叔子.基于神经网络的结构动力模型修改和破损诊断研究.振动工程学报,1997(01):8-12.

[422] 高宝成,时良平,史铁林,杨叔子.基于小波分析的简支梁裂缝识别方法研究.振动工程学报,1997(01):81-85.

[423] 阎明印,蔡振江,杨晶栾,江南,杨叔子.适于专家系统的机器学习模型与过程.沈阳工业学院学报,1997(01):59-63.

[424] 杨叔子,史铁林,李东晓.分布式监测诊断系统的开发与设计.振动、测试与诊断,1997(01):1-6.

[425] 易传云,孙国正,杨叔子,杨元山.共轭齿面鼓形齿联轴器诱导法曲率和相对滑动系数分析.武汉交通科技大学学报,1997(01):117-123.

[426] 李斌,柳庆,杨叔子.根据参考轨迹学习模糊控制规则.华中理工大学学报,1997(05):22-24.

[427] 李东晓,刘世元,史铁林,杨叔子.分布式监测诊断系统的研究及应用设计.计算机工程,1997(03):64-68.

[428] 叶伯生,杨叔子.基于参数方程的抛物线插补方法.制造技术与机床,1997(06):35-37.

[429] 杨叔子.欢庆香港回归组诗.世界科技研究与发展,1997(03):13.

[430] 刘克明,杨叔子.《周礼》中的图学记载及其有关问题的探讨(续).地图,1997(02):28-31.

[431] 张海霞,赵英俊,杨叔子.脉冲磁场传感器的理论计算与检测.仪表技术与传感器,1997(06):6-9.

[432] 朱向阳,钟秉林,熊有伦,杨叔子.双线性时间序列模型的参数辨识.振动工程学报,1997(2):109-113.

[433] 张海霞,赵英俊,杨叔子.非晶态合金传感器技术的现状及展望.测控技术,1997(3):5-8.

[434] 阎明印,杜玉波,蔡振江,杨晶,杨叔子.机械设备智能诊断系统的机器学习机制.沈阳工业学院学报,1997(02):3-7.

[435] 左力,程涛,刘艳明,杨叔子.直接作用式神经网络控制器研究.华中理工大学学报,1997(7):49-52.

[436] 张中民,卢文祥,杨叔子,张英堂.变速箱振动信号的分解及在故障诊断中的应用.华中理工大学学报,1997(7):63-66.

[437] 胡以怀,刘永长,杨叔子.柴油机排气门漏气故障诊断的试验研究.内燃机工程,1997(3):51-58.

[438] 杨叔子,易传云.现代工程技术发展的态势与我们的策略.世界科技研究与发展,1997(4):29-36.

[439] 韩西京,史铁林,李录平,杨叔子.故障诊断中几种征兆自动获取技术研究.华中理工大学学报,1997(8):54-56.

[440] 陈维克,杨叔子,郭予康.基于关系数据库的CAPP专家系统的开发与应用.组合机床与自动化加工技术,1997(8):26-30.

[441] 刘辉,吴昌林,杨叔子,胡海青.参数化啮合斜齿轮三维有限元网格的自动生成.华中理工大学学报,1997(9):14-16.

[442] 黄其柏,徐斌,师汉民,杨叔子.离心风机紊流次声理论与控制的研究.华中理工大学学报,1997(9):22-25.

[443] 何岭松,杨叔子.包络检波的数字滤波算法.振动工程学报,1997(3):

116-121.

[444] 朱向阳,钟秉林,杨叔子.基于模拟退火算法的子集 AR 模型辨识方法及其应用.应用科学学报,1997(3):259-264.

[445] 韩西京,李录平,史铁林,杨叔子.旋转机械轴心轨迹的自动识别.振动、测试与诊断,1997(3):22-27.

[446] 李东晓,史铁林,杨叔子.分布式监测诊断系统的研究及应用设计.计算机工程,1997(3):64-68.

[447] 杨叔子.历史之交,使命神圣.高等工程教育研究,1997(4):5-7.

[448] 杨叔子.永必求真 今应重善.世界科技研究与发展,1997(5):22-29.

[449] 汪朝军,刘艳明,杨叔子.铣削加工过程的自适应最优控制.华中理工大学学报,1997(10):8-10.

[450] 李东晓,史铁林,杨叔子.分布式监测诊断系统的设计思想及实现.华中理工大学学报,1997(11):32-33.

[451] 胡春华,吴波,杨叔子.智能制造系统中的合作与协调.航空制造工程,1997(12):25-28.

[452] 胡以怀,杨叔子,刘永长.柴油机气阀间隙异常振动诊断方法的改进.上海海运学院学报,1997(4):29-36.

[452] 阎明印,栾江南,杨叔子.机械设备智能诊断系统的机器学习机制.沈阳工业学院学报,1997(04):63-71.

[454] Li Hongsheng, Shi Tielin and Yang Shuzi. *Method of surveying converter lining eroding state based on laser measure technique*. Proceedings of SPIE - The International Society for Optical Engineering, v 3558, p 439-444 (1998).

[455] Li Hongsheng, Yang Xiaofei, Shi Tielin and Yang Shuzi. *Detecting laser range finding signals in surveying converter lining based on wavelet transform*. Proceedings of SPIE - The International Society for Optical Engineering, v 3558, p 445-450 (1998).

[456] Lei Ming, Yang Xiaohong, Tseng, M. M and Yang Shuzi. *A CORBA-based agent-driven design for distributed intelligent manufacturing systems*. Journal of Intelligent Manufacturing, v 9, n 5, p 457-465 (1998).

[457] Zhang Haixia, Zhao Yingjun, Yang Shuzi and Li Hejun. *A Novel*

Co-based amorphous magnetic field sensor. Sensors and Actuators, A: Physical, v 69, n 2, p 121-125 (1998).

[458] Lei Ming, Yang Xiaohong, Tseng Mitchell M. and Yang Shuzi. *Design an intelligent machine center - strategy and practice*. Mechatronics, v 8, n 3, p 271-285 (1998).

[459] 叶文,杨叔子.基于过程代数的智能机器人系统建模.机器人,1998(1):10-15.

[460] 胡以怀,杨叔子,刘永长,周轶尘.柴油机磨损故障振动诊断机理的研究.内燃机学报,1998(1):53-64.

[461] 李录平,韩守木,黄树红,杨叔子.旋转机械故障特征的定性提取.华中理工大学学报,1998(1):101-103.

[462] 李录平,韩西京,韩守木,杨叔子.从振动频谱中提取旋转机械故障特征的方法.汽轮机技术,1998(1):13-16.

[463] 徐宜桂,周轶尘,杨叔子.结构动态模型修改与动态诊断的智能方法(英文).武汉交通科技大学学报,1998(1):44-48.

[464] 左力,程涛,刘艳明,杨叔子.基于神经网络模糊控制器的铣削过程智能控制.华中理工大学学报,1998(2):41-44.

[465] 李锡文,杜润生,杨叔子.铣刀磨损过程中铣削力与磨损面积分析.华中理工大学学报,1998(2):45-48.

[466] 武新军,康宜华,卢文祥,杨叔子.非接触式霍尔位移传感器的研制及应用.华中理工大学学报,1998(2):53-54.

[467] 武新军,康宜华,谢月云,杨叔子.升船机钢丝绳断丝检测方法的研究.华中理工大学学报,1998(2):55-57.

[468] 左朝晖,吴波,周杰韩,杨叔子.应用虚拟制造技术的产品设计策略.航空制造工程,1998(02):36-37.

[469] 张桂才,史铁林,杨叔子.机械信号分析中消噪方法的研究.振动、测试与诊断,1998(1):32-36.

[470] 周杰韩,吴波,左朝晖,杨叔子.虚拟制造环境面向对象建模.华中理工大学学报,1998(3):39-41.

[471] 叶伯生,杨叔子.CNC系统中三次B-样条曲线的高速插补方法研究.中国机械工程,1998(3):42-43.

[472] 张中民,卢文祥,杨叔子,张英堂,张培林,郑海起.基于小波系数包络谱的

滚动轴承故障诊断.振动工程学报,1998(1):68-72.

[473] 金建华,康宜华,武新军,卢文祥,杨叔子.录井钢丝裂纹定量检测系统及应用.石油机械,1998(4):32-34.

[474] 汪朝军,刘艳明,杨叔子.变搜索域遗传算法及其在铣削加工参数优化中的应用.中国机械工程,1998(4):33-34.

[475] 张海霞,赵英俊,杨叔子.一种新型非晶态合金磁场传感器的设计与优化.传感技术学报,1998(2):7-12.

[476] 江汉红,赵英俊,杨叔子.基于SCF的程控低通滤波电路研究.仪表技术与传感器,1998(5):25-28.

[477] 徐宜桂,史铁林,杨叔子,周轶尘.BP神经网络及其在结构动力分析中的应用研究.计算力学学报,1998(2):87-93.

[478] 王立平,杨叔子,杨兆军,王立江.低频振动钻削提高微小钻头寿命机理的研究.工具技术,1998(5):3-5.

[479] 李斌,吴波,罗欣,杨叔子.设备自律控制器的结构及其实现技术.中国机械工程,1998(5):42-44.

[480] 杨叔子.下学上达 文质相宜.东南学术,1998(3):6-14.

[481] 来五星,轩建平,史铁林,杨叔子.分布式监测诊断系统中历史数据库系统开发的设计.振动、测试与诊断,1998(2):129-133.

[482] 杨叔子.下学上达文质相宜.山东工业大学学报(社会科学版),1998(2):5-12.

[483] 刘世元,李锡文,杜润生,杨叔子.内燃机气缸压力的振动信号倒谱识别方法.华中理工大学学报,1998(6):80-82.

[484] 胡春华,吴波,杨叔子.基于多自主体的分布式智能制造系统研究.中国机械工程,1998(7):54-57.

[485] 徐宜桂,马西庚,史铁林,杨叔子,周轶尘.基于神经网络的动力学反解算法及其应用研究.机械工程学报,1998(4):107-111.

[486] 左力,程涛,李锡文,刘艳明,杨叔子.一种新颖铣床智能数控系统的体系结构及其硬件实现.工业控制计算机,1998(4):36-37.

[487] 王立平,杨兆军,王立江,杨叔子.多元变参数振动钻床的数控系统.机械工业自动化,1998(04):30-32.

[488] 周杰韩,吴波,左朝晖,杨叔子.开放式虚拟制造开发工具框架.华中理工大学学报,1998(8):18-19.

[489] 王立平,杨兆军,王立江,杨叔子.用有限元法研究振动钻头寿命.华中理工大学学报,1998(8):29-31.

[490] 轩建平,张桂才,史铁林,杨叔子.大型机电监测系统传感器在线配置技术研究.仪表技术与传感器,1998(9):35-38.

[491] 杨叔子,吴波.基于Agent的生产制造模式.航空制造工程,1998(08):3-5.

[492] 王立平,杨叔子,杨兆军,王立江.振动钻削减小微小孔出口毛刺的试验研究.汽车工艺与材料,1998(9):15-16.

[493] 易传云,孙国正,崔可润,杨叔子.弹性共轭曲面运动基本方程.机械科学与技术,1998(5):5-7.

[494] 轩建平,张桂才,史铁林,杨叔子.大型机电监测系统传感器在线配置技术研究.测控技术,1998(5):18-20.

[495] 刘克明,杨叔子.中国古代机械设计思想初探.机械技术史,1998(10):86-93.

[496] 轩建平,来五星,史铁林,杨叔子,罗友桥,翁小雄.汽车行驶模式智能控制系统的研究.内燃机工程,1998(4):26-31.

[497] 杨叔子.分层集成组态式设备故障诊断系统.中国设备管理,1998(11):5-8.

[498] 王立平,杨叔子,赵宏伟,王立江.变参数振动钻削提高微小孔加工精度的研究.工具技术,1998(11):3-5.

[499] 李晓峰,史铁林,杨叔子,严新平.柴油机磨损故障诊断系统的研究.华中理工大学学报,1998(11):60-62.

[500] 刘辉,吴昌林,杨叔子.支承系统变形对斜齿轮承载特性的影响.华中理工大学学报,1998(11):73-75.

[501] 刘辉,吴昌林,杨叔子.基于有限元法的斜齿轮体模态计算与分析.华中理工大学学报,1998(11):76-78.

[502] 李晓峰,严新平,史铁林,杨叔子.柴油机主要摩擦副磨损型式的识别.润滑与密封,1998(6):24-26.

[503] 张桂才,轩建平,史铁林,杨叔子.基于小波包的转子轴心轨迹提纯方法.华中理工大学学报,1998(12):9-10.

[504] 王立平,杨兆军,王立江,杨叔子.低频振动钻削时微小钻头钻入横向偏移的研究.中国机械工程,1998(12):65-68.

[505] 何岭松,王峻峰,杨叔子.设备故障诊断网上开放实验室建设.振动工程学报,1998(4):110-114.

[506] Lei Ming, Yang Xiaohong and Yang Shuzi. Tool wear length estimation with a self-learning fuzzy inference algorithm in finish milling. International Journal of Advanced Manufacturing Technology, v 15, n 8, p 537-545 (1999).

[507] 刘世元,杜润生,杨叔子.利用神经网络诊断内燃机失火故障的研究.内燃机学报,1999(1):70-73.

[508] 吴崇健,杨叔子,骆东平,朱英富.WPA法计算多支承弹性梁的动响应和动应力.华中理工大学学报 1999(01):69-71.

[509] 程涛,左力,刘艳明,杨叔子.数控机床切削加工过程智能自适应控制研究.中国机械工程,1999(1):34-39.

[510] 江汉红,赵英俊,杜润生,杨叔子.大动态范围PGA电路的设计与实现.工业仪表与自动化装置,1999(1):10-13.

[511] 康宜华,武新军,杨叔子.磁性无损检测技术的分类.无损检测,1999(2):58-60.

[512] 周传宏,孙健利,杨叔子.滚动直线导轨副的振动模型研究.机械设计,1999(2):22-25.

[513] 武新军,康宜华,杨叔子.无损检测在役钢丝绳标准的研究.起重运输机械,1999(2):35-38.

[514] 吴崇健,骆东平,杨叔子,马运义.WPA法分析带动力吸振器多支承桅杆动态特性.华中理工大学学报,1999(02):22-24.

[515] 刘克明,杨叔子.中国古代机械设计思想的科学成就.中国机械工程,1999(2):199-202.

[516] 师汉民,杨叔子.知识经济条件下的制造产业——兼论我国制造企业的改革之路.模具技术,1999(02):90-95.

[517] 张桂才,史铁林,杨叔子,基于高阶统计量的机械故障特征提取方法研究.华中理工大学学报,1999(03):6-8.

[518] 王立平,杨叔子,杨兆军,王立江.振动钻削的国内外研究状况和发展趋势.机械设计与制造工程,1999(2):6-9.

[519] 轩建平,来五星,史铁林,杨叔子.500测点汽轮发电机组分布式状态监测系统关键技术的研究.中国机械工程,1999(3):78-81.

[520] 何岭松,王峻峰,杨叔子.基于因特网的设备故障远程协作诊断技术.中国机械工程,1999(3):104-106.

[521] 王立平,李晓峰,史铁林,杨叔子.连续转子系统的非线性动力学模型.汽轮机技术,1999(2):34-36.

[522] 浦耿强,杨叔子,周良弼,蒋国英,白羽.小型发动机配气机构接触应力分析.汽车科技,1999(02):6-9.

[523] 杨叔子,陶绪楠.时代呼唤:让大学生走近戏剧.戏剧之家,1999(02):19-21.

[524] 王立平,李晓峰,杜润生,杨叔子,开闭裂纹转子的模型化与动态仿真.华中理工大学学报,1999(04):65-67.

[525] 康宜华,武新军,杨叔子.磁性无损检测技术中的磁化技术.无损检测,1999(5):206-209.

[526] 周传宏,孙健利,杨叔子.精密滚动直线导轨副工作台静刚度研究.机械设计,1999(5):31-33.

[527] 金建华,康宜华,王峻峰,杨叔子.抽油杆裂纹在线定量检测原理与仪器.石油矿场机械,1999(3):17-20.

[528] 王立平,杨叔子,杨兆军,王立江.三区段振动钻削动态轴向力和扭矩的计算机仿真.工具技术,1999(5):7-9.

[529] 武新军,康宜华,杨叔子.大直径钢丝绳直径连续测量方法与装置.机械与电子,1999(3):3-5.

[530] 张洁,罗欣,杜润生,杨叔子.一种基于开放式结构的数控加工系统.机械与电子,1999(3):27-29.

[531] 刘克明,杨叔子,蔡凯.中国古代工程几何作图的科学成就.中国科学基金,1999(3):37-41.

[532] 浦耿强,杨叔子,金国栋,蒋国英,白羽,发动机顶置配气凸轮与摇臂的润滑分析.汽车科技,1999(03):29-34.

[533] 金建华,康宜华,卢文祥,杨叔子.漏磁场法在线定量检测钢丝裂纹的研究.仪器仪表学报,1999(3):285-286.

[534] 王立平,杨叔子,王立江,振动钻削工艺的发展概况及应用前景.工具技术,1999(06):3-7.

[535] 刘克明,杨叔子.中国古代工程制图的数学基础.成都大学学报(自然科学版),1999(2):16-23.

[536] 李晓峰,王立平,史铁林,杨叔子.考虑非线性油膜力的转子系统稳态响应的研究.华中理工大学学报,1999(06):57-59.

[537] 李晓峰,王立平,史铁林,杨叔子.连续不对称转子系统不平衡响应的研究.华中理工大学学报,1999(07):43-62.

[538] 刘世元,杜润生,杨叔子.内燃机缸盖振动信号的特性与诊断应用研究.华中理工大学学报,1999(7):49-51.

[539] 康宜华,武新军,杨叔子.磁性无损检测技术中磁信号测量技术.无损检测,1999(8):340-343.

[540] 程涛,左静,刘艳明,杨叔子.基于单个神经元的自适应模糊控制器及其应用.控制理论与应用,1999(4):621-624.

[541] 刘世元,杜润生,杨叔子.基于小波包分析的内燃机振动诊断方法研究.华中理工大学学报,1999(8):7-9.

[542] 轩建平,来五星,史铁林,杨叔子.隔离测量器在火电厂分布式网络化状态监测故障诊断系统中的应用.仪表技术与传感器,1999(8):22-25.

[543] 郭兴,杜润生,杨叔子.铣刀破损振动位移监测方法的研究.安徽机电学院学报(自然科学版),1999(3):5-10.

[544] 黎洪生,何岭松,史铁林,杨叔子,李知践.基于B/S的远程故障诊断专家系统研究.武汉工业大学学报,1999(04):39-41.

[545] 杨叔子,吴波.依托基金项目 开展创新研究——国家自然科学基金重点项目"智能制造技术基础"研究综述.中国机械工程,1999(9):35-38.

[546] 刘克明,杨叔子.中国古代图学对现代工程图学的贡献.工程图学学报,1999(3):116-124.

[547] 程涛,胡春,吴波,杨叔子.分布式网络化制造系统构想.中国机械工程,1999(11):1234-1238.

[548] 张桂才,史铁林,轩建平,杨叔子.高阶统计量与RBF网络结合用于齿轮故障分类.中国机械工程,1999(11):1250-1252.

[549] 崔汉锋,马维垠,杨叔子.反求工程中的多曲面光滑重建.机械与电子,1999(6):32-35.

[550] 吴崇健,骆东平,杨叔子,朱英富,马运义.离散分布式动力吸振器的设计及在船舶工程中的应用.振动工程学报,1999(4):584-588.

[551] 李斌,师汉民,胡春华,吴波,杨叔子.基于Agent分布式网络化制造模式的研究.中国机械工程,1999(12):1358-1362.

[552] Li Hongsheng, Shi Tielin, Yang Shuzi, Tao Yunfeng, Li Zhijian and Chen Wenwu. *Integrated ANN diagnostic model-based remote diagnosis approach on Internet*. 16th World Computer Congress 2000. Proceedings of Conference on Intelligent Information Processing, p 232-237 (2000).

[553] Liu Shiyuan, Du Runsheng and Yang Shuzi. *Fuzzy pattern recognition for on-line detection of engine misfire by measurement of crankshaft angular velocity*. Proceedings of SPIE - The International Society for Optical Engineering, v 4077, p 432-436 (2000).

[554] Li Hongsheng, Shi Tielin, Yang Shuzi, Li Zhijian, Tao Yunfeng and Chen Wenwu. *Internet-based remote diagnosis: Concept, system architecture and prototype*. Proceedings of the World Congress on Intelligent Control and Automation (WCICA), v 1, p 719-723 (2000).

[555] Li Hongsheng, Shi Tielin and Yang Shuzi. *An approach of sampling computing for wavelet analysis and its Web-based implementation*. 2000 5th International Conference on Signal Processing Proceedings. 16th World Computer Congress 2000, v 1, p 395-398 (2000).

[556] Li Hongsheng, Li Zhijian, Tao Yunfeng, Chen Wenwu, Shi Tielin and Yang Shuzi. *Research on Internet-based remote diagnosis system model*. 16th World Computer Congress 2000. Proceedings of Conference on Intelligent Information Processing, p 320-326 (2000).

[557] 吴伟蔚,杨叔子,吴今培. 远程协作故障诊断系统的设计及实现. 计算机工程与应用,2000(1):139-141.

[558] 金建华,康宜华,杨叔子. 用微机的增强并行口设计数据采集卡. 电子技术,2000(1):29-31.

[559] 周杰韩,吴波,贾瑜,杨叔子. 虚拟柔性加工单元物流开发与实践. 航空工程与维修,2000(1):35-36.

[560] 刘世元,杜润生,杨叔子. 小波包改进算法及其在柴油机振动诊断中的应用. 内燃机学报,2000(1):11-15.

[561] 周杰韩,吴波,杜润生,杨叔子. 微软密码应用编程接口在网络化柔性加工单元中的应用研究. 计算机应用研究,2000:16-17.

[562] 周杰韩,吴波,杜润生,杨叔子. 虚拟柔性加工单元数据库管理系统的设计. 华中理工大学学报,2000(1):10-11.

[563] 杨叔子,吴波,胡春华,程涛.网络化制造与企业集成.中国机械工程,2000:45-48.

[564] 吴伟蔚,杨叔子,吴今培.基于智能 agent 的故障诊断系统研究.模式识别与人工智能,2000(1):78-81.

[565] 崔汉锋,马维垠,林奕鸿,杨叔子.多个 B 样条曲面光滑连接的研究.华中理工大学学报,2000(3):4-6.

[566] 黎洪生,何岭松,史铁林,杨叔子.基于因特网远程故障诊断系统架构.华中理工大学学报(自然科学版)2000(3):13-15.

[567] 刘世元,杜润生,杨叔子.利用模糊模式识别诊断内燃机失火故障的研究.振动工程学报 2000(1):32-45.

[568] 朱庆华,程涛,胡春华,吴波,杨叔子.CORBA 规范在分布式制造系统中的应用.中国机械工程,2000(3):307-316.

[569] 刘世元,杜润生,杨叔子.内燃机转速波动信号的测量方法及其应用研究.振动、测试与诊断,2000(1):13-18.

[570] 来五星,李巍华,史铁林,杨叔子.基于事件驱动的矢量监测方法.华中理工大学学报 2000(4):4-6.

[571] 程涛,胡春华,吴波,杨叔子.基于 CORBA 的分布式多自主体系统研究.中国机械工程,2000(4):441-445.

[572] 肖锡武,徐鉴,李誉,杨叔子.具有非轴对称刚度转轴的分岔.力学学报,2000(3):360-365.

[573] 来五星,李巍华,史铁林,杨叔子.远程监测诊断系统中延时问题处理.计算机应用,2000(5):43-44.

[574] 黎洪生,史铁林,杨叔子.IPC 电子盘及在实时测控系统中的应用.华中理工大学学报,2000(5):21-23.

[575] 李锡文,张洁,杜润生,杨叔子.小直径立铣刀后刀面磨损带的研究.工具技术,2000(6):7-10.

[576] 康宜华,武新军,杨叔子.磁性无损检测技术中的信号处理技术.无损检测,2000(6):15-19.

[577] 周杰韩,黄正波,杨叔子.分布式对象技术与柔性制造系统分布式布局应用.计算机应用,2000(6):17-19.

[578] 王立平,杜润生,史铁林,杨叔子.带有轴承间隙的裂纹转子分叉与混沌特性.振动工程学报,2000(2):85-90.

[579] 李锡文,杜润生,杨叔子.铣削力模型的频域特性研究.工具技术,2000(7):3-6.

[580] 刘世元,杜润生,杨叔子.利用转速波动信号诊断内燃机失火故障的研究(1)——诊断模型方法.内燃机学报,2000(3):95-99.

[581] 刘世元,杜润生,杨叔子.利用转速波动信号诊断内燃机失火故障的研究(2)——波形分析方法.内燃机学报,2000(3):100-103.

[582] 李锡文,杜润生,张洁,杨叔子.立铣刀切削力模型的时域特性研究.华中理工大学学报,2000(7):46-51.

[583] 王立平,杜润生,史铁林,汪劲松,杨叔子.转子开闭裂纹的计算机仿真与特性分析.中国机械工程,2000(7):3-11.

[584] 张洁,罗欣,杜润生,杨叔子.智能化数控加工单元的远程操作与控制系统.中国机械工程,2000(7):41-44.

[585] 沈轶,赵勇,廖晓昕,杨叔子.具有可变时滞的Hopfield型随机神经网络的指数稳定性.数学物理学报,2000(3):400-404.

[586] 李锡文,杜润生,杨叔子.神经网络融合法在铣刀磨损监测中的应用.华中理工大学学报,2000(8):48-51.

[587] 周杰韩,杜润生,吴波,杨叔子.虚拟柔性加工单元网络信息安全的数据加密与数字签名方案.计算机工程与应用,2000(9):117-123.

[588] 刘世元,杜润生,杨叔子.利用转速波动信号在线识别内燃机气缸压力的研究.内燃机工程,2000(3):39-45.

[589] 熊良才,何岭松,史铁林,杨叔子.基于Intranet的轧机状态监测与故障诊断系统的研制.冶金自动化,2000(5):17-19.

[590] 熊良才,何岭松,史铁林,杨叔子.基于WEB的轧机状态监测与故障诊断系统研究.钢铁,2000(9):67-79.

[591] 吴伟蔚,杨叔子,吴今培.故障诊断Agent研究.振动工程学报,2000(3):73-79.

[592] 周杰韩,吴波,杨叔子.虚拟制造系统综述.中国科学基金,2000(5):25-29.

[593] 熊良才,史铁林,杨叔子.机电系统中传感器故障诊断的控制图法.机械与电子,2000(5):7-8.

[594] 裘法祖,杨叔子,崔崑.院士寄语新同学——创新靠人才.高等工程教育研究,2000(4):16-17.

[595] 刘世元,杜润生,杨叔子.利用转速波动信号诊断内燃机失火故障的研究(3)——多特征综合方法.内燃机学报,2000(4):63-66.

[596] 胡阳,康宜华,卢文祥,杨叔子.钢丝绳无损检测中的一些算法——信号的预处理和特征提取.无损检测,2000(11):483-488.

[597] 张洁,罗欣,杜润生,杨叔子.基于NC特征的智能加工监督控制系统.华中理工大学学报,2000(11):92-94.

[598] 张洁,罗欣,杜润生,杨叔子.NC特征的概念及自动生成方法研究,华中理工大学学报,2000(11):95-97.

[599] 金建华,康宜华,杨叔子,等.油管缺陷在线检测仪的研究.仪表技术与传感器,2000(11):18-20.

[600] 李锡文,杜润生,杨叔子.立铣刀渐进磨损过程中主电机功率信号的时域特性研究.工具技术.2000(12):7-9.

[601] 刘世元,杜润生,杨叔子.柴油机缸盖振动信号的小波包分解与诊断方法研究.振动工程学报.2000(4):85-92.

[602] Cheng Tao, Zhang Jie, Hu Chunhua, Wu Bo and Yang Shuzi. *Intelligent machine tools in a distributed network manufacturing mode environment.* International Journal of Advanced Manufacturing Technology, v 17, n 3, p 221-232 (2001).

[603] 武新军,王峻峰,杨叔子.斜拉桥缆索缺陷检测系统的研制.机械科学与技术,2001(6):901-903.

[604] 金建华,康宜华,杨叔子.用磁桥路法测量油管壁厚的研究.无损检测,2001(2):55-58.

[605] 饶贵安,康宜华,武新军,杨叔子.电涡流方法在输电线损伤检测中的应用.传感器技术,2001(2):14-16.

[606] 胡春华,朱庆华,张智勇,吴波,杨叔子.基于CORBA的分布式网络化制造系统建模.机械与电子,2001(2):3-6.

[607] 杨叔子,刘克明.从"道艺合一"到"道通为一"——庄子技术思想初探.哈尔滨工业大学学报(社会科学版),2001(1):3-7.

[608] 刘克明,杨叔子.中国古代机械制造中的数理设计方法及其应用.机械研究与应用,2001(1):56-58.

[609] 饶贵安,康宜华,杨叔子.高压输电线破损检测传感器研究.仪表技术,2001(5):47-49.

[610] 武新军,王峻峰,杨叔子,等.数字化钻杆井口无损检测仪的研制.石油机械,2001(4):42-44.

[611] 李锡文,杜润生,杨叔子.基于后刀面磨损带面积的铣刀磨损模型的建立.华中科技大学学报,2001(4):53-56.

[612] 王立平,杜润生,史铁林,汪劲松,杨叔子.考虑轴承间隙时裂纹转子的故障状态及仿真分析.中国机械工程,2001(4):18-20.

[613] 刘克明,杨叔子.《老子》中有关机械的论述及其思想.中国机械工程,2001(4):84-87.

[614] 张智勇,朱庆华,程涛,吴波,杨叔子.分布式网络化制造系统中的工作流管理.制造业自动化,2001(4):5-9.

[615] 肖锡武,杨叔子.非对称刚度转轴的参激共振和分叉分析.机械工程学报,2001(6):43-48.

[616] 申戌,黄树红,韩守木,杨叔子.旋转机械振动信号的信息熵特征.机械工程学报,2001(6):94-98.

[617] 周杰韩,吴波,杜润生,杨叔子.虚拟制造系统多任务仿真方案设计与实现.中国机械工程,2001(7):30-32.

[618] 胡春华,张智勇,程涛,吴波,杨叔子.智能制造环境下的企业集成.中国科学基金,2001(4):29-32.

[619] 李晓峰,史铁林,杨叔子.考虑定子支撑弹性的碰摩转子系统非线性特性研究.振动工程学报,2001(3):59-64.

[620] 轩建平,史铁林,杨叔子.AR模型参数的Bootstrap方差估计,华中科技大学学报,2001(9):81-83.

[621] Liao Xiaoxin, Yang Shizi, Cheng Shijie and Fu Yuli. *Stability of general neural networks with reaction-diffusion*. Science in China (Series F: Information Sciences),2001(10),11-14.

[622] 金建华,康宜华,杨叔子.漏磁场法测量油管壁厚的研究.仪器仪表学报,2001(5):469-472.

[623] 饶贵安,康宜华,杨叔子.高压输电线破损检测传感器研究.仪表技术,2001(5):47-49.

[624] 吴家洲,吴波,杨叔子.UG软件的二次开发.机床与液压,2001(5):89-91.

[625] 刘世元,杜润生,史铁林,杨叔子.汽车发动机各缸工作不均匀性的一种在

线监测方法. 中国机械工程,2001(10):74-78.

[626] 周杰韩,熊光楞,吴波,杨叔子. 虚拟制造系统的制造哲理,制造业自动化. 2001(10):1-3.

[627] 吴家洲,杨叔子. 基于实例推理的齿轮快速设计 CAD 系统研究. 机械工程师,2001(11):15-16.

[628] 吴家洲,杨叔子. 球笼式万向节快速设计 CAD 系统研究. 机械工人 冷加工,2001:47-48.

[629] 吴家洲,吴波,杨叔子. 球笼式万向节快速设计 CAD 系统研究. 组合机床与自动化加工技术,2001(11):27-28.

[630] Xiong Liangcai, Shi Tielin, Yang Shuzi. *Bispectrum Analysis in Fault Diagnosis of Gears*. Journal of Southwest Jiaotong University,2001(2):147-151.

[631] 熊良才,史铁林,杨叔子. 基于双谱分析的齿轮故障诊断研究. 华中科技大学学报,2001(11):4-5.

[632] 武新军,王峻峰,杨叔子. 斜拉桥缆索缺陷检测系统的研制. 机械科学与技术,2001(6):901-903.

[633] 武新军,康宜华,吴义峰,杨叔子. 连续油管椭圆度在线磁性检测原理与方法. 石油矿场机械,2001(6):12-14.

[634] 吴家洲,吴波,杨叔子. 基于实例推理的齿轮快速设计 CAD 系统研究. 汽车工艺与材料,2001(11):35-36.

[635] 肖健华,樊可清,吴今培,杨叔子. 应用于故障诊断的 SVM 理论研究. 振动、测试与诊断,2001(4):26-30.

[636] 周杰韩,熊光楞,吴波,杨叔子. 面向对象模型量化评价方法. 中国机械工程,2001(12):71-76.

[637] 胡春华,张智勇,程涛,吴波,杨叔子. 基于 Petri 网的智能制造系统建模. 中国机械工程,2001(12):99-103.

[638] 熊良才,史铁林,杨叔子. 基于 Intranet 状态监测与故障诊断系统的设计及实现. 中国机械工程,2001:148-150.

[639] 吴家洲,吴波,杨叔子. 基于实例推理的齿轮快速设计 CAD 系统研究. 制造业自动化,2001(12):24-26.

[640] Xiong Liangcai, Shi Tielin, Yang Shuzi and Rao R. B. K. N. *A novel application of wavelet-based bispectrum analysis to diagnose faults in*

gears. International Journal of COMADEM, v 5, n 3, p 31-38 (2002).

[641] Lai Wuxing, Zhang Guicai, Shi Tielin and Yang Shuzi. *Classification of gear faults using higher-order statistics and support vector machines*. Chinese Journal of Mechanical Engineering (English Edition), v 15, n 3, p 243-247 (2002).

[642] Jiang Hanhon, Wu Ying, Zhang Xiaofeng, Kong Li, Shi Tielin and Yang Shuzi. *Research on embedded system of remote monitoring for vessel mechanical-electronic equipment*. 6th World Multiconference on Systemics, Cybernetics and Informatics. Proceedings, v 14, p 195-198 (2002).

[643] 张智勇,刘世荣,程涛,吴波,杨叔子.网络化制造系统中资源快速重组的策略研究,中国机械工程,2002(1):82-86.

[644] 吴家洲,杨叔子.公共路灯远程监控系统研究,机电一体化,2002(1):74-75.

[645] 吴家洲,姚远,杨叔子.基于电力线载波通讯的公共路灯远程监控系统实现的研究.计算机自动测量与控制,2002(1):41-43.

[646] 杨叔子,刘克明.从"道艺合一"到"道通为一"———庄子技术思想初探.湖北大学学报(哲学社会科学版),2002(1):42-47.

[647] 王立平,叶佩青,汪劲松,杨叔子.转子碰摩故障非线性特征的数值仿真分析.汽轮机技术,2002(1):31-33.

[648] 廖晓昕,杨叔子,程时杰,沈轶.具有反应扩散的广义神经网络的稳定性.中国科学 E 辑:技术科学,2002(1):89-96.

[649] 吴家洲,波吴,杨叔子.UG 软件的二次开发.精密制造与自动化,2002(1):29-31.

[650] 肖健华,吴今培,杨叔子.基于启发式知识的属性约简方法及其在评价体系中的应用.系统工程,2002(1):92-96.

[651] 吴家洲,杨叔子.基于实例推理的齿轮快速设计 CAD 系统研究.机械,2002(1):32-33.

[652] 李晓峰,卫国爱,史铁林,杨叔子.裂纹轴转子系统故障特征分析.华中科技大学学报(自然科学版),2002(2):21-23.

[653] 吴家洲,吴波,杨叔子.球笼式万向节快速设计 CAD 系统研究.机床与液压,2002(1):58-59.

[654] 王立平,叶佩青,汪劲松,杨叔子.基于实验的碰摩转子系统故障的双谱特征.振动工程学报,2002(1):66-71.

[655] 李晓峰,高燕,史铁林,杨叔子.转子系统碰摩故障特征分析.中国机械工程,2002(6):7-10.

[656] 胡友民,杜润生,杨叔子.制造系统数据采集技术研究.制造业自动化,2002(3):23-27.

[657] 王立平,郑浩峻,李铁民,汪劲松,杨叔子.转子碰摩故障非线性特征的实验研究.汽轮机技术,2002(2):101-102.

[658] 周杰韩,熊光楞,张和明,杨叔子.面向环境的产品全生命周期工程.制造业自动化,2002(4):30-33.

[659] 胡友民,李锡文,杜润生,杨叔子.基于PLC高可靠性工业过程远程监控系统.华中科技大学学报(自然科学版),2002(4):13-15.

[660] 胡友民,杜润生,杨叔子.生产过程远程监控与诊断技术研究.华中科技大学学报(自然科学版),2002(4):16-18.

[661] 肖锡武,肖光华,杨叔子.不对称转子系统的参激强迫振动.振动工程学报,2002(3):71-74.

[662] 李晓峰,许平勇,史铁林,杨叔子.带有涡动的裂纹转子故障特征研究.应用数学和力学,2002(6):643-652.

[663] 周杰韩,熊光楞,杨叔子.知识型制造业和制造元框架.现代制造工程,2002(6):5-8.

[664] 胡友民,杜润生,杨叔子.网络化监测与诊断技术研究.测控技术,2002(7):45-47.

[665] 胡友民,杜润生,杨叔子.液压系统运行状态监测.液压与气动,2002(8):35-37.

[666] 肖健华,吴今培,杨叔子.基于SVM的综合评价方法研究.计算机工程,2002(8):36-38.

[667] 周杰韩,曾庆良,熊光楞,张和明,杨叔子.制造业知识管理研究.计算机集成制造系统——CIMS,2002(8):669-672.

[668] 周杰韩,熊光楞,范文慧,柴旭东,杨叔子.一种创新性解决问题的方法——TRIZ的研究与进展.制造业自动化,2002(8):24-27.

[669] 肖锡武,肖光华,杨叔子.不对称转子系统的参激强迫振动.振动工程学报,2002(3):71-74.

[670] 金建华,杨叔子.一种新型油管缺陷磁性检测传感器.传感技术学报,2002(3):238-242.

[671] 杨叔子.科学人文交融·育人创新.北京观察,2002(9):10-12.

[672] 胡友民,杜润生,杨叔子.大型混合机液压系统分析与设计.液压与气动,2002(9):8-10.

[673] 胡友民,李锡文,杜润生,杨叔子.大型机电设备的高可靠性监控系统.机电一体化,2002(5):35-38.

[674] 金建华,杨叔子.基于磁性传感器信息融合的油管损伤在线检测技术.无损检测,2002(9):375-380.

[675] 周杰韩,熊光楞,吴波,杨叔子.虚拟制造系统分布式应用研究.清华大学学报(自然科学版),2002(9):1265-1268.

[676] 武新军,康宜华,杨叔子,程诗斌,孟庆鑫.数字化油管井口无损检测系统.无损检测,2002(10):422-424.

[677] 金建华,阙沛文,杨叔子.油管杆状磨损缺陷的建模与定量检测.石油机械,2002(11):8-10.

[678] 金建华,杨叔子.油管壁厚测量数据的一致性加权融合估计算法.仪表技术与传感器,2002(11):43-45.

[679] 金建华,阙沛文,杨叔子.油管腐蚀缺陷的在线检测技术.计算机自动测量与控制,2002(11):18-20.

[680] 胡友民,李锡文,杜润生,杨叔子.双谱分析在铣刀状态监测中的应用.振动、测试与诊断,2002(4):10-15.

[681] 李巍华,廖广兰,史铁林,杨叔子.基于核函数主元分析的机械设备状态识别.华中科技大学学报(自然科学版),2002(12):71-74.

[682] Li Weihua, Zhang Guicai, Shi Tielin and Yang Shuzi. *Gear crack early diagnosis using bispectrum diagonal slice*. Chinese Journal of Mechanical Engineering (English Edition), v 16, n 2, p 193-196 (2003).

[683] Li Weihua, Shi Tielin, Liao Guanglan and Yang Shuzi. *Feature extraction and classification of gear faults using principal component analysis*. Journal of Quality in Maintenance Engineering, v 9, n 2, p 132-143 (2003).

[684] Chen Yonghui, Shi Tielin and Yang Shuzi. *A model for studying properties of the mode-coupling type instability in friction induced*

oscillations on 4-h cold rolling mills. Key Engineering Materials, v 245-346, p 123-130 (2003).

[685] Liu Zhiping, Kang Yihua, Wu Xinjun and Yang Shuzi. *Recent developments in magnetic flux leakage technology for petrochemical tank floor inspections*. Materials Performance, v 42, n 12, p 24-27 (2003).

[686] Xiao Xiwu, Yang Shuzi, Huang Yuying. *1/2 subharmonic resonance of a shaft with unsymmetrical stiffness*. Chinese Journal of Mechanical Engineering (English Edition), v 16, n 1, p 25-27+30 (2003).

[687] 熊良才,史铁林,杨叔子. Choi-Williams 分布参数优化及其应用. 华中科技大学学报(自然科学版),2003(1):103-104.

[688] 陈勇辉,廖广兰,史铁林,杨叔子等. 四辊冷带轧机摩擦型颤振机理的研究. 华中科技大学学报(自然科学版),2003(1):105-107.

[689] 肖健华,吴今培,樊可清,杨叔子. 基于支持向量机的齿轮故障诊断方法. 中国制造业信息化,2003:107-109.

[690] 胡友民,杨叔子,杜润生. 基于网络的状态监测与诊断技术研究,. 机械工人. 冷加工,2003(2):11-13.

[691] 张智勇,吴波,杨叔子. 网络协同制造系统的构想与实现. 中国科学基金,2003(1):18-22.

[692] 胡友民,杜润生,杨叔子. 集中式监控系统的可靠性分析. 振动、测试与诊断,2003(01):6-8.

[693] 胡友民,杜润生,杨叔子. 立式混合机温控系统分析与设计. 微计算机信息,2003(05):16-17.

[694] 陈勇辉,史铁林,杨叔子. 四辊冷带轧机非线性参激振动的研究. 机械工程学报,2003(04):56-60.

[695] 刘志平,康宜华,杨叔子,李涛,李春树. 储罐罐底板漏磁检测仪的研制. 无损检测 2003(05):234-236.

[696] 陈勇辉,李巍华,史铁林,杨叔子. 四辊冷带轧机五倍频再生颤振机理的研究. 华中科技大学学报(自然科学版),2003(05):55-57.

[697] 李巍华,陈勇辉,史铁林,杨叔子. 基于核的动态聚类算法用于机械故障模式分类. 华中科技大学学报(自然科学版),2003(05):58-61.

[698] 李晓峰,高燕,史铁林,杨叔子. 带有油膜轴承的裂纹轴转子系统故障特征分析. 华中科技大学学报(自然科学版),2003(05):62-64.

[699] 肖健华,吴今培,樊可清,杨叔子.粗糙主成分分析在齿轮故障特征提取中的应用.振动工程学报,2003(02):166-170.

[700] 来五星,轩建平,史铁林,杨叔子.Wigner-Ville 时频分布研究及其在齿轮故障诊断中的应用.振动工程学报,2003(02):247-250.

[701] 胡友民,杜润生,杨叔子.智能化状态监测技术研究.中国机械工程,2003(11):946-950.

[702] 郗庆路,罗欣,杨叔子.基于蚂蚁算法的混流车间动态调度研究.计算机集成制造系统-CIMS,2003(06)456-475.

[703] 刘克明,杨叔子.中国古代工程图学及其现代意义.哈尔滨工业大学学报(社会科学版),2003(02):1-6.

[704] 郗庆路,罗欣,杨叔子.分布式制造系统的控制协调及其方法分析.信息与控制,2003(04):295-230.

[705] 刘志平,康宜华,武新军,杨叔子.大面积钢板局部磁化的三维有限元分析.华中科技大学学报(自然科学版),2003(08):10-12.

[706] 胡友民,杜润生,杨叔子.冗余式分层监测系统可靠性分析.机械工程学报,2003(08):110-115.

[707] 杨叔子.治学育人 必正其风.高等教育研究,2003(05):13-19.

[708] 熊良才,史铁林,杨叔子.基于滤波器组的时频分布及其在故障诊断中的应用.中国机械工程,2003(19):1640-1642.

[709] 王伏林,易传云,王涛,杨叔子.数字化齿面加工误差评定基准研究.中国机械工程,2003(19),1687-1691.

[710] 刘志平,康宜华,杨叔子.漏磁检测信号的反演.无损检测,2003(10):531-535.

[711] 杨叔子,吴波.先进制造技术及其发展趋势.机械工程学报,2003(10):73-78.

[712] 王峻峰,何岭松,杨叔子.基于 Web 的设备工况视频监视.机械与电子,2003(06):56-59.

[713] 王伏林,易传云,杨叔子,刘洪贵.二维数字化齿面加工误差分析.华中科技大学学报(自然科学版),2003(11):55-57.

[714] Cheng Tao,Guan Zailin,Liu Liming,Wu Bo and Yang Shuzi. *A CORBA-based multi-agent system integration framework*. Proceedings. Ninth IEEE International Conference on Engineering of Complex Computer

Systems, p 191-198 (2004).

[715] 金建华,康宜华,杨叔子.油管损伤的磁性检测法及其实现技术.无损检测,2004(01):13-17.

[716] 杨叔子,李斌,吴波.先进制造技术发展与展望.机械制造与自动化,2004(01):1-6.

[717] 杨叔子.8大趋势引领"现代制造".人民政协报,2004-03-02.

[718] 胡友民,杨叔子,杜润生.制造系统分布式柔性可重组状态监测与诊断技术研究.机械工程学报,2004(03):40-44.

[719] 程涛,管在林,吴波,杨叔子,黄心汉.分布式虚拟制造资源中心.中国制造业信息化,2004(04)87-90.

[720] 程涛,吴波,杨叔子,黄心汉.支持分布式网络化制造的智能数控系统的研究.中国机械工程,2004(08):688-694.

[721] 程涛,管在林,吴波,杨叔子,黄心汉.网络经济下的一种新型生产组织模式——虚拟制造组织.机械制造,2004(06):7-10.

[722] 王峻峰,康宜华,杨叔子.抽油杆裂纹信号的特征分析及处理方法.华中科技大学学报(自然科学版),2004(9):69-70.

[723] 来五星,轩建平,史铁林,杨叔子.微制造光刻工艺中光刻胶性能的比较.半导体技术,2004(11):22-25.

[724] 程涛,吴波,杨叔子,等.一种支持企业集成的基于Agent的企业信息模型.组合机床与自动化加工技术,2004(11):93-96.

[725] 李孟清,张春良,杨叔子,等.质量管理图中趋势模态及阶跃模态的模糊神经网络识别.中国机械工程,2004(22):1998-2001.

[726] 刘志平,康宜华,武新军,杨叔子.储罐底板漏磁检测传感器设计.无损检测.2004(12):612-615.

[727] 王伏林,易传云,刘洪贵,杨叔子.数字化齿面加工轮廓误差两种分析模型.机床与液压,2004(12):14-16.

[728] Yi Pengxing, Yang Shuzi, Du Runsheng, Wu Bo and Liu Shiyuan. *Distributed monitoring system reliability estimation with consideration of statistical uncertainty*. Chinese Journal of Mechanical Engineering (English Edition), v 18, n 4, p 519-524 (2005).

[729] Gao Qinglu, Luo Xin and Yang Shuzi. *Stigmergic cooperation mechanism for shop floor control system*. International Journal of Advanced

Manufacturing Technology,v 25,n 7-8,p 743-753 (2005).

[730] 饶贵安,康宜华,陈龙驹,陈铁红,刘双海,杨叔子.一种新的实时小波分析.仪器仪表学报,2005(2):181-183.

[731] 李巍华,史铁林,杨叔子.基于非线性判别分析的故障分类方法研究.振动工程学报,2005(2):133-138.

[732] 易朋兴,杜润生,杨叔子,等.基于 Markov 模型的分布式监测系统可靠性研究.机械工程学报,2005(6):143-148.

[733] 曾赤梅,杨叔子.对话杨叔子.今日湖北,2005(6):18-21.

[734] 程涛,吴波,管在林,杨叔子.虚拟制造组织中服务定位机制研究.工业工程,2005(3):6-11.

[735] 陈美霞,骆东平,杨叔子.壳间连接形式对双层壳声辐射性能的影响.振动与冲击,2005(5):80-83.

[736] 陈美霞,关珊珊,骆东平,杨叔子,曹钢.具有内部甲板的环肋柱壳振动和声学试验分析.舰船科学技术,2005(06):19-25.

[737] Wang Fulin, Yi Chuanyun, Wang Tao, Yang Shuzi and Zhao Gang. *A generating method for digital gear tooth surfaces*. International Journal of Advanced Manufacturing Technology,v 28,n 5-6,p 474-485 (2006).

[738] Lai Wuxing, Shi Tielin, Xuan Jianping and Yang Shuzi. *Strip steel quality evaluation technique based on support vector machine decision methods*. WSEAS Transactions on Circuits and Systems,v 5,n 9,p 1438-1445 (2006).

[739] Wang Junfeng, Shi Tielin, He Lingsong and Yang Shuzi. *Frequency overlapped signal identification using blind source separation*. Chinese Journal of Mechanical Engineering (English Edition),v 19,n 2,p 286-289 (2006).

[740] 杨叔子,吴波,李斌.再论先进制造技术及其发展趋势.机械工程学报,2006(1):1-5.

[741] 易朋兴,刘世元,杜润生,杨叔子.分布式监测系统可靠性贝叶斯评估.华中科技大学学报(自然科学版),2006(3):42-45.

[742] 来五星,廖广兰,史铁林,杨叔子.反应离子刻蚀加工工艺技术的研究.半导体技术,2006(6):414-417.

[743] 蔡洪涛,杜润生,杨叔子.偏心圆筒混合机的运动及混合机理分析.华中科

技大学学报(自然科学版).2006(6):84-86.

[744] 来五星,史铁林,杨叔子.嵌入式激光器能量实时控制系统的开发.半导体光电.2006(3):328-330.

[745] 王林鸿,王平,吴波,杨叔子.液压缸动态特性的混沌特征.矿山机械,2006(9):118-122.

[746] 王林鸿,郭俊杰,吴波,杨叔子.液压缸运行动态特性的关联维数分析.机床与液压,2006(9):99-102.

[747] 李巍华,史铁林,杨叔子.基于核函数估计的转子故障诊断方法.机械工程学报,2006(9):76-82.

[748] 王林鸿,郭俊杰,吴波,杨叔子.液压缸低速爬行原因新探.液压与气动,2006(9):14-18.

[749] 王林鸿,杜润生,吴波,杨叔子.液压缸低速运动的动态分析.中国机械工程,2006(20):2098-2101.

[750] 陈厚桂,何利勇,康宜华,杨叔子.钢丝绳断丝检测仪器评估中标样制作方法研究.计量学报,2007(2):150-153.

[751] 黄荣杰,吴波,杨叔子.DNC通讯接口模式在网络数控系统中的应用分析.组合机床与自动化加工技术,2007(5):43-45.

[752] 谢守勇,李锡文,杨叔子,等.基于PLC的模糊控制灌溉系统的研制.农业工程学报,2007(6):208-210.

[753] 杨叔子,余东升.文化素质教育与通识教育之比较.高等教育研究.2007(6):1-7.

[754] 李锡文,杨明金,谢守勇,等.基于时域特性的铣刀磨损状态信息提取.中国机械工程.2007(13):1513-1517.

[755] 杨叔子.踏平坎坷,成人成才——在深圳大学的演讲.深圳大学学报(人文社会科学版),2007(4):132-134.

[756] 李锡文,杨明金,谢守勇,杨叔子.基于神经网络信息融合的铣刀磨损状态监测.农业机械学报,2007(7):160-163.

[757] 杨叔子.科学求真·人文为善·艺术致美·工业设计彰和谐.节能环保和谐发展——2007中国科协年会论文集(二).2007(9):28.

[758] 杨叔子.发挥优势 放眼未来.2007中国科协年会院士专家座谈会论文集.2007(9):9.

[759] 杨叔子.工业设计务彰和谐,科技导报,2007(17):1.

[760] 李锡文,杨明金,杜润生,杨叔子. 模糊模式识别在铣刀磨损监测中的应用. 机械科学与技术. 2007(9):1113-1117.

[761] 李锡文,杨明金,谢守勇,杨叔子. 大型立式混合装备设计研究. 西南大学学报(自然科学版),2007(9):158-162.

[762] 谢守勇,李锡文,杨叔子,等. 一种三轴伺服平台运动控制研究. 西南大学学报(自然科学版),2007(9):169-172.

[763] 李锡文,杨明金,谢守勇,杨叔子. 铣削加工刀具磨损过程双谱分析. 农业机械学报,2007(9):143-146.

[764] 易朋兴,胡友民,崔峰,杜润生,杨叔子. 立式捏合机捏合间隙影响CFD分析. 化工学报,2007(10):2680-2684.

[765] 谢守勇,李锡文,杨叔子. 提升机输送带纠偏研究. 起重运输机械,2007(10):55-57.

[766] 来五星,轩建平,史铁林,杨叔子. 用自回归最小二乘消噪法提纯轴承故障信号. 振动、测试与诊断,2007(4):292-294.

[767] 王林鸿,吴波,杜润生,杨叔子. 液压缸运动的非线性动态特征. 机械工程学报,2007(12):12-19.

[768] Gao Qinglu, Luo Xin, Wu Bo and Yang Shuzi. *Collaborative plan exchange: significant reinforcement for contract net based distributed manufacturing systems*. International Journal of Advanced Manufacturing Technology, v 37, n 9-10, p 1042-1050 (2008).

[769] Yang Mingjin, Li Xiwen, Li Shaoping and Yang Shuzi. *Reliability analysis of the control system of large-scale vertical mixing equipment*. Frontiers of Mechanical Engineering in China, v 3, n 2, p 133-138 (2008).

[770] Wang Linhong, Wu Bo, Du Runsheng and Yang Shuzi. *Dynamic characteristics of NC table with SVD*. Frontiers of Mechanical Engineering in China, v 3, n 4, p 385-391 (2008).

[771] 杨叔子. 制造、先进制造技术的发展及其趋势(上). 装备制造,2008(4):52-55.

[772] 杨叔子. 寻根寄语. 寻根,2008(2):1.

[773] 杨叔子. 制造、先进制造技术的发展及其趋势(下). 装备制造,2008(5):38-41.

[774] 王伏林,易传云,杨叔子,等.二维数字化齿面的插削展成加工方法.机械工程学报,2008(6):88-94.

[775] 杨叔子,史铁林.以人为本——树立制造业发展的新观念,机械工程学报.2008(7):1-5.

[776] 易朋兴,崔峰,胡友民,杨叔子.立式捏合机搅拌桨螺旋角影响数值分析.固体火箭技术,2008(4):381-385.

[777] Wang Linhong, Wu Bo, Du Runsheng and Yang Shuzi. Dynamic characteristics of an NC table with phase space reconstruction. Frontiers of Mechanical Engineering in China, v 4, n 2, p 179-183 (2009).

[778] 杨叔子,史铁林.走向"制造服务"一体化的和谐制造.机械制造与自动化,2009(1):1-5.

[779] 杨明金,李锡文,谢守勇,杨叔子.立式捏合机混合釜内流场遍历性研究.农机化研究,2009(4):20-23.

[780] 王林鸿,杜润生,吴波,杨叔子.数控工作台的非线性动态特性.中国机械工程,2009(13):1513-1519.

[781] 王林鸿,吴波,杜润生,杨叔子.数控工作台动态特性的混沌特征.中国机械工程,2009(14):1656-1659.

[782] Xie Shouyong, Li Xiwen, Yang Mingjin and Yang Shuzi. *Control algorithm of a servo platform*. Frontiers of Mechanical Engineering in China, v 5, n 3, p 353-355 (2010).

[783] Yang Mingjin, Li Xiwen, Shi Tielin and Yang Shuzi. *Performance analysis and parameter optimization of a planetary mixer*. Applied Mechanics and Materials, v 37-38, p 858-861 (2010).

[784] He Ruibo, Zhao Yingjun, Yang Shunian and Yang Shuzi. *Kinematic-parameter identification for serial-robot calibration based on POE formula*. IEEE Transactions on Robotics, v 26, n 3, p 411-423 (2010).

[785] Guo Guang, Wu Bo and Yang Shuzi. *A job-insertion heuristic for minimizing the mean flowtime in dynamic flowshops*. Frontiers of Mechanical Engineering, v 6, n 2, p 197-202(2011).

[786] Wang Erhua, Wu Bo, Yan Wenhui and Yang Shuzi. *Dynamic simulation and optimization design of polished rod hoisting system based on VB combined with Matlab and Ansys*. 2011 2nd International Conference on

Mechanic Automation and Control Engineering, MACE 2011 - Proceedings, p 245-251(2011).

[787] Yang Mingjin, Li Xiwen, Shi Tielin and Yang Shuzi. *Time series modeling and analysis of cutter wear in milling operation*. Proceedings - 3rd International Conference on Measuring Technology and Mechatronics Automation, ICMTMA 2011, v 3, p 1023-1026 (2011).

[788] 何锐波,赵英俊,韩奉林,杨曙年,杨叔子.基于指数积公式的串联机构运动学参数辨识实验.机器人,2011(1):35-39,45.

[789] 杨叔子,丁汉,李斌.高端制造装备关键技术的科学问题.机械制造与自动化,2011(1):1-5.

[790] 杨明金,李锡文,史铁林,杨叔子.高黏度流体混合机混合均匀度试验(英文).农业工程学报,2011(6):137-142.

[791] 王林鸿,吴波,杜润生,杨叔子.用奇异谱和奇异熵研究数控工作台动态特征.振动、测试与诊断,2012(1):116-119.